INTRODUCTORY READINGS IN THE
PHILOSOPHY OF SCIENCE

INTRODUCTORY READINGS IN THE
PHILOSOPHY OF SCIENCE

EDITED BY

E. D. Klemke
Robert Hollinger
David Wÿss Rudge

WITH A. David Kline

 **Prometheus Books**

59 John Glenn Drive
Amherst, New York 14228-2197

Published 1998 by Prometheus Books

02 01 00 99 98   5 4 3 2 1

Library of Congress Cataloging-in-Publication Data

Introductory readings in the philosophy of science / edited by E. D. Klemke, Robert
    Hollinger, David Wÿss Rudge. — 3rd. ed.
        p.   cm.
    Includes bibliographical references.
    ISBN 1–57392–240–4 (alk. paper)
        1. Science—Philosophy. I. Klemke, E. D., 1926–   II. Hollinger, Robert.
III. Rudge, David Wÿss.
Q175.I64   1998
501—dc21   98–8293
                                                                                                        CIP

Printed by Bookcrafters in the United States of America on acid-free paper

Contents

PART 1. SCIENCE AND PSEUDOSCIENCE

PART 2. THE NATURAL AND SOCIAL SCIENCES

PART 3. EXPLANATION AND LAW

PART 6. SCIENCE AND VALUES

APPENDIX

Preface to the Third Edition

The main revision for the third edition is the addition of a section on the natural and the social sciences. This complements the first part, science vs. nonscience, and resonates with issues about explanation, confirmation, science and values, and the role of theory. Part 2 focuses on the way in which these issues generate debates about the nature of the social sciences, and comparisons and contrasts with the natural sciences.

The new readings in part 2 provide an integrated set of papers which address each other, either explicitly (Taylor vs. Kuhn) or implicitly (Rosenberg vs. Machlup). They extend the issues of the other parts into debates about the social and behavioral sciences. These readings also anticipate and expand upon the papers in part 6 (Science and Values), and also throw additional light on Kuhn's views. The readings in the newly revised section on Science and Values (part 6) now include an essay on feminism and science (Giere), which discusses feminism and Positivism, Popper, Kuhn, realism and antirealism. Even the topic of part 1, science vs. nonscience, is discussed in the context of these new essays. Finally, the Hollinger essay, "From Weber to Habermas," is included in the newly revised part 6, to fill a gap in the readings.

We believe that the new part 2 and the revised and expanded part 6 adequately cover material in the old part 6, Science and Culture. We have therefore eliminated this section, except for the essay by Hollinger, and revised the section on Science and Values accordingly. The new material is also more current, since it deals with feminism, postmodernism, and (in the expanded editorial introduction to part 6) the so-called science wars and recent versions of the sociology of science (mainly in the form known as Science and Technology Studies [STS]). These are all topics that are of great interest to the general reading public, as well as to university professors and students.

This book, in its revised and expanded third edition, can be used in standard one-semester courses in the philosophy of science, two-semester courses which spend time on issues in the philosophy of social science and science and culture, and in philosophy of social science courses. We have therefore included outlines for several alternative course syllabuses at the end of the book.

In addition, David Rudge has made significant revisions in part 3 (on explanation). He has added editorial material to explain some of the nuances of Hempel's views about explanation, and has in fact rewritten the editor's introduction to part 3. He has also replaced the Brody and Dray readings with three exciting pieces on explanation, by van Fraassen, Kitcher, and Salmon. The material on explanation is now much more current (adding papers on the pragmatics of explanation and explanation as unification); and part 3 also has more organic unity in this edition. Rudge has also updated the Reading List and Study Questions for part 3. (The reading lists and study questions for part 4 have also been completely revised.)

We have also added "Case Studies" to each part, just before the study questions. These case studies are designed to be specific illustrations of the main topics and issues in each of the book's parts. They connect the theoretical issues discussed in the readings with a particular problem or issue of some sort. Professor Rudge, the only one of the editors who regularly teaches a standard one-semester course in the philosophy of science, has added a pedagogical essay that is addressed to both instructors and students. We think this will help the reader optimize the way in which this book can be used in courses.

E. D. Klemke
Robert Hollinger
David Wÿss Rudge

Preface to the Revised Edition

This revised edition, like its predecessor, is a collection of informative, lively, and accessible readings, organized under six main issues, and designed to provide a solid introduction to the problems of the philosophy of science.

We have been gratified by the response of both instructors and students to the first edition of this text. For this new edition we have made a number of what we hope are improvements.

Probably the most often expressed concern with regard to the first edition was the need for a better selection expressing the views of Thomas Kuhn. We not only remedied this but included a paper by Carl Hempel that examines some of the important themes raised by Kuhn.

The first edition did not contain a version of the standard account of theories given that most presentations were far too technical. The now-included essay by Rudolf Carnap is a nice nontechnical representation of the view. For similar reasons we only included a very qualitative discussion of confirmation in the first edition. The selection on the logic of confirmation, now included, was especially written for introductory students by Ronald Giere. It may require some additional help from the instructor but we believe it will be well worth the effort.

The opening part, "Science and Nonscience," has been expanded by one essay. This topic is especially exciting to introductory students. Our earlier suggestion, that one may want to wait until other topics have been covered to take up this issue, still stands.

The final two parts, on science and values and science and culture, which also have proven to be quite interesting to a variety of undergraduates, have been significantly revised. We eliminated essays that were too difficult or repetitive. The new essays also relate more directly to issues in the previous sections.

We would like to express our deep gratitude to all those people who helped us, in various ways, to prepare this book. In addition to the many who aided us with the first edition (see the preface to that edition, reprinted in this volume), for this new edition we are grateful to Carl Matheson, Warren Asher, Bernice Power, Edna Wiser, Ray Amsler, and Steven L. Mitchell and the entire staff of Prometheus Books.

Finally, we hope that instructors and students who have comments or suggestions for improving the volume or who wish to convey their evaluation of it will contact the editors.

E. D. Klemke
Robert Hollinger
A. David Kline

Preface to the First Edition

It is a truism that in our time science has become a highly respected and venerated enterprise. It is widely deemed to be a Good Thing. There are many who hold it to be the most successful pursuit of knowledge in the history of the world. Others consider it to be the noblest intellectual achievement of mankind. It is also true that (in recent years, especially) many have found science to be undeserving of such high praise. Not only has science been attacked by proponents of ESP, astrology, and other pseudosciences, and by critics of modern technology; it has even been found wanting by some scientists and ex-scientists and others who know and write about science. (See Paul Feyerabend's essay in Part I of this volume.) Who is right, or more nearly correct, in this controversy?

It is our hope that the reader of this volume will be able to arrive at his or her own answer to this question. However, in our view, one cannot attempt to answer such a question without a clear understanding of what science is and what it is not. One of our major goals has been to provide that understanding and to provide it in such a manner that it is accessible to the beginning student and the layman.

Hence we would like to stress that this is an *introductory* work in the philosophy of science. Our principles of selection for the readings included have been these: The selection should be intrinsically interesting. It should be comprehensible to a beginning student. It should serve to provoke discussion and criticism. We have also tried to stimulate a kind of dialogue among the authors by selecting works which present varying and even conflicting points of view.

We have presented the topics in the order we have found to be most desirable. However, there may be one exception. Part I could be used either first or last—or, perhaps best of all, both. Otherwise we believe that the order of presentation should be followed. (However, it is not necessary that every reading within each part be used.)

A few other comments are in order. First, although all three of us worked closely together on all of the material included, the general introduction and the introductions to the six parts were written individually, not collectively. (The initials at the end of each indicate primary authorship.) Second, the Study Questions at the end of each of the six parts were composed by the author of the introduction to that part. Third, the bibliographies—one is given at the end of each part (and at the end of the general introduction)—are not exhaustive. They are intended to provide further sources which deal with some of the major issues discussed in the volume. Fourth, although we have spent much time in revising our format and the selection of readings, we may have overlooked some items which ought to have been included. If so, we shall be grateful to hear from instructors who use the volume and to receive their suggestions with regard to this—or any other—issue.

We would like to express our gratitude to all those who helped in various ways with regard to the preparation of our initial proposal and the final manuscript. We are especially grateful to: Rowena Wright, Bernice Power, Annette Van Cleave, Richard Kniseley, David Hauser, and Steven Isaacson. We would also like to express our appreciation to the members of the Philosophy Department of Iowa State University for their constant help and encouragement.

E. D. Klemke
Robert Hollinger
A. David Kline

Acknowledgments

The editors gratefully acknowledge the kind permission of authors, editors, and publishers that has enabled us to print the essays included in this book.

PART 1

1. Sir Karl Popper, "Science: Conjectures and Refutations." From *Conjectures and Refutations,* pp. 33–41, 52-59. New York: Harper and Row, 1963. Also published by Basic Books, New York, and Routledge and Kegan Paul, London. Copyright 1963 by Karl Popper.

2. John Ziman, "What Is Science?" *Public Knowledge,* pp. 5–27. New York: Cambridge University Press, 1968. Copyright 1968 by Cambridge University Press.

3. Paul Feyerabend, "How to Defend Society against Science." *Radical Philosophy,* no. 11 (1975), pp. 3–8. Copyright 1975 by Paul Feyerabend.

4. Paul R. Thagard, "Why Astrology Is a Pseudoscience." *Proceedings of Philosophy of Science Association,* 1978, Volume One, pp. 223–24. Edited by P. D. Asquith and I. Hacking. East Lansing: Philosophy of Science Association, 1978. Copyright 1978 by Philosophy of Science Association.

5. Philip Kitcher, "Believing Where We Cannot Prove." In Kitcher, *Abusing Science: The Case against Creationism.* Cambridge: The MIT Press, 1982, pp. 30–54. Copyright 1982 by Philip Kitcher. Reprinted by permission of the publisher.

PART 2

6. Charles Taylor, "Interpretation and the Sciences of Man." Originally published in the *Review of Metaphysics* 25, no. 1 (September 1971), reprinted with permission.

7. Thomas Kuhn, "The Natural and the Social Sciences." In Hiley, D. R., Bohman, J. F., and Shusterman, R., eds., *The Interpretive Turn: Philosophy, Science, and Culture.* Ithaca, NY: Cornell University Press, 1991, pp. 17–24. Copyright 1991 by Cornell University Press.

8. Fritz Machlup, "Are the Social Sciences Really Inferior?" *Southern Economic Journal* 17 (1961): 173–84. Copyright 1961 by the Southern Economic Association and the University of

North Carolina, Chapel Hill, NC. This article appears with permission of the Southern Economic Association.

9. Alexander Rosenberg, "If Economics Isn't Science, What Is It?" *Philosophical Forum* 14 (1983): 296–314. Copyright 1983 by the philosophy department, Boston University, Boston, MA.

10. Brian C. Fay and J. Donald Moon, "What Would an Adequate Philosophy of Social Science Look Like?" *Philosophy of Social Science* 7 (1977): 209–27. Copyright 1977 by Sage Publications, Brian Fay, and J. Donald Moon.

PART 3

11. Carl G. Hempel, "Studies in the Logic of Explanation." In Hempel, *Aspects of Scientific Explanation.* Baltimore, MD: Williams and Wilkins Co., 1948, pp. 135–75. Copyright 1948, The William and Wilkins Company. Reproduced, in edited form, by permission.

12. Karel Lambert & Gordon Britten, "Laws and Conditional Statements." In Lambert and Britten, *An Introduction to the Philosophy of Science,* Englewood Cliffs, NJ: Prentice Hall, 1970, pp. 37–45. Reprinted by permission of the authors.

13. Nancy Cartwright, "The Truth Doesn't Explain Much." © Nancy Cartwright 1983. Reprinted from *How the Laws of Physics Lie* by Nancy Cartwright (1983) by permission of Oxford University Press.

14. Wesley Salmon, "Scientific Explanation: How We Got from There to Here." From *Causality and Explanation* by Wesley Salmon. Copyright © 1998 by Wesley Salmon. Used by permission of Oxford University Press, Inc.

15. Bas van Fraassen, "VII. The Pragmatics of Explanation." *American Philosophical Quarterly* 14, no. 2 (1977): 143–50. Copyright 1977 by University of Pittsburgh Press, Pittsburgh, PA.

16. Philip Kitcher, "Explanatory Unification." *Philosophy of Science* 48 (1981): 507–31. Copyright 1981 by University of Chicago Press, Chicago, IL.

PART 4

17. Rudolf Carnap, "The Nature of Theories." *Philosophical Foundations of Physics: An Introduction to the Philosophy of Science,* by Rudolf Carnap, ed. Martin Gardner, pp. 225–46. New York: Basic Books, 1966. Copyright 1966 by Basic Books, Inc. Reprinted by permission of the publisher.

18. Hilary Putnam, "What Theories Are Not." Excerpted from *Logic, Methodology, and Philosophy of Science,* edited by Ernest Nagel, Patrick Suppes, and Alfred Tarski, with the permission of the publishers, Stanford University Press. Copyright © 1977 by the Board of Trustees of the Leland Stanford Junior University.

19. N. R. Hanson, "Observation." From "Observation," *Patterns of Discovery,* New York: Cambridge University Press, pp. 4–11, 15–19. Copyright 1958 by Cambridge University Press.

20. W. T. Stace, "Science and the Physical World." Reprinted from *Man against Darkness* by W. T. Stace by permission of the University of Pittsburgh Press. Copyright 1967 by the University of Pittsburgh Press.

21. Stephen Toulmin, "Do Sub-Microscopic Entities Exist?" *The Philosophy of Science,* pp. 134–39. London: The Hutchinson Publishing Group, Ltd. Copyright 1953, The Hutchinson Publishing Group, Ltd. Reprinted by permission of the publisher.

22. Grover Maxwell, "The Ontological Status of Theoretical Entities." *Minnesota Studies in*

the Philosophy of Science, Vol. III, pp. 3–14. Edited by H. Feigl and G. Maxwell. Minneapolis, University of Minnesota Press, 1962. Copyright 1964 by University of Minnesota Press.

23. Carl Matheson and A. David Kline, "Is There a Significant Observational Theoretical Distinction?" Original essay for this volume.

PART 5

24. W. V. Quine and J. S. Ullian, "Hypothesis." From *The Web of Belief,* 2d ed., by W. V. Quine and J. S. Ullian, pp. 64–82. Copyright 1970, 1978 by W. V. Quine.

25. Ronald Giere, "Justifying Scientific Theories." Excerpts from *Understanding Scientific Reasoning,* 2d ed. by Ronald N. Giere. Copyright © 1984 by Holt, Rinehart, and Winston, reprinted by permission of the publisher.

26. Thomas S. Kuhn, "Objectivity, Value Judgment, and Theory Choice," in *The Essential Tension,* pp. 320–39. Chicago: University of Chicago Press, 1977. Copyright 1977 by Thomas S. Kuhn. Reprinted by permission of the publisher and the author.

27. Carl G. Hempel, "Scientific Rationality: Analytic vs. Pragmatic Perspectives." *Rationality, To-Day* edited by Theodore F. Geracts, pp. 46–58. Ottawa: The University of Ottawa Press, 1979.

28. Philipp G. Frank, "The Variety of Reasons for the Acceptance of Scientific Theories." *Scientific Monthly* 79 (September 1954): 139–45. Copyright 1954 by the American Association for the Advancement of Science.

PART 6

29. Richard Rudner, "The Scientist *Qua* Scientist Makes Value Judgments." *Philosophy of Science* 20 (1953): 1–6. Copyright 1953 by Williams and Wilkins Co., Baltimore, MD.

30. Carl G. Hempel, "Science and Human Values." *Social Control in a Free Society,* ed. by R. E. Spiller, pp. 39–64. Philadelphia: University of Pennsylvania Press, 1960. Copyright 1960 by University of Pennsylvania Press.

31. Ernan McMullin, "Values in Science." Copyright 1983 by the Philosophy of Science Association, East Lansing, MI, 18 Morrill Hall, Dept. of Philosophy, Michigan State University, East Lansing, MI 48824. From *PSA 1982,* volume two, ed. P. D. Asquinth and T. Nickles, 1983, pp. 3–28.

32. Robert Hollinger, "From Weber to Habermas." Original essay for this volume.

33. Ronald Giere, "The Feminist Question in Philosophy of Science." In L. H. Nelson and J. Nelson, eds. *Feminism, Science, and the Philosophy of Science,* 1996, pp. 3–15, Dordrecht, Holland: Kluwer Press. Copyright 1996 by Kluwer Press.

Introduction

What Is Philosophy of Science?

Most readers of this volume probably have some familiarity with science—or with one or more of the sciences. But the following question may come to mind: Just what is philosophy of science? How does it differ from science? How is it related to other areas of philosophy? We shall here attempt to provide answers to these and related questions.[1]

I. WHAT PHILOSOPHY OF SCIENCE IS NOT

Let us begin with a discussion of what philosophy of science is *not*.

(1) Philosophy of science is not the history of science. The history of science is a valuable pursuit for both scientists and nonscientists. But it must not be confused with the philosophy of science. This is not to deny that the two disciplines may often be interrelated. Indeed, some have held that certain problems within the philosophy of science cannot be adequately dealt with apart from the context of the history of science. Nevertheless, it is generally held that we must distinguish between the two.

(2) Philosophy of science is not metaphysical cosmology or "philosophy of nature." The latter attempts to provide cosmological or ethical speculations about the origin, nature, and purpose of the universe, or generalizations about the universe as a whole. As examples we may cite the views of Hegel and Marx, that the universe is dialectical in character; or the view of Whitehead, that it is organismic. Such cosmologies are often imaginative, metaphorical, and anthropomorphic constructions. They frequently involve interpreted extrapolations from science. Again, certain problems within the philosophy of science may aid the construction of or involve a consideration of such cosmological theories. But here, too, there is wide agreement that they must be distinguished.

(3) Philosophy of science is not the psychology or sociology of science.

19

The latter disciplines constitute a study of science as an activity, as one social phenomenon among many. Some of the topics that fall within such an inquiry are: scientists' motives for doing what they do; the behavior and activity of scientists; how (in fact) they make discoveries; what the impact of such discoveries is on society; and the sorts of governmental structures under which science has flourished. Again, certain problems in the philosophy of science may on occasion be related to such issues. But once more, it is reasonable to hold that these inquiries must be distinguished.

For the purposes of our study, the philosophy of science will not primarily mean or apply to any of the above. We will not try to comprehend the history of science. We will not present any grand cosmological speculations. We will not try to understand the scientific enterprise in terms of human or social needs. However, with regard to the latter, it is desirable to make a distinction. It is one thing to present a psychological or sociological account of science. This we will not do. It is another thing to examine philosophically the relationship of science and culture and generally of science and values. The last part of this volume will be devoted to these issues.

II. WHAT PHILOSOPHY OF SCIENCE IS

Let us attempt now to see what the philosophy of science *is*. By one widely held conception, philosophy of science is the attempt to understand the meaning, method, and logical structure of science by means of a logical and methodological analysis of the aims, methods, criteria, concepts, laws, and theories of science. Let us accept this as a preliminary characterization.

In order to illustrate or apply this characterization, let us focus on the matter of the concepts of science.

(1) There are numerous concepts that are used in many sciences but not investigated by any particular science. For example, scientists often use such concepts as: causality, law, theory, and explanation. Several questions arise: What is meant by saying that one event is the cause of another? That is, what is the correct analysis of the concept of cause? What is a law of nature? How is it related to other laws? What is the nature of a scientific theory? How are laws related to theories? What are description and explanation in science? How is explanation related to prediction? To answer such questions is to engage in logical and methodological analysis. Such an analysis is what philosophy of science, in part, is (according to this conception).

(2) There are many concepts used in the sciences that differ from the ones mentioned above. Scientists often speak of ordinary things—such as beakers, scales, pointers, tables. Let us call these observables. But they also often speak of unobservables: electrons, ions, genes, psi-functions, and so on. Several questions then arise: How are these entities (if they are entities)

related to things in the everyday world? What does a word such as "positron" mean in terms of things we can see, hear, and touch? What is the logical justification for introducing these words which (purport to) refer to unobservable entities? To answer such questions by means of logical and methodological analysis constitutes another part or aspect of what philosophy of science is (according to the conception we are considering).

Now, with regard to the kinds of concepts mentioned in (2), one might ask: Why analyze these concepts? Don't scientists know how to use them? Yes, they certainly know how to use terms such as "electron," "friction coefficient," and so on. And often they pretty much agree about whether statements employing such expressions are true or false. But a philosopher, on the other hand, might be puzzled by such terms. Why? Well no one has ever directly seen a certain subatomic particle, or a frictionless body, or an ideal gas. Now we generally agree that we see physical objects and some of their properties—spatial relations, and so on. The philosopher of science asks (among other things) whether it is possible that a term such as "positron" can be "defined" so that all the terms occurring in the definition (except logical terms, such as "not," "and," "all") refer to physical objects and their properties. He attempts to reduce or trace such "theoretical constructs" to a lower level in the realm of the observable. Why? Because unless this is done, the doors all open to arbitrarily postulating entities such as gremlins, vital forces, and whatnot.

As we can see, throughout such conceptual investigations as those mentioned above, the standpoint adopted by the philosopher of science is often a commonsense standpoint. Thus certain questions which may be asked by other divisions of philosophy (such as epistemology) are not asked here, for example, whether a table really exists. If one wants to say that this means that philosophy of science has certain limitations, then we must agree. But not much follows from admitting this, for those other questions can always be raised later when we turn to other kinds of philosophical problems. Hence for the philosophy of science, we do not need to raise them. We may use the standpoint of common sense.

III. SOME MAIN TOPICS IN PHILOSOPHY OF SCIENCE

The characterization of philosophy of science we have given in the preceding section does not adequately cover all of the kinds of issues and problems generally recognized as falling within the scope of philosophy of science. Hence it is perhaps best to resist trying to find a single formula or "definition" of philosophy of science and to turn to a different task.

Let us now briefly consider some of the main specific topics and questions with which philosophy of science is concerned. (In this volume, we will be able to focus on only some of these issues.)

(1) The formal sciences: logic and mathematics. Logic and math are often referred to as sciences. In what sense, if any, are they sciences? How do we know logical and mathematical truths? What, if anything, are they true of? What is the relation of mathematics to empirical science?

(2) Scientific description. What constitutes an adequate scientific description? What is the "logic" of concept formation which enters into such description?

(3) Scientific explanation. What is meant by saying that science explains? What is a scientific explanation? Are there other kinds of explanations? If so, how are they related to those of science?

(4) Prediction. We say that science predicts. What makes this possible? What is the relation of prediction to explanation? What is the relation of testing to both?

(5) Causality and law. We sometimes hear it said that science explains by means of laws. What are scientific laws? How do they serve to explain? Further, we sometimes speak of explaining laws. How can that be? Many laws are known as causal laws. What does that mean? Are there noncausal laws? If so, what are they?

(6) Theories, models, and scientific systems. We also hear it said that science explains by means of theories. What are theories? How are they related to laws? How do they function in explanation? What is meant by a "model" in science? What role do models play in science?

(7) Determinism. Discussions of lawfulness lead to the question of determinism. What is meant by determinism in science? Is the deterministic thesis (if it is a thesis) true? Or what reason, if any, do we have for thinking it to be true?

(8) Philosophical problems of physical science. The physical sciences have, in recent years, provided a number of philosophical problems, For example, some have held that relativity theory introduces a subjective component into science. Is this true? Others have said that quantum physics denies or refutes determinism. Is this true or false?

(9) Philosophical problems of biology and psychology. First, are these sciences genuinely distinct? If so, why? If not, why not? Further, are these sciences ultimately reducible to physics, or perhaps to physics and chemistry? This gets us into the old "vitalism/mechanism" controversy.

(10) The social sciences. There are some who deny that the social sciences are genuine sciences. Why? Are they right or wrong? Is there any fundamental difference between the natural sciences and the social sciences?

(11) History. Is history a science? We often speak of historical laws. Are there really any such laws? Or are there only general trends? Or neither?

(12) Reduction and the unity of science. We have already briefly referred to this issue. The question here is whether it is possible to reduce one science to another and whether all of the sciences are ultimately reducible to a single

science or a combination of fundamental sciences (such as physics and chemistry).

(13) Extensions of science. Sometimes scientists turn into metaphysicians. They make "radical" statements about the universe—e.g., about the ultimate heat-death, or that it is imbued with moral progress. Is there any validity in these claims?

(14) Science and values. Does science have anything to say with regard to values? Or is it value-neutral?

(15) Science and religion. Do the findings and conclusions of science have any implications for traditional religious or theological commitments? If so, what are they?

(16) Science and culture. Both religion and the domain of values may be considered to be parts or aspects of culture. But surely the term culture also refers to other activities and practices. What is the relationship of science to these?

(17) The limits of science. Are there limits of science? If so, what are they? By what criteria, if any, can we establish that such limits are genuine?

IV. PHILOSOPHY OF SCIENCE AND SCIENCE

We hope that by now the reader has a fair grasp of what philosophy of science is. In order to provide further understanding, let us examine one way by which one might contrast science with the philosophy of science. We may best do this by focusing on the activities and concerns of scientists and of philosophers of science. There are many ways in which these differ. Let us look at just a couple of them. According to one widely held view:

(1) Scientists (among other things, and not necessarily in this order): (a) observe what happens in the world and note regularities; (b) experiment— i.e., manipulate (some) things so that they can be observed under special circumstances; (c) discover (or postulate) laws of nature which are intended to explain regularities; (d) combine laws of nature into theories or subsume those laws under theories. Philosophers of science do none of the above things. Rather, they ask questions such as: What is a law of nature? What is a scientific (vs. a nonscientific or unscientific) theory? What are the criteria (if any) by which to distinguish or demarcate those theories which are genuinely scientific from those which are not? Furthermore, according to this view:

(2) Scientists, like almost everyone else, make deductions. For example, they often construct a certain theory from various laws and observations and then from it deduce other theories or laws, or even certain specific occurrences which serve to test a theory. Philosophers of science do not do that. Rather they clarify the nature of deduction (and how it differs from other inferences or reasoning), and they describe the role deduction plays in science. For example, they ask how deduction is involved in the testing of theories.

From these examples of (some of) the activities of scientists and philosophers of science, we may see that (according to the view we are considering): Whereas science is largely empirical, synthetic, and experimental, philosophy of science is largely verbal, analytic, and reflective. To be sure, in the works of some scientists—especially those who are in the more "theoretical" sciences—verbal, analytic, and reflective features may be found. But the converse is not generally true. The activities of philosophers of science are, for the most part, not empirical or experimental, and they do not add to our store of factual knowledge of the actual world. And even in those cases where the more "philosophical" activities are found in science, they are usually not pursued with the same rigor or toward the same ends as they are by philosophers of science.

We may roughly see the difference by examining the following table:

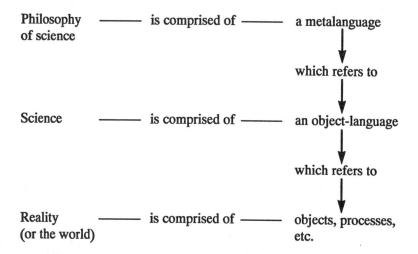

Philosophy ——— is comprised of ——— a metalanguage
of science

which refers to

Science ——— is comprised of ——— an object-language

which refers to

Reality ——— is comprised of ——— objects, processes,
(or the world)   etc.

Thus we may see that, whereas science uses (an object-) language to talk about the objects of the world, philosophy of science (or at least a large "part" of it) uses (a meta-) language to talk about the language of science. In short, as a slight oversimplification, we may say: Science is talk about the world (a certain *kind* of talk, of course). Philosophy is talk about language (again, a certain kind of talk, of course).

To summarize the view we have considered: (1) The sciences consist of such things as listings of data, generalizations from them, the formulation of laws or trends, theoretical interpretations of data or laws, and arguments and evidence in favor of them. (2) Philosophy of science, to a large extent, consists of remarks about the language of science: the analysis of concepts, methods, and arguments of the various sciences; and also the analysis of the principles underlying science.

It is hoped now that our earlier characterization of the philosophy of science may be more readily understood and appreciated. Once again, according to that characterization, philosophy of science is the attempt to understand the meaning, method, and logical structure of science by means of a logical and methodological analysis of the aims, methods, criteria, concepts, laws, and theories of science.

One might reasonably object: But this view of philosophy of science does not do justice or apply to the list of topics in the philosophy of science provided in the preceding section. We are sympathetic to such an objection. Whereas our initial characterization does apply to many of the problems and concerns found in that list, it does not apply to others—for example, the topics of science and religion, or science and culture. Hence we propose that our initial characterization be modified in order to take such matters into account. We propose the following as an amended characterization of the philosophy of science. Philosophy of science is the attempt (a) to understand the method, foundations, and logical structure of science and (b) to examine the relations and interfaces of science and other human concerns, institutions, and quests, by means of (c) a logical and methodological analysis both of the aims, methods, and criteria of science and of the aims, methods, and concerns of various cultural phenomena in their relations to science.

V. The Scope of This Book

As we have mentioned, we cannot within a single volume do justice to all of the topics which fall within the domain of philosophy of science. We have therefore chosen six topics which (a) are crucial ones in philosophy of science, (b) are intrinsically interesting to the layperson as well as to the scientist or philosopher, and (c) are accessible to the beginning student. Similarly, the readings we have selected reflect those features. The topics are:

1. Science and Pseudoscience
2. The Natural and Social Sciences
3. Explanation and Law
4. Theory and Observation
5. Confirmation and Acceptance
6. Science and Values

Since we have provided discussions of these topics in the introductions to the parts of the book, we shall not make further comments about them at this point.

We truly hope that the readers of this volume will derive as much enjoyment from the book as we have had in our production of it. We urge that the Study Questions at the end of each part be utilized. For further reading we have provided selected bibliographies.

E.D.K.

NOTE

1. Many of the views regarding science and the philosophy of science presented in this introduction and in the introduction to Part 1 stem from the lectures and writings of Herbert Feigl, May Brodbeck, John Hospers, and Sir Karl Popper.

SELECTED BIBLIOGRAPHY*

[1] Brody, B., ed. *Readings in the Philosophy of Science*. Englewood Cliffs, N.J.: Prentice-Hall, 1970.

[2] Carnap, R. *An Introduction to the Philosophy of Science*. New York: Basic Books, 1966. [Good for the beginner.]

[3] Danto, A. and Morgenbesser, S. eds. *Philosophy of Science*. New York: Meridian, 1960.

[4] Gale, G. *Theory of Science*. New York: McGraw-Hill, 1979.

[5] Hempel, C. *Philosophy of Natural Science*. Englewood Cliffs, N.J.: Prentice-Hall, 1966. [A short but excellent work.]

[6] Hocking, I. *Representing and Intervening*. Cambridge, Mass.: Cambridge University Press, 1983. [One of the few philosophical studies of experimentation.]

[7] Kuhn, T. *The Structure of Scientific Revolutions*. Chicago: University of Chicago Press, 1962. [Classical critique of positivist and other notions of scientific progress.]

[8] Michalos, A., ed. *Philosophical Problems of Science and Technology*. Rockleigh, N.J.: Allyn & Bacon, 1974.

[9] Morgenbesser, S., ed. *Philosophy of Science Today*. New York: Basic Books, 1967.

[10] Niddith, P. H., ed. *Philosophy of Science*. New York: Oxford University Press, 1968.

[11] Quine, W. V., and Ullian, J. *The Web of Belief*, 2nd ed. New York: Random House, 1978. [Relates philosophy of science to more general issues in theory of knowledge.]

[12] Shapere, D., ed. *Philosophical Problems of Natural Science*. New York: Macmillan, 1964.

[13] Toulmin, S. *The Philosophy of Science*. New York: Harper & Row, 1960. [Still worth reading; author takes a novel approach to science.]

*[1], [3], [8], [9], [10], and [12] are anthologies that deal with various issues in the philosophy of science, including many discussed in this volume.

Part 1

Science and Pseudoscience

Introduction

The major topics we shall discuss in this essay are: the aims of science; the criteria of science, or the criteria for distinguishing that which is scientific from that which is nonscientific; the question "What is science?"; and the central issues of the readings which follow. But, first, let us begin by making some distinctions.

I. SOME DISTINCTIONS

Before turning to the topics above it will be helpful to consider some ways of classifying the various sciences. Among these, the following should be noted.

(1) Pure sciences versus applied sciences. It is widely held that we must distinguish: (A) science as a field of knowledge (or set of cognitive disciplines) from (B) the applications of science. It is common to refer to these as the pure and applied sciences. (A) Among the pure sciences we may distinguish: (a) the formal sciences, logic, and mathematics; and (b) the factual or empirical sciences. Among the latter we may also distinguish: (b1) the natural sciences, which include the physical sciences, physics, chemistry, and so on, and the life and behavioral sciences, such as biology and psychology; and (b2) the social sciences, such as sociology and economics. (B) The applied sciences include the technological sciences—such as engineering and aeronautics, medicine, agriculture, and so on.

It should be noted that there are at least two levels of application among the various sciences. There is, first, the application of the formal sciences to the pure, factual sciences. Since the factual sciences must have logical form and usually utilize some mathematics, such application is often held to be essential for the development of the pure factual sciences. Different from this is, second, the application of the factual sciences to the applied sciences. Here the findings of the pure, empirical sciences are applied (in a different sense of "applied") to disciplines which fulfill various social, human purposes, such as building houses or roads and health care.

(2) Law-finding sciences versus fact-finding sciences. We recognize that such sciences as chemistry and physics attempt to discover universal laws which are applicable everywhere at all times, whereas such sciences as geography, history (if it is a science), and perhaps economics are concerned with local events. It is often said that the subject matter of the latter consists of particular facts, not general laws. As a result, there are some who wish to limit the term "science" to the law-finding sciences. Upon the basis of the criteria of science (such as those which will be presented later, or others), we believe that we may say that both the law-finding disciplines and the fact-finding disciplines are capable of being sciences if those (or other) criteria are met. Furthermore, one might argue that there are no purely fact-finding sciences. If so, to speak of law-finding versus fact-finding may, in many cases, indicate an artificial disjunction.

(3) Natural sciences versus social sciences. Related to (2), we find that some would limit the giving of scientific status to the natural sciences alone. Sometimes the reason given is the distinction referred to above—that the natural sciences are primarily law-finding, whereas the social sciences are predominantly fact-finding. But sometimes the distinction is based on subject matter. Hence it is held by some that natural phenomena constitute the field of science but cultural phenomena constitute the field of scholarship and require understanding, *verstehen*, and empathy. But there are points at which the classification does not hold up. First, there are some predominantly fact-finding natural sciences, such as geography, geology, and paleontology. And there are some law-finding social sciences, such as sociology and linguistics. Second, the distinction according to subject matter is not a clear-cut one. Hence we shall take a "liberal" view of science and allow the use of the term "science" to apply to both the natural and the social sciences—with the recognition that there are some differences.

It is widely held that distinctions (2) and (3) do not hold up but that (1) is an acceptable distinction. However, as we shall see in the readings which follow, some have even raised doubts about the significance of (1). Here, as always, we urge the reader to reflect upon these matters.

II. THE AIMS OF SCIENCE

Let us now turn to the question "What are the aims of science?" Using the above distinction between pure (empirical) and applied science, we may then cite the following as some of the aims of science.

(1) The aims of applied science include: control, planning, technological progress; the utilization of the forces of nature for practical purposes. Obvious examples are: flood control, the construction of sturdier bridges, and the improvement of agriculture. Since this is all fairly obvious, no further elaboration is needed.

(2) The aims of the pure, factual sciences may be considered from two standpoints. (a) Psychologically considered, the aims of the pure, empirical sciences are: the pursuit of knowledge; the attainment of truth (or the closest possible realization of truth); the satisfaction of using our intellectual powers to explain and predict accurately. Scientists, of course, derive enjoyment from rewards, prestige, and competing with others. But they often achieve a genuine inner gratification which goes with the search for truth. In some ways this is similar in quality to artistic satisfaction. It is seen, for example, in the enjoyment one derives from the solution of a difficult problem.

(b) Logically considered, the aims of the pure, factual sciences are often held to be: description, explanation, and prediction. (b1) Description includes giving an account of what we observe in certain contexts, the formulation of propositions which apply to (or correspond to) facts in the world. (b2) Explanation consists of accounting for the facts and regularities we observe. It involves asking and answering "Why?" or "How come?" This may be done by subsuming facts under laws and theories. (b3) Prediction is closely related to explanation. It consists in deriving propositions which refer to events which have not yet happened, the deducing of propositions from laws and theories and then seeing if they are true, and hence provide a testing of those laws and theories. (b4) We might also mention post- or retrodiction, the reconstruction of past events. This process is also inferential in character. Since these issues will be discussed in subsequent parts of this volume, we shall not elaborate upon them at this time. (See some of the readings in part 1 and those in parts 2, 3, and 4.) However, we might mention that, here again, there is not unanimity with regard to the aims characterized above. Once more we urge the reader to think about these (and other) issues.

III. THE CRITERIA OF SCIENCE

In this section we shall state and discuss one view with regard to what are the essential criteria of science, that is, those criteria which may be used for at least two purposes: first, to distinguish science from commonsense knowledge (without claiming that the two are radically disjunctive—in some cases they may differ only in degree, not in kind); second, to distinguish that which is scientific, on the one hand, from that which is either nonscientific or unscientific, on the other—for example, to distinguish between theories which are genuinely scientific and those which are not. It has been maintained that any enterprise, discipline, or theory is scientific if it is characterized by or meets those criteria.

Before turning to the view which we have selected for consideration, let us consider an example. It is quite likely that most scientists and others who reflect upon science would hold that (say) Newton's theory of gravitation is scientific (even if it had to be modified), whereas (say) astrology is not scientific. Perhaps the reader would agree. But just what is it that allows us to rule in Newton's theory and to rule out astrology? In order to stimulate the reader's reflection, we shall consider one view of what the criteria for making such distinctions are. These criteria have been stated by Professor Herbert Feigl in various lectures and in writing. Our discussion of them corresponds fairly closely to the discussion given by Professor Feigl.

The five criteria are:

(1) Intersubjective testability. This refers to the possibility of being, in principle, capable of corroboration or "check-up" by anyone. Hence: *inter*-subjective. (Hence, private intuitions and so forth must be excluded.)

(2) Reliability. This refers to that which, when put to a test, turns out to be true, or at least to be that which we can most reasonably believe to be true. Testing is not enough. We want theories which, when tested, are found to be true.

(3) Definiteness and precision. This refers to the removal of vagueness and ambiguity. We seek, for example, concepts which are definite and delimited. We are often helped here by measurement techniques and so forth.

(4) Coherence or systematic character. This refers to the organizational aspect of a theory. A set of disconnected statements is not as fruitful as one which has systematic character. It also refers to the removal of, or being free from, contradictoriness.

(5) Comprehensiveness or scope. This refers to our effort to attain a continual increase in the completeness of our knowledge and also to our seeking theories which have maximum explanatory power—for example, to account for things which other theories do not account for.

Let us consider these criteria in greater detail.

(1) Intersubjective testability. (a) Testability. We have noted that in science we encounter various kinds of statements: descriptions, laws, theoretical explanations, and so on. These are put forth as knowledge-claims. We must (if possible) be able to tell whether evidence speaks for or against such knowledge-claims. If the propositions which express those claims are not capable of tests, we cannot call those propositions true or false or even know how to go about establishing their truth or falsity. It should be noted that the criterion is one of testability, not tested. For example, at a given point in time, "There are mountains on the far side of the moon" was testable though not tested.

(b) Intersubjective. "Intersubjectivity" is often employed as a synonym for "objectivity." And the latter term has various meanings. Some of these are: (i) A view or belief is said to be objective if it is not based on illusions, hallucinations, deceptions, and so on. (ii) Something is referred to as objective if it is not merely a state of mind but is really "out there" in the external world. (iii) We often use "objective" to indicate the absence of bias and the presence of disinterestedness and dispassionateness. (iv) "Objectivity" also refers to the possibility of verification by others, and hence excludes beliefs, which stem from private, unique, unrepeatable experiences. Science strives for objectivity in all of these senses. Hence we take "intersubjective" to include all of them.

(c) Intersubjective testability. It is often held that (according to the view we are considering) in order for a proposition or theory to be judged scientific it must meet this first requirement. Indeed, many of the other criteria presuppose intersubjective testability. We cannot even begin to talk of reliability or precision unless this first criterion has been met.

(2) Reliability. Science is not merely interested in hypotheses which are intersubjectively testable. It is also interested in those which are true or at least have the greatest verisimilitude or likelihood of being true. Hence the need arises for the criterion of reliability. Whereas the first criterion stressed the possibility of finding assertions which are true *or* false, the second stresses the end result of that process. We judge a claim or body of knowledge to be reliable if it contains not merely propositions which are capable of being true *or* false but rather those which *are true* or which have the greatest verisimilitude. We find such propositions to be true (or false) by means of confirmation. Complete verification, and hence complete certainty, cannot be achieved in the factual sciences.

It should be noted that, first, the reliability of scientific assertions make them useful for prediction; second, although the assertions of many enterprises are testable (for example, those of astrology as much as those of astronomy), only some of them are reliable. And we reject some of them precisely because they are unreliable. The evidence is against them; we do not attain truth by means of them.

(3) Definiteness and precision. The terms "definiteness" and "precision" may be used in at least two related senses. First, they refer to the delimitation of our concepts and to the removal of ambiguity or vagueness. Second, they refer to a more rigid or exact formulation of *laws*. For example, "It is more probable than not that X causes disease Y" is less desirable than "The probability that X causes Y is 98%."

(4) Coherence or systematic character. In the sciences, we seek not merely disorganized or loosely related facts but a well-connected account of the facts. It has been held by many that we achieve this via what has been called the hypothetico-deductive procedure of science. This procedure includes: (a) our beginning with a problem (which pertains to some realm of phenomena); (b) the formulation of hypotheses, laws, and theories by which

to account for those phenomena or by which to resolve the problem; (c) the deriving (from (b)) of statements which refer to observable facts; (d) the testing of those deduced assertions to see if they hold up. Thus we seek an integrated, unified network, not merely a congeries of true statements.

But, of course, we also seek theories which are consistent, which are free from self-contradictions. The reason for insisting upon such coherence is obvious; hence there is no need for elaboration.

(5) Comprehensiveness or scope. The terms "comprehensiveness," and "scope" are also used in two senses, both of which are essential in science. First, a theory is said to be comprehensive if it possesses maximum explanatory power. Thus Newton's theory of gravitation was ranked high partly because it accounted not only for the laws of falling bodies but also for the revolution of the heavenly bodies and for the laws of tides. Second, by "comprehensive" we often refer to the completeness of our knowledge. This of course does not mean finality. We do not think of the hypotheses of the empirical sciences as being certain for all time. Rather we must be ready to modify them or even, on occasion, to abandon them.

To summarize: According to the view we have presented, we judge a law, hypothesis, theory, or enterprise to be scientific if it meets all five of the above criteria. If it fails to meet all five, it is judged to be unscientific or at least non-scientific. To return to our earlier example, it seems clear that Newton's theory thus passes the test. Astrological theory or Greek mythology does not.

It should be noted that, in presenting Professor Feigl's criteria for the reader's consideration, we do not claim that they are correct or free from defects. Indeed, as we shall see in the readings which follow, many writers have rejected some (or all) of those criteria. The reader should once again attempt to seek an acceptable criterion or set of criteria, if such can be found.

IV. WHAT IS SCIENCE?

A common characterization of science (or sometimes of scientific method) runs as follows. Science is knowledge obtained by: (1) making observations as accurate and definite as possible; (2) recording these intelligibly; (3) classifying them according to the subject matter being studied; (4) extracting from them, by induction, general statements (laws) which assert regularities; (5) deducing other statements from these; (6) verifying those statements by further observation; and (7) propounding theories which connect and so account for the largest possible number of laws. It is further maintained that this process runs from (1) through (7) *in that order.*

The conception of science has been challenged in recent years. Its most severe critic is Sir Karl Popper. (See the selection in part 1 of this volume.) We shall not repeat Popper's criticisms. Instead we offer a characterization of science which some believe to be more adequate than the one mentioned above and which they deem to be free from the defects it possesses.

According to this view, the following is at least a minimal characterization of (factual) science (or of a science).

Science is a body of knowledge which consists of the following, coherently organized in a systematic way:

(a) Statements which record and classify observations which are relevant for the solution of a problem in as accurate and definite a way as possible.

(b) General statements—laws or hypotheses—which assert regularities among certain classes of observed or observable phenomena.

(c) Theoretical statements which connect and account for the largest possible number of laws.

(d) Other general or specific statements which are deducible from the initial descriptions and from laws and theories and which are confirmed by further observation and testing.

At least two things should be noted about this characterization. First, it indicates the role of the formal sciences in the empirical sciences. Mathematics is important for (a); logic is important for (d). Second, nothing is said in this characterization about the *method* of obtaining knowledge or of obtaining laws. It may be induction, but it may also be a guess, intuition, hunch, or whatever.

Since a number of the readings in part 1 deal with the question "What is science?" we shall not attempt to provide a "final" answer. Instead, we encourage the student to come up with the best answer possible, based on his or her reading and reflection.

V. THE READINGS IN PART 1

Since the essays contained in part 1 are clearly written and since they are accessible to the beginning student or ordinary reader, no detailed summaries will be presented here. We urge the student to prepare his or her own summaries and to make use of the Study Questions at the end. However, a few brief remarks may be helpful.

Throughout many of his works, Sir Karl Popper has been concerned with the problem of how to distinguish between science and pseudoscience (or nonscience). He claims to have solved that problem by having provided a criterion of demarcation, a criterion by which to distinguish theories which are genuinely scientific from those which are not. By means of this criterion—of falsifiability or refutability—he attempts to show that Einstein's theory of gravitation satisfies the criterion (and hence is scientific) whereas astrology, the Marxist theory of history, and various psychoanalytic theories—for varying reasons—are not scientific. He also wishes to separate the problem of demarcation from the problem of meaning and maintains that the latter is a pseudoproblem. (The reader should reflect upon why he holds that it is a pseudoproblem and whether he has succeeded in showing that it is.)

In the middle sections of Popper's essay, he claims that the problem of demarcation has provided a key for solving a number of philosophical problems, especially the problem of induction. Since this issue does not pertain to the main topics of part 1, most of those sections of the essay have been deleted here. The problem of induction is: How, if at all, can we justify our knowledge-claims concerning matters of fact which we have not yet experienced or are not now experiencing? In the eighteenth century, David Hume maintained that we cannot provide any rational justification. Popper agrees with Hume's logical refutation of induction but disagrees with his psychological explanation of induction (in terms of custom or habit).

The selection by John Ziman consists of extracts from his book on science. In the first part he discusses and rejects various definitions of science which have been held. And he attempts to formulate a more accurate and tenable characterization based on what he takes to be the goal or objective of science, namely, consensus of rational opinion "over the widest possible field." In the second part, he provides his answer to the question "What distinguishes science from nonscience?" The reader should attempt to decide whether his "criterion of demarcation" is an improvement over Popper's and, if so, why. Since this selection is unusually clear and readable, no further comments are required.

Feyerabend's essay is, no doubt, one of the most controversial ones in this volume. Feyerabend claims that he wishes to defend society and its inhabitants from all ideologies, including science. He likens them (again, including science) to fairytales "which have lots of interesting things to say but which also contain wicked lies." He goes on to consider an argument designed to defend the exceptional status which science has in society today. According to this argument "(1) science has finally found the correct method for achieving results and (2) . . . there are many results to prove the excellence of the method." In the next sections he argues against both (1) and (2). He concludes his essay with a provocative discussion of education and myth. We urge the reader to reflect seriously upon Feyerabend's somewhat unorthodox views and to ask whether Feyerabend has adequately defended them.

Paul R. Thagard's essay constitutes both a further discussion of some of the above-mentioned topics (such as the criterion of demarcation) and an example of the application of them. Most scientists and philosophers agree that astrology is a pseudoscience. Thagard attempts to show why it is. After presenting a brief description of astrology, he attempts to show that the major objections which have been provided do not show that it is a pseudoscience. Thagard then proposes his principle of demarcation and, upon the basis of it, claims to show that and why astrology is unscientific.

In his important essay, Philip Kitcher specifies various criteria which must be met before a view or a criticism can be scientific. He then applies this to the views of Creationists and also to their criticisms of evolutionary theory. He attempts to show that their views and criticisms are either fallacious or totally unsupported.

There is a kind of dialogue which runs through the essays in this part. We urge the reader to critically evaluate the various positions presented and attempt to come to his or her own conclusion with regard to the questions "What is science?" "By what criteria can we distinguish science from nonscience or pseudoscience?" and so on. The Study Questions should provide assistance in gauging the reader's understanding of the selections and in grappling with these and related questions.

E. D. K.

1

Science: Conjectures and Refutations*

Sir Karl Popper

Mr. Turnbull had predicted evil consequences, . . . and was now doing the best in his power to bring about the verification of his own prophecies.

ANTHONY TROLLOPE

I

When I received the list of participants in this course and realized that I had been asked to speak to philosophical colleagues I thought, after some hesitation and consultation, that you would probably prefer me to speak about those problems which interest me most, and about those developments with which I am most intimately acquainted. I therefore decided to do what I have never done before: to give you a report on my own work in the philosophy of science, since the autumn of 1919 when I first began to grapple with the problem, *"When should a theory be ranked as scientific?"* or *"Is there a criterion for the scientific character or status of a theory?"*

The problem which troubled me at the time was neither, "When is a theory true?" nor, "When is a theory acceptable?" My problem was different. I *wished to distinguish between science and pseudoscience*; knowing very well that science often errs, and that pseudoscience may happen to stumble on the truth.

I knew, of course, the most widely accepted answer to my problem: that science is distinguished from pseudoscience—or from "metaphysics"—by

---

*A lecture given at Peterhouse, Cambridge, in Summer 1953, as part of a course on developments and trends in contemporary British philosophy, organized by the British Council, originally published under the title "Philosophy of Science. a Personal Report" in *British Philosophy in Mid-Century*, ed. C. A. Mace, 1957. [Portions have been deleted by the editors for this publication.]

its empirical method, which is essentially *inductive,* proceeding from observation or experiment. But this did not satisfy me. On the contrary, I often formulated my problem as one of distinguishing between a genuinely empirical method and a nonempirical or even a pseudoempirical method—that is to say, a method which, although it appeals to observation and experiment, nevertheless does not come up to scientific standards. The latter method may be exemplified by astrology with its stupendous mass of empirical evidence based on observation—on horoscopes and on biographies.

But as it was not the example of astrology which led me to my problem I should perhaps briefly describe the atmosphere in which my problem arose and the examples by which it was stimulated. After the collapse of the Austrian Empire there had been a revolution in Austria: the air was full of revolutionary slogans and ideas, and new and often wild theories. Among the theories which interested me Einstein's theory of relativity was no doubt by far the most important. Three others were Marx's theory of history, Freud's psychoanalysis, and Alfred Adler's so-called "individual psychology."

There was a lot of popular nonsense talked about these theories, and especially about relativity (as still happens even today), but I was fortunate in those who introduced me to the study of this theory. We all—the small circle of students to which I belonged—were thrilled with the result of Eddington's eclipse observations which in 1919 brought the first important confirmation of Einstein's theory of gravitation. It was a great experience for us, and one which had a lasting influence on my intellectual development.

The three other theories I have mentioned were also widely discussed among students at that time. I myself happened to come into personal contact with Alfred Adler, and even to cooperate with him in his social work among the children and young people in the working-class districts of Vienna where he had established social guidance clinics.

It was during the summer of 1919 that I began to feel more and more dissatisfied with these three theories—the Marxist theory of history, psychoanalysis, and individual psychology; and I began to feel dubious about their claims to scientific status. My problem perhaps first took the simple form, "What is wrong with Marxism, psychoanalysis, and individual psychology? Why are they so different from physical theories, from Newton's theory, and especially from the theory of relativity?"

To make this contrast clear I should explain that few of us at the time would have said that we believed in the truth of Einstein's theory of gravitation. This shows that it was not my doubting the truth of those other three theories which bothered me, but something else. Yet neither was it that I merely felt mathematical physics to be more exact than the sociological or psychological type of theory. Thus what worried me was neither the problem of truth, at that stage at least, nor the problem of exactness or measurability. It was rather that I felt that these other three theories, though posing as sciences, had in fact more in common with primitive myths than with science; that they resembled astrology rather than astronomy.

I found that those of my friends who were admirers of Marx, Freud, and Adler, were impressed by a number of points common to these theories, and especially by their apparent explanatory power. These theories appeared to be able to explain practically everything that happened within the fields to which they referred. The study of any of them seemed to have the effect of an intellectual conversion or revelation, opening your eyes to a new truth hidden from those not yet initiated. Once your eyes were thus opened you saw confirming instances everywhere: the world was full of *verifications* of the theory. Whatever happened always confirmed it. Thus its truth appeared manifest; and unbelievers were clearly people who did not want to see the manifest truth; who refused to see it, either because it was against their class interest, or because of their repressions which were still "unanalyzed" and crying aloud for treatment.

The most characteristic element in this situation seemed to me the incessant stream of confirmations, of observations which "verified" the theories in question; and this point was constantly emphasized by their adherents. A Marxist could not open a newspaper without finding on every page confirming evidence for his interpretation of history; not only in the news, but also in its presentation—which revealed the class bias of the paper—and especially of course in what the paper *did not* say. The Freudian analysts emphasized that their theories were constantly verified by their "clinical observations." As for Adler, I was much impressed by a personal experience. Once, in 1919, I reported to him a case which to me did not seem particularly Adlerian, but which he found no difficulty in analyzing in terms of his theory of inferiority feelings, although he had not even seen the child. Slightly shocked, I asked him how he could be so sure. "Because of my thousandfold experience," he replied; whereupon I could not help saying: "And with this new case, I suppose, your experience has become thousand-and-one-fold."

What I had in mind was that his previous observations may not have been much sounder than this new one; that each in its turn had been interpreted in the light of "previous experience," and at the same time counted as additional confirmation. What, I asked myself, did it confirm? No more than that a case could be interpreted in the light of the theory. But this meant very little, I reflected, since every conceivable case could be interpreted in the light of Adler's theory, or equally of Freud's. I may illustrate this by two very different examples of human behavior: that of a man who pushes a child into the water with the intention of drowning it; and that of a man who sacrifices his life in an attempt to save the child. Each of these two cases can be explained with equal ease in Freudian and in Adlerian terms. According to Freud the first man suffered from repression (say, of some component of his Oedipus complex), while the second man had achieved sublimation. According to

Adler the first man suffered from feelings of inferiority (producing perhaps the need to prove to himself that he dared to commit some crime), and so did the second man (whose need was to prove to himself that he dared to rescue the child). I could not think of any human behavior which could not be interpreted in terms of either theory. It was precisely this fact—that they always fitted, that they were always confirmed—which in the eyes of their admirers constituted the strongest argument in favor of these theories. It began to dawn on me that this apparent strength was in fact their weakness.

With Einstein's theory the situation was strikingly different. Take one typical instance—Einstein's prediction, just then confirmed by the findings of Eddington's expedition. Einstein's gravitational theory had led to the result that light must be attracted by heavy bodies (such as the sun), precisely as material bodies were attracted. As a consequence it could be calculated that light from a distant fixed star whose apparent position was close to the sun would reach the earth from such a direction that the star would seem to be slightly shifted away from the sun; or, in other words, that stars close to the sun would look as if they had moved a little away from the sun, and from one another. This is a thing which cannot normally be observed since such stars are rendered invisible in daytime by the sun's overwhelming brightness; but during an eclipse it is possible to take photographs of them. If the same constellation is photographed at night one can measure the distances on the two photographs, and check the predicted effect.

Now the impressive thing about this case is the *risk* involved in a prediction of this kind. If observation shows that the predicted effect is definitely absent, then the theory is simply refuted. The theory is *incompatible with certain possible results of observation*—in fact with results which everybody before Einstein would have expected.[1] This is quite different from the situation I have previously described, when it turned out that the theories in question were compatible with the most divergent human behavior, so that it was practically impossible to describe any human behavior that might not be claimed to be a verification of these theories.

These considerations led me in the winter of 1919–20 to conclusions which I may now reformulate as follows.

(1) It is easy to obtain confirmations, or verifications, for nearly every theory—if we look for confirmations.

(2) Confirmations should count only if they are the result of *risky predictions*; that is to say, if, unenlightened by the theory in question, we should have expected an event which was incompatible with the theory—an event which would have refuted the theory.

(3) Every "good" scientific theory is a prohibition: it forbids certain things to happen. The more a theory forbids, the better it is.

(4) A theory which is not refutable by any conceivable event is nonscientific. Irrefutability is not a virtue of theory (as people often think) but a vice.

(5) Every genuine *test* of a theory is an attempt to falsify it, or to refute it. Testability is falsifiability; but there are degrees of testability; some theories are more testable, more exposed to refutation, than others; they take, as it were, greater risks.

(6) Confirming evidence should not count *except when it is the result of a genuine test of the theory*; and this means that it can be presented as a serious but unsuccessful attempt to falsify the theory. (I now speak in such cases of "corroborating evidence.")

(7) Some genuinely testable theories, when found to be false, are still upheld by their admirers—for example by introducing ad hoc some auxiliary assumption, or by reinterpreting the theory ad hoc in such a way that it escapes refutation. Such a procedure is always possible, but it rescues the theory from refutation only at the price of destroying, or at least lowering, its scientific status. (I later described such a rescuing operation as a "*conventionalist twist*" or a "*conventionalist stratagem.*")

One can sum up all this by saying that the *criterion of the scientific status of a theory is its falsifiability, or refutability, or testability.*

II

I may perhaps exemplify this with the help of the various theories so far mentioned. Einstein's theory of gravitation clearly satisfied the criterion of falsifiability. Even if our measuring instruments at the time did not allow us to pronounce on the results of the tests with complete assurance, there was clearly a possibility of refuting the theory.

Astrology did not pass the test. Astrologers were greatly impressed, and misled, by what they believed to be confirming evidence—so much so that they were quite unimpressed by any unfavorable evidence. Moreover, by making their interpretations and prophecies sufficiently vague they were able to explain away anything that might have been a refutation of the theory had the theory and the prophecies been more precise. In order to escape falsification they destroyed the testability of their theory. It is a typical soothsayer's trick to predict things so vaguely that the predictions can hardly fail: that they become irrefutable.

The Marxist theory of history, in spite of the serious efforts of some of its founders and followers, ultimately adopted this soothsaying practice. In some of its earlier formulations (for example in Marx's analysis of the character of the "coming social revolution") their predictions were testable, and in fact falsified.[2] Yet instead of accepting the refutations the followers of Marx reinterpreted both the theory and the evidence in order to make them agree. In this way they rescued the theory from refutation; but they did so at the price of adopting a device which made it irrefutable. They thus gave a "conventionalist twist" to the theory; and by this stratagem they destroyed its much advertised claim to scientific status.

The two psychoanalytic theories were in a different class. They were simply nontestable, irrefutable. There was no conceivable human behavior which could contradict them. This does not mean that Freud and Adler were not seeing certain things correctly: I personally do not doubt that much of what they say is of considerable importance, and may well play its part one day in a psychological science which is testable. But it does mean that those "clinical observations" which analysts naively believe confirm their theory cannot do this any more than the daily confirmations which astrologers find in their practice.[3] And as for Freud's epic of the ego, the superego, and the id, no substantially stronger claim to scientific status can be made for it than for Homer's collected stories from Olympus. These theories describe some facts, but in the manner of myths. They contain most interesting psychological suggestions, but not in a testable form.

At the same time I realized that such myths may be developed, and become testable; that historically speaking all—or very nearly all—scientific theories originate from myths, and that a myth may contain important anticipations of scientific theories. Examples are Empedocles' theory of evolution by trial and error, or Parmenides' myth of the unchanging block universe in which nothing ever happens and which, if we add another dimension, becomes Einstein's block universe (in which, too, nothing ever happens, since everything is, four-dimensionally speaking, determined and laid down from the beginning). I thus felt that if a theory is found to be nonscientific, or "metaphysical" (as we might say), it is not thereby found to be unimportant, or insignificant, or "meaningless," or "nonsensical."[4] But it cannot claim to be backed by empirical evidence in the scientific sense—although it may easily be, in some genetic sense, the "result of observation."

(There were a great many other theories of this prescientific or pseudoscientific character, some of them, unfortunately, as influential as the Marxist interpretation of history; for example, the racialist interpretation of history—another of those impressive and all-explanatory theories which act upon weak minds like revelations.)

Thus the problem which I tried to solve by proposing the criterion of falsifiability was neither a problem of meaningfulness or significance, nor a problem of truth or acceptability. It was the problem of drawing a line (as well as this can be done) between the statements, or systems of statements, of the empirical sciences, and all other statements—whether they are of a religious or of a metaphysical character, or simply pseudoscientific. Years later—it must have been in 1928 or 1929—I called this first problem of mine the *"problem of demarcation."* The criterion of falsifiability is a solution to this problem of demarcation, for it says that statements or systems of statements, in order to be ranked as scientific, must be capable of conflicting with possible, or conceivable, observations. . . .

III

Let us now turn from our logical criticism of the *psychology of experience* to our real problem—the problem of the *logic of science*. Although some of the things I have said may help us here, insofar as they may have eliminated certain psychological prejudices in favor of induction, my treatment of the *logical problem of induction* is completely independent of this criticism, and of all psychological considerations. Provided you do not dogmatically believe in the alleged psychological fact that we make inductions, you may now forget my whole story with the exception of two logical points: my logical remarks on testability or falsifiability as the criterion of demarcation; and Hume's logical criticism of induction.

From what I have said it is obvious that there was a close link between the two problems which interested me at that time: demarcation, and induction or scientific method. It was easy to see that the method of science is criticism, i.e., attempted falsifications. Yet it took me a few years to notice that the two problems—of demarcation and of induction—were in a sense one. . . .

I recently came across an interesting formulation of this belief in a remarkable philosophical book by a great physicist—Max Born's *Natural Philosophy of Cause and Chance.*[5] He writes: "Induction allows us to generalize a number of observations into a general rule: that night follows day and day follows night. . . . But while everyday life has no definite criterion for the validity of an induction, . . . science has worked out a code, or rule of craft, for its application." Born nowhere reveals the contents of this inductive code (which, as his wording shows, contains a "definite criterion for the validity of an induction"); but he stresses that "there is no logical argument" for its acceptance: "it is a question of faith"; and he is therefore "willing to call induction a metaphysical principle." But why does he believe that such a code of valid inductive rules must exist? This becomes clear when he speaks of the "vast communities of people ignorant of, or rejecting, the rule of science, among them the members of antivaccination societies and believers in astrology. It is useless to argue with them; I cannot compel them to accept the same criteria of valid induction in which I believe: the code of scientific rules." This makes it quite clear that *"valid induction" was here meant to serve as a criterion of demarcation between science and pseudoscience.*

But it is obvious that this rule or craft of "valid induction" is not even metaphysical: it simply does not exist. No rule can ever guarantee that a generalization inferred from true observations, however often repeated, is true. (Born himself does not believe in the truth of Newtonian physics, in spite of its success, although he believes that it is based on induction.) And the success of science is not based upon rules of induction, but depends upon luck, ingenuity, and the purely deductive rules of critical argument.

I may summarize some of my conclusions as follows:

(1) Induction, i.e., inference based on many observations, is a myth. It is neither a psychological fact, nor a fact of ordinary life, nor one of scientific procedure.

(2) The actual procedure of science is to operate with conjectures: to jump to conclusions—often after one single observation (as noticed for example by Hume and Born).

(3) Repeated observations and experiments function in science as tests of our conjectures or hypotheses, i.e., as attempted refutations.

(4) The mistaken belief in induction is fortified by the need for a criterion of demarcation which, it is traditionally but wrongly believed, only the inductive method can provide.

(5) The conception of such an inductive method, like the criterion of verifiability, implies a faulty demarcation.

(6) None of this is altered in the least if we say that induction makes theories only probable rather than certain.

IV

If, as I have suggested, the problem of induction is only an instance or facet of the problem of demarcation, then the solution to the problem of demarcation must provide us with a solution to the problem of induction. This is indeed the case, I believe, although it is perhaps not immediately obvious.

For a brief formulation of the problem of induction we can turn again to Born, who writes: "... no observation or experiment, however extended, can give more than a finite number of repetitions"; therefore, "the statement of a law—B depends on A—always transcends experience. Yet this kind of statement is made everywhere and all the time, and sometimes from scanty material."[6]

In other words, the logical problem of induction arises from (a) Hume's discovery (so well expressed by Born) that it is impossible to justify a law by observation or experiment, since it "transcends experience"; (b) the fact that science proposes and uses laws "everywhere and all the time." (Like Hume, Born is struck by the "scanty material," i.e., the few observed instances upon which the law may be based.) To this we have to add (c) *the principle of empiricism* which asserts that in science, only observation and experiment may decide upon the *acceptance or rejection* of scientific statements, including laws and theories.

These three principles, (a), (b), and (c), appear at first sight to clash; and this apparent clash constitutes the *logical problem of induction.*

Faced with this clash, Born gives up (c), the principle of empiricism (as Kant and many others, including Bertrand Russell, have done before him), in favor of what he calls a "metaphysical principle," a metaphysical principle which he does not even attempt to formulate; which he vaguely describes as a "code or rule of craft," and of which I have never seen any formulation which even looked promising and was not clearly untenable.

But in fact the principles (a) to (c) do not clash. We can see this the moment we realize that the acceptance by science of a law or of a theory is *tentative only*; which is to say that all laws and theories are conjectures, or tentative *hypotheses* (a position which I have sometimes called "hypotheticism"); and that we may reject a law or theory on the basis of new evidence, without necessarily discarding the old evidence which originally led us to accept it.[7]

The principles of empiricism (c) can be fully preserved, since the fate of a theory, its acceptance or rejection, is decided by observation and experiment by the result of tests. So long as a theory stands up to the severest tests we can design, it is accepted; if it does not, it is rejected. But it is never inferred, in any sense, from the empirical evidence. There is neither a psychological nor a logical induction. *Only the falsity of the theory can be inferred from empirical evidence, and this inference is a purely deductive one.*

Hume showed that it is not possible to infer a theory from observation statements; but this does not affect the possibility of refuting a theory by observation statements. The full appreciation of the possibility makes the relation between theories and observations perfectly clear.

This solves the problem of the alleged clash between the principles (a), (b), and (c), and with it Hume's problem of induction. . . .

NOTES

1. This is a slight oversimplification, for about half of the Einstein effect may be derived from the classical theory, provided we assume a ballistic theory of light.

2. See, for example, my *Open Society and Its Enemies,* ch. 15, section iii, and notes 13–14.

3. "Clinical observations," like all other observations, are *interpretations in the light of theories*; and for this reason alone they are apt to seem to support those theories in the light of which they were interpreted. But real support can be obtained only from observations undertaken as tests (by "attempted refutations"); and for this purpose *criteria of refutation* have to be laid down beforehand: it must be agreed which observable situations, if actually observed, mean that the theory is refuted. But what kind of clinical responses would refute to the satisfaction of the analyst not merely a particular analytic diagnosis but psychoanalysis itself? And have such criteria ever been discussed or agreed upon by analysts? Is there not, on the contrary, a whole family of analytic concepts, such as "ambivalence" (I do not suggest that there is no such thing as ambivalence), which would make it difficult, if not impossible, to agree upon such criteria? Moreover, how much headway has been made in investigating the question of the extent to which the (conscious or unconscious) expectations and theories held by the analyst influence the "clinical responses" of the patient? (To say nothing about the conscious attempts to influence the patient by proposing interpretations to him, etc.) Years ago I introduced the term "Oedipus effect" to describe the influence of a theory or expectation or prediction *upon*

the event which it predicts or describes: it will be remembered that the causal chain leading to Oedipus' parricide was started by the oracle's prediction of this event. This is a characteristic and recurrent theme of such myths, but one which seems to have failed to attract the interest of the analysts, perhaps not accidentally. (The problem of confirmatory dreams suggested by the analyst is discussed by Freud, for example in *Gesammelte Schriften*, III, 1925, where he says on page 314: "If anybody asserts that most of the dreams which can be utilized in an analysis ... owe their origin to [the analyst's] suggestion, then no objection can be made from the point of view of analytic theory. Yet there is nothing in this fact," he surprisingly adds, "which would detract from the reliability of our results.")

4. The case of astrology, nowadays a typical pseudoscience, may illustrate this point. It was attacked, by Aristotelians and other rationalists, down to Newton's day, for the wrong reason—for its now accepted assertion that the planets had an "influence" upon terrestrial ("sublunar") events. In fact Newton's theory of gravity, and especially the lunar theory of the tides, was historically speaking an offspring of astrological lore. Newton, it seems, was most reluctant to adopt a theory which came from the same stable as for example the theory that "influenza" epidemics are due to an astral "influence." And Galileo, no doubt for the same reason, actually rejected the lunar theory of the tides; and his misgivings about Kepler may easily be explained by his misgivings about astrology.

5. Max Born, *Natural Philosophy of Cause and Chance,* Oxford, 1949, p. 7.

6. *Natural Philosophy of Cause and Chance,* p. 6.

7. I do not doubt that Born and many others would agree that theories are accepted only tentatively. But the widespread belief in induction shows that the far-reaching implications of this view are rarely seen.

2

What Is Science?

John Ziman

To answer the question "What is science?" is almost as presumptuous as to try to state the meaning of life itself. Science has become a major part of the stock of our minds; its products are the furniture of our surroundings. We must accept it, as the good lady of the fable is said to have agreed to accept the Universe.

Yet the question is puzzling rather than mysterious. Science is very clearly a conscious artifact of mankind, with well-documented historical origins, with a definable scope and content, and with recognizable professional practitioners and exponents. The task of defining poetry, say, whose subject matter is by common consent ineffable, must be self-defeating. Poetry has no rules, no method, no graduate schools, no logic: the bards are self-anointed and their spirit bloweth where it listeth. Science, by contrast, is rigorous, methodical, academic, logical, and practical. The very facility that it gives us, of clear understanding, of seeing things sharply in focus, makes us feel that the instrument itself is very real and hard and definite. Surely we can state, in a few words, its essential nature.

It is not difficult to state the order of being to which science belongs. It is one of the categories of the intellectual commentary that man makes on his world. Amongst its kith and kin we would put religion, art, poetry, law, philosophy, technology, etc.—the familiar divisions or "faculties" of the academy or the multiversity.

At this stage I do not mean to analyze the precise relationship that exists between science and each of these cognate modes of thought; I am merely asserting that they are on all fours with one another. It makes some sort of sense (though it may not always be stating a truth) to substitute these words for one another, in phrases like "Science teaches us . . ." or "The Spirit of Law is . . ." or "*Technology* benefits mankind by . . ."or "He is a student of *Philosophy.*" The famous "conflict between Science and Religion" was truly a battle between combatants of the same species—between David and Goliath if you will—and not, say, between the Philistine army and a Dryad, or between a point of order and a postage stamp.

Science is obviously like religion, law, philosophy, and so forth in being a more or less coherent set of ideas. In its own technical language, science is information; it does not act directly on the body; it speaks to the mind. Religion and poetry, we may concede, speak also to the emotions, and the statements of art can seldom be written or expressed verbally—but they all belong in the nonmaterial realm.

But in what ways are these forms of knowledge *unlike* one another? What are the special attributes of science? What is the criterion for drawing lines of demarcation about it, to distinguish it from philosophy, or from technology, or from poetry?

*   *   *

One can be zealous for science, and a splendidly successful research worker, without pretending to a clear and certain notion of what science really is. In practice it does not seem to matter.

Perhaps this is healthy. A deep interest in theology is not welcome in the average churchgoer, and the ordinary taxpayer should not really concern himself about the nature of sovereignty or the merits of bicameral legislatures. Even though church and state depend, in the end, upon such abstract matters, we may reasonably leave them to the experts if all goes smoothly. The average scientist will say that he knows from experience and common sense what he is doing, and so long as he is not striking too deeply into the foundations of knowledge he is content to leave the highly technical discussion of the nature of science to those self-appointed authorities the philosophers of science. A rough and ready conventional wisdom will see him through.

Yet in a way this neglect of—even scorn for—the philosophy of science by professional scientists is strange. They are, after all, engaged in a very difficult, rather abstract, highly intellectual activity and need all the guidance they can get from general theory. We may agree that the general principles may not in practice be very helpful, but we might have thought that at least they would be taught to young scientists in training, just as medical students are taught physiology and budding administrators were once encouraged to acquaint themselves with Plato's *Republic*. When the student graduates and goes into a laboratory, how will he know what to do to make scientific discoveries if he has not been taught the distinction between a scientific theory and a nonscientific one? Making all allowances for the initial prejudice of scientists against speculative philosophy, and for the outmoded assumption that certain general ideas would communicate themselves to the educated

and cultured man without specific instruction, I find this an odd and significant phenomenon.

The fact is that scientific investigation, as distinct from the theoretical
content of any given branch of science, is a practical art. It is not learnt out
of books, but by imitation and experience. Research workers are trained by
apprenticeship, by working for their Ph.D.'s under the supervision of more
experienced scholars, not by attending courses in the metaphysics of physics.
The graduate student is given his "problem": "You might have a look at the
effect of pressure on the band structure of the III-V compounds; I don't think
it has been done yet, and it would be interesting to see whether it fits into the
pseudopotential theory." Then, with considerable help, encouragement, and
criticism, he sets up his apparatus, makes his measurements, performs his
calculations, et cetera and in due course writes a thesis and is accounted a
qualified professional. But notice that he will not at any time have been made
to study formal logic, nor will he be expected to defend his thesis in a step
by step deductive procedure. His examiners may ask him why he had made
some particular assertion in the course of his argument, or they may enquire
as to the reliability of some particular measurement. They may even ask him
to assess the value of the "contribution" he has made to the subject as a
whole. But they will not ask him to give any opinion as to whether physics
is ultimately *true*, or whether he is justified now in believing in an external
world, or in what sense a theory is verified by the observation of favorable
instances. The examiners will assume that the candidate shares with them the
common language and principles of their discipline. No scientist really
doubts that theories are verified by observation, any more than a common
law judge hesitates to rule that hearsay evidence is inadmissible.

What one finds in practice is that scientific argument, written or spoken,
is not very complex or logically precise. The terms and concepts that are used
may be extremely subtle and technical, but they are put together in quite
simple logical forms, with expressed or implied relations as the machinery of
deduction. It is very seldom that one uses the more sophisticated types of
proof used in mathematics, such as asserting a proposition by proving that its
negation implies a contradiction. Of course actual mathematical or numerical
analysis of data may carry the deduction through many steps, but the symbolic machinery of algebra and the electronic circuits of the computer are
then relied on to keep the argument straight.[1] In my own experience, one
more often detects elementary non sequiturs in the verbal reasoning than
actual mathematical mistakes in the calculations that accompany them. This
is not said to disparage the intellectual powers of scientists; I mean simply
that the reasoning used in scientific papers is not very different from what we
should use in an everyday careful discussion of an everyday problem.

. . . [This point] is made to emphasize the inadequacy of the "logico-
inductive" metaphysic of science. How can this be correct, when few scien-

tists are interested in or understand it, and none ever uses it explicitly in his work? But then if science is distinguished from other intellectual disciplines neither by a particular style or argument nor by a definable subject matter, what is it?

The answer proposed in this essay is suggested by its title: *Science is Public Knowledge.* This is, of course, a very cryptic definition, with almost the suggestion of a play upon words. What I mean is something along the following lines. Science is not merely *published* knowledge or information. Anyone may make an observation, or conceive a hypothesis, and if he has the financial means, get it printed and distributed for other persons to read. Scientific knowledge is more than this. Its facts and theories must survive a period of critical study and testing by other competent and disinterested individuals, and must have been found so persuasive that they are almost universally accepted. The objective of science is not just to acquire information nor to utter all noncontradictory notions; its goal is a consensus of rational opinion over the widest possible field.

In a sense, this is so obvious and well-known that it scarcely needs saying. Most educated and informed people agree that science is true, and therefore impossible to gainsay. But I assert my definition much more positively; this is the basic principle upon which science is founded. It is not a subsidiary consequence of the "scientific method"; it *is* the scientific method itself.

The defect of the conventional philosophical approach to science is that it considers only two terms in the equation. The scientist is seen as an individual, pursuing a somewhat one-sided dialogue with taciturn nature. *He* observes phenomena, notices regularities, arrives at generalizations, deduces consequences, et cetera and eventually, Hey Presto! a law of nature springs into being. But it is not like that at all. The scientific enterprise is corporate. It is not merely, in Newton's incomparable phrase, that one stands on the shoulders of giants, and hence can see a little farther. Every scientist sees through his own eyes—and also through the eyes of his predecessors and colleagues. It is never one individual that goes through all the steps in the logico-inductive chain; it is a group of individuals, dividing their labor but continuously and jealously checking each other's contributions. The cliché of scientific prose betrays itself "Hence we arrive at the conclusion that . . ." The audience to which scientific publications are addressed is not passive; by its cheering or booing, its bouquets or brickbats, it actively controls the substance of the communications that it receives.

In other words, scientific research is a social activity. Technology, art, and religion are perhaps possible for Robinson Crusoe, but law and science are not. To understand the nature of science, we must look at the way in which scientists behave toward one another, how they are organized, and how information passes between them. The young scientist does not study formal logic, but he learns by imitation and experience a number of conventions that embody strong social relationships. In the language of sociology, he learns to play his *role* in a system by which knowledge is acquired, sifted, and eventually made public property.

It has, of course, long been recognized that science is peculiar in its origins to the civilization of Western Europe. The question of the social basis of science, and its relations to other organizations and institutions of our way of life, is much debated. Is it a consequence of the "Bourgeois Revolution," or of Protestantism—or what? Does it exist despite the Church and the universities, or because of them? Why did China, with its immense technological and intellectual resources, not develop the same system? What should be the status of the scientific worker in an advanced society; should he be a paid employee, with a prescribed field of study, or an aristocratic dilettante? How should decisions be taken about expenditure on research? And so on.

These problems, profoundly sociological, historical, and political though they may be, are not quite what I have in mind. Only too often the element in the argument that gets the least analysis is the actual institution about which the whole discussion hinges—scientific activity itself. To give a contemporary example, there is much talk nowadays about the importance of creating more effective systems for storing and indexing scientific literature, so that every scientist can very quickly become aware of the relevant work of every other scientist in his field. This recognizes that publication is important, but the discussion usually betrays an absence of careful thought about the part that conventional systems of scientific communication play in sifting and sorting the material that they handle. Or again, the problem of why Greek science never finally took off from its brilliant taxying runs is discussed in terms of, say, the aristocratic citizen despising the servile labor of practical experiment, when it might have been due to the absence of just such a communications system between scholars as was provided in the Renaissance by alphabetic printing. The internal sociological analysis of science itself is a necessary preliminary to the study of the sociology of knowledge in the secular world.

The present essay cannot pretend to deal with all such questions. The "science of science" is a vast topic, with many aspects. The very core of so many difficulties is suggested by my present argument—that science stands in the region where the intellectual, the psychological, and the sociological coordinate axes intersect. It is knowledge, therefore intellectual, conceptual, and abstract. It is inevitably created by individual men and women, and therefore has a strong psychological aspect. It is public, and therefore molded and determined by the social relations between individuals. To keep all these aspects in view simultaneously, and to appreciate their hidden connections, is not at all easy.

It has been put to me that one should in fact distinguish carefully between science as a body of knowledge, science as what scientists do, and science as a social institution. This is precisely the sort of distinction that one must not make; in the language of geometry, a solid object cannot be reconstructed from its projections upon the separate Cartesian planes. By assigning the intellectual aspects of science to the professional philosophers we make of it an arid exercise in logic; by allowing the psychologists to take possession of the personal dimension we overemphasize the mysteries of "creativity" at the expense of rationality and the critical power of well-ordered argument; if the social aspects are handed over to the sociologists, we get a description of research as an N-person game, with prestige points for stakes and priority claims as trumps. The problem has been to discover a unifying principle for science in all its aspects. The recognition that scientific knowledge must be public and *consensible* (to coin a necessary word) allows one to trace out the complex inner relationships between its various facets. Before one can distinguish and discuss separately the philosophical, psychological, or sociological dimensions of science, one must somehow have succeeded in characterizing it as a whole.[2]

In an ordinary work of science one does well not to dwell too long on the hypothesis that is being tested, trying to define and describe it in advance of reporting the results of the experiments or calculations that are supposed to verify or negate it. The results themselves indicate the nature of the hypothesis, its scope and limitations. The present essay is organized in the same manner. Having sketched a point of view in this chapter, I propose to turn the discussion to a number of particular topics that I think can be better understood when seen from this new angle. To give a semblance of order to the argument, the various subjects have been arranged according to whether they are primarily *intellectual*—as, for example, some attempt to discriminate between scientific and nonscientific disciplines; *psychological*—e.g., the role of education, the significance of scientific creativity; *sociological*—the structure of the scientific community and the institutions by which it maintains scientific standards and procedures. Beyond this classification, the succession of topics is likely to be pretty haphazard; or, as the good lady said, "How do I know what I think until I have heard what I have to say?". . .

NOTES

1. This point I owe to Professor Körner.
2. "Hence a true philosophy of science must be a philosophy of scientists and laboratories as well as one of waves, particles, and symbols." Patrick Meredith in *Instruments of Communication*, p. 40.

3

How to Defend Society against Science*

Paul Feyerabend

Practitioners of a strange trade, friends, enemies, ladies, and gentlemen: Before starting with my talk, let me explain to you, how it came into existence.

About a year ago I was short of funds. So I accepted an invitation to contribute to a book dealing with the relation between science and religion. To make the book sell I thought I should make my contribution a provocative one and the most provocative statement one can make about the relation between science and religion is that science is a religion. Having made the statement the core of my article I discovered that lots of reasons, lots of excellent reasons, could be found for it. I enumerated the reasons, finished my article, and got paid. That was stage one.

Next I was invited to a Conference for the Defense of Culture. I accepted the invitation because it paid for my flight to Europe. I also must admit that I was rather curious. When I arrived in Nice I had no idea what I would say. Then while the conference was taking its course I discovered that everyone thought very highly of science and that everyone was very serious. So I decided to explain how one could defend culture from science. All the reasons collected in my article would apply here as well and there was no need to invent new things. I gave my talk, was rewarded with an outcry about my "dangerous and ill-considered ideas," collected my ticket and went on to Vienna. That was stage number two.

Now I am supposed to address you. I have a hunch that in some respect you are very different from my audience in Nice. For one, you look much younger. My audience in Nice was full of professors, businessmen, and television executives, and the average age was about 58½. Then I am quite sure that most of you are considerably to the left of some of the people in Nice.

---

*[This] article is a revised version of a talk given to the Philosophy Society at Sussex University in November 1974.

54

As a matter of fact, speaking somewhat superficially I might say that you are a leftist audience while my audience in Nice was a rightist audience. Yet despite all these differences you have some things in common. Both of you, I assume, respect science and knowledge. Science, of course, must be reformed and must be made less authoritarian. But once the reforms are carried out, it is a valuable source of knowledge that must not be contaminated by ideologies of a different kind. Secondly, both of you are serious people. Knowledge is a serious matter, for the Right as well as for the Left, and it must be pursued in a serious spirit. Frivolity is out, dedication and earnest application to the task at hand is in. These similarities are all I need for repeating my Nice talk to you with hardly any change. So, here it is.

FAIRYTALES

I want to defend society and its inhabitants from all ideologies, science included. All ideologies must be seen in perspective. One must not take them too seriously. One must read them like fairytales which have lots of interesting things to say but which also contain wicked lies, or like ethical prescriptions which may be useful rules of thumb but which are deadly when followed to the letter.

Now, is this not a strange and ridiculous attitude? Science, surely, was always in the forefront of the fight against authoritarianism and superstition. It is to science that we owe our increased intellectual freedom vis-à-vis religious beliefs; it is to science that we owe the liberation of mankind from ancient and rigid forms of thought. Today these forms of thought are nothing but bad dreams—and this we learned from science. Science and enlightenment are one and the same thing—even the most radical critics of society believe this. Kropotkin wants to overthrow all traditional institutions and forms of belief, with the exception of science. Ibsen criticizes the most intimate ramifications of nineteenth-century bourgeois ideology, but he leaves science untouched. Levi-Strauss has made us realize that Western Thought is not the lonely peak of human achievement it was once believed to be, but he excludes science from his relativization of ideologies. Marx and Engels were convinced that science would aid the workers in their quest for mental and social liberation. Are all these people deceived? Are they all mistaken about the role of science? Are they all the victims of a chimera?

To these questions my answer is a firm *Yes and No.*

Now, let me explain my answer.

My explanation consists of two parts, one more general, one more specific.

The general explanation is simple. Any ideology that breaks the hold a comprehensive system of thought has on the minds of men contributes to the liberation of man. Any ideology that makes man question inherited beliefs is an aid to enlightenment. A truth that reigns without checks and balances is a tyrant who must be overthrown, and any falsehood that can aid us in the overthrow of this tyrant is to be welcomed. It follows that seventeenth- and eighteenth-century science indeed *was* an instrument of liberation and enlightenment. It does not follow that science is bound to *remain* such an instrument. There is nothing inherent in science or in any other ideology that makes it *essentially* liberating. Ideologies can deteriorate and become stupid religions. Look at Marxism. And that the science of today is very different from the science of 1650 is evident at the most superficial glance.

For example, consider the role science now plays in education. Scientific "facts" are taught at a very early age and in the very same manner in which religious "facts" were taught only a century ago. There is no attempt to waken the critical abilities of the pupil so that he may be able to see things in perspective. At the universities the situation is even worse, for indoctrination is here carried out in a much more systematic manner. Criticism is not entirely absent. Society, for example, and its institutions, are criticized most severely and often most unfairly and this already at the elementary school level. But science is excepted from the criticism. In society at large the judgement of the scientist is received with the same reverence as the judgement of bishops and cardinals was accepted not too long ago. The move towards "demythologization," for example, is largely motivated by the wish to avoid any clash between Christianity and scientific ideas. If such a clash occurs, then science is certainly right and Christianity wrong. Pursue this investigation further and you will see that science has now become as oppressive as the ideologies it had once to fight. Do not be misled by the fact that today hardly anyone gets killed for joining a scientific heresy. This has nothing to do with science. It has something to do with the general quality of our civilization. Heretics in science are still made to suffer from the *most severe* sanctions this relatively tolerant civilization has to offer.

But—is this description not utterly unfair? Have I not presented the matter in a very distorted light by using tendentious and distorting terminology? Must we not describe the situation in a very different way? I have said that science has become rigid, that it has ceased to be an instrument of *change* and *liberation,* without adding that it has found the *truth,* or a large part thereof. Considering this additional fact we realize, so the objection goes, that the rigidity of science is not due to human wilfulness. It lies in the nature of things. For once we have discovered the truth—what else can we do but follow it?

This trite reply is anything but original. It is used whenever an ideology wants to reinforce the faith of its followers. "Truth" is such a nicely neutral word. Nobody would deny that it is commendable to speak the truth and wicked to tell lies. Nobody would deny that—and yet nobody knows what such an attitude amounts to. So it is easy to twist matters and to change allegiance to truth in one's everyday affairs into allegiance to the Truth of an ideology which is nothing but the dogmatic defense of that ideology. And it is of course *not* true that we *have* to follow the truth. Human life is guided by many ideas. Truth is one of them. Freedom and mental independence are others. If Truth, as conceived by some ideologists, conflicts with freedom, then we have a *choice*. We may abandon freedom. But we may also abandon Truth. (Alternatively, we may adopt a more sophisticated idea of truth that no longer contradicts freedom; that was Hegel's solution.) My criticism of modern science is that it inhibits freedom of thought. If the reason is that it has found the truth and now follows it, then I would say that there are better things than first finding, and then following such a monster.

This finishes the general part of my explanation.

There exists a more specific argument to defend the exceptional position science has in society today. Put in a nutshell the argument says (1) that science has finally found the correct *method* for achieving results and (2) that there are many *results* to prove the excellence of the method. The argument is mistaken—but most attempts to show this lead into a dead end. Methodology has by now become so crowded with empty sophistication that it is extremely difficult to perceive the simple errors at the basis. It is like fighting the hydra—cut off one ugly head, and eight formalizations take its place. In this situation the only answer is superficiality: when sophistication loses content then the only way of keeping in touch with reality is to be crude and superficial. This is what I intend to be.

AGAINST METHOD

There is a method, says part (1) of the argument. What is it? How does it work?

One answer which is no longer as popular as it used to be is that science works by collecting facts and inferring theories from them. The answer is unsatisfactory as theories never *follow from* facts in the strict logical sense. To say that they may yet be *supported* from facts assumes a notion of support that (a) does not show this defect and (b) is sufficiently sophisticated to permit us to say to what extent, say, the theory of relativity is supported by the facts. No such notion exists today, nor is it likely that it will ever be found (one of the problems is that we need a notion of support in which grey ravens can be said to support "all ravens are black"). This was realized by conventionalists and transcendental idealists who pointed out that theories *shape* and *order* facts and can therefore be retained come what may. They can be retained because the human mind either consciously or unconsciously carries out its ordering function. The trouble with these views is that they assume for the mind what they want to explain for the world, viz., that it works in a regular fashion. There is only one view which overcomes all these difficulties. It was invented twice in the nineteenth century, by Mill, in his immortal essay *On Liberty,* and by some Darwinists who extended Darwinism to the battle of ideas. This view takes the bull by the horns: theories cannot be justified and their excellence cannot be shown without reference to other theories. We may explain the *success* of a theory by reference to a more comprehensive theory (we may explain the success of Newton's theory by using the general theory of relativity); and we may explain our *preference* for it by comparing it with other theories.

Such a comparison does not establish the intrinsic excellence of the theory we have chosen. As a matter of fact, the theory we have chosen may be pretty lousy. It may contain contradictions, it may conflict with well-known facts, it may be cumbersome, unclear, ad hoc in decisive places, and so on. But it may still be better than any other theory that is available at the time. It may in fact be the best lousy theory there is. Nor are the standards of judgment chosen in an absolute manner. Our sophistication increases with every choice we make, and so do our standards. Standards compete just as theories compete and we choose the standards most appropriate to the his-torical situation in which the choice occurs. The rejected alternatives (theo-ries; standards; "facts") are not eliminated. They serve as correctives (after all, we may have made the wrong choice) and they also explain the content of the preferred views (we understand relativity better when we understand the structure of its competitors; we know the full meaning of freedom only when we have an idea of life in a totalitarian state, of its advantages—and there are many advantages—as well as of its disadvantages). Knowledge so conceived is an ocean of alternatives channeled and subdivided by an ocean of standards. It forces our mind to make imaginative choices and thus makes it grow. It makes our mind capable of choosing, imagining, criticizing.

Today this view is often connected with the name of Karl Popper. But there are some very decisive differences between Popper and Mill. To start with, Popper developed his view to solve a special problem of episte-mology—he wanted to solve "Hume's problem." Mill, on the other hand, is interested in conditions favorable to human growth. His epistemology is the result of a certain theory of man, and not the other way around. Also Popper, being influenced by the Vienna Circle, improves on the logical form of a theory before discussing it, while Mill uses every theory in the form in which it occurs in science. Thirdly, Popper's standards of comparison are rigid and fixed, while Mill's standards are permitted to change with the historical situ-ation. Finally, Popper's standards eliminate competitors once and for all: the-ories that are either not falsifiable or falsifiable and falsified have no place in science. Popper's criteria are clear, unambiguous, precisely formulated; Mill's criteria are not. This would be an advantage if science itself were clear, unambiguous, and precisely formulated. Fortunately, it is not.

To start with, no new and revolutionary scientific theory is ever formulated in a manner that permits us to say under what circumstances we must regard it as endangered: many revolutionary theories are unfalsifiable. Falsifiable versions do exist, but they are hardly ever in agreement with accepted basic statements: every moderately interesting theory is falsified. Moreover, theories have formal flaws, many of them contain contradictions, ad hoc adjustments, and so on and so forth. Applied resolutely, Popperian criteria would eliminate science without replacing it by anything comparable. They are useless as an aid to science. In the past decade this has been realized by various thinkers, Kuhn and Lakatos among them. Kuhn's ideas are interesting but, alas, they are much too vague to give rise to anything but lots of hot air. If you don't believe me, look at the literature. Never before has the literature on the philosophy of science been invaded by so many creeps and incompetents. Kuhn encourages people who have no idea why a stone falls to the ground to talk with assurance about scientific method. Now I have no objection to incompetence but I do object when incompetence is accompanied by boredom and self-righteousness. And this is exactly what happens. We do not get interesting false ideas, we get boring ideas or words connected with no ideas at all. Secondly, wherever one tries to make Kuhn's ideas more definite one finds that they are *false*. Was there ever a period of normal science in the history of thought? No—and I challenge anyone to prove the contrary.

Lakatos is immeasurably more sophisticated than Kuhn. Instead of theories he considers research programs which are sequences of theories connected by methods of modification, so-called heuristics. Each theory in the sequence may be full of faults. It may be beset by anomalies, contradictions, ambiguities. What counts is not the shape of the single theories, but the tendency exhibited by the sequence. We judge historical developments and achievements over a period of time, rather than the situation at a particular time. History and methodology are combined into a single enterprise. A research program is said to progress if the sequence of theories leads to novel predictions. It is said to degenerate if it is reduced to absorbing facts that have been discovered without its help. A decisive feature of Lakatos's methodology is that such evaluations are no longer tied to methodological rules which tell the scientist either to retain or to abandon a research program. Scientists may stick to a degenerating program; they may even succeed in making the program overtake its rivals and they therefore proceed rationally whatever they are doing (provided they continue calling degenerating programs degenerating and progressive programs progressive). This means that Lakatos offers *words* which *sound* like the elements of a methodology; he does not offer a methodology. There is no method according to the most advanced and sophisticated methodology in existence today. This finishes my reply to part (1) of the specific argument.

AGAINST RESULTS

According to part (2), science deserves a special position because it has produced *results*. This is an argument only if it can be taken for granted that nothing else has ever produced results. Now it may be admitted that almost everyone who discusses the matter makes such an assumption. It may also be admitted that it is not easy to show that the assumption is false. Forms of life different from science either have disappeared or have degenerated to an extent that makes a fair comparison impossible. Still, the situation is not as hopeless as it was only a decade ago. We have become acquainted with methods of medical diagnosis and therapy which are effective (and perhaps even more effective than the corresponding parts of Western medicine) and which are yet based on an ideology that is radically different from the ideology of Western science. We have learned that there are phenomena such as telepathy and telekinesis which are obliterated by a scientific approach and which could be used to do research in an entirely novel way (earlier thinkers such as Agrippa of Nettesheim, John Dee, and even Bacon were aware of these phenomena). And then—is it not the case that the Church saved souls while science often does the very opposite? Of course, nobody now believes in the ontology that underlies this judgment. Why? Because of ideological pressures identical with those which today make us listen to science to the exclusion of everything else. It is also true that phenomena such as telekinesis and acupuncture may eventually be absorbed into the body of science and may therefore be called "scientific." But note that this happens only after a long period of resistance during which a science *not yet* containing the phenomena wants to get the upper hand over forms of life that contain them. And this leads to a further objection against part (2) of the specific argument. The fact that science has results counts in its favor only if these results were achieved by science alone, and without any outside help. A look at history shows that science hardly ever gets its results in this way. When Copernicus introduced a new view of the universe, he did not consult *scientific* predecessors, he consulted a crazy Pythagorean such as Philolaos. He adopted his ideas and he maintained them in the face of all sound rules of scientific method. Mechanics and optics owe a lot to artisans, medicine to midwives and witches. And in our own day we have seen how the interference of the state can advance science: when the Chinese communists refused to be intimidated by the judgment of experts and ordered traditional medicine back into universities and hospitals there was an outcry all over the world that science would now be ruined in China. The very opposite occurred: Chinese science advanced and Western science learned from it. Wherever we look we see that great scientific advances are due to outside interference which is made to prevail in the face of the most basic and most "rational" methodological rules. The lesson is plain: there does not exist a single argument that could be used to support the exceptional role which science today plays in society. Science has done many things, but so have other ideologies. Science often proceeds systematically, but so do other ideologies (just consult the records of the many doctrinal debates that took place in the Church) and, besides, there are no overriding rules which are adhered to under any circumstances; there is no "scientific methodology" that can be used to separate science from the rest. *Science is just one of the many ideologies that propel society and it should be treated as such* (this statement applies even to the most progressive and most dialectical sections of science). What consequences can we draw from this result?

The most important consequence is that there must be a *formal separation between state and science* just as there is now a formal separation between state and church. Science may influence society but only to the extent to which any political or other pressure group is permitted to influence society. Scientists may be consulted on important projects but the final judgement must be left to the democratically elected consulting bodies. These bodies will consist mainly of laymen. Will the laymen be able to come to a correct judgment? Most certainly, for the competence, the complications and the successes of science are vastly exaggerated. One of the most exhilarating experiences is to see how a lawyer, who is a layman, can find holes in the testimony, the technical testimony, of the most advanced expert and thus prepare the jury for its verdict. Science is not a closed book that is understood only after years of training. It is an intellectual discipline that can be examined and criticized by anyone who is interested and that looks difficult and profound only because of a systematic campaign of obfuscation carried out by many scientists (though, I am happy to say, not by all). Organs of the state should never hesitate to reject the judgment of scientists when they have reason for doing so. Such rejection will educate the general public, will make it more confident, and it may even lead to improvement. Considering the sizeable chauvinism of the scientific establishment we can say: the more Lysenko affairs, the better (it is not the *interference* of the state that is objectionable in the case of Lysenko, but the *totalitarian* interference which kills the opponent rather than just neglecting his advice). Three cheers to the fundamentalists in California who succeeded in having a dogmatic formulation of the theory of evolution removed from the textbooks and an account of Genesis included. (But I know that they would become as chauvinistic and totalitarian as scientists are today when given the chance to run society all by themselves. Ideologies are marvelous when used in the companies of other ideologies. They become boring and doctrinaire as soon as their merits lead to the removal of their opponents.) The most important change, however, will have to occur in the field of *education.*

EDUCATION AND MYTH

The purpose of education, so one would think, is to introduce the young into life, and that means: into the *society* where they are born and into the *physical* universe that surrounds the society. The method of education often consists in the teaching of some *basic myth*. The myth is available in various versions. More advanced versions may be taught by initiation rites which firmly implant them into the mind. Knowing the myth, the grownup can explain almost everything (or else he can turn to experts for more detailed information). He is the master of Nature and of Society. He understands them both and he knows how to interact with them. However, *he is not the master of the myth that guides his understanding.*

Such further mastery was aimed at, and was partly achieved, by the Presocratics. The Presocratics not only tried to understand the *world.* They also tried to understand, and thus to become the masters of, the *means of understanding the world.* Instead of being content with a single myth they developed many and so diminished the power which a well-told story has over the minds of men. The sophists introduced still further methods for reducing the debilitating effect of interesting, coherent, "empirically adequate," et cetera, et cetera tales. The achievements of these thinkers were not appreciated and they certainly are not understood today. When teaching a myth we want to increase the chance that it will be understood (i.e., no puzzlement about any feature of the myth), believed, *and accepted.* This does not do any harm when the myth is counterbalanced by other myths: even the most dedicated (i.e., totalitarian) instructor in a certain version of Christianity cannot prevent his pupils from getting in touch with Buddhists, Jews, and other disreputable people. It is very different in the case of science, or of rationalism where the field is almost completely dominated by the believers. In this case it is of paramount importance to strengthen the minds of the young, and "strengthening the minds of the young" means strengthening them *against* any easy acceptance of comprehensive views. What we need here is an education that makes people *contrary, countersuggestive,* without making them incapable of devoting themselves to the elaboration of any single view. How can this aim be achieved?

It can be achieved by protecting the tremendous imagination which children possess and by developing to the full the spirit of contradiction that exists in them. On the whole children are much more intelligent than their teachers. They succumb, and give up their intelligence because they are bullied, or because their teachers get the better of them by emotional means. Children can learn, understand, and keep separate two to three different languages ("children" and by this I mean three- to five-year-olds, *not* eight-year-olds who were experimented upon quite recently and did not come out too well; why? because they were already loused up by incompetent teaching at an earlier age). Of course, the languages must be introduced in a more interesting way than is usually done. There are marvelous writers in all languages who have told marvelous stories—let us begin our language teaching with *them* and not with "der Hund hat einen Schwanz" and similar inanities. Using stories we may of course also introduce "scientific" accounts, say, of the origin of the world and thus make the children acquainted with science as well. But science must not be given any special position except for pointing out that there are lots of people who believe in it. Later on the stories which have been told will be supplemented with "reasons," where by reasons I mean further accounts of the kind found in the tradition to which the story belongs. And, of course, there will also be contrary reasons. Both reasons and contrary reasons will be told by the experts in the fields and so the young generation becomes acquainted with all kinds of sermons and all types of wayfarers. It becomes acquainted with them, it becomes acquainted with their stories, and every individual can make up his mind which way to go. By now everyone knows that you can earn a lot of money and respect and perhaps even a Nobel Prize by becoming a scientist, so many will become scientists. They will *become* scientists *without having been taken in by the ideology of science,* they will *be* scientists *because they have made a free choice.* But has not much time been wasted on unscientific subjects and will this not detract from their competence once they have become scientists? Not at all! The progress of science, of good science depends on novel ideas and on intellectual freedom: science has very often been advanced by outsiders (remember that Bohr and Einstein regarded themselves as outsiders). Will not many people make the wrong choice and end up in a dead end? Well, that depends on what you mean by a "dead end." Most scientists today are devoid of ideas, full of fear, intent on producing some paltry result so that they can add to the flood of inane papers that now constitutes "scientific progress" in many areas. And, besides, what is more important? To lead a life which one has chosen with open eyes, or to spend one's time in the nervous attempt of avoiding what some not so intelligent people call "dead ends"? Will not the number of scientists decrease so that in the end there is nobody to run our precious laboratories? I do not think so. Given a choice many people may choose science, for a science that is run by free agents looks much more attractive than the science of today which is run by slaves, slaves of institutions and slaves of "reason." And if there is a temporary shortage of scientists the situation may always be remedied by various kinds of incentives. Of course, scientists will not play any predominant role in the society I envisage. They will be more than balanced by magicians, or priests, or astrologers. Such a situation is unbearable for many people, old and young, right and left. Almost all of you have the firm belief that at least some kind of truth has been found, that it must be preserved, and that the method of teaching I advocate and the form of society I defend will dilute it and make it finally disappear. You have this firm belief; many of you may even have reasons. *But what you have to consider is that the absence of good contrary reasons is due to a historical accident*; it does *not* lie in the nature of things. Build up the kind of society I recommend and the views you now despise (without knowing them, to be sure) will return in such splendor that you will have to work hard to maintain your own position and will perhaps be entirely unable to do so. You do not believe me? Then look at history. Scientific astronomy was firmly founded on Ptolemy and Aristotle, two of the greatest minds in the history of Western Thought. Who upset their well-argued, empirically adequate, and precisely formulated system? Philolaos the mad and antediluvian Pythagorean. How was it that Philolaos could stage such a comeback? Because he found an able defender: Copernicus. Of course, you may follow your intuitions as I am following mine. But remember that your intuitions are the result of your "scientific" training where by science I also mean the science of Karl Marx. My training, or, rather, my nontraining, is that of a journalist who is interested in strange and bizarre events. Finally, is it not utterly irresponsible, in the present world situation, with millions of people starving, others enslaved, downtrodden, in abject misery of body and mind, to think luxurious thoughts such as these? Is not freedom of choice a luxury under such circumstances? Is not the flippancy and the humor I want to see combined with the freedom of choice a luxury under such circumstances? Must we not give up all self-indulgence and *act*? Join together, and *act*? This is the most important objection which today is raised against an approach such as the one recommended by me. It has tremendous appeal, it has the appeal of unselfish dedication. Unselfish dedication—to what? Let us see!

We are supposed to give up our selfish inclinations and dedicate ourselves to the liberation of the oppressed. And selfish inclinations are what? They are our wish for maximum liberty of thought in the society in which we live *now*, maximum liberty not only of an abstract kind, but expressed in appropriate institutions and methods of teaching. This wish for concrete intellectual and physical liberty in our own surroundings is to be put aside, for the time being. This assumes, first, that we do not need this liberty for our task. It assumes that we can carry out our task with a mind that is firmly closed to some alternatives. It assumes that the correct way of liberating others *has always been found* and that all that is needed is to carry it out. I am sorry, I cannot accept such doctrinaire self-assurance in such extremely important matters. Does this mean that we cannot act at all? It does not. But it means that *while acting we have to try to realize as much of the freedom I have recommended so that our actions may be corrected in the light of the ideas we get while increasing our freedom.* This will slow us down, no doubt, but are we supposed to charge ahead simply because some people tell us that they have found an explanation for all the misery and an excellent way out of it? Also we want to liberate people not to make them succumb to a new kind of slavery, *but to make them realize their own wishes,* however different these wishes may be from our own. Self-righteous and narrow-minded liberators cannot do this. As a rule they soon impose a slavery that is worse, because more systematic, than the very sloppy slavery they have removed. And as regards humor and flippancy the answer should be obvious. Why would anyone want to liberate anyone else? Surely not because of some *abstract* advantage of liberty but because liberty is the best way to free development *and thus to happiness.* We want to liberate people so that *they can smile.* Shall we be able to do this if we ourselves have forgotten how to smile and are frowning on those who still remember? Shall we then not spread another disease, comparable to the one we want to remove, the disease of puritanical self-righteousness? Do not object that dedication and humor do not go together—Socrates is an excellent example to the contrary. *The hardest task needs the lightest hand or else its completion will not lead to freedom but to a tyranny much worse than the one it replaces.*

4

Why Astrology Is a Pseudoscience

Paul R. Thagard

Most philosophers and historians of science agree that astrology is a pseudoscience, but there is little agreement on *why* it is a pseudoscience. Answers range from matters of verifiability and falsifiability, to questions of progress and Kuhnian normal science, to the different sort of objections raised by a large panel of scientists recently organized by *The Humanist* magazine. Of course there are also Feyerabendian anarchists and others who say that no demarcation of science from pseudoscience is possible. However, I shall propose a complex criterion for distinguishing disciplines as pseudoscientific; this criterion is unlike verificationist and falsificationist attempts in that it introduces social and historical features as well as logical ones.

I begin with a brief description of astrology. It would be most unfair to evaluate astrology by reference to the daily horoscopes found in newspapers and popular magazines. These horoscopes deal only with sun signs, whereas a full horoscope makes reference to the "influences" also of the moon and the planets, while also discussing the ascendant sign and other matters.

Astrology divides the sky into twelve regions, represented by the familiar signs of the Zodiac: Aquarius, Libra, and so on. The sun sign represents the part of the sky occupied by the sun at the time of birth. For example, anyone born between September 23 and October 22 is a Libran. The ascendant sign, often assumed to be at least as important as the sun sign, represents the part of the sky rising on the eastern horizon at the time of birth, and therefore changes every two hours. To determine this sign, accurate knowledge of the time and place of birth is essential. The moon and the planets (of which there are five or eight depending on whether Uranus, Neptune, and Pluto are taken into account) are also located by means of charts on one of the parts of the Zodiac. Each planet is said to exercise an influence in a special sphere of human activity; for example, Mars governs drive, courage, and daring, while Venus governs love and artistic endeavor. The immense number of combinations of sun, ascendant, moon, and planetary influences allegedly determines human personality, behavior, and fate.

Astrology is an ancient practice, and appears to have its origins in Chaldea, thousands of years B.C.E. By 700 B.C.E., the Zodiac was established, and a few centuries later the signs of the Zodiac were very similar to current ones. The conquests of Alexander the Great brought astrology to Greece, and the Romans were exposed in turn. Astrology was very popular during the fall of the Republic, with many notables such as Julius Caesar having their horoscopes cast. However, there was opposition from such men as Lucretius and Cicero.

Astrology underwent a gradual codification culminating in Ptolemy's *Tetrabiblos* [20], written in the second century C.E. This work describes in great detail the powers of the sun, moon, and planets, and their significance in people's lives. It is still recognized as a fundamental textbook of astrology. Ptolemy took astrology as seriously as he took his famous work in geography and astronomy; this is evident from the introduction to the *Tetrabiblos*, where he discusses two available means of making predictions based on the heavens. The first and admittedly more effective of these concerns the relative movements of the sun, moon and planets, which Ptolemy had already treated in his celebrated *Almagest* [19]. The secondary but still legitimate means of prediction is that in which we use the "natural character" of the aspects of movement of heavenly bodies to "investigate the changes which they bring about in that which they surround" ([20], p. 3). He argues that this method of prediction is possible because of the manifest effects of the sun, moon, and planets on the earth, for example on weather and the tides.

The European Renaissance is heralded for the rise of modern science, but occult arts such as astrology and alchemy flourished as well. Arthur Koestler has described Kepler's interest in astrology: not only did astrology provide Kepler with a livelihood, he also pursued it as a serious interest, although he was skeptical of the particular analyses of previous astrologers ([13], pp. 244–248). Astrology was popular both among intellectuals and the general public through the seventeenth century. However, astrology lost most of this popularity in the eighteenth century, when it was attacked by such figures of the Enlightenment as Swift [24] and Voltaire [29]. Only since the 1930s has astrology again gained a huge audience: most people today know at least their sun signs, and a great many believe that the stars and planets exercise an important influence on their lives.

In an attempt to reverse this trend, Bart Bok, Lawrence Jerome, and Paul Kurtz drafted in 1975 a statement attacking astrology; the statement was signed by 192 leading scientists, including 19 Nobel Prize winners. The statement raises three main issues: astrology originated as part of a magical world view, the planets are too distant for there to be any physical foundation for astrology, and people believe it merely out of longing for comfort ([21], pp. 9f.). None of these objections is ground for condemning astrology as pseudoscience. To show this, I shall briefly discuss articles written by Bok [1] and Jerome [12] in support of the statement.

According to Bok, to work on statistical tests of astrological predictions is a waste of time unless it is demonstrated that astrology has some sort of physical foundation ([1], p. 31). He uses the smallness of gravitational and radiative effects of the stars and planets to suggest that there is no such foundation. He also discusses the psychology of belief in astrology, which is the result of individuals' desperation in seeking solutions to their serious personal problems. Jerome devotes most of his article to the origins of astrology in the magical principle of correspondences. He claims that astrology is a system of magic rather than science, and that it fails "not because of any inherent inaccuracies due to precession or lack of exact knowledge concerning time of birth or conception, but rather because its interpretations and predictions are grounded in the ancients' magical world view" ([12], p. 46). He does however discuss some statistical tests of astrology, which I shall return to below.

These objections do not show that astrology is a pseudoscience. First, origins are irrelevant to scientific status. The alchemical origins of chemistry ([11], pp. 10–18) and the occult beginnings of medicine [8] are as magical as those of astrology, and historians have detected mystical influences in the work of many great scientists, including Newton and Einstein. Hence astrology cannot be condemned simply for the magical origins of its principles. Similarly, the psychology of popular belief is also in itself irrelevant to the status of astrology: people often believe even good theories for illegitimate reasons, and even if most people believe astrology for personal, irrational reasons, good reasons may be available.[1] Finally the lack of a physical foundation hardly marks a theory as unscientific ([22], p. 2). Examples: when Wegener [31] proposed continental drift, no mechanism was known, and a link between smoking and cancer has been established statistically [28] though the details of carcinogenesis remain to be discovered. Hence the objections of Bok, Jerome, and Kurtz fail to mark astrology as pseudoscience.

Now we must consider the application of the criteria of verifiability and falsifiability to astrology. Roughly, a theory is said to be verifiable if it is possible to deduce observation statements from it. Then in principle, observations can be used to confirm or disconfirm the theory. A theory is scientific only if it is verifiable. The vicissitudes of the verification principle are too well known to recount here ([9], ch. 4). Attempts by A. J. Ayer to articulate the principle failed either by ruling out most of science as unscientific, or by ruling out nothing. Moreover, the theory/observation distinction has increasingly come into question. All that remains is a vague sense that testability somehow is a mark of scientific theories ([9], ch. 4; [10], pp. 30–32).

Well, astrology *is* vaguely testable. Because of the multitude of influences resting on tendencies rather than laws, astrology is incapable of making precise predictions. Nevertheless, attempts have been made to test the reality of these alleged tendencies, using large-scale surveys and statistical evaluation. The pioneer in this area was Michel Gauquelin, who examined the careers and times of birth of 25,000 Frenchmen. Astrology suggests that people born under certain signs or planets are likely to adopt certain occupations: for example, the influence of the warlike planet Mars tends to produce soldiers or athletes, while Venus has an artistic influence. Notably, Gauquelin found *no significant correlation* between careers and either sun sign, moon sign, or ascendant sign. However, he did find some statistically interesting correlations between certain occupations of people and the position of certain planets at the time of their birth ([5], ch. 11; [6]). For example, just as astrology would suggest, there is a greater than chance association of athletes and Mars, and a greater than chance association of scientists and Saturn, where the planet is rising or at its zenith at the moment of the individual's birth.

These findings and their interpretation are highly controversial, as are subsequent studies in a similar vein [7]. Even if correct, they hardly verify astrology, especially considering the negative results found for the most important astrological categories. I have mentioned Gauquelin in order to suggest that through the use of statistical techniques astrology is at least *verifiable*. Hence the verification principle does not mark astrology as pseudoscience.

Because the predictions of astrologers are generally vague, a Popperian would assert that the real problem with astrology is that it is not falsifiable: astrologers cannot make predictions which if unfulfilled would lead them to give up their theory. Hence because it is unfalsifiable, astrology is unscientific.

But the doctrine of falsifiability faces serious problems as described by Duhem [4], Quine [21], and Lakatos [15]. Popper himself noticed early that no observation ever guarantees falsification: a theory can always be retained by introducing or modifying auxiliary hypotheses, and even observation statements are not incorrigible ([17], p. 50). Methodological decisions about what can be tampered with are required to block the escape from falsification. However, Lakatos has persuasively argued that making such decision in advance of tests is arbitrary and may often lead to overhasty rejection of a sound theory which *ought* to be saved by antifalsificationist stratagems ([15], pp. 112 ff). Falsification only occurs when a better theory comes along. Then falsifiability is only a matter of replaceability by another theory, and since astrology is in principle replaceable by another theory, falsifiability provides no criterion for rejecting astrology as pseudoscientific. We saw in the discussion of Gauquelin that astrology can be used to make predictions about statistical regularities, but the nonexistence of these regularities does not falsify astrology; but here astrology does not appear worse than the best of scientific theories, which also resist falsification until alternative theories arise.[2]

Astrology cannot be condemned as pseudoscientific on the grounds proposed by verificationists, falsificationists, or Bok and Jerome. But undoubtedly astrology today faces a great many unsolved problems ([32], ch. 5). One is the negative result found by Gauquelin concerning careers and signs. Another is the problem of the precession of the equinoxes, which astrologers generally take into account when heralding the "Age of Aquarius" but totally neglect when figuring their charts. Astrologers do not always agree on the significance of the three planets, Neptune, Uranus, and Pluto, that were discovered since Ptolemy. Studies of twins do not show similarities of personality and fate that astrology would suggest. Nor does astrology make sense of mass disasters, where numerous individuals with very different horoscopes come to similar ends.

But problems such as these do not in themselves show that astrology is either false or pseudoscientific. Even the best theories face unsolved problems throughout their history. To get a criterion demarcating astrology from science, we need to consider it in a wider historical and social context.

A demarcation criterion requires a matrix of three elements: [theory, community, historical context]. Under the first heading, "theory," fall familiar matters of structure, prediction, explanation, and problem solving. We might also include the issue raised by Bok and Jerome about whether the theory has a physical foundation. Previous demarcationists have concentrated on this theoretical element, evident in the concern of the verification and falsification principles with prediction. But we have seen that this approach is not sufficient for characterizing astrology as pseudoscientific.

We must also consider the *community* of advocates of the theory, in this case the community of practitioners of astrology. Several questions are important here. First, are the practitioners in agreement on the principles of the theory and on how to go about solving problems which the theory faces? Second, do they care, that is, are they concerned about explaining anomalies and comparing the success of their theory to the record of other theories? Third, are the practitioners actively involved in attempts at confirming and disconfirming their theory?

The question about comparing the success of a theory with that of other theories introduces the third element of the matrix, historical context. The historical work of Kuhn and others has shown that in general a theory is rejected only when (1) it has faced anomalies over a long period of time and (2) it has been challenged by another theory. Hence under the heading of historical context we must consider two factors relevant to demarcation: the record of a theory over time in explaining new facts and dealing with anomalies, and the availability of alternative theories.

We can now propose the following principles of demarcation:

A theory or discipline which purports to be scientific is pseudoscientific if and only if:

(1) It has been less progressive than alternative theories over a long period of time, and faces many unsolved problems; but

(2) the community of practitioners makes little attempt to develop the theory towards solutions of the problems, shows no concern for attempts to evaluate the theory in relation to others, and is selective in considering confirmations and disconfirmations.

Progressiveness is a matter of the success of the theory in adding to its set of facts explained and problems solved ([15], p. 118; cf. [26], p. 83).

This principle captures, I believe, what is most importantly unscientific about astrology. First, astrology is dramatically unprogressive, in that it has changed little and has added nothing to its explanatory power since the time of Ptolemy. Second, problems such as the precession of equinoxes are outstanding. Third, there are alternative theories of personality and behavior available: one need not be an uncritical advocate of behaviorist, Freudian, *or* Gestalt theories to see that since the nineteenth century psychological theories have been expanding to deal with many of the phenomena which astrology explains in terms of heavenly influences. The important point is not that any of these psychological theories is established or true, only that they are growing alternatives to a long-static astrology. Fourth and finally, the community of astrologers is generally unconcerned with advancing astrology to deal with outstanding problems or with evaluating the theory in relation to others.[3] For these reasons, my criterion marks astrology as pseudoscientific.

This demarcation criterion differs from those implicit in Lakatos and Kuhn. Lakatos has said that what makes a series of theories constituting a research program scientific is that it is progressive: each theory in the series has greater corroborated content than its predecessor ([15], p. 118). While I agree with Lakatos that progressiveness is a central notion here, it is not sufficient to distinguish science from pseudoscience. We should not brand a nonprogressive discipline as pseudoscientific unless it is being maintained against more progressive alternatives. Kuhn's discussion of astrology focuses on a different aspect of my criterion. He says that what makes astrology unscientific is the absence of the paradigm-dominated puzzle-solving activity characteristic of what he calls normal science ([14], p. 9). But as Watkins has suggested, astrologers are in some respects model normal scientists: they concern themselves with solving puzzles at the level of individual horoscopes, unconcerned with the foundations of their general theory or paradigm ([30], p. 32). Hence that feature of normal science does not distinguish science from pseudoscience. What makes astrology pseudoscientific is not that it lacks periods of Kuhnian normal science, but that its proponents adopt uncritical attitudes of "normal" scientists despite the existence of more progressive alternative theories. (Note that I am not agreeing with Popper [18] that Kuhn's normal scientists are unscientific; they can become unscientific only when an alternative paradigm has been developed.) However, if one looks not at the puzzle solving at the level of particular astrological predictions, but at the level of theoretical problems such as the precession of the equinoxes, there is some agreement between my criterion and Kuhn's; astrologers do not have a paradigm-induced confidence about solving theoretical problems.

Of course, the criterion is intended to have applications beyond astrology. I think that discussion would show that the criterion marks as pseudoscientific such practices as witchcraft and pyramidology, while leaving contemporary physics, chemistry, and biology unthreatened. The current fad of biorhythms, implausibly based like astrology on date of birth, cannot be branded as pseudoscientific because we lack alternative theories giving more detailed accounts of cyclical variations in human beings, although much research is in progress.[4]

One interesting consequence of the above criterion is that a theory can be scientific at one time but pseudoscientific at another. In the time of Ptolemy or even Kepler, astrology had few alternatives in the explanation of human personality and behavior. Existing alternatives were scarcely more sophisticated or corroborated than astrology. Hence astrology should be judged as not pseudoscientific in classical or Renaissance times, even though it is pseudoscientific today. Astrology was not simply a perverse sideline of Ptolemy and Kepler, but part of their scientific activity, even if a physicist involved with astrology today should be looked at askance. Only when the historical and social aspects of science are neglected does it become plausible that pseudoscience is an unchanging category. Rationality is not a property of ideas eternally: ideas, like actions, can be rational at one time but irrational at others. Hence relativizing the science/pseudoscience distinction to historical periods is a desirable result.

But there remains a challenging historical problem. According to my criterion, astrology only became pseudoscientific with the rise of modern psychology in the nineteenth century. But astrology was already virtually excised from scientific circles by the beginning of the eighteenth. How could this be? The simple answer is that a theory can take on the appearance of an unpromising project well before it deserves the label of pseudoscience. The Copernican revolution and the mechanism of Newton, Descartes, and Hobbes undermined the plausibility of astrology.[5] Lynn Thorndike [27] has described how the Newtonian theory pushed aside what had been accepted as a universal natural law, that inferiors such as inhabitants of earth are ruled and governed by superiors such as the stars and the planets. William Stahlman [23] has described how the immense growth of science in the seventeenth century contrasted with stagnation of astrology. These developments provided good reason for discarding astrology as a promising pursuit, but they were not yet enough to brand it as pseudoscientific, or even to refute it.

Because of its social aspect, my criterion might suggest a kind of cultural relativism. Suppose there is an isolated group of astrologers in the jungles of South America, practicing their art with no awareness of alternatives. Are we to say that astrology is *for them* scientific? Or, going in the other direction, should we count as alternative theories ones which are available to extraterrestrial beings, or which someday will be conceived? This wide construal of "alternative" would have the result that our best current theories are probably pseudoscientific. These two questions employ, respectively, a too narrow and too broad view of alternatives. By an alternative theory I mean one generally available in the world. This assumes first that there is some kind of communication network to which a community has, or should have, access. Second, it assumes that the onus is on individuals and communities to find out about alternatives. I would argue (perhaps against Kuhn) that this second assumption is a general feature of rationality; it is at least sufficient to preclude ostrichism as a defense against being judged pseudoscientific.

In conclusion, I would like to say why I think the question of what constitutes a pseudoscience is important. Unlike the logical positivists, I am not grinding an antimetaphysical ax, and unlike Popper, I am not grinding an anti-Freudian or anti-Marxian one.[6] My concern is social: society faces the twin problems of lack of public concern with the advancement of science, and lack of public concern with the important ethical issues now arising in science and technology, for example around the topic of genetic engineering. One reason for this dual lack of concern is the wide popularity of pseudoscience and the occult among the general public. Elucidation of how science differs from pseudoscience is the philosophical side of an attempt to overcome public neglect of genuine science.

NOTES

1. However, astrology would doubtlessly have many fewer supporters if horoscopes tended less toward compliments and pleasant predictions and more toward the kind of analysis included in the following satirical horoscope from the December, 1977, issue of *Mother Jones*: VIRGO (Aug. 23-Sept. 22). You are the logical type and hate disorder. This nit-picking is sickening to your friends. You are cold and unemotional and sometimes fall asleep while making love. Virgos make good bus drivers.

2. For an account of the comparative evaluation of theories, see [26].

3. There appear to be a few exceptions; see [32].

4. The fad of biorhythms, now assuming a place beside astrology in the popular press, must be distinguished from the very interesting work of Frank Brown and others on biological rhythms. For a survey, see [5].

5. Plausibility is in part a matter of a hypothesis being of an appropriate kind, and is relevant even to the acceptance of a theory. See [26], p. 90, and [25].

6. On psychoanalysis see [3]. I would argue that Cioffi neglects the question of alternatives to psychoanalysis and the question of its progressiveness.

The author is grateful to Dan Hausman and Elias Baumgarten for comments.

REFERENCES

[1] Bok, Bart J. "A Critical Look at Astrology." In [2]. Pages 21–33.

[2] Bok, Bart, Lawrence E. Jerome, and Paul Kurtz. *Objections to Astrology.* Buffalo: Prometheus Books, 1975.

[3] Cioffi, Frank. "Freud and the Idea of a Pseudoscience." In *Explanation in the Behavioral Sciences,* edited by R. Borger and F. Cioffi, 471–499. Cambridge: Cambridge University Press, 1970.

[4] Duhem, P. *The Aim and Structure of Physical Theory.* Translated by P. Wiener. New York: Atheneum, 1954. (Translated from 2nd edition of *La Theorie Physique: Son Object Sa Structure.* Paris: Marcel Riviere & Cie, 1914.)

[5] Gauquelin, Michel. *The Cosmic Clocks.* Chicago: Henry Regnery, 1967.

[6] ———. *The Scientific Basis of Astrology.* New York: Stein & Day, 1969.

[7] ———. "The Zelen Test of the Mars Effect." *The Humanist* 37 (1977):30–35.

[8] Haggard, Howard W. *Mystery, Magic, and Medicine.* Garden City, N.Y.: Doubleday, Doran, & Company, 1933.

[9] Hempel, Carl. *Aspects of Scientific Explanation.* New York: The Free Press, 1965.

[10] ———. *Philosophy of Natural Science.* Englewood Cliffs, N.J.: Prentice-Hall, 1966.

[11] Ihde, Aaron, J. *The Development of Modern Chemistry.* New York: Harper & Row, 1964.

[12] Jerome, Lawrence E. "Astrology: Magic or Science?" In [2]. Pages 37–62.

[13] Koestler, Arthur. *The Sleepwalkers.* Harmondsworth: Penguin, 1964.

[14] Kuhn, T. S. "Logic of Discovery or Psychology of Research." In [16]. Pages 1–23.

[15] Lakatos, Imre. "Falsification and the Methodology of Scientific Research Programmes." In [16]. Pages 91–195.

[16] Lakatos, Imre, and Alan Musgrave, eds. *Criticism and the Growth of Knowledge.* Cambridge: Cambridge University Press, 1970.

[17] Popper, Karl. *The Logic of Scientific Discovery.* London: Hutchinson, 1959. (Originally published as *Logik der Forschung.* Vienna: J. Springer, 1935.)

[18] ———. "Normal Science and Its Dangers." In [16]. Pages 51–58.

[19] Ptolemy. *The Almagest (The Mathematical Composition).* (As printed in Hutchins, Robert Maynard, ed. *Great Books of the Western World.* Vol. 16, 1–478. Chicago: Encyclopedia Britannica, Inc., 1952.)

[20] ———. *Tetrabiblos.* Edited and translated by F. E. Robbins. Cambridge, Mass.: Harvard University Press, 1940.

[21] Quine, W. V. O. "Two Dogmas of Empiricism." In *From a Logical Point of View,* 20–46. New York: Harper & Row, 1963. (Originally published in *The Philosophical Review* 60 [1951]: 20–43.)

[22] Sagan, Carl. "Letter." *The Humanist* 36 (1976): 2.

[23] Stahlman, William D. "Astrology in Colonial America: An Extended Query." *William and Mary Quarterly* 13 (1956): 551–563.

[24] Swift, Jonathan. "The Partridge Papers." In *The Prose Works of Jonathan Swift.* Vol. 2, 139–170. Oxford: Basil Blackwell, 1940–1968.

[25] Thagard, Paul R. "The Autonomy of a Logic of Discovery." Forthcoming in the *Festschrift for T. A. Goudge.*

[26] ———. "The Best Explanation: Criteria for Theory Choice." *Journal of Philosophy* 75 (1978): 76–92.

[27] Thorndike, Lynn. "The True Place of Astrology in the History of Science." *Isis* 46 (1955): 273–278.

[28] U.S. Department of Health, Education, and Welfare. *Smoking and Health: Report of the Advisory Committee to the Surgeon General of the Public Health Service.* Washington, D.C.: U.S. Governmment Printing Office, 1964.

[29] Voltaire. "Astrologie" and "Astronomie." *Dictionnaire Philosophique.* In *Oeuvres Completes de Voltaire.* Vol. 17, 446–453. Paris: Garnier Frères, 1878–1885.

[30] Watkins, J. W. N. "Against 'Normal Science.'" In [16]. Pages 25–37.

[31] Wegener, Alfred. "Die Entstchung der Kontinente." *Petermann's Geographische Mitteilung* 58 (1912): 185–195, 235–256, 305–309.

[32] West, J. A., and J. G. Toonder, *The Case for Astrology.* Harmondsworth: Penguin, 1973.

5

Believing Where We Cannot Prove

Philip Kitcher

I. Opening Moves

Simple distinctions come all too easily. Frequently we open the way for later puzzlement by restricting the options we take to be available. So, for example, in contrasting science and religion, we often operate with a simple pair of categories. On one side there is science, proof, and certainty; on the other, religion, conjecture, and faith.

The opening lines of Tennyson's *In Memoriam* offer an eloquent statement of the contrast:

> Strong Son of God, immortal love,
> Whom we, that have not seen Thy face,
> By faith, and faith alone, embrace,
> Believing where we cannot prove.

A principal theme of Tennyson's great poem is his struggle to maintain faith in the face of what seems to be powerful scientific evidence. Tennyson had read a popular work by Robert Chambers, *Vestiges of the Natural History of Creation,* and he was greatly troubled by the account of the course of life on earth that the book contains. *In Memoriam* reveals a man trying to believe where he cannot prove, a man haunted by the thought that the proofs may be against him.

Like Tennyson, contemporary Creationists accept the traditional contrast between science and religion. But where Tennyson agonized, they attack. While they are less eloquent, they are supremely confident of their own solution. They open their onslaught on evolutionary theory by denying that it is a science. In *The Troubled Waters of Evolution,* Henry Morris characterizes evolutionary theory as maintaining that large amounts of time are required for evolution to produce "new kinds." As a result, we should not expect to see such "new kinds" emerging. Morris comments, "Creationists in turn insist that this belief is not scientific evidence but only a statement of faith. The evolutionist seems to be saying, Of course, we cannot really *prove* evolution, since this requires ages of time, and so, therefore, you should accept it as a proved fact of science! Creationists regard this as an odd type of logic, which would be entirely unacceptable in any other field of science" (Morris 1974b, 16). David Watson makes a similar point in comparing Darwin with Galileo: "So here is the difference between Darwin and Galileo: Galileo set a demonstrable *fact* against a few words of Bible poetry which the Church at that time had understood in an obviously naive way; Darwin set an unprovable *theory* against eleven chapters of straightforward history which cannot be reinterpreted in any satisfactory way" (Watson 1975, 46).

The idea that evolution is conjecture, faith, or "philosophy" pervades Creationist writings (Morris 1974a, 4–8; Morris 1974b, 22, 172; Wysong 1976, 43–45; Gish 1979, 11–13, 26, 186; Wilder-Smith 1981, 7–8). It is absolutely crucial to their case for equal time for "scientific" Creationism. This ploy has succeeded in winning important adherents to the Creationist cause. As he prepared to defend Arkansas law 590, Attorney General Steven Clark echoed the Creationist judgment. "Evolution," he said, "is just a theory." Similar words have been heard in Congress. William Dannemeyer, a congressman from California, introduced a bill to limit funding to the Smithsonian with the following words: "If the theory of evolution is just that—a theory—and if that theory can be regarded as a religion . . . then it occurs to this Member that other Members might prefer it not to be given exclusive or top billing in our Nation's most famous museum but equal billing or perhaps no billing at all."

In their attempt to show that evolution is not science, Creationists receive help from the least likely sources. Great scientists sometimes claim that certain facts about the past evolution of organisms are "demonstrated" or "indubitable" (Simpson 1953, 70, 371; also Mayr 1976, 9). But Creationists also can (and do) quote scientists who characterize evolution as "dogma" and contend that there is no conclusive proof of evolutionary theory (Matthews 1972, xi; Birch and Ehrlich 1967, 349; quoted in Gish 1979, 15–16; similar passages are quoted in Morris, 1974a, 6–8, and in Wysong 1976, 44). Evolution is not part of science because, as evolutionary biologists themselves concede, science demands proof, and, as other biologists point out, proof of evolution is not forthcoming.

The rest of the Creationist argument flows easily. We educate our children in evolutionary theory as if it were a proven fact. We subscribe officially, in our school system, to one faith—an atheistic, materialistic faith—ignoring rival beliefs. Antireligious educators deform the minds of children, warping them to accept as gospel a doctrine that has no more scientific support than the true Gospel. The very least that should be done is to allow for both alternatives to be presented.

We should reject the Creationists' gambit. Eminent scientists notwithstanding, science is not a body of demonstrated truths. Virtually all of science is an exercise in believing where we cannot prove. Yet, scientific conclusions are not embraced by faith alone. Tennyson's dichotomy was too simple.

II. INCONCLUSIVE EVIDENCE

Sometimes we seem to have conclusive reasons for accepting a statement as true. It is hard to doubt that $2 + 2 = 4$. If, unlike Lord Kelvin's ideal mathematician, we do not find it obvious that

$$\int \pm {}^{\infty}_{\infty} e^{-x^2} \, dx = \sqrt{\pi}.$$

at least the elementary parts of mathematics appear to command our agreement. The direct evidence of our senses seems equally compelling. If I see the pen with which I am writing, holding it firmly in my unclouded view, how can I doubt that it exists? The talented mathematician who has proved a theorem and the keen-eyed witness of an episode furnish our ideals of certainty in knowledge. What they tell us can be engraved in stone, for there is no cause for worry that it will need to be modified.

Yet, in another mood, one that seems "deeper" or more "philosophical," skeptical doubts begin to creep in. Is there really anything of which we are so certain that later evidence could not give us reason to change our minds? Even when we think about mathematical proof, can we not imagine that new discoveries may cast doubt on the cogency of our reasoning? (The history of mathematics reveals that sometimes what seems for all the world like a proof may have a false conclusion.) Is it not possible that the most careful observer may have missed something? Or that the witness brought preconceptions to the observation that subtly biased what was reported? Are we not *always* fallible?

I am mildly sympathetic to the skeptic's worries. Complete certainty is best seen as an ideal toward which we strive and that is rarely, if ever, attained. Conclusive evidence always eludes us. Yet even if we ignore skeptical complaints and imagine that we are sometimes lucky enough to have conclusive reasons for accepting a claim as true, we should not include scientific reasoning among our paradigms of proof. Fallibility is the hallmark of science. This point should not be so surprising. The trouble is that we frequently forget it in discussing contemporary science. When we turn to the history of science, however, our fallibility stares us in the face. The history of the natural sciences is strewn with the corpses of intricately organized theories, each of which had, in its day, considerable evidence in its favor. When we look at the confident defenders of those theories we should see anticipations of ourselves. The eighteenth-century scientists who believed that heat is a "subtle fluid," the atomic theorists who maintained that water molecules are compounded out of one atom of hydrogen and one of oxygen, the biochemists who identified protein as the genetic material, and the geologists who thought that continents cannot move were neither unintelligent nor ill informed. Given the evidence available to them, they were eminently reasonable in drawing their conclusions. History proved them wrong. It did not show that they were unjustified.

Why is science fallible? Scientific investigation aims to disclose the general principles that govern the workings of the universe. These principles are not intended merely to summarize what some select groups of humans have witnessed. Natural science is not just natural history. It is vastly more ambitious. Science offers us laws that are supposed to hold universally, and it advances claims about things that are beyond our power to observe. The nuclear physicist who sets down the law governing a particular type of radioactive decay is attempting to state a truth that holds throughout the entire cosmos and also to describe the behavior of things that we cannot even see. Yet, of necessity, the physicist's ultimate evidence is highly restricted. Like the rest of us, scientists are confined to a relatively small region of space and time and equipped with limited and imperfect senses. How is science possible at all? How are we able to have any confidence about the distant regions of the cosmos and the invisible realm that lies behind the surfaces of ordinary things? The answer is complicated. Natural science follows intricate and ingenious procedures for fathoming the secrets of the universe. Scientists devise ways of obtaining especially revealing evidence. They single out some of the things we are able to see as crucial clues to the way that nature works. These clues are used to answer questions that cannot be addressed by direct observation. Scientific theories, even those that are most respected and most successful, rest on indirect arguments from the observational evidence. New discoveries can always call those arguments into question, showing scientists that the observed data should be understood in a different way, that they have misread their evidence.

But scientists often forget the fallibility of their enterprise. This is not just absentmindedness or wishful thinking. During the heyday of a scientific theory, so much evidence may support the theory, so many observational clues may seem to attest to its truth, that the idea that it could be overthrown appears ludicrous. In addition, the theory may provide ways of identifying quickly what is inaccessible to our unaided senses. Electron microscopes and cloud chambers are obvious examples of those extensions of our perceptual system that theories can inspire. Trained biochemists will talk quite naturally of seeing large molecules, and it is easy to overlook the fact that they are presupposing a massive body of theory in describing what they "see." If that theory were to be amended, even in subtle ways, then the descriptions of the "observed characteristics" of large molecules might have to be given up. Nor should we pride ourselves that the enormous successes of contemporary science secure us against future amendments. No theory in the history of science enjoyed a more spectacular career than Newton's mechanics. Yet Newton's ideas had to give way to Einstein's.

When practicing scientists are reminded of these straightforward points, they frequently adopt what the philosopher George Berkeley called a "forlorn skepticism." From the idea of science as certain and infallible, they jump to a cynical description of their endeavors. Science is sometimes held to be a game played with arbitrary rules, an irrational acceptance of dogma, an enterprise based ultimately on faith. Once we have appreciated the fallibility of natural science and recognized its sources, we can move beyond the simple opposition of proof and faith. Between these extremes lies the vast field of cases in which we believe something on the basis of good—even excellent—but inconclusive evidence.

If we want to emphasize the fact that what scientists believe today may have to be revised in the light of observations made tomorrow, then we can describe all our science as "theory." But the description should not confuse us. To concede that evolutionary biology is a theory is not to suppose that there are alternatives to it that are equally worthy of a place in our curriculum. All theories are revisable, but not all theories are equal. Even though our present evidence does not *prove* that evolutionary biology—or quantum physics, or plate tectonics, or any other theory—is true, evolutionary biologists will maintain that the present evidence is overwhelmingly in favor of their theory and overwhelmingly against its supposed rivals. Their enthusiastic assertions that evolution is a proven fact can be charitably understood as claims that the (admittedly inconclusive) evidence we have for evolutionary theory is as good as we ever obtain for any theory in any field of science.

Hence the Creationist cry for a quick Fools' Mate can easily be avoided. Creationists attempt to draw a line between evolutionary biology and the rest of science by remarking that large-scale evolution cannot be observed. This tactic fails. Large-scale evolution is no more inaccessible to observation than nuclear reactions or the molecular composition of water. For the Creationists to succeed in divorcing evolutionary biology from the rest of science, they need to argue that evolutionary theory is less well supported by the evidence than are theories in, for example, physics and chemistry. It will come as no surprise to learn that they try to do this. To assess the merits of their arguments we need a deeper understanding of the logic of inconclusive justification. We shall begin with a simple and popular idea: Scientific theories earn our acceptance by making successful predictions.

III. PREDICTIVE SUCCESS

Imagine that somebody puts forward a new theory about the origins of hay fever. The theory makes a number of startling predictions concerning connections that we would not have thought worth investigating. For example, it tells us that people who develop hay fever invariably secrete a particular substance in certain fatty tissues and that anyone who eats rhubarb as a child never develops hay fever. The theory predicts things that initially appear fantastic. Suppose that we check up on these predictions and find that they are borne out by clinical tests. Would we not begin to believe—and believe reasonably—that the theory was *at least* on the right track?

This example illustrates a pattern of reasoning that is familiar in the history of science. Theories win support by producing claims about what can be observed, claims that would not have seemed plausible prior to the advancement of the theory, but that are in fact found to be true when we make the appropriate observations. A classic (real) example is Pascal's confirmation of Torricelli's hypothesis that we live at the bottom of an ocean of air that presses down upon us. Pascal reasoned that if Torricelli's hypothesis were true, then air pressure should decrease at higher altitudes (because at higher altitudes we are closer to the "surface" of the atmosphere, so that the length of the column of air that presses down is shorter). Accordingly, he sent his brother-in-law to the top of a mountain to make some barometric measurements. Pascal's clever working out of the observational predictions of Torricelli's theory led to a dramatic predictive success for the theory.

The idea of predictive success has encouraged a popular picture of science. (We shall see later that this picture, while popular, is not terribly accurate.) Philosophers sometimes regard a theory as a collection of claims or statements. Some of these statements offer generalizations about the features of particular, recondite things (genes, atoms, gravitational force, quasars, and the like). These statements are used to infer statements whose truth or falsity can be decided by observation. (This appears to be just what Pascal did.) Statements belonging to this second group are called the *observational consequences* of the theory. Theories are supported when we find that their observational consequences (those that we have checked) are true. The credentials of a theory are damaged if we discover that some of its observational consequences are false.

We can make the idea more precise by being clearer about the inferences involved. Those who talk of inferring observational predictions from our theories think that we can *deduce* from the statements of the theory, and from those statements alone, some predictions whose accuracy we can check by direct observation. Deductive inference is well understood. The fundamental idea of deductive inference is this: We say that a statement *S* is a valid deductive consequence of a group of statements if and only if it is *impossible* that all the statements in the group should be true and that *S* should be false; alternatively, *S* is a valid deductive consequence (or, more simply, a valid consequence) of a group of statements if and only if it would be self-contradictory to assert all the statements in the group and to deny *S*.

It will be helpful to make the idea of valid consequence more familiar with some examples. Consider the statements "All lovers of baseball dislike George Steinbrenner" and "George Steinbrenner loves baseball." The statement "George Steinbrenner dislikes himself" is a deductively valid consequence of these two statements. For it is impossible that the first two should be true and the third false. However, in claiming that this is a case of deductively valid consequence, we do not commit ourselves to maintaining that *any* of the statements is true. (Perhaps there are some ardent baseball fans who admire Steinbrenner. Perhaps Steinbrenner himself has no time for the game.) What deductive validity means is that the truth of the first two statements would guarantee the truth of the third; that is, *if* the first two *were* true, then the third would have to be true.

Another example will help rule out other misunderstandings. Here are two statements: "Shortly after noon on January 1, 1982, in the Oval Office, a jelly bean was released from rest more than two feet above any surface"; "Shortly after noon on January 1, 1982, in the Oval Office, a jelly bean fell." Is the second statement a deductively valid consequence of the first? You might think that it is, on the grounds that it would have been impossible for the unfortunate object to have been released and not to have fallen. In one sense this is correct, but that is not the sense of impossibility that deductive logicians have in mind. Strictly speaking, it is not *impossible* for the jelly bean to have been released without falling; we can imagine, for example, that the law of gravity might suddenly cease to operate. We do not *contradict* ourselves when we assert that the jelly bean was released but deny that it fell; we simply refuse to accept the law of gravity (or some other relevant physical fact).

Thus, *S* is a deductively valid consequence of a group of statements if and only if there is *absolutely no possibility* that all the statements in the group should be true and *S* should be false. This conception allows us to state the popular view of theory and prediction more precisely. Theories are collections of statements. The observational consequences of a theory are statements that have to be true if the statements belonging to the theory are all true. These observational consequences also have to be statements whose truth or falsity can be ascertained by direct observation. My initial discussion of predictive success presented the rough idea that, when we find the observational consequences of a theory to be true, our findings bring credit to the theory. Conversely discovery that some observational consequences of a theory are false was viewed as damaging. We can now make the second point much more precise. Any theory that has a false observational consequence must contain some false statement (or statements). For if all the statements in the theory were true, then, according to the standard definitions of *deductive validity* and *observational consequence,* any observational consequence would also have to be true. Hence, if a theory is found to have a false observational consequence, we must conclude that one or more statements of the theory is false.

This means that theories can be conclusively falsified through the discovery that they have false observational consequences. Some philosophers, most notably Sir Karl Popper (Popper 1959; 1963), have taken this point to have enormous significance for our understanding of science. According to Popper, the essence of a scientific theory is that it should be *falsifiable*. That is, if the theory is false, then it must be possible to show that it is false. Now, if a theory has utterly no observational consequences, it would be extraordinarily difficult to unmask that theory as false. So, to be a genuine scientific theory, a group of statements must have observational consequences. It is important to realize that Popper is not suggesting that every good theory must be false. The difference between being falsifiable and being false is like the difference between being vulnerable and actually being hurt. A good scientific theory should not be false. Rather, it must have observational consequences that could reveal the theory as mistaken if the experiments give the wrong results.

While these ideas about theory testing may seem strange in their formal attire, they emerge quite frequently in discussions of science. They also find their way into the creation-evolution debate.

IV. PREDICTIVE FAILURE

From the beginning, evolutionary theory has been charged with just about every possible type of predictive failure. Critics of the theory have argued that (a) the theory makes no predictions (it is unfalsifiable and so fails Popper's criterion for science), (b) the theory makes false predictions (it is falsified), (c) the theory does not make the kinds of predictions it ought to make (the observations and experiments that evolutionary theorists undertake have no bearing on the theory). Many critics, including several Creationists (Morris 1974a, Gish 1979; Wysong 1976), manage to advance all these objections in the same work. This is somewhat surprising, since points (a) and (b) are, of course, mutually contradictory.

The first objection is vitally important to the Creationist cause. Their opponents frequently insist that Creationism fails the crucial test for a scientific theory. The hypothesis that all kinds of organisms were separately fashioned by some "originator" is unfalsifiable (Gould 1981b). Creationists retort that they can play the same game equally well. *Any* hypothesis about the origins of life, including that advanced by evolutionary theory, is not subject to falsification. Hence we cannot justify a decision to teach evolutionary theory and not to teach Creationism by appealing to the Popperian criterion for genuine science.

The allegation that evolutionary theory fails to make any predictions is a completely predictable episode in any Creationist discussion of evolution. Often the point is made by appeal to the authority of Popper. Here are two sample passages:

The outstanding philosopher of science, Karl Popper, though himself an evolutionist, pointed out cogently that evolution, no less than creation, is untestable
and thus unprovable. (Morris 1974b, 80)

Thus, for a theory to qualify as a scientific theory, it must be supported by
events, processes or properties which can be observed, and the theory must be
useful in predicting the outcome of future natural phenomena or laboratory
experiments. An additional limitation usually imposed is that the theory must be
capable of falsification. That is, it must be possible to conceive some experiment, the failure of which would disprove the theory.
     It is on the basis of such criteria that most evolutionists insist that creation
be refused consideration as a possible explanation for origins. Creation has not
been witnessed by human observers, it cannot be tested experimentally, and as
a theory it is nonfalsifiable.
     The general theory of evolution also fails to meet all three of these criteria,
however. (Gish 1979, 13)

These passages, and many others (for example, Morris 1974a, 150; Morris
1975, 9; Moore 1974, 9; Wilder-Smith 1981, 133), draw on the picture of science sketched above. It is not clear that the Creationists really understand the
philosophical views that they attempt to apply. Gish presents the most articulate discussion of the falsifiability criterion. Yet he muddles the issue by
describing falsifiability as an "additional limitation" beyond predictive
power. (The previous section shows that theories that make predictions are
automatically falsifiable.) Nevertheless, the Creationist challenge is a serious
one, and, if it could not be met, evolutionary theory would be in trouble.
     Creationists buttress their charge of unfalsifiability with further objec-
tions. They are aware that biologists frequently look as though they are engaged in observations and experiments. Creationists would allow that researchers in biology sometimes make discoveries. What they deny is that the
discoveries support evolutionary theory. They claim that laboratory manipulations fail to teach us about evolution in nature: "Even if modern scientists
should ever actually achieve the artificial creation of life from non-life, or of
higher kinds from lower kinds, in the laboratory, this would not *prove* in any
way that such changes did, or even could, take place in the past by random
natural processes" (Morris 1974a, 6). The standards of evidence to be applied
to evolutionary biology have suddenly been raised. In this area of inquiry, it
is not sufficient that a theory yield observational consequences whose truth or
falsity can be decided in the laboratory. Creationists demand special kinds of
predictions, and will dismiss as irrelevant any laboratory evidence that evolutionary theorists produce. [In this way, they try to defend point (c).]
     Oddly enough, however, the most popular supplement to the charge that
evolutionary theory is unfalsifiable is a determined effort to falsify it [point
(b)]. Creationists cannot resist arguing that the theory is actually falsified.
Some of them, Morris and Gish, for example, recognize the tension between

the two objections. They try to paper over the problem by claiming that evolutionary theory and the Creationist account are both "models." Each "model" would "naturally" incline us to expect certain observational results. A favorite Creationist ploy is to draw up tables in which these "predictions" are compared. When we look at the tables we find that the evolutionary expectations are confounded. By contrast, the Creationist "model" leads us to anticipate features of the world that are actually there. Faced with such adverse results, the benighted evolutionary biologist is portrayed as struggling to "explain away" the findings by whatever means he can invent.

Morris's own practice of this form of evolution baiting can serve as an example. Morris constructs a table (1974a, 12; see below and facing page) whose function is to indicate "The predictions that would probably be made in several important categories" (1974a, 12). Morris admits magnanimously that "these primary models may be modified by secondary [additional] assumptions to fit certain conditions. For example, the basic evolution model may be extended to include harmful, as well as beneficial, mutations, but this is not a natural prediction of the basic concept of evolution" (1974a, 13). The idea that the "natural predictions" of the evolution "model" are at odds with the phenomena is used to suggest that evolutionary biologists are forced to desperate measures to protect their "faith." As Morris triumphantly concludes, "the data must be *explained* by the evolutionist, but they are predicted by the *creationist*" (1974a, 13).

The careful reader ought to be puzzled. If Morris really thinks that evolutionary theory has been falsified, why does he not say so? Of course, he would have to admit that the theory is falsifiable. Seemingly, however, a staunch Creationist should be delighted to abandon a relatively abstruse point about unfalsifiability in favor of a clear-cut refutation. The truth of the matter is that the alleged refutations fail. No evolutionary theorist will grant that (for example) the theory predicts that the fossil record should show "innumerable transitions." Instead, paleontologists will point out that we can deduce conclusions about what we should find in the rocks only if we make assumptions about the fossilization process. Morris makes highly dubious assumptions, hails them as "natural," and then announces that the "natural predictions" of the theory have been defeated.

(This example suggests a method for coping with Morris's "table of natural predictions." Each of these predictions can be deduced from evolutionary theory only if the theory is extended by adding extra assumptions. Morris saddles evolutionary theory with faulty additional claims. These are the source of the false predictions. Later, I shall show this in detail for some of Morris's "natural predictions" and the similar difficulties raised by other Creationists [Gish 1979, 53–54; Wysong 1975, 421–426].)

Category	Evolution Model	Creation Model
Structure of Natural Law	Constantly Changing	Invariable
Galactic Universe	Galaxies Changing	Galaxies Constant
Structure of Stars	Stars Changing into Other Types	Stars Unchanged
Other Heavenly Bodies	Building Up	Breaking Down
Types of Rock Formations	Different in Different "Ages"	Similar in All "Ages"
Appearance of Life	Life Evolving from Nonlife	Life Only from Life
Array of Organisms	Continuum of Organisms	Distinct Kinds of Organisms
Appearance of Kinds of Life	New Kinds Appearing	No New Kinds Appearing
Mutations in Organisms	Beneficial	Harmful
Natural Selection	Creative Process	Conservative Process
Age of Earth	Extremely Old	Probably Young
Fossil Record	Innumerable Transitions	Systematic Gaps
Appearance of Man	Ape-Human Intermediates	No Ape-Human Intermediates
Nature of Man	Quantitatively Superior to Animals	Qualitatively Distinct from Animals
Origin of Civilization	Slow and Gradual	Contemporaneous with Man

To make a serious assessment of these broad Creationist charges, we must begin by asking some basic methodological questions. We cannot decide whether evolutionary biologists are guilty of trying to save their theory by using ad hoc assumptions (new and implausible claims dreamed up for the sole purpose of protecting some cherished ideas) unless we have some way of deciding when a proposal is ad hoc. Similarly, we cannot make a reasoned response to the charge that laboratory experiments are irrelevant, or to the fundamental objection that evolutionary theory is unfalsifiable, unless we have a firmer grasp on the relation between theory and evidence.

V. NAIVE FALSIFICATIONISM

The time has come to tell a dreadful secret. While the picture of scientific testing sketched above continues to be influential among scientists, it has been shown to be seriously incorrect. (To give my profession its due, historians and philosophers of science have been trying to let this particular cat out of the bag for at least thirty years. See, for example, Hempel 1941; Quine 1952.) Important work in the history of science has made it increasingly clear that no major scientific theory has ever exemplified the relation between theory and evidence that the traditional model presents.

What is wrong with the old picture? Answer: Either it debars most of what we take to be science from counting as science or it allows virtually anything to count. On the traditional view of "theory," textbook cases of scientific theories turn out to be unfalsifiable. Suppose we identify Newtonian mechanics with Newton's three laws of motion plus the law of gravitation. What observational consequences can we deduce from these four statements? You might think that we could deduce that if, as the (undoubtedly apocryphal) story alleges, an apple became detached from a branch above where Newton was sitting, the apple would have fallen on his head. But this does not follow at all. To see why not, it is only necessary to recognize that the failure of this alleged prediction would not force us to deny any of the four statements of the theory. All we need do is assume that some other forces were at work that overcame the force of gravity and caused the apple to depart from its usual trajectory. So, given this simple way of applying Popper's criterion, Newtonian mechanics would be unfalsifiable. The same would go for any other scientific theory. Hence none of what we normally take to be science would count as science. (I might note that Popper is aware of this problem and has suggestions of his own as to how it should be overcome. However, what concerns me here are the *applications* of Popper's ideas, that are made by Creationists, as well as by scientists in their professional debates.)

The example of the last paragraph suggests an obvious remedy. Instead of thinking about theories in the simple way just illustrated, we might take them to be far more elaborate. Newton's laws (the three laws of motion and the law of gravitation) are *embedded* in Newtonian mechanics. They form the core of the theory, but do not constitute the whole of it. Newtonian mechanics also contains supplementary assumptions, telling us, for example, that for certain special systems the effects of forces other than gravity are negligible. This more elaborate collection of statements *does* have observational consequences and *is* falsifiable.

But the remedy fails. Imagine that we attempt to expose some self-styled spiritual teacher as an overpaid fraud. We try to point out that the teacher's central message—"Quietness is wholeness in the center of stillness"—is unfalsifiable. The teacher cheerfully admits that, taken by itself, this profound doctrine yields no observational consequences. He then points out that, by themselves, the central statements of scientific theories are also incapable of generating observational consequences. Alas, if all that is demanded is that a doctrine be embedded in a group of statements with observational consequences, our imagined guru will easily slither off the hook. He replies, "You have forgotten that my doctrine has many other claims. For example, I believe that if quietness is wholeness in the center of stillness, then flowers bloom in the spring, bees gather pollen, and bunkered defenders of so-called science raise futile objections to the world's spiritual benefactors. You will see that these three predictions are borne out by experience. Of course, there are countless others. Perhaps when you see how my central message yields so much evident truth, you will recognize the wealth of evidence behind my claim. Quietness is wholeness in the center of stillness."

More formally, the trouble is that *any* statement can be coupled with other statements to produce observational consequences. Given any doctrine *D,* and any statement *O* that records the results of an observation, we can enable *D* to "predict" *O* by adding the extra assumption, "if *D,* then *O.*" (In the example, *D* is "Quietness is wholeness in the center of stillness"; examples of *O* would be statements describing the blooming of particular flowers in the spring, the pollen gathering of specific bees, and so forth.)

The falsifiability criterion adopted from Popper—which I shall call the naive *falsificationist* criterion—is hopelessly flawed. It runs aground on a fundamental fact about the relation between theory and prediction: On their own, individual scientific laws, or the small groups of laws that are often identified as theories, do not have observational consequences. This crucial point about theories was first understood by the great historian and philosopher of science Pierre Duhem. Duhem saw clearly that individual scientific claims do not, and cannot, confront the evidence one by one. Rather, in his picturesque phrase, "Hypotheses are tested in bundles." Besides ruling out the possibility of testing an individual scientific theory (read, small group of laws), Duhem's insight has another startling consequence. We can only test relatively large bundles of claims. What this means is that when our experiments go awry we are not logically compelled to select any particular claim as the culprit. We can always save a cherished hypothesis from refutation by rejecting (however implausibly) one of the other members of the bundle. Of course, this is exactly what I did in the illustration of Newton and the apple above. Faced with disappointing results, I suggested that we could abandon the (tacit) additional claim that no large forces besides gravity were operating on the apple.

Creationists wheel out the ancient warhorse of naive falsificationism so that they can bolster their charge that evolutionary theory is not a science. The (very) brief course in deductive logic plus the whirlwind tour through naive falsificationism and its pitfalls enable us to see what is at the bottom of this seemingly important criticism. Creationists can appeal to naive falsificationism to show evolution is not a science. But, given the traditional picture of theory evidence I have sketched, one can appeal to naive falsificationism to show that *any* science is not a science. So, as with the charge that evolutionary change is unobservable, Creationists have again failed to find some "fault" of evolution not shared with every other science. (And, as we shall see, Creationists like some sciences, especially thermodynamics.) Consistent application of naive falsificationism can show that anybody's favorite science (whether it be quantum physics, molecular biology, or whatever) is not science. Of course, what this shows is that the naive falsificationist criterion is a very poor test of genuine science. To be fair, this point can cut both ways. Scientists who charge that "scientific" Creationism is unfalsifiable are not insulting the theory as much as they think.

VI. SUCCESSFUL SCIENCE

Despite the inadequacies of naive falsificationism, there is surely something right in the idea that a science can succeed only if it can fail. An invulnerable "science" would not be science at all. To achieve a more adequate understanding of how a science can succeed and how it runs the risk of failure, let us look at one of the most successful sciences and at a famous episode in its development.

Newtonian celestial mechanics is one of the star turns in the history of science. Among its numerous achievements were convincing explanations of the orbits of most of the known planets. Newton and his successors viewed the solar system as a collection of bodies subject only to gravitational interactions; they used the law of gravitation and the laws of motion to compute the orbits. (Bodies whose effects were negligible in any particular case would be disregarded. For example, the gravitational attraction due to Mercury would not be considered in working out the orbit of Saturn.) The results tallied beautifully with astronomical observations. But one case proved difficult. The outermost known planet, Uranus, stubbornly followed an orbit that diverged from the best computations. By the early nineteenth century it was clear that something was wrong. Either astronomers erred in treating the solar system as a Newtonian gravitational system or there was some particular difficulty in applying the general method to Uranus.

Perhaps the most naive of falsificationists would have recommended that the central claim of Newtonian mechanics—the claim that the solar system is a Newtonian gravitational system—be abandoned. But there was obviously a more sensible strategy. Astronomers faced one problematical planet, and they asked themselves what made Uranus so difficult. Two of them, John Adams and Urbain Leverrier, came up with an answer. They proposed (independently) that there was a hitherto unobserved planet beyond Uranus. They computed the orbit of the postulated planet and demonstrated that the anomalies of the motion of Uranus could be explained if a planet followed this path. There was a straightforward way to test their proposal. Astronomers began to look for the new planet. Within a few years, the planet—Neptune—was found.

I will extract several morals from this success story. The first concerns an issue we originally encountered in Morris's "table of natural predictions": What is the proper use of auxiliary hypotheses? Adams and Leverrier saved the central claim of Newtonian celestial mechanics by offering an auxiliary hypothesis. They maintained that there were more things in the heavens than had been dreamed of in previous natural philosophy. The anomalies in the orbit of Uranus could be explained on the assumption of an extra planet. Adams and Leverrier worked out the exact orbit of that planet so that they could provide a detailed account of the perturbations—and so that they could tell their fellow astronomers where to look for Neptune. Thus, their auxiliary hypothesis was *independently testable.* The evidence for Neptune's existence was not just the anomalous motion of Uranus. The hypothesis could be checked independently of any assumptions about Uranus or about the correctness of Newtonian celestial mechanics—by making telescopic observations.

Since hypotheses are always tested in bundles, this method of checking presupposed other assumptions, in particular, the optical principles that justify the use of telescopes. The crucial point is that, while hypotheses are always tested in bundles, they can be tested in *different* bundles. An auxiliary hypothesis ought to be testable independently of the particular problem it is introduced to solve, independently of the theory it is designed to save.

While it is obvious in retrospect—indeed it was obvious at the time—that the problem with Uranus should not be construed as "falsifying" celestial mechanics, it is worth asking explicitly why scientists should have clung to Newton's theory in the face of this difficulty. The answer is not just that nothing succeeds like success, and that Newton's theory had been strikingly successful in calculating the orbits of the other planets. The crucial point concerns the way in which Newton's successes had been achieved. Newton was no opportunist, using one batch of assumptions to cope with Mercury, and then moving on to new devices to handle Venus. Celestial mechanics was a remarkably *unified* theory. It solved problems by invoking the same pattern of reasoning, or *problem-solving strategy,* again and again: From a specification of the positions of the bodies under study, use the law of gravitation to calculate the forces acting; from a statement of the forces acting, use the laws of dynamics to compute the equations of motion; solve the equations of motion to obtain the motions of the bodies. This single pattern of reasoning was applied in case after case to yield conclusions that were independently found to be correct.

At a higher level, celestial mechanics was itself contained in a broader theory. Newtonian physics, as a whole, was remarkably unified. It offered a strategy for solving a diverse collection of problems. Faced with *any* question about motion, the Newtonian suggestion was the same: Find the forces acting, from the forces and the laws of dynamics work out the equations of motion, and solve the equations of motion. The method was employed in a broad range of cases. The revolutions of planets, the motions of projectiles, tidal cycles, and pendulum oscillations—all fell to the same problem-solving strategy.

We can draw a second moral. A science should be *unified.* A thriving science is not a gerrymandered patchwork but a coherent whole. Good theories consist of just one problem-solving strategy, or a small family of problem-solving strategies, that can be applied to a wide range of problems. The theory succeeds as it is able to encompass more and more problem areas. Failure looms when the basic problem-solving strategy (or strategies) can resolve almost none of the problems in its intended domain without the "aid" of untestable auxiliary hypotheses.

Despite the vast successes of his theory, Newton hoped for more. He envisaged a time when scientists would recognize other force laws, akin to the law of gravitation, so that other branches of physics could model themselves after celestial mechanics. In addition, he suggested that many physical questions that are not ostensibly about motion—questions about heat and about chemical combination, for example—could be reduced to problems of motion. *Principia,* Newton's masterpiece, not only offered a theory; it also advertised a program:

> I wish we could derive the rest of the phenomena of Nature by the same kind of reasoning from mechanical principles, for I am induced by many reasons to suspect that they may all depend upon certain forces by which the particles of bodies, by some causes hitherto unknown, are either mutually impelled towards one another, and cohere in regular figures, or are repelled and recede from one another. These forces being unknown, philosophers have hitherto attempted the search of Nature in vain; but I hope the principles here laid down will afford some light either to this or some truer method of philosophy. (Newton 1687/Motte-Cajori 1960, xviii)

Newton's message was clear. His own work only began the task of applying an immensely fruitful, unifying idea.

Newton's successors were moved, quite justifiably, to extend the theory he had offered. They attempted to show how Newton's main problem-solving strategy could be applied to a broader range of physical phenomena. During the eighteenth and nineteenth centuries, the search for understanding of the forces of nature was carried into hydrodynamics, optics, chemistry, and the studies of heat, elasticity, electricity, and magnetism. Not all of these endeavors were equally successful. Nevertheless, Newton's directive fostered the rise of some important new sciences.

The final moral I want to draw from this brief look at Newtonian physics concerns *fecundity*. A great scientific theory, like Newton's, opens up new areas of research. Celestial mechanics led to the discovery of a previously unknown planet. Newtonian physics as a whole led to the development of previously unknown sciences. Because a theory presents a new way of looking at the world, it can lead us to ask new questions, and so to embark on new and fruitful lines of inquiry. Of the many flaws with the earlier picture of theories as sets of statements, none is more important than the misleading presentation of sciences as static and insular. Typically, a flourishing science is incomplete. At any time, it raises more questions than it can currently answer. But incompleteness is no vice. On the contrary, incompleteness is the mother of fecundity. Unresolved problems present challenges that enable a theory to flower in unanticipated ways. They also make the theory hostage to future developments. A good theory should be productive; it should raise new questions and presume that those questions can be answered without giving up its problem-solving strategies.

I have highlighted three characteristics of successful science. *Independent testability* is achieved when it is possible to test auxiliary hypotheses independently of the particular cases for which they are introduced. *Unification* is the result of applying a small family of problem-solving strategies to a broad class of cases. *Fecundity* grows out of incompleteness when a theory opens up new and profitable lines of investigation. Given these marks of successful science, it is easy to see how sciences can fall short, and how some doctrines can do so badly that they fail to count as science at all. A scientific theory begins to wither if some of its auxiliary assumptions can be saved from refutation only by rendering them untestable; or if its problem-solving strategies become a hodgepodge, a collection of unrelated methods, each designed for a separate recalcitrant case; or if the promise of the theory just fizzles, the few questions it raises leading only to dead ends.

When does a doctrine fail to be a science? If a doctrine fails sufficiently abjectly as a science, then it fails to be a science. Where bad science becomes egregious enough, pseudoscience begins. The example of Newtonian physics shows us how to replace the simple (and incorrect) naive falsificationist criterion with a battery of tests. Do the doctrine's problem-solving strategies encounter recurrent difficulties in a significant range of cases? Are the problem-solving strategies an opportunistic collection of unmotivated and unrelated methods? Does the doctrine have too cozy a relationship with auxiliary hypotheses, applying its strategies with claims that can be "tested" only in their applications? Does the doctrine refuse to follow up on unresolved problems, airily dismissing them as "exceptional cases"? Does the doctrine restrict the domain of its methods, forswearing excursions into new areas of investigation where embarrassing questions might arise? If all, or many, of these tests are positive, then the doctrine is not a poor scientific theory. It is not a scientific theory at all.

The account of successful science that I have given not only enables us to replace the naive falsificationist criterion with something better. It also provides a deeper understanding of how theories are justified. Predictive success is one important way in which a theory can win our acceptance. But it is not the only way. In general, theories earn their laurels by solving problems—providing answers that can be independently recognized as correct—and by their fruitfulness. Making a prediction is answering a special kind of question. The astronomers who used celestial mechanics to predict the motion of Mars were answering the question of where Mars would be found. Yet, very frequently, our questions do not concern *what* occurs, but *why* it occurs. We already know that something happens and we want an explanation. Science offers us explanations by setting the phenomena within a unified framework. Using a widely applicable problem-solving strategy, together with independently confirmed auxiliary hypotheses, scientists show that what happened was to be expected. It was known before Newton that the orbits of the planets are approximately elliptical. One of the great achievements of Newton's celestial mechanics was to apply its problem-solving strategy to deduce that the orbit of any planet will be approximately elliptical, thereby explaining the shape of the orbits. In general, science is at least as concerned with reducing the number of unexplained phenomena as it is with generating correct predictions.

The most global Creationist attack on evolutionary theory is the claim that evolution is not a science. If this claim were correct, then the dispute about what to teach in high school science classes would be over. In earlier parts of this chapter, we saw how Creationists were able to launch their broad criticisms. If one accepts the idea that science requires proof, or if one adopts the naive falsificationist criterion, then the theory of evolution—and every other scientific theory—will turn out not to be a part of science. So Creationist standards for science imply that there is no science to be taught.

However, we have seen that Creationist standards rest on a very poor understanding of science. In light of a clearer picture of the scientific enterprise, I have provided a more realistic group of tests for good science, bad science, and pseudoscience. Using this more sophisticated approach, I now want to address seriously the global Creationist questions about the theory of evolution. Is it a pseudoscience? Is it a poor science? Or is it a great science? These are very important questions, for the appropriateness of granting equal time to Creation "science" depends, in part, on whether it can be regarded as the equal of the theory of evolution.

VII. DARWIN'S DARING

The heart of Darwinian evolutionary theory is a family of problem-solving strategies, related by their common employment of a particular style of historical narrative. A *Darwinian history* is a piece of reasoning of the following general form. The first step consists in a description of an ancestral population of organisms. The reasoning proceeds by tracing the modification of the population through subsequent generations, showing how characteristics were selected, inherited, and became prevalent. (. . . Natural selection is taken to be the primary—but not the only—force of evolutionary change.)

Reasoning like this can be used to answer a host of biological questions. Suppose that we want to know why a contemporary species manifests a particular trait. We can answer that question by supplying a Darwinian history that describes the emergence of that trait. Equally, we can use Darwinian histories to answer questions about relationships among groups of organisms. One way to explain why two species share a common feature is to trace their descent from a common ancestor. Questions of biogeography can be addressed in a similar way. We can explain why we find contemporary organisms where we do by following the course of their historical modifications and migrations. Finally, we can tackle problems about extinction by showing how characteristics that had enabled organisms to thrive were no longer advantageous when the environment (or the competition) changed. In all these cases, we find related strategies for solving problems. The history of the development of populations, understood in terms of variation, competition, selection, and inheritance, is used to shed light on broad classes of biological phenomena.

The questions that evolutionary theory has addressed are so numerous that any sample is bound to omit important types. The following short selection undoubtedly reflects the idiosyncrasy of my interests: Why do orchids have such intricate internal structures? Why are male birds of paradise so brightly colored? Why do some reptilian precursors of mammals have enormous "sails" on their backs? Why do bats typically roost upside down? Why are the hemoglobins of humans and apes so similar? Why are there no marsupial analogues of seals and whales? Why is the mammalian fauna of Madagascar so distinctive? Why did the large, carnivorous ground birds of South America become extinct? Why is the sex ratio in most species one to one (although it is markedly different in some species of insects)? Answers to these questions, employing Darwinian histories, can be found in works written by contemporary Darwinian biologists. Those works contain answers to a myriad of questions of the same general types. Darwinian histories are constructed again and again to illuminate the characteristics of contemporary organisms, to account for the similarities and differences among species, to explain why the forms preserved in the fossil record emerged and became extinct, to cast light on the geographical distribution of animals and plants.

We can see the history in action by taking a brief look at one of these examples. The island of Madagascar, off the east coast of Africa, supports a peculiar group of mammals. Many of these mammals are endemic. Among them is a group of relatively small insectivorous mammals, the *tenrecs*. All tenrecs share certain features that mark them out as relatively primitive mammals. They have very poor vision, their excretory system is rudimentary, the testes in the male are carried within the body, their capacity for regulating their body temperature is poor compared with that of most mammals. Yet, on their simple and rudimentary body plan, specialized characteristics have often been imposed. Some tenrecs have the hedgehog's method of defense. Others have the forelimbs characteristic of moles. There are climbing tenrecs that resemble the shrews, and there are tenrecs that defend themselves by attempting to stick their quills into a would-be predator. Hedgehogs, moles, tree shrews, and porcupines do not inhabit Madagascar. But they seem to have their imitators. (These are examples of convergent evolution, cases in which unrelated organisms take on some of the same characteristics.) Why are these peculiar animals found on Madagascar, and nowhere else?

A straightforward evolutionary story makes sense of what we observe. In the late Mesozoic or early Cenozoic, small, primitive, insectivorous mammals rafted across the Mozambique channel and colonized Madagascar. Later the channel widened and Madagascar became inaccessible to the more advanced mammals that evolved on the mainland. Hence the early colonists developed without competition from advanced mainland forms and without pressure from many of the normal predators who make life difficult for small mammals. The tenrecs have been relatively protected. In the absence of rigorous competition, they have preserved their simple body plan, and they have exploited unoccupied niches, which are filled elsewhere by more advanced creatures. Tenrecs have gone up the trees and burrowed in the ground because those are good ways to make a living and because they have had nobody but one another to contend with.

The same kind of story can be told again and again to answer all sorts of questions about all sorts of living things. Evolutionary theory is unified because so many diverse questions—questions as various as those I listed— can be addressed by advancing Darwinian histories. Moreover, these narratives constantly make claims that are subject to independent check. Here are four examples from the case of the triumphant tenrecs. (1) The explanation presupposes that Madagascar has drifted away from the east coast of Africa. That is something that can be checked by using geological criteria for the movement of land-masses, criteria that are independent of biology. (2) The account claims that the tenrecs would have been able to raft across the Mozambique channel, but that the present channel constitutes a barrier to more advanced mammals (small rodents). These claims could be tested by looking to see whether the animals in question can disperse across channels of the appropriate sizes. (3) The narrative assumes that the specialized methods of defense offered advantages against the predators that were present in Madagascar. Studies of animal interactions can test whether the particular defenses are effective against local predators. (4) Central to the explanatory account is the thesis that the tenrecs are related. If this is so, then studies of the minute details of tenrec anatomy should reveal many common features, and the structures of proteins ought to be similar. In particular, the tenrecs ought to be much more like one another than they are like hedgehogs, shrews, or moles.

Looking at one example, or even at a small number of examples, does not really convey the strength of evolutionary theory. The same patterns of reasoning can be applied again and again, in book after book, monograph after monograph, article after article. Yet the particular successes in dealing with details of natural history, numerous though they are, do not exhaust the accomplishments of the theory. Darwin's original theory—the problem-solving strategies advanced in the *Origin*, which are, in essence, those just described—gave rise to important new areas of scientific investigation. Evolutionary theory has been remarkably fruitful.

Darwin not only provided a scheme for unifying the diversity of life. He also gave a structure to our ignorance. After Darwin, it was important to resolve general issues about the presuppositions of Darwinian histories. The way in which biology should proceed had been made admirably plain, and it was clear that biologists had to tackle questions for which they had, as yet, no answers. How do new characteristics arise in populations? What are the mechanisms of inheritance? How do characteristics become fixed in populations? What criteria decide when a characteristic confers some advantage on its possessor? What interactions among populations of organisms affect the adaptive value of characteristics? With respect to all of these questions, Darwin was forced to confess ignorance. By raising them, his theory pointed the way to its further articulation.

Since Darwin's day, biologists have contributed parts of evolutionary theory that help to answer these important questions. Geneticists have advanced our understanding of the transmission of characteristics between generations and have enabled us to see how new characteristics can arise. Population geneticists have analyzed the variation present in populations of organisms; they have suggested how that variation is maintained and have specified ways in which characteristics can be fixed or eliminated. Workers in morphology and physiology have helped us to see how variations of particular kinds might yield advantages in particular environments. Ecologists have studied the ways in which interactions among populations can affect survival and fecundity.

The moral is obvious. Darwin gambled. He trusted that the questions he left open would be answered by independent biological sciences and that the deliverances of these sciences would be consistent with the presuppositions of Darwinian histories. Because of the breadth of his vision, Darwin made his theory vulnerable from a number of different directions. To take just one example, it could have turned out the mechanisms of heredity would have made it impossible for advantageous variations to be preserved and to spread. Indeed, earlier in this century, many biologists felt that the emerging views about inheritance did not fit into Darwin's picture, and the fortunes of Darwinian evolutionary theory were on the wane.

When we look at the last 120 years of the history of biology, it is impossible to ignore the fecundity of Darwin's ideas. Not only have inquiries into the presuppositions of Darwinian histories yielded new theoretical disciplines (like population genetics), but the problem-solving strategies have been extended to cover phenomena that initially appeared troublesome. One recent triumph has been the development of explanations for social interactions among animals. Behavior involving one animal's promotion of the good of others seems initially to pose a problem for evolutionary theory. How can we construct Darwinian histories for the emergence of such behavior? W. D. Hamilton's concept of inclusive fitness, and the deployment of game-theoretic ideas by R. L. Trivers and John Maynard Smith, revealed how the difficulty could be resolved by a clever extension of traditional Darwinian concepts.

Yet puzzles remain. One problem is the existence of sex. When an organism forms gametes (sperm cells or egg cells) there is a meiotic division, so that in sexual reproduction only half of an organism's genes are transmitted to each of its progeny. Because of this "cost of meiosis," it is hard to see how genotypes for sexual reproduction might have become prevalent. (Apparently, they will spread only half as fast as their asexual rivals.) So why is there sex? We do not have a compelling answer to the question. Despite some ingenious suggestions by orthodox Darwinians (notably G. C. Williams 1975; John Maynard Smith 1978), there is no convincing Darwinian history for the emergence of sexual reproduction. However, evolutionary theorists believe that the problem will be solved without abandoning the main Darwinian insights—just as early nineteenth-century astronomers believed that the problem of the motion of Uranus could be overcome without major modification of Newton's celestial mechanics.

The comparison is apt. Like Newton's physics in 1800, evolutionary theory today rests on a huge record of successes. In both cases, we find a unified theory whose problem-solving strategies are applied to illuminate a host of diverse phenomena. Both theories offer problem solutions that can be subjected to rigorous independent checks. Both open up new lines of inquiry and have a history of surmounting apparent obstacles. The virtues of successful science are clearly displayed in both.

There is a simple way to put the point. Darwin is the Newton of biology. Evolutionary theory is not simply an area of science that has had some success at solving problems. It has unified biology and it has inspired important biological disciplines. Darwin himself appreciated the unification achieved by his theory and its promise of further development (Darwin 1859/Mayr

1964, 188, 253–254, 484–486). Over a century later, at the beginning of his authoritative account of current views of species and their origins, Ernst Mayr explained how that promise has been fulfilled: "The theory of evolution is quite rightly called the greatest unifying theory in biology. The diversity of organisms, similarities and differences between kinds of organisms, patterns of distribution and behavior, adaptation and interaction, all this was merely a bewildering chaos of facts until given meaning by the evolutionary theory" (Mayr 1970, 1). Dobzhansky put the point even more concisely: "Nothing in biology makes sense except in the light of evolution" (Dobzhansky, 1973).

REFERENCES

Birch, L. C., and Ehrlich, P. R. 1967. "Evolutionary History and Population Biology." *Nature* 214:349–352.

Darwin, C. 1859/Mayr, E. 1964. *The Origin of the Species*. Facsimile of the first edition: Cambridge: Harvard University Press.

Dobzhansky, T. 1973. "Nothing in Biology Makes Sense Except in the Light of Evolution." *American Biology Teacher* 35:125–129.

Gish. D. T. 1979. *Evolution? The Fossils Say No!* San Diego: Creation Life Publishers.

Gould, S. J. 1981. *The Mismeasure of Man*. New York: Norton.

Hempel, C. G. 1951. "Problems and Changes in the Empiricist Criterion of Meaning." In *Aspects of Scientific Exploration*. Glencoe, Ill.: The Free Press, 1965.

Matthews, L. H. 1971. *Introduction to The Origin of the Species*. London: Dent.

Maynard Smith, J. 1978. *The Evolution of Sex*. Cambridge: Cambridge University Press.

Mayr, E. 1970. *Population, Species, and Evolution*. Cambridge, Mass.: Harvard University Press.

———. 1976. *Evolution and the Diversity of Life*. Cambridge, Mass.: Harvard University Press.

Moore, J. 1974. *Should Evolution Be Taught?* San Diego: Creation Life Publishers.

Morris, H. Ma. 1974a. *Scientific Creationism* (general ed.). San Diego: Creation Life Publishers.

——— 1974b. *The Troubled Waters of Evolution*. San Diego: Creation Life Publishers.

——— 1975. *Introducing Creationism in the Public Schools*. San Diego: Creation Life Publishers.

Popper, K. R. 1959. *The Logic of Scientific Discovery*. London: Hutchinson.

———. 1968. *Conjectures and Refutations*. New York: Harper and Row.

Quine, W. V. O. 1952. *The Dogmas of Empiricism*. Reprinted from "A Logical Point of View," Cambridge, Mass.: Harvard University Press.

Simpson, C. G. 1953. *The Major Features of Evolution*. New York: Columbia University Press.

Watson, D. C. C. 1976. *The Great Brain Robbery*. Chicago: Moody Press.

Wilder-Smith, A. E. 1981. *The Natural Sciences Know Nothing of Evolution*. San Diego: Creation Life Publishers.

Williams, G. C. 1975. *Sex and Evolution*. Princeton, N.J.: Princeton University.

Wysong, R. L. 1976. *The Creation Evolution Controversy*. Midland, Mich.: Inquiry Press.

CASE STUDY FOR PART 1

The following is a letter which was received by the editor of a science journal.

Dear Sir:

I am taking the liberty of calling upon you to be the judge in a dispute between me and an acquaintance who is no longer a friend. The question at issue is this: Is my creation, umbrellaology, a science? Allow me to explain. . . . For the past eighteen years assisted by a few faithful disciples, I have been collecting materials on a subject hitherto almost wholly neglected by scientists, the umbrella. The results of my investigations to date are embodied in the nine volumes which I am sending to you under a separate cover. Pending their receipt, let me describe to you briefly the nature of their contents and the method I pursued in compiling them. I began on the Island of Manhattan. Proceeding block by block, house by house, family by family, and individual by individual, I ascertained (1) the number of umbrellas possessed, (2) their size, (3) their weight, (4) their color. Having covered Manhattan after many years, I eventually extended the survey to the other boroughs of the City of New York, and at length completed the entire city. Thus I was ready to carry forward the work to the rest of the state and indeed the rest of the United States and the whole known world.

It was at this point that I approached my erstwhile friend. I am a modest man, but I felt I had the right to be recognized as the creator of a new science. He, on the other hand, claimed that umbrellaology was not a science at all. First, he said, it was silly to investigate umbrellas. Now this argument is false, because science scorns not to deal with any object, however humble and lowly, even to the "hind leg of a flea." Then why not umbrellas? Next, he said that umbrellaology could not be recognized as a science because it was of no use or benefit to mankind. But is not the truth the most precious thing in life? Are not my nine volumes filled with the truth about my subject? Every word in them is true. Every sentence contains a hard, cold fact. When he asked me what was the object of umbrellaology I was proud to say, "To seek and discover the truth is object enough for me." I am a pure scientist; I have no ulterior motives. Hence it follows that I am satisfied with truth alone. Next, he said my truths were dated and that any one of my finds might cease to be true tomorrow. But this, I pointed out, is not an argument against umbrellaology, but rather an argument for keeping it up to date, which exactly is what I propose. Let us have surveys monthly, weekly, or even daily, to keep our knowledge abreast of the changing facts. His next contention was that umbrellaology had entertained no hypotheses and had developed no theories or laws. This is a great error. In the course of my investigations, I employed innumerable hypotheses. Before entering each new block and each new section of the city, I entertained an hypothesis as regards the number and characteristics of the umbrellas that would be found there, which hypotheses were either verified or nullified by my subsequent observations, in accordance with proper scientific procedure, as explained in authoritative texts. (In fact, it is of interest to note that I can substantiate and document every one of my replies to these objections by numerous quotations from standard works, leading journals, public speeches of eminent scientists, and the like.) As for theories and laws, my work represents an abundance of them. I will here mention only a few, by way of illustration. There is the Law of Color Variation Relative to Ownership by Sex. (Umbrellas owned by women tend to great variety of color, whereas those owned by men are almost all black.) To this law I have given exact statistical formulation. (See vol. 6, Appendix 1, Table 3, p. 582.) There are the curiously interrelated Laws of Individual Ownership of Plurality of Umbrellas, and Plurality of Ownership of Individual Umbrellas. The interrelationship assumes the form, in the first law, of almost direct ratio to annual income, and in the second, in almost inverse ratio to annual income. (For an exact statement of the modifying circumstances, see vol. 8, p. 350.) There is also the Law of Tendency Toward Acquisition of Umbrellas in Rainy Weather. To this law I have given experimental verification in chapter 3 of volume 3. In the same way I have performed numerous other experiments in connection with my generalizations.

Thus I feel that my creation is in all respects a genuine science, and I appeal to you for substantiation of my opinion. . . .

We invite you to compose a reply to this letter.

STUDY QUESTIONS FOR PART 1

1. According to Popper, what is the problem of demarcation? State Popper's solution of that problem.

2. Show why, in Popper's views, psychoanalytic theories do not meet his criterion of demarcation whereas Einstein's theory of gravitation does.

3. In addition to the examples which he already gives, list several theories, statements, or enterprises which do *not* meet Popper's criterion of demarcation. Show *why* they do not do so.

4. Try to think of some theories or truths which (in your view) should be accorded scientific status but which do *not* meet Popper's criterion for being scientific. How would you revise—or replace—Popper's criterion in order to account for them?

5. Defend or attack Ziman's thesis that the distinction between pure science and technological knowledge is difficult to make.

6. According to Feyerabend, *how* should one defend society against science? *Why* should one do so? Do you agree with him? Why or why not?

7. Why does Feyerabend reject Popper's criterion of falsifiability? Has he provided *good reasons* for rejecting it? Why or why not?

8. What thesis, if any, in Feyerabend's essay do you disagree with or find fault with? Why?

9. What is Thagard's criterion for distinguishing science from pseudoscience? Compare and contrast with the views of Popper, Ziman, and Feyerabend. Which of the criteria formulated by these four authors do you find most acceptable? Why?

10. On the basis of Thagard's criterion, why does he claim that astrology is a pseudoscience? Has he (in your view) successfully shown that it is a pseudoscience? Why or why not?

11. How do (explicitly or implicitly) Popper, Ziman, Feyerabend, and Thagard answer the question "What is science?" Which answer seems preferable to you? Why?

12. What are the central theses of Kitcher's view of scientific theories?

13. Show how, according to Kitcher, the Creationist account does not live up to the requirements of being scientific.

14. Show how Darwinian evolutionary theory does live up to them.

SELECTED BIBLIOGRAPHY

A. What Is Science?

[1] Campbell, N. *What Is Science?* New York: Dover Publications, 1952. [Elementary but significant answer to the question.]

[2] Cohen. M. R., and E. Nagel. *An Introduction to Logic and Scientific Method.* Bk 2. New York: Harcourt, Brace, 1934. [Old but worthwhile.]

[3] Conant, J. B. On Understanding Science. New Haven, Conn.: Yale University Press, 1951. [Very readable; good introduction.]

[4] Feyerabend, P. *Against Method.* New York: Shocken Books, 1977. [Very critical of all prevailing views of science.]

[5] Kemeny, J. G. *A Philosopher Looks at Science.* Princeton, N. J.: D. Van Nostrand, 1959. [One of many good introductions.]

[6] Kuhn, T. S. *The Structure of Scientific Revolutions.* 2d ed. Chicago, Ill.: University of Chicago Press, 1971. [A classic.]

[7] Nagel, E. *The Structure of Science.* New York: Harcourt, Brace, & World, 1961. [Difficult in spots, but excellent.]

[8] Popper, K. *Conjectures and Refutations.* New York: Harper & Row, 1968. [A seminal and classic collection of essays.]

[9] ———. *The Logic of Scientific Discovery.* New York: Harper & Row, 1968. [A major and important work.]

[10] Toulmin, *S. Foresight and Understanding.* London: Hutchinson, 1961. [Contains interesting insights; very readable.]

B. Science and Nonscience

[1] Flew, A. "ESP." *The New York Review of Books.* 8 October 1964, pp. 15–16. [Short but worthwhile.]

[2] Gardner, M. *Fads and Fallacies in the Name of Science.* New York: Dover Publications, 1957. (See also the detailed appendix and notes to this volume.) [The best book of its kind.]

[3] ———. "Funny Coincidence." *The New York Review of Books*. 26 May 1966. [One of many essays by Gardner.]

[4] ———. "What Hath Hoova Wrought?" *The New York Review of Books*. 16 May 1974, pp. 18–19.

[5] ———. "Paranosense." *The New York Review of Books*. 30 October 1975, pp. 12–14. [See 3, above.]

[6] ———. "Supergull." *The New York Review of Books*. 17 March 1977, pp. 18–20. [See 3, above.]

[7] Hansel, C. E. M. *ESP: A Scientific Evaluation*. New York: Scribner's, 1966. [See 3, above.]

[8] Kline, A. D. "Theories, Facts, and Gods." In *Did the Devil Make Darwin Do It?*, edited by David Wilson. Ames, Iowa: Iowa State University Press, 1983. [All of the essays in this volume are worth reading.]

[9] Popper, K. *Conjectures and Refutations*. New York: Harper & Row, 1968. [See A8.]

[10] ———. *The Logic of Scientific Discovery*. New York: Harper & Row, 1968. [See A8.]

[11] Toulmin, S. "Contemporary Scientific Mythology." In *Metaphysical Beliefs*, by S. Toulmin et al. London: SCM Press, 1957. [Excellent exposure; a must.]

Part 2

The Natural and Social Sciences

Introduction

The main arguments for a general distinction between the natural and the social sciences can be stated briefly. First, human beings are not like billiard balls. Since Newtonian science was the dominant scientific paradigm until recently, attempts to establish a science or sciences of human behavior and society took it for granted that any human or social science must be based upon the assumptions and canons of Newtonian physics. Advocates of this view were often positivists. Their view about the social sciences is often called "naturalism." Writers who rejected this view, because they believed human life cannot adequately be studied by extending the methods and doctrines of physics to human beings and society, have come to be called "antinaturalists."

The second argument for antinaturalism is that the methods of the natural sciences are not appropriate to the study of human behavior. Antinaturalists thus deny the "unity of method" doctrine of naturalists: that the same concepts, methods, and laws that are used in physics can be applied to the social and behavioral sciences to study human behavior (at both the individual and social levels). We must make use of literature and literary methods, history and historical methods, interpretive and philological methods for studying texts, and not experimental methods such as those used in physics.

Third, human beings have free will and human life is pervaded by values and significance, both "subjective" and cultural. But this is not the case with the physical world. Hence, we miss the essence of the human if we accept the doctrines and methods of naturalism.

Generally speaking, the social and behavioral sciences developed at least in the English-speaking world have, until recently, been dominated by the naturalist paradigm. The unity of method hypothesis; the idea that the social sciences are objective and value free; the notion that all of the social sciences will eventually be reduced to the laws of physics ("Psychology is a branch of physics"; the social sciences reduce to psychology; psychology reduces to biology; biology reduces to physics); the development of the policy and applied sciences as applications of social scientific (deterministic) laws are the hallmarks of much twentieth-century social science. This attitude was reinforced by the social, political, and economic prestige of the natural sciences in academic culture. Social scientists became obsessed with the question "Is it a science?" where "science" meant Newtonian physics and classical mechanics. Even the now discredited idea that biology reduces to physics played a role. Indeed, a kind of culture lag prevented social scientists from moving away from the coercive and corrosive influence of a Newtonian paradigm until relatively recently.

Psychology and political science were dominated by either behaviorism or materialism. Economics and political science were (and still are) dominated by rational choice theory, another form of the sort of theory of human nature (we are all hard wired to pursue our rational self-interests) that underlies psychology. Sociology alternated between various forms of naturalism (usually positivism) and antinaturalism, or infrequent attempts to combine them.

American social scientists were influenced by "systems" (usually Newtonian) models of society and the quest for academic respectability. Cultural anthropology moved back and forth between positivist models, often charged with being Eurocentric, and antinaturalist models which seemed to lack provable results and reliable methods, and which (to critics) smacked too much of relativism and irrationalism. Historians have been unable to decide whether they are social scientists or humanists for one hundred years, although interesting history is usually done despite these historiographical debates.

Other approaches, often antinaturalistic, were dismissed or marginalized to the various humanistic disciplines, pushed off on philosophers and literary theorists or historians, who took them more seriously.

In recent years, efforts to reduce the social sciences to evolutionary biology, genetics, and the brain sciences have all become fashionable in academic circles.

Yet there is also the impression that many of the social sciences (even economics and psychology, the most mathematical and quantitative, and therefore the most "respectable" from the point of view of naturalism) do little more than dress up well-known truths in jargon and numbers, or involve idealizations (such as "Rational Economic Man") that seem to have little to do with human comings and goings. This spurred various writers to develop antinaturalistic approaches to the social and behavioral sciences, by making use of distinctions such as "bodily movements" vs. "meaningful actions," "causes" vs. "reasons" or "rules," "the agent's (subjective, participatory) point of view" vs. "the scientist's (objective, spectator) point of view," together with a rejection of the use of modern science and standards of rationality as an Archimedean standard for understanding all people. From a historical viewpoint, the old debates between naturalism and antinaturalism continued (and in some ways continue) unabated. Whether this is because the social sciences are "really inferior," or "haven't found their Newton," or don't have one paradigm, or seek to ape Newtonian physics, which is not even the dominant paradigm in physics (let alone biology or chemistry), may not make much difference at this late date.

For much of the twentieth century, antinaturalism was an unpopular view, dismissed as implausible and even reactionary, or at least antiscientific. A case in point is Peter Winch, whose *Idea of a Social Science and Its Relation to Philosophy* was published in 1958, one year before the first edition of Kuhn's *Structure of Scientific Revolutions.* Winch used Wittgenstein's views about language, reality, and the cultural basis of human practices and "language games" to argue that the social sciences cannot succeed by using any naturalistic models. Human behavior is meaningful because it is governed by cultural rules and norms. The naturalist's use of causal theories and other tools to study human behavior is totally irrelevant to understanding human action. At best, naturalistic social science can supplement, but never replace, the everyday, culture-bound understandings we have of ourselves and our social realities. Because Winch's view rested upon controversial points in Wittgenstein's philosophy, and seemed much too extreme—it appears to rule out any social science—it was not taken very seriously, except as a means of dismissing antinaturalism out of hand.

But one result of Kuhn's book, which complements Winch's view, was to raise the question as to whether or not the natural sciences were not themselves similar to, if not identical with, *Geisteswissenschaften.* Some forms of antirealism seem to suggest this approach, as have recent developments in the sociology of science, feminist studies, cultural studies, science and technology studies (including various forms of "social constructivism") and various forms of postmodernism. (See part 6 for several essays addressing these issues.) Indeed, some believe that Kuhn himself was moving in this direction, despite his strong misgivings about the social sciences (owing to the lack of agreed-upon paradigms) in *The Structure of Scientific Revolution.*

For these and other reasons, it is, to begin with, much less obvious today that there are sharp distinctions between the natural and the social sciences; or that the distinctions between, say, psychology and sociology are any less significant between either of them and physics. Traditional criteria for making the distinction, such as those discussed above, have become problematic for a variety of reasons. The classical unity-of-science thesis (with everything reducible to psychology, then biology, then physics) has few, if any, adherents. Cultural changes (feminism, multiculturalism, postmodernism, radical changes in social structures and institutions) are moving the social and behavioral sciences in new directions, often amidst basic and often venomous intra- and interdisciplinary debates. The natural sciences, as they become less tuned into basic research and more into commercialized research, are also being altered in significant ways. Claims about the relations between knowledge and power, charges that the sciences are Eurocentric and male biased, are causing all of the sciences to be rethought. Perhaps hitherto unthought-of developments will allow all of the social sciences to make real progress without being hung up by debates about naturalism and antinaturalism.

To be sure, dominant traditions in the social and cognitive sciences are still, broadly speaking, naturalistic. But other traditions, for example, qualitative research, interviewing techniques, and ethnographic approaches, are both more significant and more often combined with quantitative (usually naturalistic) approaches. Here the influence of writers such as the sociologist Emile Durkheim (1858–1917), who argued that naturalistic methods could deal with social phenomena which were reducible neither to physics nor psychology, may be making progress. Of course, many prominent research programs, cognitive psychology, sociobiology, evolutionary psychology, rational choice theory in economics and political science, are squarely within the naturalist and positivist (if not the Newtonian) traditions. Yet many writers, some influenced by Kuhn, others by postmodernism, want to "blur the genres," including the boundaries between literature or history and the social sciences. These competing research paradigms will no doubt spawn others. Debates and conflicts will no doubt continue. But in a post-Kuhnian, postpositivist climate, where "science" is no longer equivalent to Newtonian physics, and where "being scientific" is not as uniformly honorific as it once was, many of the worries that inspired earlier debates may either take other forms or become altogether marginal.

Instead, most writers, such as the authors in this part, are more interested in examining what is going in the specific sciences (e.g., economics, sociology, and psychology) and in developing adequate philosophical accounts of the particular social sciences. General models for the philosophy of the social sciences are still being explored, but without the usual presumption that the social sciences are inferior, or have to prove they are "real" sciences (against the presumed universality of Newtonian models). This is all to the good, since it will allow us to get clearer on what is going on in these disciplines, and how they increase our understanding of the various sciences and the varieties of ways we can study the world and ourselves scientifically, without being needlessly worried about labels and hierarchies.

More specifically, current debates renew the attention paid to issues about the role of explanation and laws, theory and observation, and criteria of confirmation and acceptance in the social and natural sciences (and within specific areas of these larger domains), but within a postpositivist, post-Kuhnian framework. From this vantage point, it is not assumed that explanations in, say, sociology or history must be deductive-nomological. By the same token, the antinaturalist conception of understanding as interpretation is no longer confined to the expression of some version of "subjective" interpretation, as was often the case. The current climate provides more options for both naturalists and antinaturalists. Indeed, even the idea that the social sciences must be either "objective" or "subjective," by reference to the assumption that physics is "objective," needs to be abandoned or modified. As Charles Taylor suggests in his article, the connection between the "reality" of a social practice and the way it is understood is very close indeed. Voting practices constitute what voting (really) is in a given social context. So notions about what count as legitimate standards of acceptance and confirmation of an interpretation of human behavior may not easily fit pre-

Kuhnian naturalist models. Even Popper, a staunch defender of the unity of methods thesis in his book *The Poverty of Historicism,* allows that there are significant differences between and within the natural and the social sciences, and that it may not be obvious what the "unified" methods are.

Moreover, as we shall see in part 6, new debates about the nature and role of values, particularly within and about science, will require still more complexities in any adequate account of the sciences in general, and the social sciences in particular. At the same time, such complexities will allow us further to enrich our understanding of values and human practices in all of the sciences.

Finally, recent insights about rationality and the construction and justification of scientific theories (by feminists, postmodernists, sociologists of science, cognitive scientists, and others) will need to be systematically interwoven with these new insights about the nature of science and human values. Postpositivist philosophy of science is only just emerging in all its potential richness.

THE READINGS IN PART 2

Taylor defends a naturalistic view of the natural sciences and an antinaturalistic view of the human or social sciences. According to Taylor, the nature of human life and social reality makes the naturalistic approach to human behavior irrelevant. Understanding behavior is more like reading a text than performing experiments.

Kuhn argues that Taylor misinterprets the natural sciences. They, too, have an interpretive (or "hermeneutical") component, even though they are not interpretive sciences. Kuhn claims that some social sciences (but probably not all of them) may come to resemble, say, physics, which has interpretive elements but also naturalistic empirical research programs.

Machlup surveys the ways that the social sciences are supposed to be "inferior" to the natural sciences, and asks what "inferior" is supposed to mean. He concludes by asking why it matters whether, and in what senses, the social sciences are or are not inferior.

Fay and Moon criticize both naturalistic and antinaturalistic approaches to the social sciences, since none of them can answer three basic questions that any adequate philosophy of the social sciences must be able to answer. But instead of rejecting naturalism and antinaturalism, they call for a combination of them that would be adequate to answering the questions.

Rosenberg argues that economics is a formal science, not an empirical science. It is basically trying to develop models for prediction that apply mathematics and statistics. Rosenberg claims all of the main problems in economics stem from its attempt to be an empirical science. Some general conclusions about all of empirical social science are implicit in his argument.

R. H.

6

Interpretation and the Sciences of Man*

Charles Taylor

I

i

Is there a sense in which interpretation is essential to explanation in the sciences of man? The view that it is, that there is an unavoidably "hermeneutical" component in the sciences of man, goes back to Dilthey. But recently the question has come again to the fore, for instance, in the work of Gadamer,[1] in Ricoeur's interpretation of Freud,[2] and in the writings of Habermas.[3]

Interpretation, in the sense relevant to hermeneutics, is an attempt to make clear, to make sense of an object of study. This object must, therefore, be a text, or a text-analogue, which in some way is confused, incomplete, cloudy, seemingly contradictory—in one way or another unclear. The interpretation aims to bring to light an underlying coherence or sense.

This means that any science which can be called "hermeneutical," even in an extended sense, must be dealing with one or another of the confusingly interrelated forms of meaning. Let us try to see a little more clearly what this involves.

We need, first, an object or a field of objects, about which we can speak in terms of coherence or its absence, or making sense or nonsense.

Second, we need to be able to make a distinction, even if only a relative one, between the sense of coherence made, and its embodiment in a particular field of carriers or signifiers. For otherwise, the task of making clear what is fragmentary or confused would be radically impossible. No sense could be given to this idea. We have to be able to make for our interpretations claims of the order: the meaning confusedly present in this text or text-ana-

---

*[This selection is abridged from the original. It reprints sections I and IV of Charles Taylor's (1971) "Interpretation and the Sciences of Man," *The Review of Metaphysics* 25 (1: 3–51, i.e., pp. 3–17 and 45–51.]

logue is clearly expressed here. The meaning, in other words, is one which admits of more than one expression, and in this sense a distinction must be possible between meaning and expression.

The point of the above qualification, that this distinction may be only relative, is that there are cases where no clear, unambiguous, nonarbitrary line can be drawn between what is said and its expression. It can be plausibly argued (I think convincingly, although there is no space to go into it here) that this is the normal and fundamental condition of meaningful expression, that exact synonymy, or equivalence of meaning, is a rare and localized achievement of specialized languages or uses of civilization. But this, if true (and I think it is), does not do away with the distinction between meaning and expression. Even if there is an important sense in which a meaning reexpressed in a new medium cannot be declared identical, this by no means entails that we can give no sense to the project of expressing a meaning in a new way. It does of course raise an interesting and difficult question about what can be meant by expressing it in a clearer way: what is the "it" which is clarified if equivalence is denied? I hope to return to this in examining interpretation in the sciences of man.

Hence the object of a science of interpretation must be describable in terms of sense and nonsense, coherence and its absence; and must admit of a distinction between meaning and its expression.

There is also a third condition it must meet. We can speak of sense or coherence, and of their different embodiments, in connection with such phenomena as gestalts, or patterns in rock formations, or snow crystals, where the notion of expression has no real warrant. What is lacking here is the notion of a subject for whom these meanings are. Without such a subject, the choice of criteria of sameness and difference, the choice among the different forms of coherence which can be identified in a given pattern, among the different conceptual fields in which it can be seen, is arbitrary.

In a text or text-analogue, on the other hand, we are trying to make explicit the meaning expressed, and this means expressed by or for a subject or subjects. The notion of expression refers us to that of a subject. The identification of the subject is by no means necessarily unproblematical, as we shall see further on; it may be one of the most difficult problems, an area in which prevailing epistemological prejudice may blind us to the nature of our object of study. I think this has been the case, as I will show below. And moreover, the identification of a subject does not assure us of a clear and absolute distinction between meaning and expression as we saw above. But any such distinction, even a relative one, is without any anchor at all, is totally arbitrary, without appeal to a subject.

The object of a science of interpretation must thus have sense, distinguishable from its expression, which is for or by a subject.

ii

Before going on to see in what way, if any, these conditions are realized in the sciences of man, I think it would be useful to set out more clearly what rides on this question, why it matters whether we think of the sciences of man as hermeneutical, what the issue is at stake here.

The issue here is at root an epistemological one. But it is inextricable from an ontological one, and hence, cannot but be relevant to our notions of science and of the proper conduct of inquiry. We might say that it is an onto-logical issue which has been argued ever since the seventeenth century in terms of epistemological considerations which have appeared to some to be unanswerable.

The case could be put in these terms: what are the criteria of judgment in a hermeneutical science? A successful interpretation is one which makes clear the meaning originally present in a confused, fragmentary, cloudy form. But how does one know that this interpretation is correct? Presumably because it makes sense of the original text: what is strange, mystifying, puz-zling, contradictory is no longer so, is accounted for. The interpretation appeals throughout to our understanding of the "language" of expression, which understanding allows us to see that this expression is puzzling, that it is in contradiction to that other, etc., and that these difficulties are cleared up when the meaning is expressed in a new way.

But this appeal to our understanding seems to be crucially inadequate. What if someone does not "see" the adequacy of our interpretation, does not accept our reading? We try to show him how it makes sense of the original non- or partial sense. But for him to follow us he must read the original lan-guage as we do, he must recognize these expressions as puzzling in a certain way, and hence be looking for a solution to our problem. If he does not, what can we do? The answer, it would seem, can only be more of the same. We have to show him through the reading of other expressions why this expres-sion must be read in the way we propose. But success here requires that he follow us in these other readings, and so on, it would seem, potentially for-ever. We cannot escape an ultimate appeal to a common understanding of the expressions, of the "language" involved. This is one way of trying to express what has been called the "hermeneutical circle." What we are trying to estab-lish is a certain reading of text or expressions, and what we appeal to as our grounds for this reading can only be other readings. The circle can also be put in terms of part-whole relations: we are trying to establish a reading for the whole text, and for this we appeal to readings of its partial expressions; and yet because we are dealing with meaning, with making sense, where expres-sions only make sense or not in relation to others, the readings of partial expressions depend on those of others, and ultimately of the whole.

Put in forensic terms, as we started to do above, we can only convince an interlocutor if at some point he shares our understanding of the language concerned. If he does not, there is no further step to take in rational argument; we can try to awaken these intuitions in him or we can simply give up; argument will advance us no further. But of course the forensic predicament can be transferred into my own judging: if I am this ill equipped to convince a stubborn interlocutor, how can I convince myself? how can I be sure? Maybe my intuitions are wrong or distorted, maybe I am locked into a circle of illusion.

Now one, and perhaps the only sane response to this would be to say that such uncertainty is an ineradicable part of our epistemological predicament. That even to characterize it as "uncertainty" is to adopt an absurdly severe criterion of "certainty," which deprives the concept of any sensible use. But this has not been the only or even the main response of our philosophical tradition. And it is another response which has had an important and far-reaching effect on the sciences of man. The demand has been for a level of certainty which can only be attained by breaking beyond the circle.

There are two ways in which this breakout has been envisaged. The first might be called the "rationalist" one and could be thought to reach a culmination in Hegel. It does not involve a negation of intuition, or of our understanding of meaning, but rather aspires to attainment of an understanding of such clarity that it would carry with it the certainty of the undeniable. In Hegel's case, for instance, our full understanding of the whole in "thought" carries with it a grasp of its inner necessity, such that we see how it could not be otherwise. No higher grade of certainty is conceivable. For this aspiration the word "breakout" is badly chosen; the aim is rather to bring understanding to an inner clarity which is absolute.

The other way, which we can call "empiricist," is a genuine attempt to go beyond the circle of our own interpretations, to get beyond subjectivity. The attempt is to reconstruct knowledge in such a way that there is no need to make final appeal to readings or judgments which cannot be checked further. That is why the basic building block of knowledge on this view is the impression, or sense datum, a unit of information which is not the deliverance of a judgment, which has by definition no element in it of reading or interpretation, which is a brute datum. The highest ambition would be to build our knowledge from such building blocks by judgments which could be anchored in a certainty beyond subjective intuition. This is what underlies the attraction of the notion of the association of ideas, or if the same procedure is viewed as a method, induction. If the original acquisition of the units of information is not the fruit of judgment or interpretation, then the constatation that two such elements occur together need not either be the fruit of interpretation, of a reading or intuition which cannot be checked. For if the occurrence of a single element is a brute datum, then so is the co-occurrence of two such elements. The path to true knowledge would then repose crucially on the correct recording of such co-occurrences.

This is what lies behind an ideal of verification which is central to an important tradition in the philosophy of science, whose main contemporary protagonists are the logical empiricists. Verification must be grounded ultimately in the acquisition of brute data. By "brute data," I mean here and throughout data whose validity cannot be questioned by offering another interpretation or reading, data whose credibility cannot be founded or undermined by further reasoning.[4] If such a difference of interpretation can arise over given data, then it must be possible to structure the argument so as to distinguish the basic, brute data from the inferences made on the basis of them.

The inferences themselves, of course, to be valid must similarly be beyond the challenge of a rival interpretation. Here the logical empiricists added to the armory of traditional empiricism which set great store by the method of induction, the whole domain of logical and mathematical inference which had been central to the rationalist position (with Leibniz at least, although not with Hegel), and which offered another brand of unquestionable certainty.

Of course, mathematical inference and empirical verification were combined in such a way that two theories or more could be verified within the same domain of facts. But this was a consequence to which logical empiricism was willing to accommodate itself. As for the surplus meaning in a theory which could not be rigorously coordinated with brute data, it was considered to be quite outside the logic of verification.

As a theory of perception, this epistemology gave rise to all sorts of problems, not least of which was the perpetual threat of skepticism and solipsism inseparable from a conception of the basic data of knowledge as brute data, beyond investigation. As a theory of perception, however, it seems largely a thing of the past, in spite of a surprising recrudescence in the Anglo-Saxon world in the thirties and forties. But there is no doubt that it goes marching on, among other places, as a theory of how the human mind and human knowledge actually function.

In a sense, the contemporary period has seen a better, more rigorous statement of what this epistemology is about in the form of computer-influenced theories of intelligence. These try to model intelligence as consisting of operations on machine-recognizable input which could themselves be matched by programs which could be run on machines. The machine criterion provides us with our assurance against an appeal to intuition or interpretations which cannot be understood by fully explicit procedures operating on brute data—the input.[5]

The progress of natural science has lent great credibility to this epistemology, since it can be plausibly reconstructed on this model, as has been done, for instance, by the logical empiricists. And of course the temptation has been overwhelming to reconstruct the sciences of man on the same model; or rather to launch them in lines of inquiry that fit this paradigm, since they are constantly said to be in their "infancy." Psychology, where an earlier vogue of behaviorism is being replaced by a boom of computer-based models, is far from the only case.

The form this epistemological bias—one might say obsession—takes is different for different sciences. Later I should like to look at a particular case, the study of politics, where the issue can be followed out. But in general, the empiricist orientation must be hostile to a conduct of inquiry which is based on interpretation, and which encounters the hermeneutical circle as this was characterized above. This cannot meet the requirements of intersubjective, nonarbitrary verification which it considers essential to science. And along with the epistemological stance goes the ontological belief that reality must be susceptible to understanding and explanation by science so understood. From this follows a certain set of notions of what the sciences of man must be.

On the other hand, many, including myself, would like to argue that these notions about the sciences of man are sterile, that we cannot come to understand important dimensions of human life within the bounds set by this epistemological orientation. This dispute is of course familiar to all in at least some of its ramifications. What I want to claim is that the issue can be fruitfully posed in terms of the notion of interpretation as I began to outline it above.

I think this way of putting the question is useful because it allows us at once to bring to the surface the powerful epistemological beliefs that underlie the orthodox view of the sciences of man in our academy, and to make explicit the notion of our epistemological predicament implicit in the opposing thesis. This is in fact rather more way-out and shocking to the tradition of scientific thought than is often admitted or realized by the opponents of narrow scientism. It may not strengthen the case of the opposition to bring out fully what is involved in a hermeneutical science as far as convincing waverers is concerned, but a gain in clarity is surely worth a thinning of the ranks—at least in philosophy.

iii

Before going on to look at the case of political science, it might be worth asking another question: why should we even pose the question whether the sciences of man are hermeneutical? What gives us the idea in the first place that men and their actions constitute an object or a series of objects which meet the conditions outlined above?

The answer is that on the phenomenological level or that of ordinary speech (and the two converge for the purposes of this argument) a certain notion of meaning has an essential place in the characterization of human behavior. This is the sense in which we speak of a situation, an action, a demand, a prospect having a certain meaning for a person.

Now it is frequently thought that "meaning" is used here in a sense that is a kind of illegitimate extension from the notion of linguistic meaning.

Whether it can be considered an extension is another matter; it certainly differs from linguistic meaning. But it would be very hard to argue that it is an illegitimate use of the term.

When we speak of the "meaning" of a given predicament, we are using a concept which has the following articulation. (1) Meaning is for a subject: it is not the meaning of the situation *in vacuo,* but its meaning for a subject, a specific subject, a group of subjects, or perhaps what its meaning is for the human subject as such (even though particular humans might be reproached with not admitting or realizing this). (2) Meaning is of something; that is, we can distinguish between a given element—situation, action, or whatever— and its meaning. But this is not to say that they are physically separable. Rather we are dealing with two descriptions of the element, in one of which it is characterized in terms of its meaning for the subject. But the relations between the two descriptions are not symmetrical. For, on the one hand, the description in terms of meaning cannot be unless descriptions of the other kind apply as well; or put differently, there can be no meaning without a substrate. But on the other hand, it may be that the same meaning may be borne by another substrate—e.g., a situation with the same meaning may be realized in different physical conditions. There is a necessary role for a potentially substitutable substrate; or all meanings are of something.

And (3) things only have meaning in a field, that is, in relation to the meanings of other things. This means that there is no such thing as a single, unrelated meaningful element; and it means that changes in the other meanings in the field can involve changes in the given element. Meanings cannot be identified except in relation to others, and in this way resemble words. The meaning of a word depends, for instance, on those words with which it contrasts, on those that define its place in the language (e.g., those defining "determinable" dimensions, like color, shape), on those that define the activity or "language game" it figures in (describing, invoking, establishing communion), and so on. The relations between meanings in this sense are like those between concepts in a semantic field.

Just as our color concepts are given their meaning by the field of contrast they set up together, so that the introduction of new concepts will alter the boundaries of others, so the various meanings that a subordinate's demeanor can have for us, as deferential, respectful, cringing, mildly mocking, ironical, insolent, provoking, downright rude, are established by a field of contrast; and as with finer discrimination on our part, or a more sophisticated culture, new possibilities are born, so other terms of this range are altered. And as the meaning of our terms "red," "blue," "green" is fixed by the definition of a field of contrast through the determinable term "color," so all these alternative demeanors are only available in a society which has, among other types, hierarchical relations of power and command. And corresponding to the underlying language game of designating colored objects is the set of social practices which sustain these hierarchical structures and are fulfilled in them.

Meaning in this sense—let us call it experiential meaning—thus is for a subject, of something, in a field. This distinguishes it from linguistic meaning which has a four- and not a three-dimensional structure. Linguistic meaning is for subjects and in a field, but it is the meaning of signifiers and it is about a world of referents. Once we are clear about the likenesses and differences there should be little doubt that the term "meaning" is not a misnomer, the product of an illegitimate extension into this context of experience and behavior.

There is thus a quite legitimate notion of meaning which we use when we speak of the meaning of a situation for an agent. And that this concept has a place is integral to our ordinary consciousness and hence speech about our actions. Our actions are ordinarily characterized by the purpose sought and explained by desires, feelings, emotions. But the language by which we describe our goals, feelings, desires is also a definition of the meaning things have for us. The vocabulary defining meaning—words like "terrifying," "attractive"—is linked with that describing feeling—"fear," "desire"—and that describing goals—"safety," "possession."

Moreover, our understanding of these terms moves inescapably in a hermeneutical circle. An emotion term like "shame," for instance, essentially refers us to a certain kind of situation, the "shameful," or "humiliating," and a certain mode of response, that of hiding oneself, of covering up, or else "wiping out" the blot. That is, it is essential to this feeling's identification as shame that it be related to this situation and give rise to this type of disposition. But this situation in its turn can only be identified in relation to the feelings it provokes; and the disposition is to a goal that can similarly not be understood without reference to the feelings experienced: the "hiding" in question is one which will cover up my shame; it is not the same as hiding from an armed pursuer; we can only understand what is meant by "hiding" here if we understand what kind of feeling and situation is being talked about. We have to be within the circle.

An emotion term like "shame" can only be explained by reference to other concepts which in turn cannot be understood without reference to shame. To understand these concepts we have to be in on a certain experience, we have to understand a certain language, not just of words, but also a certain language of mutual action and communication, by which we blame, exhort, admire, esteem each other. In the end we are in on this because we grow up in the ambit of certain common meanings. But we can often experience what it is like to be on the outside when we encounter the feeling, action, and experiential meaning language of another civilization. Here there is no translation, no way of explaining in other, more accessible concepts. We can only catch on by getting somehow into their way of life, if only in imagination. Thus if we look at human behavior as action done out of a background of desire, feeling, emotion, then we are looking at a reality which must be characterized in terms of meaning. But does this mean that it can be the object of a hermeneutical science as this was outlined above?

There are, to remind ourselves, three characteristics that the object of a science of interpretation has: it must have sense or coherence; this must be distinguishable from its expression; and this sense must be for a subject.

Now insofar as we are talking about behavior as action, hence in terms of meaning, the category of sense or coherence must apply to it. This is not to say that all behavior must "make sense," if we mean by this be rational, avoid contradiction, confusion of purpose, and the like. Plainly a great deal of our action falls short of this goal. But in another sense, even contradictory, irrational action is "made sense of," when we understand why it was engaged in. We make sense of action when there is a coherence between the actions of the agent and the meaning of his situation for him. We find his action puzzling until we find such a coherence. It may not be bad to repeat that this coherence in no way implies that the action is rational: the meaning of a situation for an agent may be full of confusion and contradiction, but the adequate depiction of this contradiction makes sense of it.

Making sense in this way through coherence of meaning and action, the meanings of action and situation, cannot but move in a hermeneutical circle. Our conviction that the account makes sense is contingent on our reading of action and situation. But these readings cannot be explained or justified except by reference to other such readings, and their relation to the whole. If an interlocutor does not understand this kind of reading, or will not accept it as valid, there is nowhere else the argument can go. Ultimately, a good explanation is one which makes sense of the behavior; but then to appreciate a good explanation one has to agree on what makes good sense; what makes good sense is a function of one's readings; and these in turn are based on the kind of sense one understands.

But how about the second characteristic, that sense should be distinguishable from its embodiment? This is necessary for a science of interpretation because interpretation lays a claim to make a confused meaning clearer; hence there must be some sense in which the "same" meaning is expressed, but differently.

This immediately raises a difficulty. In talking of experiential meaning above, I mentioned that we can distinguish between a given element and its meaning, between meaning and substrate. This carried the claim that a given meaning *may be* realized in another substrate. But does this mean that we can *always* embody the same meaning in another situation? Perhaps there are some situations, standing before death, for instance, which have a meaning which cannot be embodied otherwise.

But fortunately this difficult question is irrelevant for our purposes. For here we have a case in which the analogy between text and behavior implicit in the notion of a hermeneutical science of man only applies with important modifications. The text is replaced in the interpretation by another text, one which is clearer. The text-analogue of behavior is not replaced by another such text-analogue. When this happens we have revolutionary theater, or terroristic acts designed to make propaganda of the deed, in which the hidden relations of a society are supposedly shown up in a dramatic confrontation. But this is not scientific understanding, even though it may perhaps be based on such understanding, or claim to be.

But in science the text-analogue is replaced by a text, an account. Which might prompt the question, how we can even begin to talk of interpretation here, of expressing the same meaning more clearly, when we have two such utterly different terms of comparison, a text and a tract of behavior? Is the whole thing not just a bad pun?

This question leads us to open up another aspect of experiential meaning which we abstracted from earlier. Experiential meanings are defined in fields of contrast, as words are in semantic fields.

But what was not mentioned above is that these two kinds of definition are not independent of each other. The range of human desires, feelings, emotions, and hence meanings is bound up with the level and type of culture, which in turn is inseparable from the distinctions and categories marked by the language people speak. The field of meanings in which a given situation can find its place is bound up with the semantic field of the terms characterizing these meanings and the related feelings, desires, predicaments.

But the relationship involved here is not a simple one. There are two simple types of models of relation which could be offered here, but both are inadequate. We could think of the feeling vocabulary as simply describing preexisting feelings, as marking distinctions that would be there without them. But this is not adequate, because we often experience in ourselves or others how achieving, say, a more sophisticated vocabulary of the emotions makes our emotional life more sophisticated and not just our descriptions of it. Reading a good, powerful novel may give me the picture of an emotion which I had not previously been aware of. But we cannot draw a neat line between an increased ability to identify and an altered ability to feel emotions which this enables.

The other simple inadequate model of the relationship is to jump from the above to the conclusion that thinking makes it so. But this clearly won't do either, since not just any new definition can be forced on us, nor can we force it on ourselves; and some that we do gladly take up can be judged inauthentic, in bad faith, or just wrong-headed by others. These judgments may be wrong, but they are not in principle illicit. Rather we make an effort to be lucid about ourselves and our feelings, and admire a man who achieves this.

Thus, neither the simple correspondence view is correct, nor the view that thinking makes it so. But both have prima facie warrant. There is such a thing as self-lucidity, which points us to a correspondence view; but the achievement of such lucidity means moral change, that is, it changes the object known. At the same time, error about oneself is not just an absence of correspondence; it is also in some form inauthenticity, bad faith, self-delusion, repression of one's human feelings, or something of the kind; it is a matter of the quality of what is felt just as much as what is known about this, just as self-knowledge is.

If this is so, then we have to think of man as a self-interpreting animal. He is necessarily so, for there is no such thing as the structure of meanings for him independently of his interpretation of them; one is woven into the other. But then the text of our interpretation is not that heterogeneous from what is interpreted, for what is interpreted is itself an interpretation: a self-interpretation which is embedded in a stream of action. It is an interpretation of experiential meaning which contributes to the constitution of this meaning. Or to put it in another way, that of which we are trying to find the coherence is itself partly constituted by self-interpretation.

Our aim is to replace this confused, incomplete, partly erroneous self-interpretation by a correct one. And in doing this we look not only to the self-interpretation but to the stream of behavior in which it is set, just as in interpreting a historical document we have to place it in the stream of events which it relates to. But of course the analogy is not exact, for here we are interpreting the interpretation and the stream of behavior in which it is set together, and not just one or the other.

There is thus no utter heterogeneity of interpretation to what it is about; rather there is a slide in the notion of interpretation. Already to be a living agent is to experience one's situation in terms of certain meanings, and this in a sense can be thought of as a sort of proto-"interpretation." This is in turn interpreted and shaped by the language in which the agent lives these meanings. This whole is then at a third level interpreted by the explanation we proffer of his actions.

In this way the second condition of a hermeneutical science is met. But this account poses in a new light the question mentioned at the beginning whether the interpretation can ever express the same meaning as the interpreted. And in this case, there is clearly a way in which the two will not be congruent. For if the explanation is really clearer than the lived interpretation, then it will be such that it would alter in some way the behavior if it came to be internalized by the agent as his self-interpretation. In this way a hermeneutical science that achieves its goal, that is, attains greater clarity than the immediate understanding of agent or observer, must offer us an interpretation that is in this way crucially out of phase with the explicandum.

Thus human behavior seen as action of agents who desire and are moved, who have goals and aspirations, necessarily offers a purchase for descriptions in terms of meaning—what I have called "experiential meaning." The norm of explanation which it posits is one that "makes sense" of the behavior, that shows a coherence of meaning. This "making sense of" is the proffering of an interpretation, and we have seen that what is interpreted meets the conditions of a science of interpretation: first, that we can speak of its sense or coherence; and second, that this sense can be expressed in another form, so that we can speak of the interpretation as giving clearer expression to what is only implicit in the explicandum. The third condition, that this sense be for a subject, is obviously met in this case, although who this subject is, is by no means an unproblematical question as we shall see later on.

This should be enough to show that there is a good prima facie case to the effect that men and their actions are amenable to explanation of a hermeneutical kind. There is therefore some reason to raise the issue and challenge the epistemological orientation that would rule interpretation out of the sciences of man. A great deal more must be said to bring out what is involved in the hermeneutical sciences of man. But before getting on to this, it might help to clarify the issue with a couple of examples drawn from a specific field, that of politics.

In the last pages, I have presented some hypotheses which may appear very speculative; and they may indeed turn out to be without foundation, even without much interest. But their aim was mainly illustrative. My principal claim is that we can only come to grips with this phenomenon of breakdown by trying to understand more clearly and profoundly the common and intersubjective meanings of the society in which we have been living. For it is these which no longer hold us, and to understand this change we have to have an adequate grasp of these meanings. But this we cannot do as long as we remain within the ambit of mainstream social science, for it will not recognize intersubjective meaning, and is forced to look at the central ones of our society as though they were the inescapable background of all political action. Breakdown is thus inexplicable in political terms; it is an outbreak of irrationality which must ultimately be explained by some form of psychological illness.

Mainstream science may thus venture into the area explored by the above hypotheses, but after its own fashion, by forcing the psychohistorical facts of identity into the grid of an individual psychology, in short, by reinterpreting all meanings as subjective. The result might be a psychological theory of emotional maladjustment, perhaps traced to certain features of family background, analogous to the theories of the authoritarian personality and the California F-scale. But this would no longer be a political or a social theory. We would be giving up the attempt to understand the change in social reality at the level of its constitutive intersubjective meanings.

IV

It can be argued, then, that mainstream social science is kept within certain limits by its categorical principles which are rooted in the traditional epistemology of empiricism; and second, that these restrictions are a severe handicap and prevent us from coming to grips with important problems of our day, which should be the object of political science. We need to go beyond the bounds of a science based on verification to one which would study the intersubjective and common meanings embedded in social reality.

But this science would be hermeneutical in the sense that has been developed in this chapter. It would not be founded on brute data; its most primitive data would be readings of meanings, and its object would have the three properties mentioned above: the meanings are for a subject in a field or fields; they are, moreover, meanings which are partially constituted by self-definitions, which are in this sense already interpretations, and which can thus be reexpressed or made explicit by a science of politics. In our case, the subject may be a society or community; but the intersubjective meanings, as we saw, embody a certain self-definition, a vision of the agent and his society, which is that of the society or community.

But then the difficulties which the proponents of the verification model foresee will arise. If we have a science that has no brute data, that relies on readings, then it cannot but move in a hermeneutical circle. A given reading of the intersubjective meanings of a society, or of given institutions or practices, may seem well founded, because it makes sense of these practices or the development of that society. But the conviction that it does make sense of this history itself is founded on further related readings. Thus, what I said above on the identity crisis which is generated by our society makes sense and holds together only if one accepts this reading of the intersubjective meanings of our society, and if one accepts this reading of the rebellion against our society by many young people (sc. the reading in terms of identity crisis). These two readings make sense together, so that in a sense the explanation as a whole reposes on the readings, and the readings in their turn are strengthened by the explanation as a whole.

But if these readings seem implausible, or even more, if they are not understood by our interlocutor, there is no verification procedure that we can fall back on. We can only continue to offer interpretations; we are in an interpretative circle.

But the ideal of a science of verification is to find an appeal beyond differences of interpretation. Insight will always be useful in discovery, but should not have to play any part in establishing the truth of its findings. This ideal can be said to have been met by our natural sciences. But a hermeneutic science cannot but rely on insight. It requires that one have the sensibility and understanding necessary to be able to make and comprehend the readings by which we can explain the reality concerned. In physics we might argue that if someone does not accept a true theory, then either he has not been shown enough (brute data) evidence (perhaps not enough is yet available), or he cannot understand and apply some formalized language. But in the sciences of man conceived as hermeneutical, the nonacceptance of a true or illuminating theory may come from neither of these, indeed is unlikely to be due to either of these, but rather from a failure to grasp the meaning field in question, an inability to make and understand readings of this field.

In other words, in a hermeneutical science, a certain measure of insight is indispensable, and this insight cannot be communicated by the gathering of brute data, or initiation in modes of formal reasoning or some combination of these. It is unformalizable. But this is a scandalous result according to the authoritative conception of science in our tradition, which is shared even by many of those who are highly critical of the approach of mainstream psychology, or sociology, or political science. For it means that this is not a study in which anyone can engage, regardless of their level of insight; that some claims of the form, "If you don't understand, then your intuitions are at fault, are blind or inadequate," some claims of this form will be justified; that some differences will be nonarbitrable by further evidence, but that each side can only make appeal to deeper insight on the part of the other. The superiority of one position over another will thus consist in this, that from the more adequate position one can understand one's own stand and that of one's opponent, but not the other way around. It goes without saying that this argument can only have weight for those in the superior position.

Thus, a hermeneutical science encounters a gap in intuitions, which is the other side, as it were, of the hermeneutical circle. But the situation is graver than this; for this gap is bound up with our divergent options in politics and life.

We speak of a gap when some cannot understand the kind of self-definition which others are proposing as underlying a certain society or set of institutions. Thus some positivistically minded thinkers will find the language of identity theory quite opaque: and some thinkers will not recognize any theory which does not fit with the categorical presuppositions of empiricism. But self-definitions are not only important to us as scientists who are trying to understand some, perhaps distant, social reality, As men we are self-defining beings, and we are partly what we are in virtue of the self-definitions which we have accepted, however we have come by them. What self-definitions we understand and what ones we do not understand are closely linked with the self-definitions that help to constitute what we are. If it is too simple to say that one only understands an "ideology" which one subscribes to, it is nevertheless hard to deny that we have great difficulty grasping definitions whose terms structure the world in ways that are utterly different from, incompatible with, our own.

Hence the gap in intuitions does not just divide different theoretical positions; it also tends to divide different fundamental options in life. The practical and the theoretical are inextricably joined here. It may not just be that to understand a certain explanation one has to sharpen one's intuitions; it may be that one has to change one's orientation—if not in adopting another orientation, at least in living one's own in a way which allows for greater comprehension of others. Thus, in the sciences of man insofar as they are hermeneutical there can be a valid response to, "I don't understand," which takes the form, not only, "develop your intuitions," but more radically, "change yourself." This puts an end to any aspiration to a value-free or "ideology-free" science of man. A study of the science of man is inseparable from an examination of the options between which men must choose.

This means that we can speak here not only of error, but of illusion. We speak of "illusion" when we are dealing with something of greater substance than error, error that in a sense builds a counterfeit reality of its own. But errors of interpretation of meaning, which are also self-definitions of those who interpret and hence inform their lives, are more than errors in this sense: they are sustained by certain practices of which they are constitutive. It is not implausible to single out as examples two rampant illusions in our present society. One is that of the proponents of the bargaining society who can recognize nothing but either bargaining gambits or madness in those who rebel against this society. Here the error is sustained by the practices of the bargaining culture, and given a semblance of reality by the refusal to treat any protests on other terms; it hence acquires the more substantive reality of illusion. The second example is provided by much "revolutionary" activity in our society which in desperate search for an alternative mode of life purports to see its situation in that of an Andean guerrilla or Chinese peasants. Lived out, this passes from the stage of laughable error to tragic illusion. One illusion cannot recognize the possibility of human variation; the other cannot see any limits to mankind's ability to transform itself. Both make a valid science of man impossible.

In face of all this, we might be so scandalized by the prospect of such a hermeneutical science that we will want to go back to the verification model. Why can we not take our understanding of meaning as part of the logic of discovery, as the logical empiricists suggest for our unformalizable insights, and still found our science on the exactness of our predictions? Our insightful understanding of the intersubjective meanings of our society will then serve to elaborate fruitful hypotheses, but the proof of these puddings will remain in the degree to which they enable us to predict.

The answer is that if the epistemological views underlying the science of interpretation are right, such exact prediction is radically impossible—this, for three reasons of ascending order of fundamentalness.

The first is the well-known "open system" predicament, one shared by human life and meteorology, that we cannot shield a certain domain of human events, the psychological, economic, political, from external interference; it is impossible to delineate a closed system.

The second, more fundamental, is that if we are to understand men by a science of interpretation, we cannot achieve the degree of fine exactitude of a science based on brute data. The data of natural science admit of measurement to virtually any degree of exactitude. But different interpretations cannot be judged in this way. At the same time different nuances of interpretation may lead to different predictions in some circumstances, and these different outcomes may eventually create widely varying futures. Hence it is more than easy to be wide of the mark.

But the third and most fundamental reason for the impossibility of hard prediction is that man is a self-defining animal. With changes in his self-definition go changes in what man is, such that he has to be understood in different terms. But the conceptual mutations in human history can and frequently do produce conceptual webs which are incommensurable, that is, where the terms cannot be defined in relation to a common stratum of expressions. The entirely different notions of bargaining in our society and in some primitive ones provide an example. Each will be glossed in terms of practices, institutions, ideas in each society which have nothing corresponding to them in the other.

The success of prediction in the natural sciences is bound up with the fact that all states of the system, past and future, can be described in the same range of concepts, as values, say, of the same variables. Hence all future states of the solar system can be characterized, as past ones are, in the language of Newtonian mechanics. This is far from being a sufficient condition of exact prediction, but it is a necessary one in this sense, that only if past and future are brought under the same conceptual net can one understand the states of the latter as some function of the states of the former, and hence predict.

This conceptual unity is vitiated in the sciences of man by the fact of conceptual innovation, which in turn alters human reality. The very terms in which the future will have to be characterized if we are to understand it properly are not all available to us at present. Hence we have such radically unpredictable events as the culture of youth today, the Puritan rebellion of the sixteenth and seventeenth centuries, the development of Soviet society, and so forth.

And thus, it is much easier to understand after the fact than it is to predict. Human science is largely ex post understanding. Or often one has the sense of impending change, of some big reorganization, but is powerless to make clear what it will consist in: one lacks the vocabulary. But there is a clear asymmetry here, which there is not (or not supposed to be) in natural science, where events are said to be predicted from the theory with exactly the same ease with which one explains past events and by exactly the same process. In human science this will never be the case.

Of course, we strive ex post to understand the changes, and to do this we try to develop a language in which we can situate the incommensurable webs of concepts. We see the rise of Puritanism, for instance, as a shift in man's stance to the sacred; and thus, we have a language in which we can express both stances—the earlier medieval Catholic one and the Puritan rebellion—as "glosses" on this fundamental term. We thus have a language in which to talk of the transition. But think how we acquired it. This general category of the sacred is acquired not only from our experience of the shift that came in the Reformation, but from the study of human religion in general, including primitive religion, and with the detachment that came with secularization. It would be conceivable, but unthinkable, that a medieval Catholic could have this conception—or for that matter a Puritan. These two protagonists only had a language of condemnation for each other: "heretic," "idolater." The place for such a concept was preempted by a certain way of living the sacred. After a big change has happened, and the trauma has been resorbed, it is possible to try to understand it, because one now has available the new language, the transformed meaning world. But hard prediction before just makes one a laughingstock. Really to be able to predict the future would be to have explicated so clearly the human condition that one would already have preempted all cultural innovation and transformation. This is hardly in the bounds of the possible.

Sometimes men show amazing prescience: the myth of Faust, for instance, which is treated several times at the beginning of the modern era. There is a kind of prophesy here, a premonition. But what characterizes these bursts of foresight is that they see through a glass darkly, for they see in terms of the old language: Faust sells his soul to the devil. They are in no sense hard predictions. Human science looks backward. It is inescapably historical.

There are thus good grounds both in epistemological arguments and in their greater fruitfulness for opting for hermeneutical sciences of man. But we cannot hide from ourselves how greatly this option breaks with certain commonly held notions about our scientific tradition. We cannot measure such sciences against the requirements of a science of verification: we cannot judge them by their predictive capacity. We have to accept that they are founded on intuitions which all do not share, and what is worse, that these intuitions are closely bound up with our fundamental options. These sciences cannot be *wertfrei*; they are moral sciences in a more radical sense than the eighteenth century understood. Finally, their successful prosecution requires a high degree of self-knowledge, a freedom from illusion, in the sense of error which is rooted and expressed in one's way of life; for our incapacity to understand is rooted in our own self-definitions, hence in what we are. To say this is not to say anything new: Aristotle makes a similar point in Book I of the *Ethics*. But it is still radically shocking and unassimilable to the mainstream of modern science.

NOTES

1. See, e.g., H. G. Gadamer, *Wahrheit und Methode* (Tübingen, 1960).
2. See Paul Ricoeur, *De l'interprétation* (Paris, 1965).
3. See, e.g., J. Hambermas, *Erkenntnis und Interesse* (Frankfurt, 1968).
4. The notion of brute data here has some relation to, but is not at all the same as, the brute facts discussed by Elizabeth Anscombe, "On Brute Facts," *Analysis* 18 (1957–1958): 69–72, and John Searle, *Speech Acts: An Essay in the Philosophy of Language* (Cambridge, 1969), 50–53. For Anscombe and Searle, brute facts are contrasted to what may be called "institutional facts," to use Searle's term, facts which presuppose the existence of certain institutions. Voting would be an example. But as we shall see below, some institutional facts, such as Xs have voted Liberal, can be verified as brute data in the sense used here, and thus find a place in the category of political behavior. What cannot as easily be described in terms of brute data are the institutions themselves.
5. See the discussion in M. Minsky, *Computation* (Englewood Cliffs, N.J., 1967), 104–107, where Minsky explicitly argues that an effective procedure that no longer requires intuition or interpretation is one which can be realized by a machine.

7

The Natural and the Human Sciences*

Thomas S. Kuhn

Let me begin with a fragment of autobiography. Forty years ago, when I first began to develop heterodox ideas about the nature of natural science, especially physical science, I came upon a few pieces of the Continental literature on the methodology of social science. In particular, if memory serves, I read a couple of Max Weber's methodological essays, then recently translated by Talcott Parsons and Edward Shils, as well as some relevant chapters from Ernst Cassirer's *Essay on Man.* What I found in them thrilled and encouraged me. These eminent authors were describing the social sciences in ways that closely paralleled the sort of description I hoped to provide for the physical sciences. Perhaps I really was onto something worthwhile.

My euphoria was, however, regularly damped by the closing paragraphs of these discussions, which reminded readers that their analyses applied only to the *Geisteswissenschaften,* the social sciences. "Die Naturwissenschaften," their authors loudly proclaimed, "sind ganz anders." (The natural sciences are entirely different.) What then followed was a relatively standard, quasi-positivist, empiricist account of natural science, just the image that I hoped to set aside.

Under those circumstances, I promptly returned to my own knitting, the materials for which were the physical sciences in which I had taken my Ph.D. Then and now, my acquaintance with the social sciences was extremely limited. My present topic—the relation of the natural and human sciences—is not one I have thought a great deal about, nor do I have the background to do so. Nevertheless, though maintaining my distance from the social sciences, I've from time to time encountered other papers to which I reacted as I had to Weber's and Cassirer's. Brilliant, penetrating essays on the social or human sciences, they seemed to me, but papers that apparently needed to define their

---

*Remarks delivered on 11 February 1989 at a panel discussion held at LaSalle University and sponsored by the Greater Philadelphia Philosophy Consortium.

position by using as foil an image of the natural sciences to which I remain deeply opposed. One such essay supplies the reason for my presence here.

That paper is Charles Taylor's "Interpretation and the Sciences of Man."[1] For me it's a special favorite: I've read it often, learned a great deal from it, and used it regularly in my teaching. As a result, I took particular pleasure in the opportunity to participate with its author in an NEH Summer Institute on Interpretation held during the summer of 1988. The two of us had not had the opportunity to talk together before, but we quickly started a spirited dialogue, and we undertook to continue it before this panel. As I planned my introductory contribution, I was confident of a lively and fruitful exchange to follow. Professor Taylor's forced withdrawal has been correspondingly disappointing, but by the time it occurred, it was too late for a radical change of plans. Though I'm reluctant to talk about Professor Taylor behind his back, I've had no alternative but to play a role close to the one for which I was originally cast.

To avoid confusion, I shall start by locating what Taylor and I, in our discussions at the 1988 institute, primarily differed about. It was not the question whether the human and natural sciences were of the same kind. He insisted they were not, and I, though a bit of an agnostic, was inclined to agree. But we did differ, often sharply, about how the line between the two enterprises might be drawn. I did not think his way would do at all. But my notions of how to replace it—about which I shall later have just a bit to say—remained extremely vague and uncertain.

To make our difference more concrete, let me start from a too simple version of what most of you know. For Taylor, human actions constitute a text written in behavioral characters. To understand the actions, recover the meaning of the behavior, requires hermeneutic interpretation, and the interpretation appropriate to a particular piece of behavior will, Taylor emphasizes, differ systematically from culture to culture, sometimes even from individual to individual. It is this characteristic—the intentionality of behavior—that, in Taylor's view, distinguishes the study of human actions from that of natural phenomena. Early in the classic paper to which I previously referred, he says, for example, that even objects like rock patterns and snow crystals, though they have a coherent pattern, have no meaning, nothing that they express. And later in the same essay he insists that the heavens are the same for all cultures, say for the Japanese and for us. Nothing like hermeneutic interpretation, Taylor insists, is required to study objects like these. If they can properly be said to have meaning, those meanings are the same for all. They are, as he has more recently put it, absolute, independent of interpretation by human subjects.

That viewpoint seems to me mistaken. To suggest why, I shall also use the example of the heavens, which, as it happens, I had used also in the set of manuscript lectures that provided my primary text at the 1988 institute. It is not perhaps, the most conclusive example, but it is surely the least complex and thus the most suitable for brief presentation. I did not and cannot compare our heavens with those of the Japanese, but I did and will here insist that ours are different from the ancient Greeks'. More particularly, I want to emphasize that we and Greeks divided the population of the heavens into different kinds, different categories of things. Our celestial taxonomies are systematically distinct. For the Greeks, heavenly objects divided into three categories: stars, planets, and meteors. We have categories with those names, but what the Greeks put into theirs was very different from what we put into ours. The sun and moon went into the same category as Jupiter, Mars, Mercury, Saturn, and Venus. For them these bodies were like each other, and unlike members of the categories "star" and "meteor." On the other hand, they placed the Milky Way, which for us is populated by stars, in the same category as the rainbow, rings round the moon, shooting stars, and other meteors. There are other similar classificatory differences. Things like each other in one system were unlike in the other. Since Greek antiquity, the taxonomy of the heavens, the patterns of celestial similarity and difference, have systematically changed.

Many of you will, I know, wish to join Charles Taylor in telling me that these are merely differences in beliefs about objects that themselves remained the same for the Greeks as for us—something that could be shown, for example, by getting observers to point at them or to describe their relative positions. This is not the place for me to try very seriously to talk you out of that plausible position. But given more time, I would certainly make the attempt, and I want here to indicate what the structure of my argument would be.

It would begin with some points about which Charles Taylor and I agree. Concepts—whether of the natural or social world—are the possession of communities (cultures or subcultures), At any given time they are largely shared by members of the community, and their transmission from generation to generation (sometimes with changes) plays a key role in the process by which the community accredits new members. What I take "sharing a concept" to be must here remain mysterious, but I am at one with Taylor in vehemently rejecting a long-standard view. To have grasped a concept—of planets or stars, on the one hand, of equity or negotiation, on the other—is not to have internalized a set of features that provide necessary and sufficient conditions for the concept's application. Though anyone who understands a concept must know *some* salient features of the objects or situations that fall under it, those features may vary from individual to individual, and no one of them need be shared to permit the concept's proper application. Two people could, that is, share a concept without sharing a single belief about the feature or features of the objects or situations to which it applied. I don't suppose that often occurs, but in principle it could.

This much is largely common ground for Taylor and me. We part company, however, when he insists that, though social concepts shape the world to which they are applied, concepts of the natural world do not. For him but not for me, the heavens are culture-independent. To make that point, he would, I believe, emphasize that an American or European can, for example, point out planets or stars to a Japanese but cannot do the same for equity or negotiation. I would counter that one can point only to individual exemplifications of a concept—to this star or that planet, this episode of negotiation or that of equity—and that the difficulties involved in doing so are of the same nature in the natural and social worlds.

For the social world Taylor has himself supplied the arguments. For the natural world the basic arguments are supplied by David Wiggins in, among other places, *Sameness and Substance.*[2] To point usefully, informatively, to a particular planet or star, one must be able to point to it more than once, to pick out the same individual object again. And this one cannot do unless one has already grasped the sortal concept under which the individual falls. Hesperus and Phosphorus are the same *planet,* but it is only under that description, only as planets, that they can be recognized as one and the same. Until identity can be made out, there is nothing to be learned (or taught) by pointing. As in the case of equity or negotiation, neither the presentation nor the study of examples can begin until the concept of the object to be exemplified or studied is available. And what makes it available, whether in the natural or the social sciences, is a culture, within which it is transmitted by exemplification, sometimes in altered form, from one generation to the next.

I do, in short, really believe some—though by no means all—of the nonsense attributed to me. The heavens of the Greeks were irreducibly different from ours. The nature of the difference is the same as that Taylor so brilliantly describes between the social practices of different cultures. In both cases the difference is rooted in conceptual vocabulary. In neither can it be bridged by description in a brute data, behavioral vocabulary. And in the absence of a brute data vocabulary, any attempt to describe one set of practices in the conceptual vocabulary, the meaning system, used to express the other, can only do violence. That does not mean that one cannot, with sufficient patience and effort, discover the categories of another culture or of an earlier stage of one's own. But it does indicate that discovery is required and that hermeneutic interpretation—whether by the anthropologist or the historian—is how such discovering is done. No more in the natural than in the human sciences is there some neutral, culture-independent, set of categories within which the population—whether of objects or of actions—can be described.

Most of you will long since have recognized these theses as redevelopments of themes to be found in my *Structure of Scientific Revolutions* and related writings. Letting a single example serve for all, the gap that I have here described as separating the Greek heavens from our own is the sort that could only have resulted from what I earlier called a scientific revolution. The violence and misrepresentation consequent on describing their heavens in the conceptual vocabulary required to describe our own is an example of what I then called incommensurability. And the shock generated by substituting their conceptual spectacles for our own is the one I ascribed, however inadequately, to their living in a different world. Where the social world of another culture is at issue, we have learned, against our own deep-seated ethnocentric resistance, to take shock for granted. We can, and in my view must, learn to do the same for their natural worlds.

What does all of this, supposing it cogent, have to tell us about the natural and human sciences? Does it indicate that they are alike except perhaps in their degree of maturity? Certainly it reopens that possibility but it need not force that conclusion. My disagreement with Taylor was not, I remind you, about the existence of a line between natural and human sciences, but rather about the way in which that line may be drawn. Though the classic way to draw it is unavailable to those who take the viewpoint developed here, another way to draw the line emerges clearly. What I'm uncertain about is not whether differences exist, but whether they are principled or merely a consequence of the relative states of development of the two sets of fields.

Let me therefore conclude these reflections with a few tentative remarks about this alternate way of line-drawing. My argument has so far been that the natural sciences of any period are grounded in a set of concepts that the current generation of practitioners inherit from their immediate predecessors. That set of concepts is a historical product, embedded in the culture to which current practitioners are initiated by training, and it is accessible to nonmembers only through the hermeneutic techniques by which historians and anthropologists come to understand other modes of thought. Sometimes I have spoken of it as the hermeneutic basis for the science of a particular period, and you may note that it bears a considerable resemblance to one of the senses of what I once called a paradigm. Though I seldom use that term these days, having totally lost control of it, I shall for brevity sometimes use it here.

If one adopts the viewpoint I've been describing toward the natural sciences, it is striking that what their practitioners mostly do, given a paradigm or hermeneutic basis, is not ordinarily hermeneutic. Rather, they put to use the paradigm received from their teachers in an endeavor I've spoken of as normal science, an enterprise that attempts to solve puzzles like those of improving and extending the match between theory and experiment at the advancing forefront of the field. The social sciences, on the other hand—at least for scholars like Taylor, for whose view I have the deepest respect—appear to be hermeneutic, interpretive, through and through. Very little of what goes on in them at all resembles the normal puzzle-solving research of the natural sciences. Their aim is, or should be in Taylor's view, to understand behavior, not to discover the laws, if any, that govern it. That difference has a converse that seems to me equally striking. In the natural sciences the practice of research does occasionally produce new paradigms, new ways of understanding nature, of reading its texts. But the people responsible for those changes were not looking for them. The reinterpretation that resulted from their work was involuntary, often the work of the next generation. The people responsible typically failed to recognize the nature of what they had done. Contrast that pattern with the one normal to Taylor's social sciences. In the latter, new and deeper interpretations are the recognized object of the game.

The natural sciences, therefore, though they may require what I have called a hermeneutic base, are not themselves hermeneutic enterprises. The human sciences, on the other hand, often are, and they may have no alternative. Even if that's right, however, one may still reasonably ask whether they are restricted to the hermeneutic, to interpretation. Isn't it possible that here and there, over time, an increasing number of specialties will find paradigms that can support normal, puzzle-solving research?

About the answer to that question, I am totally uncertain. But I shall venture two remarks, pointing in opposite directions. First, I'm aware of no principle that bars the possibility that one or another part of some human science might find a paradigm capable of supporting normal, puzzle-solving research. And the likelihood of that transition's occurring is for me increased by a strong sense of déjà vu. Much of what is ordinarily said to argue the impossibility of puzzle-solving research in the human sciences was said two centuries ago to bar the possibility of a science of chemistry and was repeated a century later to show the impossibility of a science of living things. Very probably the transition I'm suggesting is already under way in some current specialties within the human sciences. My impression is that in parts of economics and psychology, the case might already be made.

On the other hand, in some major parts of the human sciences there is a strong and well-known argument against the possibility of anything quite like normal, puzzle-solving research. I earlier insisted that the Greek heavens were different from ours. I should now also insist that the transition between them was relatively sudden, that it resulted from research done on the prior version of the heavens, and that the heavens remained the same while that research was under way. Without that stability, the research responsible for the change could not have occurred. But stability of that sort cannot be expected when the unit under study is a social or political system. No lasting base for normal, puzzle-solving science need be available to those who investigate them; hermeneutic reinterpretation may constantly be required. Where that is the case, the line that Charles Taylor seeks between the human and the natural sciences may be firmly in place. I expect that in some areas it may forever remain there.

NOTES

1. Charles Taylor, "Interpretation and the Sciences of Man," in his *Philosophy and the Human Sciences* (Cambridge: Cambridge University Press, 1985).

2. David Wiggins, *Sameness and Substance* (Cambridge, Mass.: Harvard University Press, 1980).

8

Are the Social Sciences Really Inferior?

Fritz Machlup

If we ask whether the "social sciences" are "really inferior," let us first make sure that we understand each part of the question.

"*Inferior*" to what? Of course to the natural sciences. "Inferior" in what respect? It will be our main task to examine all the "respects," all the scores on which such inferiority has been alleged. I shall enumerate them presently.

The adverb "*really*" which qualifies the adjective "inferior" refers to allegations made by some scientists, scholars, and laymen. But it refers also to the "inferiority complex" which I have noted among many social scientists. A few years ago I wrote an essay entitled "The Inferiority Complex of the Social Sciences."[1] In that essay I said that "an inferiority complex may or may not be justified by some 'objective' standards," and I went on to discuss the consequences which "the *feeling* of inferiority"—conscious or subconscious—has for the behavior of the social scientists who are suffering from it. I did not then discuss whether the complex has an objective basis, that is, whether the social sciences are "really" inferior. This is our question today.

The subject noun would call for a long disquisition. What is meant by "*social sciences*," what is included, what is not included? Are they the same as what others have referred to as the "moral sciences," the "*Geisteswissenschaften*," the "cultural sciences," the "behavioral sciences"? Is geography, or that part of it that is called "human geography," a social science? Is history a social science—or perhaps even the social science *par excellence*, as some philosophers have contended? We shall not spend time on this business of defining and classifying. A few remarks may later be necessary in connection with some points of methodology, but by and large we shall not bother here with a definition of "social sciences" and with drawing boundary lines around them.

THE GROUNDS OF COMPARISON

The social sciences and the natural sciences are compared and contrasted on many scores, and the discussions are often quite unsystematic. If we try to review them systematically, we shall encounter a good deal of overlap and unavoidable duplication. Nonetheless, it will help if we enumerate in advance some of the grounds of comparison most often mentioned, grounds on which the social sciences are judged to come out "second best":

1. Invariability of observations
2. Objectivity of observations and explanations
3. Verifiability of hypotheses
4. Exactness of findings
5. Measurability of phenomena
6. Constancy of numerical relationships
7. Predictability of future events
8. Distance from everyday experience
9. Standards of admission and requirements

We shall examine all these comparisons.

INVARIABILITY OF OBSERVATIONS

The idea is that you cannot have much of a science unless things recur, unless phenomena repeat themselves. In nature we find many factors and conditions "invariant." Do we in society? Are not conditions in society changing all the time, and so fast that most events are unique, each quite different from anything that has happened before? Or can one rely on the saying that "history repeats itself" with sufficient invariance to permit generalizations about social events?

There is a great deal of truth, and important truth, in this comparison. Some philosophers were so impressed with the invariance of nature and the variability of social phenomena that they used this difference as the criterion in the definitions of natural and cultural sciences. Following Windelband's distinction between generalizing ("nomothetic") and individualizing ("ideographic") propositions, the German philosopher Heinrich Rickert distinguished between the generalizing sciences of nature and the individualizing sciences of cultural phenomena; and by individualizing sciences he meant historical sciences.[2] In order to be right, he redefined both "nature" and "history" by stating that reality is "nature" if we deal with it in terms of the *general* but becomes "history" if we deal with it in terms of the *unique*. To him, geology was largely history, and economics, most similar to physics, was a natural science. This implies a rejection of the contention that all fields which are normally called social sciences suffer from a lack of invariance; indeed, economics is here considered so much a matter of immutable laws of nature that it is handed over to the natural sciences.

This is not satisfactory, nor does it dispose of the main issue that natural phenomena provide *more* invariance than social phenomena. The main difference lies probably in the number of factors that must be taken into account in explanations and predictions of natural and social events. Only a small number of reproducible facts will normally be involved in a physical explanation or prediction. A much larger number of facts, some of them probably unique historical events, will be found relevant in an explanation or prediction of economic or other social events. This is true, and methodological devices will not do away with the difference. But it is, of course, only a difference in degree.

The physicist Robert Oppenheimer once raised the question whether, if the universe is *a unique* phenomenon, we may assume that *universal* or *general* propositions can be formulated about it. Economists of the Historical School insisted on treating each "stage" or phase of economic society as a completely unique one, not permitting the formulation of universal propositions. Yet, in the physical world, phenomena are not quite so homogeneous as many have liked to think; and in the social world, phenomena are not quite so heterogeneous as many have been afraid they are. (If they were, we could not even have generalized concepts of social events and words naming them.) In any case, where reality seems to show a bewildering number of variations, we construct an ideal world of abstract models in which we create enough homogeneity to permit us to apply reason and deduce the implied consequences of assumed constellations. This artificial homogenization of types of phenomena is carried out in natural and social sciences alike.

There is thus no difference in invariance in the sequences of events in nature and in society as long as we theorize about them—because in the abstract models homogeneity is assumed. There is only a difference of degree in the variability of phenomena of nature and society if we talk about the real world—as long as heterogeneity is not reduced by means of deliberate "controls." There is a third world, between the abstract world of theory and the real unmanipulated world, namely, the artificial world of the experimental laboratory. In this world there is less variability than in the real world and more than in the model world. But this third world does not exist in most of the social sciences (nor in all natural sciences). We shall see later that the mistake is often made of comparing the artificial laboratory world of manipulated nature with the real world of unmanipulated society.

We conclude on this point of comparative invariance, that there is indeed a difference between natural and social sciences, and that the difference—

apart from the possibility of laboratory experiments—lies chiefly in the number of relevant factors, and hence of possible combinations, to be taken into account for explaining or predicting events occurring in the real world.

OBJECTIVITY OF OBSERVATIONS AND EXPLANATIONS

The idea behind a comparison between the "objectivity" of observations and explorations in the natural and social sciences may be conveyed by an imaginary quotation: "Science must be objective and not affected by value judgments; but the social sciences are inherently concerned with values and, hence, they lack the disinterested objectivity of science." True? Frightfully muddled. The trouble is that the problem of "subjective value," which is at the very root of the social sciences, is quite delicate and has in fact confused many, including some fine scholars.

To remove confusion one must separate the different meanings of "value" and the different ways in which they relate to the social sciences, particularly economics. I have distinguished eleven different kinds of value-reference in economics, but have enough sense to spare you this exhibition of my pedagogic dissecting zeal. But we cannot dispense entirely with the problem and overlook the danger of confusion. Thus, I offer you a bargain and shall reduce my distinctions from eleven to four. I am asking you to keep apart the following four meanings in which value judgment may come into our present discussion: (1) The analyst's judgment may be biased for one reason or another, perhaps because his views of the social "Good" or his personal pecuniary interests in the practical use of his findings interfere with the proper scientific detachment. (2) Some normative issues may be connected with the problem under investigation, perhaps ethical judgments which may color some of the investigator's incidental pronouncements—*obiter dicta*—without however causing a bias in his reported findings of his research. (3) The interest in solving the problems under investigation is surely affected by values since, after all, the investigator selects his problems because he believes that their solution would be of value. (4) The investigator in the social sciences has to explain his observations as results of human actions which can be interpreted only with reference to motives and purposes of the actors, that is, to values entertained by them.

With regard to the first of these possibilities, some authorities have held that the social sciences may more easily succumb to temptation and may show obvious biases. The philosopher Morris Cohen, for example, spoke of "the subjective difficulty of maintaining scientific detachment in the study of human affairs. Few human beings can calmly and with equal fairness consider both sides of a question such as socialism, free love, or birth-control."[3] This is quite true, but one should not forget similar difficulties in the natural sciences. Remember the difficulties which, in deference to religious values, biologists had in discussions of evolution and, going further back, the troubles of astronomers in discussions of the heliocentric theory and of geologists in discussions of the age of the earth. Let us also recall that only 25 years ago [1936] German mathematicians and physicists rejected "Jewish" theorems and theories, including physical relativity, under the pressure of nationalistic values, and only ten years ago [1951] Russian biologists stuck to a mutation theory which was evidently affected by political values. I do not know whether one cannot detect in our own period here in the United States an association between political views and scientific answers to the question of the genetic dangers from fallout and from other nuclear testing.

Apart from political bias, there have been cases of real cheating in science. Think of physical anthropology and its faked Piltdown Man. That the possibility of deception is not entirely beyond the pale of experimental scientists can be gathered from a splendid piece of fiction, a recent novel, *The Affair,* by C. P. Snow, the well-known Cambridge don.

. Having said all this about the possibility of bias existing in the presentation of evidence and findings in the natural sciences, we should hasten to admit that not a few economists, especially when concerned with current problems and the interpretation of recent history, are given to "lying with statistics." It is hardly a coincidence if labor economists choose one base year and business economists choose another base year when they compare wage increases and price increases; or if for their computations of growth rates expert witnesses for different political parties choose different statistical series and different base years. This does not indicate that the social sciences are in this respect "superior" or "inferior" to the natural sciences. Think of physicists, chemists, medical scientists, psychiatrists, et cetera, appearing as expert witnesses in court litigation to testify in support of their clients' cases. In these instances the scientists are in the role of analyzing concrete individual events, of interpreting recent history. If there is a difference at all between the natural and social sciences in this respect, it may be that economists these days have more opportunities to present biased findings than their colleagues in the physical sciences. But even this may not be so. I may underestimate the opportunities of scientists and engineers to submit expert testimonies with paid-for bias.

The second way in which value judgments may affect the investigator does not involve any bias in his findings or his reports on his findings. But ethical judgments may be so closely connected with his problems that he may feel impelled to make evaluative pronouncements on the normative issues in question. For example, scientists may have strong views about vivisection, sterilization, abortion, hydrogen bombs, biological warfare, et cetera, and may express these views in connection with their scientific work. Likewise, social scientists may have strong views about the right to privacy, free enterprise, free markets, equality of income, old-age pensions, socialized medicine, segregation, education, et cetera, and they may express these views in connection with the results of their research. Let us repeat that this need not imply that their findings are biased. There is no difference on this score between the natural and the social sciences. The research and its results may be closely connected with values of all sorts, and value judgments may be expressed, and yet the objectivity of the research and of the reports on the findings need not be impaired.

The third way value judgments affect research is in the selection of the project, in the choice of the subject for investigation. This is unavoidable and the only question is what kinds of value and whose values are paramount. If research is financed by foundations or by the government, the values may be those which the chief investigator believes are held by the agencies or committees that pass on the allocation of funds. If the research is not aided by outside funds, the project may be chosen on the basis of what the investigator believes to be "social values," that is, he chooses a project that may yield solutions to problems supposed to be important for society. Society wants to know how to cure cancer, how to prevent hay fever, how to eliminate mosquitoes, how to get rid of crab grass and weeds, how to restrain juvenile delinquency, how to reduce illegitimacy and other accidents, how to increase employment, to raise real wages, to aid farmers, to avoid price inflation, and so on, and so forth. These examples suggest that the value component in the project selection is the same in the natural and in the social sciences. There are instances, thank God, in which the investigator selects his project out of sheer intellectual curiosity and does not give "two hoots" about the social importance of his findings. Still, to satisfy curiosity is a value too, and indeed a very potent one. We must not fail to mention the case of the graduate student who lacks imagination as well as intellectual curiosity and undertakes a project just because it is the only one he can think of, though neither he nor anybody else finds it interesting, let alone important. We may accept this case as the exception to the rule. Such exceptions probably are equally rare in the natural and the social sciences.

Now we come to the one real difference, the fourth of our value-references. Social phenomena are defined as results of human action, and all human action is defined as motivated action. Hence, social phenomena are explained only if they are attributed to definite types of action which are "understood" in terms of the values motivating those who decide and act. This concern with values—not values which the investigator entertains but values he understands to be effective in guiding the actions which bring about the events he studies—is the crucial difference between the social sciences and the natural sciences. To explain the motion of molecules, the fusion or fission of atoms, the paths of celestial bodies, the growth or mutation of organic matter, et cetera, the scientist will not ask why the molecules want to move about, why atoms decide to merge or to split, why Venus has chosen her particular orbit, why certain cells are anxious to divide. The social scientist, however, is not doing his job unless he explains changes in the circulation of money by going back to the decisions of the spenders and hoarders, explains company mergers by the goals that may have persuaded managements and boards of corporate bodies to take such actions, explains the location of industries by calculations of such things as transportation costs and wage differentials, and economic growth by propensities to save, to invest, to innovate, to procreate or prevent procreation, and so on. My social science examples were all from economics, but I might just as well have taken examples from sociology, cultural anthropology, political science, et cetera, to show that explanation in the social sciences regularly requires the interpretation of phenomena in terms of idealized motivations of the idealized persons whose idealized actions bring forth the phenomena under investigation.

An example may further elucidate the difference between the explanatory principles in nonhuman nature and human society. A rock does not say to us: "I am a beast,"[4] nor does it say: "I came here because I did not like it up there near the glaciers, where I used to live: here I like it fine, especially this nice view of the valley." We do not inquire into value judgments of rocks. But we must not fail to take account of valuations of humans; social phenomena must be explained as the results of motivated human actions.

The greatest authorities on the methodology of the social sciences have referred to this fundamental postulate as the requirement of "subjective interpretation," and all such interpretation of "subjective meanings" implies references to values motivating actions. This has of course nothing to do with value judgments impairing the "scientific objectivity" of the investigators or affecting them in any way that would make their findings suspect. Whether the postulate of subjective interpretation which *differentiates* the social sciences from the natural sciences should be held to make them either "inferior" or "superior" is a matter of taste.

VERIFIABILITY OF HYPOTHESES

It is said that verification is not easy to come by in the social sciences, while it is the chief business of the investigator in the natural sciences. This is true, though many do not fully understand what is involved and, consequently, are apt to exaggerate the difference.

One should distinguish between what a British philosopher has recently called "high-level hypotheses" and "low-level generalizations."[5] The former are postulated and can never be *directly* verified; a single high-level hypothesis cannot even be *indirectly verified,* because from one hypothesis standing alone nothing follows. Only a *whole system* of hypotheses can be tested by deducing from some set of general postulates and some set of specific assumptions the logical consequences, and comparing these with records of observations regarded as the approximate empirical counterparts of the specific assumptions and specific consequences.[6] This holds for both the natural and the social sciences. (There is no need for *direct* tests of the fundamental postulates in physics—such as the laws of conservation of energy, of angular momentum, of motion—or of the fundamental postulates in economics—such as the laws of maximizing utility and profits.)

While entire theoretical systems and the low-level generalizations derived from them are tested in the natural sciences, there exist at any one time many unverified hypotheses. This holds especially with regard to theories of creation and evolution in such fields as biology, geology, and cosmogony; for example (if my reading is correct), of the theory of the expanding universe, the dust-cloud hypothesis of the formation of stars and planets, of the low-temperature or high-temperature theories of the formation of the earth, of the various (conflicting) theories of granitization, et cetera. In other words, where the natural sciences deal with nonreproducible occurrences and with sequences for which controlled experiments cannot be devised, they have to work with hypotheses which remain untested for a long time, perhaps forever.

In the social sciences, low-level generalizations about recurring events are being tested all the time. Unfortunately, often several conflicting hypotheses are consistent with the observed facts and there are no crucial experiments to eliminate some of the hypotheses. But every one of us could name dozens of propositions that have been disconfirmed, and this means that the verification process has done what it is supposed to do. The impossibility of controlled experiments and the relatively large number of relevant variables are the chief obstacles to more efficient verification in the social sciences. This is not an inefficiency on the part of our investigators, but it lies in the nature of things.

EXACTNESS OF FINDINGS

Those who claim that the social sciences are "less exact" than the natural sciences often have a very incomplete knowledge of either of them, and a rather hazy idea of the meaning of "exactness." Some mean by exactness measurability. This we shall discuss under a separate heading. Others mean accuracy and success in predicting future events, which is something different. Others mean reducibility to mathematical language. The meaning of exactness best founded in intellectual history is the possibility of constructing a theoretical system of idealized models containing abstract constructs of variables and of relations between variables, from which most or all propositions concerning particular connections can be deduced. Such systems do not exist in several of the natural sciences—for example, in several areas of biology—while they do exist in at least one of the social sciences: economics.

We cannot foretell the development of any discipline. We cannot say now whether there will soon or ever be a "unified theory" of political science, or whether the piecemeal generalizations which sociology has yielded thus far can be integrated into one comprehensive theoretical system. In any case, the quality of "exactness," if this is what is meant by it, cannot be attributed to all the natural sciences nor denied to all the social sciences.

MEASURABILITY OF PHENOMENA

If the availability of numerical data were in and of itself an advantage in scientific investigation, economics would be on the top of all sciences. Economics is the only field in which the raw data of experience are already in numerical form. In other fields the analyst must first quantify and measure before he can obtain data in numerical form. The physicist must weigh and count and must invent and build instruments from which numbers can be read, numbers standing for certain relations pertaining to essentially nonnumerical observations. Information which first appears only in some such form as "relatively" large, heavy, hot, fast, is later transformed into numerical data by means of measuring devices such as rods, scales, thermometers, speedometers. The economist can begin with numbers. What he observes are prices and sums of money. He can start out with numerical data given to him without the use of measuring devices.

The compilation of masses of data calls for resources which only large organizations, frequently only the government, can muster. This, in my opinion, is unfortunate because it implies that the availability of numerical data is associated with the extent of government intervention in economic affairs, and there is therefore an inverse relation between economic information and individual freedom.

Numbers, moreover, are not all that is needed. To be useful, the numbers must fit the concepts used in theoretical propositions or in comprehensive theoretical systems. This is rarely the case with regard to the raw data of economics, and thus the economic analyst still has the problem of obtaining comparable figures by tranforming his raw data into adjusted and corrected ones, acceptable as the operational counterparts of the abstract constructs in his theoretical models. His success in this respect has been commendable, but very far short of what is needed; it cannot compare with the success of the physicist in developing measurement techniques yielding numerical data that can serve as operational counterparts of constructs in the models of theoretical physics.

Physics, however, does not stand for all natural sciences, nor economics for all social sciences. There are several fields, in both natural and social sciences, where quantification of relevant factors has not been achieved and may never be achieved. If Lord Kelvin's phrase, "Science is Measurement," were taken seriously, science might miss some of the most important problems. There is no way of judging whether nonquantifiable factors are more prevalent in nature or in society. The common reference to the "hard" facts of nature and the "soft" facts with which the student of society has to deal seems to imply a judgment about measurability. "Hard" things can be firmly gripped and measured, "soft" things cannot. There may be something to this. The facts of nature are perceived with our "senses," the facts of society are interpreted in terms of the "sense" they make in a motivational analysis. However, this contrast is not quite to the point, because the "sensory" experience of the natural scientist refers to the *data,* while the "sense" interpretation by the social scientist of the ideal-typical inner experience of the members of society refers to basic *postulates* and intervening variables.

The conclusion, that we cannot be sure about the prevalence of nonquantifiable factors in natural and social sciences, still holds.

CONSTANCY OF NUMERICAL RELATIONSHIPS

On this score there can be no doubt that some of the natural sciences have got something which none of the social sciences has got: "constants," unchanging numbers expressing unchanging relationships between measurable quantities.

The discipline with the largest number of constants is, of course, physics. Examples are the velocity of light ($c = 2.99776 \times 10^{10}$ cm/sec), Planck's constant for the smallest increment of spin or angular momentum ($h = 6.624 \times 10^{-27}$ erg sec), the gravitation constant ($G = 6.6 \times 10^{-8}$ dyne CM^2 gram $^{-2}$), the Coulomb constant ($e = 4.8025 \times 10^{-10}$ units), proton mass ($M = 1.672 \times 10^{-24}$ gram), the ratio of proton mass to electron mass ($M/m = 1836.13$), the fine-structure constant ($\alpha^{-1} = 137,0371$). Some of these constants are postulated (conventional), others (the last two) are empirical, but this makes no difference for our purposes. Max Planck contended, the postulated "universal constants" were not just "invented for reasons of practical convenience, but have forced themselves upon us irresistibly because of the agreement between the results of all relevant measurements."[7]

I know of no numerical constant in any of the social sciences. In economics we have been computing certain ratios which, however, are found to vary relatively widely with time and place. The annual income-velocity of circulation of money, the marginal propensities to consume, to save, to import, the elasticities of demand for various goods, the savings ratios, capital-output ratios, growth rates—none of these has remained constant over time or is the same for different countries. They all have varied, some by several hundred percent of the lowest value. Of course, one has found "limits" of these variations, but what does this mean in comparison with the virtually immutable physical constants? When it was noticed that the ratio between labor income and national income in some countries has varied by "only" ten percent over some twenty years, some economists were so perplexed that they spoke of the "constancy" of the relative shares. (They hardly realized that the 10 percent variation in that ratio was the same as about a 25 percent variation in the ratio between labor income and nonlabor income.) That the income velocity of circulation of money has rarely risen above 3 or fallen below 1 is surely interesting, but this is anything but a "constant." That the marginal propensity to consume cannot in the long run be above 2 is rather obvious, but in the short run it may vary between .7 and 1.2 or even more. That saving ratios (to national income) have never been above 15 percent in any country regardless of the economic system (communistic or capitalistic, regulated or essentially free) is a very important fact; but saving ratios have been known to be next to zero, or even negative, and the variations from time to time and country to country are very large indeed.

Sociologists and actuaries have reported some "relatively stable" ratios —accident rates, birth rates, crime rates, et cetera—but the "stability" is only relative to the extreme variability of other numerical ratios. Indeed, most of these ratios are subject to "human engineering," to governmental policies designed to change them, and hence they are not even thought of as constants.

The verdict is confirmed: while there are important numerical constants in the natural sciences, there are none in the social sciences.

PREDICTABILITY OF FUTURE EVENTS

Before we try to compare the success which natural and social sciences have had in correctly predicting future events, a few important distinctions should be made. We must distinguish hypothetical or conditional predictions from unconditional predictions or forecasts. And among the former we must distinguish those where all the stated conditions can be controlled, those where all the stated conditions can be either controlled or unambiguously ascertained before the event, and finally those where some of the stated conditions can neither be controlled nor ascertained early enough (if at all). A conditional prediction of the third kind is such an "iffy" statement that it may be of no use unless one can know with confidence that it would be highly improbable for these problematic conditions (uncontrollable and not ascertainable before the event) to interfere with the prediction. A different kind of distinction concerns the numerical definiteness of the prediction: one may predict that a certain magnitude (1) will change, (2) will increase, (3) will increase by at least so-and-so much, (4) will increase within definite limits, or (5) will increase by a definite amount. Similarly, the prediction may be more or less definite with respect to the time within which it is supposed to come true. A prediction without any time specification is worthless.

Some people are inclined to believe that the natural sciences can beat the social sciences on any count, in unconditional predictions as well as in conditional predictions fully specified as to definite conditions, exact degree and time of fulfilment. But what they have in mind are the laboratory experiments of the natural sciences, in which predictions have proved so eminently successful; and then they look at the poor record social scientists have had in predicting future events in the social world which they observe but cannot control. This comparison is unfair and unreasonable. The artificial laboratory world in which the experimenter tries to control all conditions as best as he can is different from the real world of nature. If a comparison is made, it must be between predictions of events in the real natural world and in the real social world.

Even for the real world, we should distinguish between predictions of events which we try to bring about by design and predictions of events in which we have no part at all. The teams of physicists and engineers who have been designing and developing machines and apparatuses are not very successful in predicting their performance when the design is still new. The record of predictions of the paths of moon shots and space missiles has been rather spotty. The so-called bugs that have to be worked out in any new contraption are nothing but predictions gone wrong. After a while predictions become more reliable. The same is true, however, with predictions concerning the performance of organized social institutions. For example, if I take an envelope, put a certain address on it and a certain postage stamp, and deposit it in a certain box on the street, I can predict that after three or four days it will be delivered at a certain house thousands of miles away. This prediction and any number of similar predictions will prove correct with a remarkably high frequency. And you don't have to be a social scientist to make such successful predictions about an organized social machinery, just as you don't have to be a natural scientist to predict the result of your pushing the electric-light switch or of similar manipulations of a well-tried mechanical or electrical apparatus.

There are more misses and fewer hits with regard to predictions of completely unmanipulated and unorganized reality. Meteorologists have a hard time forecasting the weather for the next twenty-four hours or two or three days. There are too many variables involved and it is too difficult to obtain complete information about some of them. Economists are only slightly better in forecasting employment and income, exports and tax revenues for the next six months or for a year or two. Economists, moreover, have better excuses for their failures because of unpredictable "interferences" by governmental agencies or power groups which may even be influenced by the forecasts of the economists and may operate to defeat their predictions. On the other hand, some of the predictions may be self-fulfilling in that people, learning of the predictions, act in ways which bring about the predicted events. One might say that economists ought to be able to include the "psychological" effects of their communications among the variables of their models and take full account of these influences. There are, however, too many variables, personal and political, involved to make it possible to allow for all effects which anticipations, and anticipations of anticipations, may have upon the end results. To give an example of a simple self-defeating prediction from another social science: traffic experts regularly forecast the number of automobile accidents and fatalities that are going to occur over holiday weekends, and at the same time they hope that their forecasts will influence drivers to be more careful and thus to turn the forecasts into exaggerated fears.

We must not be too sanguine about the success of social scientists in making either unconditional forecasts or conditional predictions. Let us admit that we are not good in the business of prophecy and let us be modest in our claims about our ability to predict. After all, it is not our stupidity which hampers us, but chiefly our lack of information, and when one has to make do with bad guesses in lieu of information the success cannot be great. But there is a significant difference between the natural sciences and the social sciences in this respect: Experts in the natural sciences usually do not try to do what they know they cannot do; and nobody expects them to do it. They would never undertake to predict the number of fatalities in a train wreck that might happen under certain conditions during the next year. They do not even predict next year's explosions and epidemics, floods and mountain slides, earthquakes and water pollution. Social scientists, for some strange reason, are expected to foretell the future and they feel badly if they fail.

DISTANCE FROM EVERYDAY EXPERIENCE

Science is, almost by definition, what the layman cannot understand. Science is knowledge accessible only to superior minds with great effort. What everybody can know cannot be science.

A layman could not undertake to read and grasp a professional article in physics or chemistry or biophysics. He would hardly be able to pronounce many of the words and he might not have the faintest idea of what the article was all about. Needless to say, it would be out of the question for a layman to pose as an expert in a natural science. On the other hand, a layman might read articles in descriptive economics, sociology, anthropology, social psychology. Although in all these fields technical jargon is used which he could not really understand, he might *think* that he knows the sense of the words and grasps the meanings of the sentences; he might even be inclined to poke fun at some of the stuff. He believes he is—from his own experience and from his reading of newspapers and popular magazines—familiar with the subject matter of the social sciences. In consequence, he has little respect for the analyses which the social scientists present.

The fact that social scientists use less Latin and Greek words and less mathematics than their colleagues in the natural science departments and, instead, use everyday words in special, and often quite technical, meanings may have something to do with the attitude of the layman. The sentences of the sociologist, for example, make little sense if the borrowed words are understood in their nontechnical, everyday meaning. But if the layman is told of the special meanings that have been bestowed upon his words, he gets angry or condescendingly amused.

But we must not exaggerate this business of language and professional jargon because the problem really lies deeper. The natural sciences talk about nuclei, isotopes, galaxies, benzoids, drosophilas, chromosomes, dodecahedrons, Pleistocene fossils, and the layman marvels that anyone really cares. The social sciences, however—and the layman usually finds this out—talk about—him. While he never identifies himself with a positron, a pneumococcus, a coenzyme, or a digital computer, he does identify himself with many of the ideal types presented by the social scientist, and he finds that the likeness is poor and the analysis "consequently" wrong.

The fact that the social sciences deal with man in his relations with fellow man brings them so close to man's own everyday experience that he cannot see the analysis of this experience as something above and beyond him. Hence he is suspicious of the analysts and disappointed in what he supposes to be a portrait of him.

Standards of Admission and Requirements

High school physics is taken chiefly by the students with the highest IQs. At college the students majoring in physics, and again at graduate school the students of physics, are reported to have on the average higher IQs than those in other fields. This gives physics and physicists a special prestige in schools and universities, and this prestige carries over to all natural sciences and puts them somehow above the social sciences. This is rather odd, since the average quality of students in different departments depends chiefly on departmental policies, which may vary from institution to institution. The preeminence of physics is rather general because of the requirement of calculus. In those universities in which the economics department requires calculus, the students of economics rank as high as the students of physics in intelligence, achievement, and prestige.

The lumping of all natural sciences for comparisons of student quality and admission standards is particularly unreasonable in view of the fact that at many colleges some of the natural science departments, such as biology and geology, attract a rather poor average quality of student. (This is not so in biology at universities with many applicants for a premedical curriculum.) The lumping of all social sciences in this respect is equally wrong, since the differences in admission standards and graduation requirements among departments, say between economics, history, and sociology, may be very great. Many sociology departments have been notorious for their role as refuge for mentally underprivileged undergraduates. Given the propensity to overgeneralize, it is no wonder then that the social sciences are being regarded as the poor relations of the natural sciences and as disciplines for which students who cannot qualify for the sciences are still good enough.

Since I am addressing economists, and since economics departments, at least some of the better colleges and universities, are maintaining standards as high as physics and mathematics departments, it would be unfair to level exhortations at my present audience. But perhaps we should try to convince our colleagues in all social science departments of the disservice they are doing to their fields and to the social sciences at large by admitting and keeping inferior students as majors. Even if some of us think that one can study social sciences without knowing higher mathematics, we should insist on making calculus and mathematical statistics absolute requirements—as a device for keeping away the weakest students.

Despite my protest against improper generalizations, I must admit that averages may be indicative of something or other, and that the average IQ of the students in the natural science departments is higher than that of the students in the social science department.[8] No field can be better than the men who work in it. On this score, therefore, the natural sciences would be superior to the social sciences.

THE SCORE CARD

We may now summarize the tallies on the nine scores.

1. With respect to the invariability or recurrence of observations, we found that the greater number of variables—of relevant factors—in the social sciences makes for more variation, for less recurrence of exactly the same sequences of events.

2. With respect to the objectivity of observations and explanations, we distinguished several ways in which references to values and value judgments enter scientific activity. Whereas the social sciences have a requirement of "subjective interpretation of value-motivated actions" which does not exist in the natural sciences, this does not affect the proper "scientific objectivity" of the social scientist.

3. With respect to the verifiability of hypotheses, we found that the impossibility of controlled experiments combined with the larger number of relevant variables does make verification in the social sciences more difficult than in most of the natural sciences.

4. With respect to the exactness of the findings, we decided to mean by it the existence of a theoretical system from which most propositions concerning particular connections can be deduced. Exactness in this sense exists in physics and in economics, but much less so in other natural and other social sciences.

5. With respect to the measurability of phenomena, we saw an important difference between the availability of an ample supply of numerical data and the availability of such numerical data as can be used as good counterparts of the constructs in theoretical models. On this score, physics is clearly ahead of all other disciplines. it is doubtful that this can be said about the natural sciences in general relative to the social sciences in general.

6. With respect to the constancy of numerical relationships, we entertained no doubt concerning the existence of constants, postulated or empirical, in physics and in other natural sciences, whereas no numerical constants can be found in the study of society.

7. With respect to the predictability of future events, we ruled out comparisons between the laboratory world of some of the natural sciences and the unmanipulated real world studied by the social sciences. Comparing only the comparable, the real worlds—and excepting the special case of astronomy—we found no essential differences in the predictability of natural and social phenomena.

8. With respect to the distance of scientific from everyday experience, we saw that in linguistic expression as well as in their main concerns the social sciences are so much closer to prescientific language and thought that they do not command the respect that is accorded to the natural sciences.

9. With respect to the standards of admission and requirements, we found that they are on the average lower in the social than in the natural sciences.

The last of these scores relates to the current practice of colleges and universities, not to the character of the disciplines. The point before the last, though connected with the character of the social sciences, relates only to the popular appreciation of these disciplines; it does not aid in answering the question whether the social sciences are "really" inferior. Thus the last two scores will not be considered relevant to our question. This leaves seven scores to consider. On four of the seven no real differences could be established. But on the other three scores, on "Invariance," "Verifiability," and "Numerical Constants," we found the social sciences to be inferior to the natural sciences.

THE IMPLICATIONS OF INFERIORITY

What does it mean if one thing is called "inferior" to another with regard to a particular "quality"? If this "quality" is something that is highly valued in any object, and if the absence of this "quality" is seriously missed regardless of other qualities present, then, but only then, does the noted "inferiority" have any evaluative implications. In order to show that "inferiority" sometimes means very little, I shall present here several statements about differences in particular qualities.

> "Champagne is inferior to rubbing alcohol in alcoholic content."
> "Beef steak is inferior to strawberry Jell-O in sweetness."
> "A violin is inferior to a violoncello in physical weight."
> "Chamber music is inferior to band music in loudness."
> "Hamlet is inferior to Joe Palooka in appeal to children."
> "Sandpaper is inferior to velvet in smoothness."
> "Psychiatry is inferior to surgery in ability to effect quick cures."
> "Biology is inferior to physics in internal consistency."

It all depends on what you want. Each member in a pair of things is inferior to the other in some respect. In some instances it may be precisely this inferiority that makes the thing desirable. (Sandpaper is wanted *because* of its inferior smoothness.) In other instances the inferiority in a particular respect may be a matter of indifference. (The violin's inferiority in physical weight neither adds to nor detracts from its relative value.) Again in other instances the particular inferiority may be regrettable, but nothing can be done about it and the thing in question may be wanted nonetheless. (We need psychiatry, however much we regret that in general it cannot effect quick cures; and we need biology, no matter how little internal consistency has been attained in its theoretical systems.)We have stated that the social sciences are inferior to the natural sciences in some respects, for example, in verifiability. This is regrettable. If propositions cannot be readily tested, this calls for more judgment, more patience, more ingenuity. But does it mean much else?

THE CRUCIAL QUESTION: "SO WHAT?"

What is the pragmatic meaning of the statement in question? If I learn, for example, that drug E is inferior to drug P as a cure for hay fever, this means that, if I want such a cure, I shall not buy drug E. If I am told Mr. A is inferior to Mr. B as an automobile mechanic, I shall avoid using Mr. A when my car needs repair. If I find textbook K inferior to textbook S in accuracy, organization, as well as exposition, I shall not adopt textbook K. In every one of these examples, the statement that one thing is inferior to another makes pragmatic sense. The point is that all these pairs are alternatives between which a choice is to be made.

Are the natural sciences and the social sciences alternatives between which we have to choose? If they were, a claim that the social sciences are "inferior" could have the following meanings:

1. We should not study the social sciences.
2. We should not spend money on teaching and research in the social sciences.
3. We should not permit gifted persons to study social sciences and should steer them toward superior pursuits.
4. We should not respect scholars who so imprudently chose to be social scientists.

If one realizes that none of these things could possibly be meant, that every one of these meanings would be preposterous, and that the social sciences and the natural sciences can by no means be regarded as alternatives but, instead, that both are needed and neither can be dispensed with, he can give the inferiority statement perhaps one other meaning:

5. We should do something to improve the social sciences and remedy their defects.

This last interpretation would make sense if the differences which are presented as grounds for the supposed inferiority were "defects" that can be remedied. But they are not. That there are more variety and change in social phenomena; that, because of the large number of relevant variables and the impossibility of controlled experiments, hypotheses in the social sciences cannot be easily verified; and that no numerical constants can be detected in the social world—these are not defects to be remedied but fundamental properties to be grasped, accepted, and taken into account. Because of these properties research and analysis in the social sciences hold greater complexities and difficulties, If you wish, you may take this to be a greater challenge, rather than a deterrent. To be sure, difficulty and complexity alone are not sufficient reasons for studying certain problems. But the problems presented by the social world are certainly not unimportant. If they are also difficult to tackle, they ought to attract ample resources and the best minds. Today they are getting neither. The social sciences are "really inferior" regarding the place they are accorded by society and the priorities with which financial and human resources are allocated. This inferiority is curable.

NOTES

1. Published in *On Freedom and Free Enterprise: Essays in Honor of Ludwig von Mises,* ed. Mary Sennholz, pp. 161–172.

2. H. Rickert, *Die Grenzen der naturwissenschaftlichen Begnfisbildung.*

3. M. Cohen, *Reason and Nature,* p. 348.

4. H. Kelsen, *Allgemeine Staatslehre,* p. 129. Quoted with illuminating comments in A. Schutz, *Der sinnhafte Aufbau der Sozialen Welt.*

5. R. B. Braithwaite, *Scientific Explanation: A Study of the Function of Theory, Probability, and Law in Science.*

6. F. Machlup, "The Problem of Verification in Economics," *Southern Economic Journal,* XXII, 1955, 1–21.

7. M. Planck, *Scientific Autobiography and Other Papers,* p. 173.

8. The average IQ of students receiving bachelor's degrees was, according to a 1954 study, 121 in the biological sciences, and 122 in economics, 127 in the physical sciences, and 119 in business. See D. Wolfe, *America's Resources of Specialized Talent: The Report of the Commission on Human Resources and Advanced Training,* pp. 319–322.

9

If Economics Isn't Science, What Is It?

Alexander Rosenberg

In a number of papers, and in *Microeconomic Laws,*[1] I argued that economic theory is a conceptually coherent body of causal general claims that stand a chance of being laws. My arguments elicited no great sigh of relief among economists, for they are not anxious about the scientific respectability of their discipline.[2] But others eager to adopt or adapt microeconomic theory to their own uses have appealed to these and other arguments which attempt to defend economic theory from a litany of charges that are as old as the theory itself.[3] Among these charges, the perennial ones were those that denied to economic theory the status of a contingent empirical discipline because it failed to meet one or another fashionable positivist or Popperian criterion of scientific respectability. With the waning of positivism these charges have seemed less and less serious to philosophers, although they have retained their force for the few economists still distracted by methodology.[4] But among philosophers charges that economics does not measure up to standards for being a science have run afoul of the general consensus that we have no notion of science good enough to measure candidates against. This makes it difficult to raise the question of whether economics is a science, and tends to leave economists, and their erstwhile apologists like me, satisfied with the conclusion that since there is nothing logically or conceptually incoherent about economics, it must be a respectable empirical theory of human behavior and/or its aggregate consequences.

The trouble with this attitude is that it is unwarrantably complacent. It is all well and good to say that economics is conceptually coherent, and that there are no uncontroversial standards against which economics may be found wanting, but this attitude will not make the serious anomalies and puzzles about economic theory go away. These puzzles surround its thoroughgoing predictive weakness. The ability to predict and control may be neither necessary nor sufficient criteria for cognitively respectable scientific theories. But the fact is that microeconomic theory has made no advances in the management of economic processes since its current formalism was first elaborated in the nineteenth century. And this surely undermines a complacent conviction that the credentials of economics as a science are entirely in order. For a long time after 1945 it might confidently have been said that Keynesian macroeconomics was a theory moving in the right direction: although a macro theory, it would ultimately provide the sort of explanatory and predictive satisfaction characteristic of science. But the simultaneous inflation and unemployment levels of the last decade and the economy's imperviousness to fiscal policy have eroded the layman's and the economist's confidence in the theory. Moreover the profession's reaction to the failures of Keynesian theory is even more disquieting to those who view economic theory as unimpeachably a scientific enterprise. For a large part of the response to its failures has been a return to the microeconomic theories which it was sometimes claimed to supersede. The diagnosis offered for the failure of the Keynesian theory has been that it does not accord individual agents the kind of rationality in the use of information and the satisfaction of preferences that neoclassical microeconomic theory accords them.[5] The alternative offered to Keynesian theory in the light of this result is nothing more nor less than a return to the status quo ante, to the neoclassical theory of Walras, Marshall, and the early Hicks, that Keynesianism had preempted.[6] This cycle brings economic theory right back to where it was before 1937, and it should seriously undermine the confidence of anyone's beliefs that economics is an empirical science, with aims and standards roughly identical to other empirical sciences. For the twentieth-century history of economic theory certainly does not appear to be that of an empirical science.

Of course, eighty years is not a long time in the life of a science, or even a theory, so the fact that economics has not substantially changed, either in its form or in its degree of confirmation, since Walras, or arguably since Adam Smith, is no reason to deny it scientific respectability. But it is reason to ask why economics has not moved away from the theoretical strategies that have characterized it at least since 1874, in spite of their practical inapplicability to crucial matters like the business cycle, economic development, or stagflation. On some views of proper scientific method, of course, economists have been doing just what they should be doing. Since the nineteenth century they have been pursuing a single research strategy, acting in accordance with a ubiquitous and powerful paradigm. For, economists have been steadily elaborating a theory whose *form* is identical to that of the great theoretical breakthroughs in science since the sixteenth century. Accordingly it may be argued that it would be irrational for economists to surrender this strategy short of a conclusive demonstration that it is inappropriate to the explanation of economic activity. The strategy is that of viewing the behavior economists seek to explain as reflecting forces which always move toward stable equilibria that maximize or minimize some theoretically crucial variable. In the case of microeconomics, this crucial variable is utility (or its latter-day surrogates), and the equilibrium is given by a level of price in all markets that maximizes this variable. This strategy is most impressively exemplified in Newtonian mechanics and in the Darwinian theory of natural selection. It is no surprise that a strategy which serves so well in these two signal accomplishments of science should have as strong a grip in other domains to which it seems applicable. Moreover., the constraints on theoretical and empirical developments that this strategy imposes can explain many of the greatest successes of Newtonian and Darwinian science, and much of the puzzling character of developments in economic theory.

I call this strategy the extremal strategy, because it is especially apparent in Newtonian mechanics when that theory is expressed in so-called extremal principles, according to which a system's behavior always minimizes or maximizes variables reflecting the mechanically possible states of the system. In the theory of natural selection this strategy assumes that the environment acts so as to maximize fitness. This strategy is crucial to the success of these theories because of the way it directs and shapes the research motivated by them. Thus, if we believe that a system always acts to maximize the value of a mechanical variable, for example, total energy, and our measurements of the observable value of that variable diverge from the predictions of the theory and the initial conditions, we do not infer that the system described is failing to maximize the value of the variable in question. We do not falsify the theory. We assume that we have incompletely described the constraints under which the system is actually operating. In Newtonian mechanics attempts to more completely describe the systems under study resulted in the discovery of new planets, Neptune, Uranus, and Pluto, the invention of new instruments, and eventually in the discovery of new laws, like those of thermodynamics. Similarly in biology, assuming that fitness is maximized led to the discovery of forces not previously recognized to effect genetic variation within a population, and more important, led to the discovery of genetic laws that explain the persistence in a population of apparently maladaptive traits, like sicklecell anemia, for instance. Because these theories are "extremal" ones, differential calculus may be employed to express and interrelate their leading ideas. Microeconomics is an avowedly extremal theory, asserting that the systems it describes maximize utility (or some surrogate). That is why it can be couched in the language of differential calculus. It is the extremal character of the theory, and not the fact that it deals with "quantifiable" variables, like money, that makes microeconomics a quantitively expressed theory.

More important than the fact that they all employ differential calculus, these theories are all committed *to explain everything in their domains* because of their extremal character. In virtue of the claim that systems in their domains always behave in a way which maximizes or minimizes some quantity, the theories ipso facto provide the explanation of all of their subjects' behavior by citing the determinants of all their subjects' relevant states. An extremal theory cannot be treated as only a partial account of the behavior of objects in its domain, or as enumerating just *some* of the many determinants of its subject's states; for any behavior that actually fails to maximize or minimize the value of the privileged variable simply refutes the theory *tout court*. In fact, the pervasive character of extremal theories insulates them from falsification to a degree absent from nonextremal theories. All theories are strictly unfalsifiable, simply because testing them involves auxiliary hypotheses. But extremal theories are not only insulated against strict falsification, they are also insulated against the sort of actual falsification that usually overthrows theories, instead of auxiliary hypotheses. In the case of a nonextremal theory falsification may lead us either to revise the auxiliary assumptions about test conditions, or to revise the theory by adding new antecedent clauses to its generalizations, or new qualifications to its ceteris paribus clauses. But this is not possible in the case of extremal theories. The axioms of theories like Newton's, or Darwin's, or Walras's do not embody even implicit ceteris paribus clauses. Microeconomics does not, for example, assume that agents maximize utility, ceteris paribus. With these theories the choice is always between rejecting the auxiliary hypotheses—the description of test conditions—or rejecting the theory altogether. For the only change that can be made to the theory is to deny that its subjects invariably maximize or minimize its chosen variable. This is why high-level extremal theories like Newtonian mechanics are left untouched by apparent counterinstances; why they are not simply improved by qualifications and caveats in their antecedent conditions; why they are superseded only by utterly new theories, in which the values of very different variables are maximized, or minimized.

Extremal theories are an important methodological strategy because they are so well insulated from falsification. This has enabled them to function at the core of research programs, turning what otherwise might be anomalies and counterinstances into new predictions and new opportunities for extending their domains and deepening their precision. Accordingly, it may be argued, economists' attachment to their extremal theory represents not complacency, but a well-grounded methodological conservatism. Given the fantastic successes of this approach in such diverse areas as mechanics and biology, it would be unreasonable to forgo similar strategies in the attempt to explain human behavior. So viewed, the history of attempts to make recalcitrant facts about human behavior, and the economic systems humans have constructed, fit the extremal theory of microeconomics reflects a commitment that is on a par with astronomers' attempts to make recalcitrant facts about planetary perihelions fit the demands of Newtonian mechanics; it is on a par with biologists' attempts to make the persistent genetic predisposition to malaria fit the facts of adaptation demanded by the theory of natural selection. Since these attempts do not discredit their theories as empty or unfalsifiable, it should not be inferred that there is anything improper in the economists' attempts to do the same thing. Or so it may be argued.

It is certainly correct that much of the commitment to microeconomic strategies does in fact reflect this sort of reasoning. After all, it is not just the intellectual prestige associated with the scope for differential calculus, topology, and differential geometry attending any extremal theory that explains the reluctance of economists to forgo the strategy. But this conservative rationale for the attachment of economists to extremal theories is vitiated by a crucial disanalogy between microeconomics and mechanics or evolution. Economists would indeed be well advised not to surrender their extremal research program, if only they could boast even a small part of the startling successes that other extremal research programs have achieved. But two hundred years of work in the same direction have produced nothing comparable to the physicists' discovery of new planets, or of new technologies by which to control the mechanical phenomena that Newton's laws systemized. Economists have attained no independently substantiated insight into their domain to rival the biologists' understanding of macroevolution and its underlying mechanism of adaptation and heredity. There has been no signal success of economic theory akin to these advances of extremal theory. This is a disanalogy important enough to bear explaining. Failing a satisfactory explanation, the difference is significant enough to make economists question the merits of their extremal approach, and to make us query the scientific credentials of economic theory. There is, of course, a vast literature on why economics has so little in the way of predictive content, and on how a theory so dependent on idealizations and factually false assumptions as microeconomics can nevertheless constitute a respectable scientific enterprise. This literature goes back to John Stuart Mill[7] and forward to, for example, Hal Varian.[8] The only two things clear about this literature are that economists have found it almost universally satisfying and legitimating, and noneconomists have consistently been left unsatisfied, insisting that methodological excuses are no substitute for attempting to do what economics has hitherto not done: improve its predictive content.

Having shaken free from the complacent attitude toward economic theory evinced in *Microeconomic Laws,* I have come to think that the failure of economics is not methodological, or conceptual, but very broadly empirical. Despite its conceptual integrity, microeconomics, together with all the sciences of human action and its aggregation, rests on a false but central conviction that vitiates its axioms and so bedevils the theorems deduced from them. Economic theory assumes that the categories of preference and expectation are the classes in which economic causes are to be systematized, and that the events to be explained are properly classified as actions like buying, selling, and the movements of markets, industries, and economies that these actions aggregate to. The theory has made this assumption, because of course it is an assumption we all make about human behavior; our behavior constitutes action and is caused by the joint operation of our desires and beliefs. Marginalists of the late nineteenth century like Wicksteed saw clearly that microeconomics is but the formalization of this commonsense notion, and the history of the theory of consumer behavior is the search for laws that will express the relations between desire, belief, and action, first in terms of cardinal utility and certainty, later in terms of ordinal utility, revealed preference, and expected utility under varying conditions of uncertainty and risk. The failure to find such a law or any approximation to it that actually improves our ability to predict consumer behavior any better than Adam Smith could have resulted on the one hand in a reinterpretation of the aims of economic theory away from explaining individual human action, and on the other in the tissue of apologetics with which the consumer of economic methodology is familiar.[9]

The real trouble with economics, the real source of its failure to find *improvable* laws of economic behavior, is something that has only become clear in philosophy's recent attempts to understand and improve the foundations of another science in trouble: psychology, and particularly behavioral and cognitive psychology. Philosophers have shown that the terms in which ordinary thought and the behavioral sciences describe the causes and effects of human action do not describe "natural kinds," they do not divide nature at the joints. They do not label categories of states that share the same manageably small set of causes and effects, and so cannot be brought together in causal generalizations that improve on our ordinary level of prediction and control of human actions, let alone attain the sort of continuing improvement characteristic of science. I cannot hope to do more in the present compass than identify the conclusions to which thirty-five years of work in the philosophy of psychology has arrived, and to show their bearing on economic theory. This work has been devoted to understanding "intentional" terms like "belief," "desire," "action," and their vast hoard of cognates. It is worth noting that the term "intentional" does not mean simply "purposive"; rather, it is employed by philosophers to note that the mental states like beliefs and desires are identified by their "contents," by the propositions they "contain" or the ends and objects to which they are "directed." The trouble with beliefs and desires is that when people have them the propositions they "contain" need not be true or false and the objects they are directed at need not exist or even be possible objects. So we can't decide whether a person is in a given mental state by determining either the truth or falsity of any statement open to our confirmation, or the existence or attainment of any object or end of human action. In psychology and philosophy the two leading attempts to solve this problem of other minds have been behaviorism and the so-called identity theory, the claim that each *type* of mental state ordinary language distinguishes is identical to, or is correlated with, an identifiable type of brain state. Behaviorism is false, however, just because no intentional state can be identified without making assumptions about other mental states. I can't identify your beliefs by observing your actions unless I know what your desires are; indeed I can't identify your movements as actions at all, as opposed to mere reflexes, unless I assume they are caused by the joint operation of your beliefs and desires. There is an intentional circle into which we cannot break by discoveries about mere movement, about behavior. The identity theory seems equally fruitless an approach, because the brain states neuroscience identifies just don't seem to line up in lockstep with mental states that introspection reports. This means that the intentional vocabulary of common sense, (intentional) psychology, and the social sciences are fated to remain isolated from any other conceptual scheme that identifies and systematizes the mental states and human behavior they describe and explain.[10] This isolation is an obstacle to improving the predictive power of these disciplines, because improvement in any explanatory theory requires the variables of the theory to be measured with increasing accuracy. This sort of precision is plainly impossible when the theory's variables are restricted to an interdefinable circle alone, and are not independently identifiable. But the failure of behaviorism and neurological reduction forecloses all such identifications. Accordingly we can't expect to improve our intentional explanations of action beyond their present levels of predictive power. But this level of predictive power is no higher than Plato's. The predictive weakness of theories couched in intentional vocabulary reflects the fact that the terms of this vocabulary do not correlate in a manageable way the vocabulary of other successful scientific theories; they don't divide nature at the joints, insofar as its joints are revealed in already successful theories like those of neuroscience. Some philosophers hold that we must reconcile ourselves to this state of affairs, and reduce our expectations about the possibilities of a predictively powerful theory of human behavior. Others are more optimistic, but insist that we must jettison "folk psychology" and its intentional idiom if we are to hit upon an improvable theory in the science of psychology.[11] This choice extends, of course, beyond psychology to all the other intentional sciences, of which economics, with its reliance on expectation and preference, is certainly one.[12]

Applying the new orthodoxy in the philosophy of psychology, it becomes clear that economics' predictive weakness hinges on the intentional typology of the phenomena it explains and the causes it identifies. Its failure to uncover laws of human behavior is due to its wrongly assuming that these laws will trade in desires, beliefs, or their cognates. And the system of propositions about markets and economies that economists have constructed on the basis of its assumptions about human behavior is deprived of improving explanatory and predictive power because its assumptions can't be improved in a way that transmits improved precision to their consequences. Thus the failure of economics is traced not to a conceptual mistake, or to the inappropriateness of extremal theories and their elegant mathematical apparatus to human action, but to a false assumption economists share with all other social scientists, indeed with everyone who has ever explained their own or others' behavior by appeal to the operation of desires and beliefs.

Just as economists have been given no pause by previous attacks on their discipline, they are unlikely to lay down their tools in the fact of this diagnosis either. Indeed, the persistence of economists in pursuing the extremal and intentional approach that has been conventional for well over a century suggests that nothing could make them give it up, or at any rate that nothing which would make empirical scientists give up a theory will make economists give up their theoretical strategy. But this conviction leads to the conclusion that economics is not empirical science at all. Despite its appearances, and the interest of some economists in applying their formalism to practical matters, this formalism does not any longer have the aims, nor does it make the claims, of an unequivocally empirical theory.

My diagnosis and conclusion face many potential rejoinders and several serious questions. In the remainder of this chapter I shall address one of these rejoinders and attempt to answer one of these questions. The rejoinder takes the following form:

> Surely the fact that the fundamental axioms of economic theory fail to divide nature at the joints does not vitiate the entire enterprise. After all, your explanation would have us forgo not just the abstract claims about preference orders and individual choice under uncertainty, but also the laws of supply and demand. Yet surely these are useful approximations, regularities roughly and frequently enough instanced to reflect some underlying truth about economic behavior, enough at any rate to make economics a worthwhile pursuit even if all your claims are correct. Though you may have provided an abstract explanation of why we cannot improve the current theory indefinitely, you have not given enough reason to deprive it of its usefulness, nor deny it scientific standing.

The question I want to address is closely related to this rejoinder:

> If the extremal intentional research program of economics is an empirical failure, and if economists are not about to surrender either the intentional stance or the extremal method, what is it on your view that economists are doing? What after all is economics? On your view, it does not constitute a scientific discipline. What then is it? This query is not just a rhetorical demand that I attach a label to the subject, but the reasonable request that my diagnosis must come complete with an explanation of what is really going on in economics, if as I have claimed what is going on cannot be viewed as empirical science.

The strength of the rejoinder rests on two undeniable considerations. One is that for all its infirmities, economic theory does at least sometimes *seem* to be insightful. Occasionally, qualitative predictions are borne out, and even more frequently, retrospective economic explanations of events that were unexpected, like a 15 percent reduction in the consumption of gasoline can be given. The second consideration is more abstract, but quite telling as a point in the philosophy of science: There are several scientific theories which to varying degrees fail to divide nature at the joints, and yet they are useful approximations, even if they cannot be reduced to more fundamental theories that do divide nature at the joints (or that are currently believed to do so). For instance, the Mendelian unit of inheritance cannot be reduced to the molecular gene and so does not divide its phenomena at the joints. Yet Mendel's laws are useful approximations that we would be silly to forgo. *Mutatis mutandis* for economic theory. Thus, even if my claims about the intentional vocabulary of economics are correct, they are not sufficient reason to surrender the theory.

It is quite correct that the problems I have noted for economics have parallels in other successful scientific theories. On the other hand these theories have proved successful, on standards of *improving* technological and predictive success that economics has not met. The fact that such theories do not carve nature at the joints, that they are to a degree incommensurable with their successors, or with more fundamental theories, is a problem in the philosophy of science. But it is a different problem from that which faces economics. For economics has not met with anything like the success they have met with, and what is required by this fact is an explanation of why it has not. I claim that part of the explanation is that the descriptive vocabulary fails to divide nature at the joints, so that no improvement of current economic theory can provide laws governing intentional economic activities. A better comparison for economic theory than Mendelian genetics is phlogiston theory, whose failure is traceable to its *incommensurability* with the oxygen theory that superseded it. Phlogiston theory is a scientific dead end, because there is no such thing as phlogiston, because the notion of phlogiston does not divide nature at any joint. Phlogiston is not a natural kind. This is an empirical fact about nature, and the claim that intentional notions are not natural kinds is equally a contingent claim. Thus, economics and phlogiston theory are not methodologically defective. They are simply false.

But, the rejoinder continues, what of the successes of economic theory? How can I square my arcane philosopher's argument with the evident applicability of such staples as the laws of supply and demand? After all, it is a fact about markets in all commodities that *eventually* price will influence demand and supply in the directions that microeconomic theories of economic action dictate. Surely this is an economic regularity and surely it is a consequence of individual choices, preferences, and beliefs.

Bringing what little utility there is in economic theory into harmony with my diagnosis of its ills is a task for which there are many solutions. The first thing to note is this: although the laws of supply and demand and other market-level general statements are deduced from claims about the intentional determinants of individual actions, they are logically separable from such claims, and, more important, they can be shown to follow from assumptions which are the direct denial of these general claims about rational action. From the assumption that individuals behave in purely habitual ways, always purchasing the same or the most nearly similar bundle of commodities available, no matter what the price, the law of downward sloping demand follows, as it does from the assumption that their purchases are all impulsively random.[13] The same can be shown for the choices of entrepreneurs. So surrendering the extremal intentional approach to human behavior does not logically or even theoretically oblige us to surrender these "laws." On the other hand, we cannot sharpen their applicability beyond the most qualitative or generic levels, or quantify the values of their parameters like elasticity, or improve our foresight or hindsight in the employment of these principles. Now the fact that we can usefully employ false or vacuous general statements, up to certain limits, is no mystery in the philosophy of science at all. The clearest instance of such restrictedly useful though false or vacuous general statements is Euclidean geometry. For millennia this axiomatic system was viewed as the science of space, and the great mystery which surrounded it was how we can have the apparently a priori knowledge of the nature of the world that the science of space, Euclidean geometry, gave us. Since Poincaré and Einstein this problem has been largely resolved. Prior to the twentieth century Euclidean geometry was equivocally interpreted as both a pure axiomatic system about abstract objects, one that constituted the implicit definitions of its terms, and was therefore a priori true, and as a body of claims about actual spatial relations among real objects in the world. The equivocation between these two interpretations in part caused Kant's problem of how synthetic propositions could be known a priori. Once distinguished, we came to discover that, interpreted as a theory of actual spatial relations, Euclidean geometry is false, and interpreted as a body of a priori truths implicitly defining the terms that figure in it, Euclidean geometry is vacuous. More important for our purposes, it was shown to be useless and inapplicable as a body of conventions, beyond certain values of distance and mass in space. In retrospect, we can explain why no one ever noticed these facts about geometry, and why before 1919 it proved entirely satisfactory for settling empirical questions of geography, surveying, engineering, mechanics, and astronomy. The reason, of course, is that for these questions we neither needed nor had the means to make measurements fine enough to reveal the inadequacies of Euclidean geometry. When we need to improve our measurements beyond this level of fineness, in contemporary cosmology, for instance, we must forgo Euclidean geometry in favor of one or another of its non-Euclidean alternatives. One way to describe the twentieth-century fate of

Euclidean geometry is to say that its kind terms proved not to name *natural kinds*: nature diverges from the predictions of an applied Euclidean geometry, because it does not contain examples, realizations, and instances of the kind terms of that theory. There are no Euclidean triangles. This is something we came to learn only with the advent of another theory, the general theory of relativity, which not only revealed this fact but also explained the degree of success Euclidean geometry does in fact attain when applied to small regions of space.

Of course, economic theory has attained nothing like the success of Euclidean geometry. But the apparent applicability of some of its claims is to be explained by appeal to the same factors which explain why we can employ, e.g., the Pythagorean theorem, even though there are no Euclidean triangles and no Euclidean straight lines. We can employ the laws of supply and demand, even though human beings are not economically rational agents; that is, we can employ these "laws" even though individuals do not make choices reflecting any empirical regularity governing their expectations and their intentions. We can employ them all right, but the laws of supply and demand cannot be applied with the usefulness and exactitude of the Pythagorean theorem, just because the kind terms of economic theory are different from the real kinds in which human behavior is correctly classified. And this difference is comparatively much greater than the difference between the kind terms of applied geometry and those of physics. There are no Euclidean triangles, but we know why, and we can calculate the amount of the divergence between any physical triangle and the Euclidean claims about it, because we have a physical theory to make these corrections, the very one which showed Euclidean geometry to be factually false. We can make no such improvements in the application of the laws of supply and demand; we can never do any better than apply them retrospectively or generically; we cannot specify their parameters, or their exceptions, because the axiomatic system in which they figure diverges from the facts very greatly, and because we have no associated theory that enables us to measure this divergence and make appropriate corrections for it. This is a difference in kind between Euclidean geometry and economic theory. They differ in applicability only by degree, the predicates of neither pick out natural kinds; but they differ in kind because for Euclidean geometry there is a theory, physics, that enables us to correct and improve the applicability of its implications. There is no such theory that enables us to improve on the applicability of economic theory.

Such a theory is of course logically possible, says a version of cognitive psychology, that provides bridges from economic variables like preference and expectation to independently identifiable psychological states. Such a theory might enable us to actually predict individual economic choices and to correct our microeconomic predictions of them, when these predictions go wrong. It would either enable us to improve microeconomics beyond the level at which it has been stuck for a hundred years, or it would show that the determinants of human behavior are so orthogonal to the theory's assumptions about them that microeconomics is best given up altogether. In other words, such a theory would show either that economic theory is like Euclidean geometry or that it is like phlogiston theory, or perhaps that it is somewhere in between. But the fact is that no such theory is in the offing, or on the horizon. What is worse, even if it were available, it is not likely to actually deflect practicing economists from their intentional extremal research program. And the reason is not that they are satisfied with the level of success their theory has attained, a level much closer to that of phlogiston theory than to that of Euclidean geometry. Rather, the reason is that they are not really much interested in questions of empirical applicability at all. Otherwise some of the attractive nonintentional and/or nonextremal approaches to economic behavior that are available would long ago have elicited more interest from economists than they have.[14]

My explanation for the failures of economics does economists more credit than several possible alternatives. It does not, for example, simply write off economic theory as the ideological rationalization of bourgeois capitalism; it renders the immense amount of sheer genius bestowed on the development of this theory its due. The explanation does not stigmatize the methods of economists as conceptually confused or misdirected. It isolates the failings of the theory in an empirical supposition about the determinants of human behavior, one that economists share with all of us. But this supposition is false, and so economics rests on a purely contingent, though nevertheless central, mistaken belief; just as the conviction that Euclidean geometry was the science of space rested on a purely contingent, almost equally central mistaken belief that the paths of light rays are Euclidean straight lines.

But, as I have said, the history of economic theory shows that economists cannot be expected to surrender their commitment to an extremal intentional research program, no matter what its empirical inadequacies, And this raises the question that goes together with the rejoinder: If economics does not behave the way a science does, what sort of an activity is it? If economists have not in fact been elaborating a contingent, empirical theory that successively improves our explanatory and predictive understanding of economic behavior, what has this notable intellectual achievement been aimed at? The parallel I have drawn with Euclidean geometry can help answer this question.

Euclidean geometry was once styled the science of space, but calling it a science did not make it one, and we have come to view advances in the axiomatization and extension of geometry as events not in science, but in mathematics. Economics is often defined as the science of the distribution of scarce resources, but calling it a science does not make it one. For much of their histories, since 1800, advances in both these disciplines have consisted in improvements of deductive rigor, economy, and elegance of expression, in better axiomatizations, and in the proofs of more and more general results, without much concern as to the usefulness of these results. In geometry, the fifth axiom, the postulate of the parallels, came increasingly to be the focus of attention, not because it was in doubt, but because it seemed so much more ampliative than the others. The crisis of nineteenth-century geometry was provoked by the discovery that denying the postulate of the parallels did not generate a logically inconsistent axiomatic system. Thus the question of the cognitive status of geometry became acute. Some, following Plato, held it to be an intuitively certain body of abstract truths. Some, following Mill, held it to be a body of empirical generalizations. Others, following Kant, viewed it as a body of synthetic a priori truths. Matters were settled by distinguishing between geometry as a pure axiomatic system, composed of analytic truths about abstract objects with or without real physical instances; and geometry as an applied theory about the path of light rays, which was shown to be false for reasons given in the general theory of relativity. Moreover, the abstract and apparently pointless exercises of nineteenth-century geometers in developing non-Euclidean geometries turned out to have an altogether unexpected and important empirical role to play in helping us understand the structure of space after all. For they apparently describe the real structure of space in the large. Of course pure geometry, both Euclidean and non-Euclidean, has continued to be a subject of sustained mathematical interest, and both have had applications undreamed of eighty years ago.

Compare the history of economic theory during the same time. Unlike physical theory, or for that matter the other social sciences, economics has been subject to exactly the same conceptual pigeonholing as geometry. Some have viewed it, with Lionel Robbins, as a Platonic body of intuitively obvious, idealized but nonetheless correct descriptions of human behavior. Others, following Ludwig Von Mises, have insisted it is a Kantian body of synthetic a priori truths about rationality. Others, like the geometrical conventionalists and following T. W. Hutchison, have derided it as a body of tautologies, as a pure system of implicit definitions without any grip on the real world. Still others, following Mill, have held it to be a body of idealizations of rough empirical regularities. Finally some, following Friedman, have treated it as an uninterpreted calculus in the way positivists treated geometry.[15] But most economists, like most geometers, have gone about their business proving theorems, and deriving results, without giving much thought at all to the question of economic theory's cognitive status. For them the really important question, the one which parallels the geometer's concern about the postulate of the parallels, was whether Walras's theorem that a general market clearing equilibrium exists, that it is stable and unique, follows from the axioms of microeconomic theory. Walras offered this result in 1874, as a formalization of Adam Smith's conviction about decentralized economies, but he was unable to give more than intuitive arguments for the theorem. It was only in 1934 that Abraham Wald provided an arduous and intricate satisfactory proof, and much work since his time has been devoted to producing more elegant, more intuitive, and more powerful proofs of new wrinkles on the theorem.[16] Just as geometers in the nineteenth century explored the ramifications of varying the strongest assumptions of Euclidean geometry, economists have devoted great energies to varying equally crucial assumptions about the number of agents, their expectations, returns to scale and divisibilities, and determining whether a consistent economy—a market clearing equilibrium—will still result, will be stable, and will be unique. Their interests in this formal result are quite independent of, indeed are in spite of, the fact that its assumptions about production, distribution, and information are manifestly false. The proof of general equilibrium is the crowning achievement of mathematical economics. But just as geometry as a science faced a crisis in the 1919 observations that confirmed the general theory of relativity, so too economic theory faced a crisis in the evident fact of the Great Depression. For a long time after 1929 the economists lost the *conviction* that the Walrasian general equilibrium was at least a state toward which markets must, in the long run, move. The main reaction to this crisis was, of course, Keynesianism. Insofar as this extremal theory rests on a denial of the fundamental microeconomic assumption that economic agents' expectations are rational, that they do not suffer from money illusions, that they will tailor their actions to current and future economic environments, Keynesian theory represents as much of a conceptual revolution as non-Euclidean geometry did. Keynes, of course, did not entirely win the field, even during the period when his theory appeared to explain why the market clearing general equilibrium might never be approached, let alone realized. One reason for this is that many economists continued to be interested in the purely formal questions of the conventional theory, quite regardless of its irrelevance to understanding the actual world. These economists were implicitly treating microeconomics as a pure axiomatic system, whose terms may not be instantiated in the real world, but which is of great interest, like Euclidean geometry, whether its objects actually exist. More crucially for the history of economics, there never was and is not yet a theory which can play a role for economics like the role played for geometry by physical theory. Physics enables us to choose between alternative applied geometries, and to explain the deviations from actual observation of the ones we reject. There is no such theory to serve as an auxiliary in any choice between an applied neoclassical equilibrium theory and a Keynesian equilibrium theory. When, in the 1970s, Keynesian theory foundered on empirical facts of joint unemployment and inflation, as unremitting as was the fact of the apparent nonmarket clearing equilibrium of the 1930s, the result was an eager return to the traditional theory. Economists have not forgotten the Great Depression, but their interest in it seems limited to showing that, after all, the Walrasian approach is at least logically consistent with it, something Keynes's earliest opponents could have vouchsafed them. In short, the theory and its development have been as insulated from empirical influences as geometry ever was before Einstein. All this suggests that, like geometry, economics is best viewed as a branch of mathematics somewhere on the intersection between pure and applied axiomatic systems.

Much of the mystery surrounding the actual development of economic theory—its shifts in formalism, its insulation from empirical assessment, its interest in proving purely formal, abstract possibilities, its unchanged character over a period of centuries, the controversies about its cognitive status- can be comprehended and properly appreciated if we give up the notion that economics any longer has the aims or makes the claims of an empirical science of human behavior. Rather we should view it as a branch of mathematics, one devoted to examining the formal properties of a set of assumptions about the transitivity of abstract relations: axioms that implicitly define a technical notion of "rationality," just as geometry examines the formal properties of abstract points and lines. This abstract term "rationality" may have far more potential interpretations than economists themselves realize,[17] but rather less bearing on human behavior and its consequences than we have unreasonably demanded economists reveal.

There are some important practical consequences of this answer to the question of what economics really is. If it is best viewed as more akin to a branch of mathematics on the intersection between pure axiomatization and applied geometry, then not only are several cognitive mysteries about economics solved, but more important our perspective on the bearing of economic theory must be fundamentally altered. For if this view is correct we cannot demand that it provide the reliable guide to the behavior of economic agents and the performance of economies as a whole for which the formulation of public policy looks to economics. We should neither attach much confidence to predictions made on its basis nor condemn it severely when these predictions fail. For it can no more be relied on or faulted than Euclidean geometry should be in the context of astrophysics. Admittedly this attitude leaves a vacuum in the foundations of public policy. For without economics we lose even the illusion that we understand the probable, or potential, long-term or merely possible consequences of choices that policy makers are forced to make. Of course, the caution that loss of illusions may foster is certain to be salubrious. On the other hand the vacuum may attract a really useful foundation for decisions about the economy and its improvement.

NOTES

For helpful comments on earlier drafts of this chapter I owe thanks to Jeffrey Straussman, R. J. Wolfson, and participants in a colloquium at the University of Colorado, Boulder. Research was supported by the American Council of Learned Societies and the John Simon Guggenheim Memorial Foundation.

1. *Microeconomic Laws: A Philosophical Analysis* (Pittsburgh: University of Pittsburgh Press, 1976).
2. Cf. Scott Gordon, "Should Economists Pay Attention to Philosophers?" *Journal of Political Economy,* 86 (1978). His answer is no.
3. Cf., for instance, R. A. Posner, "Some Uses and Abuses of Economics in Law," *University of Chicago Law Review,* 46 (1979), 281–306.
4. Cf. L. Boland, "A Critique of Friedman's Critics," *Journal of Economic Literature,* 17 (1979), 502–22 and M. Willes, "Rational Expectations as a Counterrevolution," *Public Interest* (1980), 81–96 in which Milton Friedman's classic, "The Methodology of Positive Economics," *Essays in Positive Economics* (Chicago: University of Chicago Press, 1953), is rehashed.
5. Exponents of "rational expectation theory" have shown that, on the basis of attributions of microeconomic rationality to individuals operating in an economy regulated in accordance with Keynesian policies, we can expect just what has happened: increasing the deficit or the money supply does nothing but increase inflation, without lowering unemployment. Cf. B. Kantor, "Rational Expectations and Economic Thought," *Journal of Economic Literature,* 17 (1979), 1427–1441.
6. Willes, op. cit., note 4, reprinted in *The Crisis in Economic Theory* (Basic Books, 1982).
7. *The System of Logic* (London, 1867). Mill's views are ably expounded in D. Hausman, "John Stuart Mill's Philosophy of Economics," *Philosophy of Science,* 48 (1981), 362–85.
8. Hal Varian and Alan Gibbard, "Economic Models," *Journal of Philosophy,* 75 (1978), 669–77.
9. Cf. A. Rosenberg, "Obstacles to Nomological Connection of Reasons and Actions," *Philosophy of Social Science,* 10 (1980), 79–91.
10. Cf. D. Dennett, *Content and Consciousness* (London: Routledge and Kegan Paul, 1966) and D. Davidson, "Mental Events," in Foster et al., *Experience and Theory* (Amherst: University of Massachusetts Press and Duckworth, 1970), pp. 79–101.
11. Cf. Paul Churchland, "Eliminative Materialism and the Propositional Attitudes," *Journal of Philosophy,* 78 (1981), 67–90.
12. Cf. A. Rosenberg, *Sociobiology and the Preemption of Social Science* (Baltimore: Johns Hopkins Press, 1980), where I expound these implications in detail.
13. Gary Becker, "Irrational Behavior and Economic Theory," *Journal of Political Economy,* 70 (1962), 1–13.
14. In this space I can mention only three such approaches, but all three undercut the rejoinder's suggestion that we cannot give up the intentional approach to economic behavior. There is first of all the behavioral approach to economic behavior associated with Herbert Simon or Richard Cyert, which forgoes both the generality and the elegance of extremal theories in favor of an attempt to uncover at least rough empirical generalizations about the actual behavior of economic units like the household and the firm. This approach involves treating the economic units as feedback systems, though not intentional ones, which respond to environmental forces as adaptive mechanisms and in turn generate outputs that affect other economic units. Still another approach associated with the work of S. G. Winter involves applying a natural selection model to attempt to account for the behavior of economic agents. Finally, N.

Georgescu-Roegen's insistence on the relevance of thermodynamic approaches to economic processes represents a clear alternative to the conventional attitude. Adopting or adapting any of these alternatives will result in a theory very different from the classical one, and they all show that the extremal intentional research program is by no means unavoidable and inevitable as an approach to economic phenomena. Moreover, so far as preserving whatever might be useful in conventional theory is concerned, these alternatives are explicitly designed to have no less applicability to market phenomena, and even to individual behavior, than neoclassical theory now has. If there is anything in the laws of supply and demand, in the possibility of stable or unique partial or general equilibria, at least some of these approaches are designed to capture it without burdening themselves by commitments to the causal force of preference and expectation operating uniformly and invariably on choice. Cf. R. Cyert and J. A. March, *A Behavioral Theory of the Firm* (Englewood Cliffs, N. J.: Prentice-Hall, 1963) and R. Cyert and G. Pottinger, "Towards a Better Microeconomic Theory," *Philosophy of Science*, 46 (1979), 204–22, and H. Simon, "A Behavioral Model of Rational Choice," *Quarterly Journal of Economics,* 69 (1955), 99–118; S. J. Winter, "Economic Natural Selection and the Theory of the Firm," *Yale Economic Essay,* 4 (1967), 224–72; N. Georgescu-Roegen, *The Entropy Law and the Economic Process* (Cambridge, Mass.: Harvard University Press, 1971); and R. Cyert and C. L. Hendrick, "Theory of the Firm," *Journal of Economic Literature,* 10 (1972), 398–414.

15. Lionel Robbins, *An Essay on the Nature and Significance of Economic Science* (London: St. Martins, 1932). T. W. Hutchison, *The Significance and Basic Postulates of Economics* (London: McMillan, 1938); L. Von Mises, *Human Action* (1949); and Milton Friedman, op. cit., note 4.

16. Leonn Walras, *Elements of Pure Economics,* trans. W. Jaffe (Homewood, Ill.; Irwin, 1954). A. Wald, "On Some Systems of Equations for Mathematical Economics," *Econometrica,* 19, 368–403. For a contemporary version of the proof, cf. Gerard Debreu, *Theory of Value* (New York: Wiley, 1959).

17. Indeed, like apparently useless arcana of nineteenth-century non-Euclidean geometry, the formalism and results of general equilibrium theory are turning out to have applications undreamed of by the economists who have proved the impressive theorems in this system. For their results are being taken over and reinterpreted by mathematical ecology: stripped of their intentional interpretation, they provide proofs and stability conditions for unique stable equilibria, that modern evolutionary biology requires in the development of its own extremal theory of balance and competition in the evolution of the biosphere. Cf., for instance, Oster and Wilson, *The Social Insects* (Cambridge, Mass.: Harvard University Press, 1978), and R. May, *Stability and Complexity in Model Eco-System* (Princeton, N.J.: Princeton University Press, 1973).

10

What Would an Adequate Philosophy of Social Science Look Like?

Brian Fay and J. Donald Moon

I

During the last twenty years an enormous literature has grown up around the question, what is the nature of social science? Two positions have dominated these discussions, the "naturalist" view which holds that social science involves no essential differences from the natural sciences, and the "humanist" view which holds that social life cannot adequately be studied "scientifically." Whole models of social science have been propounded that argue for one position and view the other as an incompatible alternative.[1] Given such a vigorous tradition of discourse, it may seem odd that anyone would now ask the question, what would an adequate philosophy of social science look like? Unfortunately, however, neither naturalism nor humanism is capable of answering the three questions which the idea of a science of behavior raises. These questions are: first, What is the *relationship between interpretation and explanation in social science*? second, What is the *nature of social scientific theory*? and third, What is *the role of critique*?

In this essay we will show why these three questions must be answered by any compelling account of social science, and why humanism and naturalism are unable to answer them. The first question will be taken up in section II, the second in section III, and the third in section IV. By showing that the dualism which dominates current philosophical thinking makes it impossible to answer these questions adequately, we will point to the need for a new synthesis in the philosophy of social science, one that transcends the antimony of humanism and naturalism.

II   *At first glance*

One way of beginning to talk about the nature of social phenomena is to invoke the now familiar prima facie distinction between human action, on the one hand, and mere bodily movements on the other—between raising one's arm and one's arm rising, to use the time-worn example. According to this distinction, actions differ from mere movements in that they are intentional and rule-governed: they are performed in order to achieve a particular purpose, and in conformity to some rules. These purposes and rules constitute what we shall call the "semantic dimension" of human behavior[2]—its symbolic or expressive aspect. An action, then, is not simply a physical occurrence, but has a certain intentional content which specifies what sort of an action it is, and which can be grasped only in terms of the system of meanings in which the action is performed. A given movement counts as a vote, a signal, a salute or an attempt to reach something, only against the background of a set of applicable rules and conventions, and the purposes of the actor involved.

To the prima facie fact that human actions are intentional events in the sense that their identity is a function of their content—what they express or the states of affairs they refer to—and, consequently, that they are characterized by invoking the rules and intentions which define them to be what they are, there are three possible responses. The first of these is to accept this prima facie fact and to try to construct a science of intentional objects in terms of it; this is the intentionalist response.[3] The second is to attempt an analysis of the concepts "intention," "meaning," and "action" in purely observational (usually behavioral) terms, so that one can use these concepts in one's science but in a purified form; this is the tack of the definitional behaviorist.[4] The third response is to accept that one cannot capture the meaning of intentional concepts without reference to mental states such as beliefs and institutional norms such as rules, and so conclude that these concepts are radically defective for scientific purposes; it therefore seeks to develop a science of behavior without using these concepts at all. This is the position of the eliminative materialists who ultimately wish to confine their accounts of language and other social behavior to a purely extensional terminology.[5]

The important thing to realize about the third response is that it requires a radically different approach from anything remotely resembling what is understood to be social science as it is practiced today. Broadly speaking, social scientists seek to offer accounts of events described in terms of their significance; thus, they want to understand why it is that a certain group is *dancing* (and not why the feet of its members are *twitching* in a manner describable in purely spatiotemporal terms), or *voting* (and not why the arms of certain bodies are rising), and they characterize the speech of people in terms of its content rather than in terms of its purely phonetic qualities. However, if the approach of the eliminative materialists came to dominate the science of behavior, it would become a sort of mechanics or neurophysiology whose explanatory concepts would be drawn from the natural sciences.

Of course, this observation does not in itself show that this approach is incoherent, or that it cannot be realized. Just as natural science abandoned intentional concepts—a strategy that was unthinkable to many at the time— so the sciences of human behavior might also be transformed in this way. The question of what might be called the conceptual solvency of a "natural science of man," as well as the problems that would arise in attempting to implement this program, are exceedingly interesting ones. However, in some sense they lie outside the boundaries of our inquiry, just because we are trying to offer an account of the many forms which *social* science now takes. It is for this reason that we feel justified in setting this position to one side.

Nor does the second response seem to be adequate. It appears to be the case that any attempt to translate intentional concepts, which involve reference to such things as rules and beliefs, into dispositional terms, which specify a set of dispositions to engage in overt movements under particular stimulus-conditions, is bound to fail. No matter how one tries to construe these concepts, it is ultimately necessary to employ another intentional concept in order to explicate its meaning.[6]

Take, for example, the statement, "Jones asked the cashier to deposit the money into his account"; the concepts "cashier," "deposited," and "money" are all prima facie intentional, in that their meanings involve certain rules (a cashier is a person who has a certain role to play in a specific institution, with certain duties and orders to follow), beliefs (in order to deposit the money, the cashier must believe that Jones has an account), and desires (Jones must want to put his money into his account in order for it to be said that he deposited it). Now, to take one of these intentional concepts, a behaviorist might argue that the beliefs involved in making a deposit can be explicated in purely dispositional terms. A line he might take is this: when a certain sound is made in the cashier's presence ("Do you think that Jones has an account?"), he will produce another ("yes"). But this construal is adequate only if the person understands the question, and understanding is an intentional state. The behaviorist, of course, may then try to give a nonintentionalist account of understanding a question—for example, that a person may be said to understand a question if he or she is able to answer it correctly most of the time. But this account also involves an intentional object, since an answer is correct only in terms of certain rules indicating what is appropriate and what is not. And so the discussion will proceed, until gradually it will become clear that what is wrong is not simply this particular attempt to reduce intentional concepts to nonintentional ones, but that there is something in principle wrong with the whole definitional behaviorist program.

Thus, we are left with the intentionalist response to the prima facie meaningful character of human actions, mental events, and social institutions. The question then arises, what implications does this have for social science? The most obvious task which an intentionalist perspective imposes on the study of human action is the need for interpretation. In order to study human behavior as meaningful performances, we must grasp the meanings expressed in speech and action, and this requires that we understand the system of concepts, rules, conventions, and beliefs which give such behavior its meaning. This is the doctrine of understanding, or *verstehen*, which figures as a prominent methodological principle in the humanist account of social science. It marks an essential methodological difference between the human sciences and the study of nature, expressing itself most clearly in the principles of concept-formation appropriate to each. Briefly, concept-formation in the natural sciences is governed by two related sets of considerations—those of theory, and those of measurement. We require that concepts be developed which permit the formation of testable laws and theories, and other issues—e.g., those deriving from ordinary language—may simply be set aside. But in the human sciences there is another set of considerations as well: the concepts we use to describe and explain human activity must be drawn from the social life that is being studied, and not from the observer's theories, at least in the first instance. Because the very identity of a particular action depends upon its meanings for the social actors, the concepts we use to describe it must capture this meaning.

Another way of putting this point is to say that concepts bear a fundamentally different relationship to social phenomena from that which they bear to natural phenomena. In the social sciences, concepts partially constitute the reality we study, while in the latter case they merely serve in describing and explaining it. As Winch has argued, something can be an "order" only if the *social actors* involved have the concept of an order, and such related concepts as obedience, authority, et cetera; but the natural event of lightning is the same whether it is conceptualized as an expression of Zeus's anger, or as an atmospheric electrical discharge: its identity is not a function of its meaning or intentional content.[7]

The interpretation of the meanings of actions, practices, and cultural objects is an extremely difficult and complicated enterprise. The basic reason for this is that, as Wittgenstein has shown, the meaning of something depends upon the role which it has in the system of which it is a part. To understand a particular action, we must grasp the beliefs and intentions which motivated it, and this further requires that we know the social contexts of practices and institutions which specify what the action in question "counts as," what sort of an action it is. To return to our check depositing example: in order for the social scientist to know what the overt movements he observes actually mean, i.e., to understand what action is being performed, it is necessary that he have an understanding of the beliefs, desires, and values of the particular people involved. But in order to understand these, he must know the vocabulary in terms of which they are expressed, and this, in turn, will require that he know the social rules and conventions which specify what a certain movement or object will count as. Moreover, in order to grasp these particular rules, he will also have to know the set of institutional practices (in this case, those of banking) of which they are a part, and how these are related to other practices of the society (in this case, the institutions of a money economy).

Nor can our scientist stop here. For the conventions of a social group, as Taylor has convincingly argued,[8] presuppose a set of fundamental conceptualizations or basic assumptions regarding man, nature, and society. These basic conceptualizations might be called the "constitutive meanings of a form of life," for they are the basic ideas or notions in terms of which the meanings of specific practices and schemes of activity must be analyzed. For example, the social practice of banking can only occur given the shared constitutive meanings of (say) some conception of property, some notion of being a unit with a particular identity, some idea of exchange value. An adequate account of the practices of a particular society, by setting out the basic ideas and conceptualizations which underlie these practices, will show how various aspects of the social order are related to each other, and how (or the extent to which) the social order constitutes a coherent whole.

The need for such a high level of interpretation may be missed if one focuses one's attention only on studies of one's own culture by other members of it. For in these situations, the scientists do not have to make explicit their interpretive scheme in order to identify and characterize the class of actions and institutions in which they are interested. They, as well as their readers, already know what banks are and what depositing funds means. However, this point should not be pushed too far, because the sorts of implicit self-understandings which we have as practitioners are generally going to be inadequate for the tasks of social science. This is the reason why some of the very best work in social science will partially consist in explicating the sets of shared rules and constitutive meanings which underlie quite ordinary, everyday practices. (Here we are thinking of such works as Beer's *Modern British Politics,* Douglas's *The Social Meanings of Suicide,* Cicourel's *The Social Organization of Juvenile Justice,* and Goffman's *Asylums.*)[9]

Impressed by the elegance and penetration of interpretive theories, humanist philosophers of social science have assumed or argued that interpretation is all there is. They have gone from the correct observation that social theories must be interpretive, to the incorrect conclusion that they can *only* be interpretive. For social phenomena do not consist in abstract structures of meanings which can be set forth and analyzed, but they consist in actions (and other events) which actually occur in particular places at particular times. And, while we cannot even approach our subject without understanding what these actions *mean,* such understanding does not, by itself, constitute an explanation of why they *occur.* To know, e.g., what someone said, and what it means, is not to know why he or she said it.

Accounts of why something happened are commonly said to be causal explanations, for they explain why it occurred by setting out what led it to happen. In the case of actions, e.g., we explain why an agent does something by pointing out the motives, or purposes which led him or her to do it. Thus, for example, Weber explained the type of behavior typical of capitalists in the sixteenth and seventeenth centuries by citing the set of religious beliefs and desires which caused certain sorts of Protestants to act in this manner.

One of the principal tenets of humanism over the last twenty years [1957-1977] has been that beliefs, purposes, values, desires, and so forth—reasons, for short—cannot be causes, and that therefore there is no real "explanation" in social science but only a further form of interpretation in which the scientist tries to uncover the rationale or warrant for the actions in question.[10] But such arguments are now generally recognized to have been inadequate because, while reasons cannot be causes (they *are* utterly different sorts of things), the having of reasons, the believing in reasons, the giving of reasons, et cetera, are all psychological events and, as such, nothing prevents them from figuring in causal explanations.[11] (We say this though we are aware that in order to actually detail the nature of these causes one needs to develop a philosophy of mental events which will do justice to their peculiar qualities, e.g., their having an intentional content, and their very close relationship to overt behavior. Unfortunately, this sort of philosophical analysis has been strangely omitted in most discussions of action theory and its relationship to the philosophy of social science, even by those who advocate a causalist position.)[12]

Moreover, social scientists are interested in explaining a great many phenomena other than actions. They want to explain why it is that people have certain beliefs and values (as in the sociology of knowledge); to account for patterns of unintended consequences of actions; to discover why a social structure arose in the first place, and why it continues to exist despite a changing membership; and so forth. In these, and in all the other questions in which a social scientist is interested, the form of explanation is causal. For in each of them what is required as an explanation is the identification of the necessary and/or sufficient condition or events which produced the phenomena in question.

We will return to the question of causal explanations in social science in a moment when we come to discuss theory in social science, but already enough has been said to demonstrate that social science is an explanatory enterprise as well as an interpretive one. And so the questions which immediately arise are: What is the relationship between interpretation and explanation in social science? How does one influence or restrain the other? How do the criteria for a good interpretation fit with the criteria for a good explanation? These questions arise just because social science is the systematic scientific study of intentional phenomena. Because humanists have failed to appreciate the explanatory task of social sciences (i.e., they have failed to see in what way these disciplines are scientific), and because naturalists have misunderstood the crucial role which interpretation plays in the social sciences (i.e., they have given insufficient or misleading analysis of what it means for a phenomenon to be intentional), both of them have neglected such questions. This is one reason why the current traditions in the analytical philosophy of social science are not only inadequate but, given their terms of reference, incapable of getting onto the right track.[13]

III

The dichotomy between humanist and naturalist also makes it impossible to answer the second question which is critical for a science of behavior: What is the nature of social-scientific theory? For many writers in the humanist tradition, particularly as represented in recent analytical philosophy, the question scarcely seems to exist; one can look in vain in the work of Louch, Winch, Taylor, or von Wright—to mention the most important humanist statements of the last fifteen years—for even a mention of social-scientific theories, let alone a discussion of them.

The reason for this is not hard to find. From the humanist perspective, there is neither a need for theories nor a place for them in the study of society. (At least this is true if we understand "theory" to refer to systematic, unified explanations of a diverse range of social phenomena.) There is no place for theories in the humanist position because its cardinal point is that social science is simply interpretive: it seeks to provide us with an understanding of the meanings of particular actions or practices of a given society. As we have already shown, such understanding may require that we grasp the worldview of the society or culture in question, and elaborate and sophisticated intellectual structures may be necessary to do so.

But an account of a society's worldview, or its intersubjective or constitutive meanings, is not a theory which explains why the society has the institutions it has, or why certain processes of social change occur, or why it is characterized by certain regularities, or why people of a certain sort perform particular kinds of actions. To explain such phenomena we need theories that are, broadly speaking, causal, and the fixation of the humanist tradition with the meaningful dimension of human action has prevented it from developing an account of this kind of social-scientific theory.

This failure to give an account of explanatory theory has proved a particular embarrassment to those espousing the humanist case because it has meant that they have failed to deal with just those aspects of social-scientific work which are of paramount importance to many of its practitioners, and which constitute some of its most conspicuous successes. The clearest example of this is Keynesian economic theory; but all the social sciences possess theories of one sort or another. Thus, kinship theory in anthropology, exchange theory in sociology, the theory of transformational grammar in linguistics, modernization theory in political science, and cognitive dissonance theory in psychology are all examples of the theoretical dimension operative in modern social science. Although humanism is popular among analytical philosophers, naturalism is still the dominant position among social scientists; one of the reasons for this is that the antitheoretical stance of the humanist model has made it appear patently deficient and even irrelevant to those actually engaged in doing substantive social scientific work.

Social science must be theoretical because one of its aims is to give causal explanations of events, and even singular causal explanations require some sort of general law or laws. To say that an event, x, causes another event, y, is to say (speaking very roughly) that x's occurrence is a necessary and/or sufficient condition for the occurrence of y. The idea that the occurrence of one event is a condition for the occurrence of the other distinguishes causal statements of the form, "Under C, x caused y," from mere statements of conjunction of the form, "Under C, x occurred and then y occurred." But this is to say that when we give causal explanations we are implicitly asserting that, whenever an x-type event occurs under conditions C, a y-type event will also occur, which is to say that causal explanations ultimately rest on general laws.

This does not mean that we must actually be able to state a law in order to offer a valid causal explanation, for we may have good reasons for believing that two events are causally related even though we cannot provide the appropriate covering law. Indeed, it may even be the case that we will not be able to state the covering law until we redescribe the events in question in the language of some theory.[14] Thus, we may be warranted in explaining the decrease in the mass of a piece of wood by its having been burned, even though we cannot state the general law upon which this explanation rests, and even though we would have to redescribe this event in terms of the theory of oxidation before we could do so. In these cases we must justify our claim by presenting reasons to believe that there is a causal law operating here. Such reasons will consist of reports of other instances in which the two events are conjoined, together with evidence that the relationship is actually a causal one. Such evidence could include our ability to manipulate the putatively causal variables so as to bring about or suppress the effects in question, and/or a specification of the causal mechanisms by which one event produces the other.

In explaining the occurrence of one event or condition in terms of another, it is not sufficient merely to offer a generalization reporting the covariance of these two events. Rather, what we require is a general statement that is lawlike in the sense that it explains its instances. Take, for example, a social scientist trying to explain why it is that in Western Europe support for totalitarian parties is inversely related to education. In the first instance, he may attempt to explain this finding with the observation that, in Western Europe, less educated people tend to have authoritarian personalities, and with the generalization that people with authoritarian personalities support authoritarian political movements. Here he tries to offer an explanation by showing that the phenomenon in question is an instance of a deeper, generally recurring pattern by embedding descriptions of it in higher-level generalizations. However, there is something problematic about this putative explanation, and that is the status of the generalizations it contains. For it immediately leads us to ask what it is about people with authoritarian personalities which leads them to support antidemocratic parties. Is it just a coincidence that they do, or is their behavior somehow necessitated by their having the kind of personality they have? If we could change a person's personality, would his or her political preferences change as well? In short, unless the general statement is not simply an empirical generalization, but what we have called a nomic generalization, so that it can support contrary-to-fact and subjunctive conditionals, then this account is not a genuine explanation. A generalization, we might say, cannot serve as the required basis for making a causal explanation unless it can explain its instances, and it cannot explain its instances unless one can give an account of *why* the generalization holds.

It is precisely at this point that theories are required, for it is in terms of theories that such an account can be forthcoming. (This is the reason why it is said that causal generalizations must be theory-impregnated.) Theories provide a systematic account of a diverse set of phenomena by showing that the events in question all result from the operation of a few basic principles. A theory goes beyond particular generalizations by showing why the generalizations hold, and it does this by specifying the basic entities which constitute the phenomena to be explained, and their modes of interaction, from which the observed generalizations can be inferred. Thus, a theory not only provides unity and coherence to a field of inquiry, but it also gives us the grounds required for asserting subjunctive conditionals, or the reasons for believing that these generalizations are, in some sense, necessary. Thus, we are inevitably led from the need to explain particular occurrences to the need for social theories.

Moreover, as a social science attempts to become more rigorously scientific, it will naturally attempt to organize and structure its various particular causal explanations and the relatively specific nomic generalizations upon which they rest by systematically interrelating them, and by subjecting them to experimental and other empirical verification. In this process, the self-conscious development of "large-scale" theory is absolutely crucial, and it is for this reason that the sciences of behavior have developed the social theories of extremely wide scope and power which we have already mentioned.

However, if the humanists fail to provide an account of social theories, we fare little better at the hands of the naturalists. Of course, the naturalists spend a great deal of time talking about scientific theories, but they analyze scientific theories in general, and give little attention to the specific problems of *social* theories. For the naturalists, the human sciences and the natural sciences share the same methodology, and so there is no need to discuss social theories apart from physical theories: what can be said of the latter applies a fortiori to the former. And since theories in the physical sciences are far more elaborate and developed than theories in the social sciences, discussion of the nature of theories is usually focused on physical theories. Moreover, when social theories are discussed in the naturalist tradition, it is often to set forth their deficiencies in terms of the naturalist ideal, rather than to analyze them in their own right.[15]

This is not a happy situation, however, because theories in the social sciences are needed to explain phenomena which are different from those in the natural world—they are intentional—and therefore we cannot assume that they will have the same structure as, or be similar in all important respects to, theories in the natural sciences. Indeed, there are at least three ways that it can be seen that theory-construction in the social sciences faces problems quite unlike theory-construction in the natural sciences.

In the first place, because intentional actions are rule-governed, they have an irreducibly normative character. Speech acts, e.g., are performed in accordance with linguistic rules, and so they can be assessed as correct or incorrect. Similarly, instrumental actions can be assessed as more or less rational depending on the extent to which they are likely to realize their intended aims. Because of this normative character of action, a distinction can be made between the *competence* of an actor and his or her *actual performance.* An actor's competence is his or her mastery of the rules (or norms of rationality) which apply to a particular area of activity; performance, on the other hand, refers to the person's actual behavior, which is determined not only by his or her competence, but also by such other factors as fatigue, inattention, misperception, learning failure, and the like.

Now, corresponding to this distinction between competence and performance is a distinction between two types of theory. A theory of competence is designed to explain the competence of an actor, or, more likely, the competence of an idealized actor who is perfectly rational, or has perfectly mastered the relevant rules. A theory of performance, on the other hand, while perhaps making use of, or presupposing, a theory of competence, is designed to explain what a person actually does, and so it would encompass all of the causal factors which bear upon behavior. In his theory of transformational grammar, for example, Chomsky attempts to set forth the basic rules or principles which generate all grammatically well-formed sentences of a language, and only such sentences. An adequate theory, then, would model the (idealized) native speaker's mastery of his or her language, his or her (potential) capacity to recognize well-formed utterances.[16] Similarly, modern economics is based upon a theory of choice which sets forth the rules which must be followed by an ideally rational actor in different kinds of choice situations.

In both linguistics and economics there is considerable controversy over the role of these theories of competence in accounting for actual performance, since it is not obvious what relevance such idealized accounts could have for explaining the behavior of any particular actor. But what is crucial for our purposes is the failure of naturalists to recognize this problem at all. Because they take social theories to have the same structure as physical theories, they are not even able to ask what the relationship is between theories which model competence and causal theories which explain overt behavior.

A second problem for theory construction that is unique to the human sciences is the relationship between the concepts and principles which the scientist uses to account for social phenomena, and those which inform the actions and beliefs of social actors. As we have already shown, because social phenomena are intentional, their very identity depends upon the concepts and self-understandings of social actors, and so in order to explain social behavior social scientists are constrained to use the actors' framework. If social scientists wish to go beyond these self-understandings by introducing concepts and principles which may be at variance with them, they face the problem of relating these new principles to those employed by the actors themselves. Failure to make this relationship would result in the scientists' failing to capture the phenomena they wish to explain, since the events in question would slip through the conceptual net the scientists had constructed.

Nor is this the problem that natural scientists face in giving empirical content to theoretical terms and principles by means of correspondence rules or bridge principles, since the concepts which the actors employ are no more "empirical" or "observable" than the concepts of the scientist: in terms of the distinction between theoretical and observational terms, concepts such as "belief" or "decision" are as "theoretical" as concepts such as "social structure" or "national income." Moreover, the problem of theoretical interpretation in the social sciences may not simply be one of developing bridge principles which specify, in part, what it is that theoretical terms refer to, or how statements employing theoretical terms can be tested. Rather, it may be a matter of establishing that different behaviors have the same or similar meanings, as when aspects of American Halloween customs are shown to be similar in meaning to the myths and rituals of the Kayapo, a people living in the Amazon basin.[17] Once again, because social theories are theories of intentional objects, they pose problems for analysis which cannot be grasped merely from an understanding of theories of physical things.

A third problem for theory construction which the naturalists have failed to discuss is the nature of paradigms or research programs in the social sciences. For a long time, of course, this was due to the dominance of positivism within the philosophy of science, and so it reflected a failure to discuss the conceptual presuppositions of scientific or theoretical work within all branches of empirical science. In the past decade or so this deficiency has been corrected, particularly with the seminal work of Kuhn and Lakatos. Lakatos's account of science in terms of the concept of a research program is of particular importance, for the touchstone of his work is not the history of actual scientific developments, as it was for Kuhn, but the logic of science itself. A research program, according to Lakatos,[18] sets out the fundamental conceptual framework or conceptualization of the phenomena we wish to explain, and the rules in accordance with which theoretical innovations or developments will be made. We require such rules, Lakatos argues, because we must have criteria which can be used to recognize adjustments to a theory which are essentially ad hoc, or unconnected with the rest of the theory. Theoretical developments which are ad hoc must then be rejected, for they do not represent genuine scientific progress.

Given that we recognize the need for research programs for theory construction in the social sciences, the question immediately arises whether the intentional nature of social phenomena constrains what can count as an adequate research program. For to identify a phenomenon as intentional is to identify it as something which was brought about for some reason: it is part of what we mean by "intentional" that it was done for a reason or purpose. And so describing something in intentional terms is implicitly to make an explanatory claim. If this argument is correct, it suggests that an adequate explanation of a social phenomenon would have to include, or be based upon, an account of the reasons or motivations which led to the behavior which brought about the phenomenon in question. If this is the case, then research programs in the social sciences would have to include a conception of human needs, purposes, rationality, et cetera, in terms of which these motivational accounts could be constructed. Research programs, such as certain versions of systems theory, which dispensed entirely with the motivations and orientations of social actors, could be dismissed as inadequate to explain intentional phenomena.[19]

The purpose of this discussion of research programs in social science is not to show that they must have some particular form. Rather, it is to point to yet another problem that is distinctive to theory construction in the social sciences, and which the naturalist tradition in the philosophy of social science cannot address, let alone solve. Until we can transcend the sterile antinomy between naturalist and humanist in the philosophy of social science, we will be completely unable to provide an adequate account of the nature of social theory.

IV

To this point we have argued that the self-understandings which people have play a causal role in bringing about the behavior in which they engage. Now this fact has often served as the basis for humanist philosophers and social scientists to make the further claim that explanations of social behavior consist solely of reconstructions of these self-understandings. Since actions are events that occur because they aɾᴇ warranted by the beliefs and desires of the actor, the task of explaining them is thought to consist of laying out the structure of reasons which justifies them. According to the humanist model, social science grasps the intelligibility of a particular form of behavior by making explicit the conceptual links that, it is hypothesized, implicitly exist between various sorts of activities, institutions, and psychological states like beliefs and desires. A good interpretation, then, is one which demonstrates the coherence which an initially unintelligible act, rule, or belief has in terms of the whole of which it is a part.

Humanists often draw two important conclusions from this construal of social science. The first is that the social scientist must assume that the beliefs, practices, and actions which he encounters are congruent with one another insofar as they are explicable. The second is that, since it is the conceptual linkages between the actors' beliefs, actions and practices which he must uncover, the explanations which the social scientist puts forward must employ essentially the same concepts which an ideal, fully informed, and articulate participant would give.[20] Both of these conclusions support a view of social life which takes it to be, by definition, rational at some level and understandable in its own terms.

Unfortunately, such a view is woefully inadequate just because it ignores crucial elements of social experience which are obviously present in social life, and which are often studied by social scientists. These include cases in which people's self-understandings are at variance with their actual situation and behavior, or in which a specific belief and action system is incompatible with other norms of the culture, or in which there are endemic conflicts as the result of conflicts in social-structural principles.[21]

In short, people may systematically misunderstand their own motives, wants, values, and actions, as well as the nature of their social order, and—given what we have said about the constitutive role of self-understandings in social life—these misunderstandings may underlie and sustain particular forms of social interaction. In these situations, the actors' ideas may mask social reality as much as reveal it, and so the social scientist cannot confine himself to explicating the way in which the actors' concepts and self-understandings form a coherent system. In order to understand these cases, the social scientist must recognize how the actors' self-understandings are incoherent, and he must show what consequences these incoherencies have.

Concrete examples of social phenomena which cannot be understood in their own terms include the idea of nobility in feudal society, and the witch craze of early modern Europe. The concept of "nobility," as Gellner has pointed out,[22] was used to legitimate rulership in feudal society: one is entitled to rule because one is "noble" or virtuous. But, at the same time, a person was a member of the ruling class, or a "noble," simply by virtue of birth, not personal merit. Thus, the concept of nobility is at best equivocal, if not thoroughly incoherent, and the failure to notice and correct this incoherency is a misunderstanding or confusion that is a condition of the feudal form of political domination.

The European witch craze is also a social phenomenon that is, in many significant respects, irrational. As Trevor-Roper has argued,[23] the belief in "witches" was not necessarily irrational in the intellectual context of the time, but the belief in witches did not cause the witch craze. What was distinctive about the witch craze, and what requires explanation going beyond the self-understandings of the actors involved, are such factors as the ferocity of the persecutions, the sudden and dramatic increase in the number of putative witches who were discovered and condemned, the geographic and social patterns of persecution, and the widespread use of torture. By focusing only on the concepts available to the actors involved, we could not explain these phenomena adequately: it would certainly not do to say that the cause of the witch craze was the fact that the number of witches has dramatically increased! Moreover, in doing so we would also fail to set the witch craze in the context of the social tensions of the time, and we would fail to see how it involved a process of scapegoating that served to deflect social discontent. By focusing only on witchcraft in terms of the system of beliefs and values of which it was a part, we would miss much that is essential to the social reality of the witch craze.

Irrational social phenomena, unfortunately, are quite common. Consider, e.g., sociologists' and psychologists' attempts to uncover the "real meaning" of neurotic behavior (like compulsive handwashing), of violent prejudicial behavior toward minority groups, of recurring self-destructive patterns of social interaction, and so forth. Moreover, in situations such as these, the particular form of irrational behavior may not be just an isolated feature of a person's life, but may instead be systematically related to a wide range of different emotions, beliefs, and actions. The very basis of a person's life—the terms in which he talks about himself in his most lucid and reflective moments, and the fears, aspirations, beliefs, passions, and values which he ascribes to himself at these times—may be fundamentally mistaken, and, as a result, he may be unable to adequately explain his behavior to himself or others. Worse than this, as a result of such misunderstanding he may pursue ends he cannot achieve, and the goals he does reach may not be satisfying. Such frustration may lead him to intensify his efforts, and so to perpetuate his misery. And just as it is possible for a person to be systematically mistaken, so whole forms of life may be based upon such self-misunderstandings, or what might be called "false consciousness." This is the picture of life that is painted for us by Rousseau, Hegel, Marx, and, more recently, by Freud, Brown, Habermas, Becker, and a host of others.

The social scientist attempts to explain such irrational phenomena by treating the actors' beliefs and desires as ciphers for something else that constitutes the actors' actual reason for acting, or the real need which they are trying to fill. Thus, according to Rousseau, people desire wealth, but what they really want is social distinction, and money is an expression of social distinction in certain societies.[24] Similarly, according to Marx, people engage in religious practices because they desire to be complete and whole human beings, and they believe that God will provide that fulfillment; but God is really nothing more than a picture of themselves fully actualized, and what would really satisfy them is to develop and exercise their productive capacities in forms of cooperative, social labor.[25] Finally, to offer a third example, Becker argues that people pursue sexual romance and contact, because sex is a cipher for everlasting life, and what they really want is to overcome the fear of their own death.[26]

Such accounts of human motivation and behavior immediately lead to the question, How is it possible for people to be so ignorant and confused about their own needs and motives, thereby leading them to engage in destructive and frustrating activities? To answer this question we must have an account of what causes people to mistake some purpose or object (wealth, God, sex), for what they really want (social distinction, happiness, eternal life), and how these delusions are maintained. Freud's notions of sublimation and repression, and Marx's notions of alienation and ideology, are examples of concepts created in order to explain the process by which an activity acquires symbolic import and with it causal power, and how this process itself is hidden from the agent's view.

Thus, systematic misunderstandings of the meanings of one's activities, reinforced by repressive mechanisms, can result in irrational behavior whose upshot is social conflict and the experience of frustration. And this is the case just because human behavior is intentional in the sense of being undertaken on the basis of the ideas, desires, and perceptions of those who perform it. But in these situations the traditional humanist goal of understanding intentional phenomena by grasping the coherence which exists among their meanings must be replaced by the need to critique these phenomena. Or better, the only way to understand such a social situation is to engage in a critique in which one lays bare the ways in which the ideas people have of themselves mask the social reality which their behavior creates, and in which one tries to demonstrate that the coherence of the relevant behavior occurs at a level so deep that it is beyond the capacity of the actors to appreciate it given the conceptual and emotional responses open to them. In doing this, the social scientist will undoubtedly have to make use of concepts and conceptual distinctions which in a basic way go beyond those operative in the social life which is being studied. It is in this way that the humanist model will be transcended.

Of course, the naturalist model will be of no help in this matter either. For though the naturalists have always been insistent that social theorists need not be confined to the categories of thought of the people they are analyzing, there is nothing in the natural sciences comparable to assessing the rationality of a particular belief system, institution, or system of actions, and deciding on a certain type of explanation depending on this assessment. Only of an intentional phenomenon can one ask: Are the factors which support it mistaken? Could it have been undertaken out of ignorance? What role does deception play in its continuation? And so forth.

The humanist cannot appreciate the role of critique in social science because he artificially confines himself to interpreting the meanings which various aspects of a social life are supposed to have by grasping the coherence which he thinks exists between these aspects understood in their own terms. By so confining himself, he not only ensures that he will fail to see the conflict, irrationality, and mechanisms of repression operative in all social orders, but also deprives himself of the means necessary to understand these phenomena, namely, a categorical scheme which allows him to speak about the relevant social order in terms radically opposed to that of the participants. The naturalist, on the other hand, cannot give an account of critique because, by neglecting the particular features of intentional phenomena, he cannot appreciate the crucial role which rationality plays in social life, or its assessment plays in social science. These inadequacies of both the humanist model and the naturalist model in elucidating the role of critique in social theory give a third reason why the dualist approach of humanism versus naturalism must be overcome if a satisfactory philosophy of social science is to be forthcoming.

V

In this essay we have not tried to set out a philosophical account of social science, but to show that neither of the two prevailing accounts is adequate. An adequate philosophy of social science must be capable of answering the three questions we have discussed: first, What is the relationship between interpretation and explanation? second, What is the nature of social scientific theory? and third, What is the role of critique in social science? Broadly speaking, these questions arise because of the conjunction of two important features of social science. In the first place, these sciences are *social*, which is to say that the phenomena they study are intentional phenomena, and so must be identified in terms of their meanings. Secondly, these sciences are *sciences,* in the sense that they try to develop systematic theories to explain the underlying causal interconnections among phenomena of a widely divergent sort. Because they each fasten on only one of these features, humanism and naturalism fail to provide an adequate account of social science.

This does not mean, however, that we reject both of these traditions of thought entirely. On the contrary, these philosophical metatheories are partial realizations of the task of giving an account of social science. What is wrong with them is not that they are false, but that they are one-sided. Indeed, as we have suggested throughout our analysis, these two positions can be reformulated in such a way as to render them compatible, and their insights complementary. By showing just where humanism and naturalism are inadequate, we hope to have contributed to the construction of the framework for a new synthesis in the philosophy of social science.

NOTES

The essay was written while the authors were receiving a Joint Research Grant from the National Endowment for the Humanities (RO-2210675-139). The authors wish to acknowledge the importance of the free time which this grant made possible, and to publicly thank the National Endowment. The argument presented here does not necessarily represent the view of NEH. An earlier version of this essay was read at Williams College.

1. See Maurice Roche, *Phenomenology, Language, and the Social Sciences,* London 1973, and G. H. von Wright, *Explanation and Understanding,* Ithaca, N.Y. 1971, for recent examples of this opposition between these two models of social science.

2. Following von Wright, p. 6.

3. The intentionalist response may be adopted by a broad spectrum of what otherwise might be strange bedfellows, including humanists (such as phenomenologists) and naturalists who are not at the same time explanatory reductionists (such as those who adopt a functionalist theory of mind). Moreover there are some who think that a simple intentionalist response is inadequate, because the science of man will ultimately be a hybrid science which employs both intentional and extensional terminology (much as computer science today employs both the language of programming and the language of electronics); for this, see Daniel Dennett, "Intentional Systems," *Journal of Philosophy,* 68, 1971. Of course, in this case the problems associated with an intentionalist analysis would still remain.

4. The classical statement of such a position is G. Ryle, *The Concept of Mind,* New York 1949. Behaviorist psychology is based on this response; see B. F. Skinner, *Verbal Behavior,* New York 1957. In *The Nature of Cultural Things,* New York 1964, Marvin Harris advocates such a program for anthropology.

5. See W. V. O. Quine, *Word and Object,* Cambridge, Mass. 1960. Actually, the third response is more varied than it might at first appear. For, on the one hand, it characterizes those who think that the task of a science of man is to discover the contingent identities between those states and events now characterized in intentional terms and these same states and events designated in purely physical terms, such that a mapping of one terminology into the other via their supposed common extension, followed by the replacement of the intentionalist vocabulary by the physicalist one, would characterize the development of human sciences. (This is the view of the Identity theorists; see D. Armstrong, *A Materialist Theory of Mind,* London 1968.) And,

on the other hand, it also characterizes those who believe that the intentionalist idiom should be abandoned altogether in the construction of a science of man. See P. K. Feyerabend, "Mental Events and the Brain," *Journal of Philosophy*, 60, 1963.

6. See Roderick Chisholm, *Perceiving*, Ithaca, N.Y. 1957, ch. 11.

7. Peter Winch, *The Idea of a Social Science*, London 1958, p. 125.

8. Charles Taylor "Interpretation and the Sciences of Man," *Review of Metaphysics*, 25, 1971.

9. It is in anthropology that the need for interpretation is most obvious, because the anthropologist does not have an "insider's" implicit understanding of the society he studies, and so he must develop an explicit scheme of the whole in his work. It is, thus, no accident that it is in anthropology that the interpretive enterprise is most highly developed.

10. See, e.g., A. I. Melden, *Free Action*, London 1961, passim.

11. The classic statement of this position is Donald Davidson, "Actions, Reasons, and Causes," *Journal of Philosophy*, 60, 1963.

12. The exception to this is Arthur Danto, *Analytical Philosophy of Action*, Cambridge 1973, and especially the writings of those who propound a functionalistic theory of mind, such as J. Fodor, *Psychological Explanation*, New York 1968; Daniel Dennett, *Content and Consciousness*, London 1969; and the essays by Hilary Putnam in his *Mind, Language, and Reality*, Cambridge 1975.

13. This was, of course, Max Weber's position, especially in his "Critical Studies in the Logic of the Cultural Science," and in this respect we feel a return to Weber would be a progressive step in the philosophy of social science. We say this even though we do not agree with his answers to our questions.

14. See Donald Davidson, "Causal Relations," *Journal of Philosophy*, 64, 1967, and his "Mental Events," in L. Foster and J. Swanson (eds.), *Experience and Theory*, Amherst, Mass. 1970. In the latter essay Davidson argues that the causal relationships which hold between the having of reasons and actions can only be stated in a nonintentionalist vocabulary.

15. Richard Rudner, e.g., in his discussion of social theory in his *Philosophy of Social Science*, Englewood Cliffs, N.J. 1966, first gives an account of physical theories, and then discusses various theoretical formulations found in the social sciences, including typologies and analytical conceptual schemata, pointing out how these fail to meet the criteria for genuine theories.

16. Noam Chomsky discusses the relevance of competence theories for linguistics in *Aspects of a Theory of Syntax*, Cambridge, Mass. 1965, part 1.

17. See Victory Turner, *The Ritual Process*, Chicago 1967, pp. 172ff.

18. Imre Lakatos, "Falsification and the Methodology of Scientific Research Programmes," in Lakatos and Musgrave (eds.), *Criticism and the Growth of Knowledge*, Cambridge 1970.

19. For further discussion of the use of the idea of a "research program" in explicating the structure of social science theories, see J. Donald Moon, "The Logic of Political Inquiry," in Fred I. Greenstein and Nelson W. Polsby (eds.), *The Handbook of Political Science*, vol. 1, Reading, Mass. 1975, pp. 192ff.

20. R. G. Collingwood, *The Idea of History*, New York 1945, pp. 308ff., and pp. 282ff.

21. For examples of studies investigating each of these three types of situations, see V. Aubert, *Sociology of the Law*, Oslo 1964; Chalmers Johnson, *Revolutionary Change*, Boston 1966; and Victor Turner, *Schism and Continuity in an African* Society, Manchester 1957, respectively.

22. E. Gellner, "Concepts and Society," reprinted in his *Cause and Meaning in the Social Sciences*, London 1973, pp. 18–46.

23. See H. R. Trevor-Roper, *The European Witch-Craze*, New York 1969, and Alasdair MacIntyre, "Rationality and the Explanation of Action," in his *Against the Self-Images of the Age*, New York 1971.

24. *Discourse on the Origins of Inequality,* ed. Masters, New York 1964, pp. 265–66.

25. See "A Contribution to the Critique of Hegel's 'Philosophy of Right,' Introduction," in Karl Marx, *Critique of Hegel's "Philosophy of Right,"* ed. Joseph O'Malley, Cambridge 1970, pp. 129ff.

26. See *The Denial of Death,* New York 1973, pp. 160–70.

CASE STUDY FOR PART 2

Several years ago, a man came home and found his wife in bed with another man; he killed her. At his trial, the judge gave the man a very light sentence after he was convicted of murder by the jury. The judge reasoned as follows: Evolutionary psychology and other disciplines have proven that human behavior is controlled by our genetic makeup and evolutionary history. In most species, including the human species, males are naturally jealous and protective of their mates, whom they need to control. The husband was therefore only doing what comes naturally. Society's laws and moral customs must conform to these biological forces. Therefore, the man should not be punished severely. And our laws and morality must be adjusted to accommodate these natural facts about human males.

Some social and cognitive scientists—naturalists—would likely find this a plausible argument. If it is, why are any social sciences necessary at all, if human behavior is explainable in biological terms? Or, at the very least, doesn't the judge's argument support the naturalistic program of explaining cultural and psychological behavior in terms of the methods and assumptions of the natural (in this case the biological) sciences?

1. How might an antinaturalist such as Taylor respond to the judge's argument?

2. How might Fay and Moon respond to the judge's argument?

3. Suppose one uses the same sort of assumptions the judge used to explain the altruistic behavior of a saint. Suppose it turns out that human beings are naturally selfish (for genetic-evolutionary reasons). Would this make the saint (or anybody who wants to distribute wealth and resources more equitably) genetically defective?

4. Suppose crime and low intelligence are largely or entirely genetic in origin. Would that make criminals genetically defective? If so, are explanations about social factors (poverty, etc.) irrelevant to the social and behavioral sciences?

5. Would these sorts of views justify male dominance and gender roles for men and women exclusively in biological terms (women are naturally suited to be mothers and nurses, not medical doctors or theoretical physicists)?

6. How do the debates about the similarities and differences between the natural and the social sciences discussed in this section relate to this scenario and questions 1–5?

STUDY QUESTIONS FOR PART 2

1. In what ways does interpretation depend upon the meanings of behavior?

2. What kinds of meaning does Taylor think relevant to the human sciences?

3. What is the hermeneutic circle? Why isn't it a vicious circle, according to Taylor?

4. Why would empiricist approaches to the social sciences regard the hermeneutical circle as vicious?

5. What does Taylor mean by claiming that human beings are self-defining or self-interpreting animals?

6. How does the fact that we are self-interpreting animals explain why the social sciences must be hermeneutical?

7. What is Kuhn's major objection to Taylor?

8. In what sense is physics interpretive, according to Kuhn?

9. Why isn't physics a hermeneutical science for Kuhn?

10. Why might economics and psychology become more like physics, on Kuhn's view?

11. Why does it matter whether the social sciences are inferior to the natural sciences, according to Machlup?

12. In what respects are the social sciences inferior?

13. Are there any respects in which the social sciences are not inferior, according to Machlup?

14. Can you think of any other grounds of comparison between the natural and social sciences? Would the latter be inferior according to any of these?

15. Why are naturalism and humanism (antinaturalism) both inadequate philosophies of the social sciences, according to Fay and Moon?

16. What are the three questions they think all philosophies of the social sciences must answer?

17. Why are these questions important?

18. Why does naturalism fail to answer these questions satisfactorily?

19. Why can't humanism or antinaturalism adequately answer these questions?

20. What is "critique" and what is its role in an adequate social science?

21 What do Fay and Moon conclude about naturalism and humanism?

22. What is the main failure of economics, according to Rosenberg?

23. What is the main source of this failure?

24. What is Rosenberg's proposal for dealing with this problem?

25. Rosenberg seems to be claiming that economics cannot (or should not) be understood realistically, but instrumentally. Why?

SELECTED BIBLIOGRAPHY
[*INDICATES GOOD BIBLIOGRAPHIES]

A. General Readers

1. Martin, Michael, and MacIntyre, Lee, C., eds. *Readings in the Philosophy of Social Science.* Cambridge: Massachusetts Institute of Technology Press, 1994. [Must reading; moderately advanced.]*
2. Rabinow, Paul and Sullivan, William, eds. *Interpretive Social Science: A Second Look,* Berkeley: University of California Press, 1987. [Developments in Hermeneutics. Accessible.]*
3. Sabia, Daniel R. and Wallulis, Jerald, eds. *Changing Social Science,* Albany: State University of New York Press, 1983. [Moderate level.]

B. General Texts

4. Braybrooke, David. *Philosophy of Social Science.* Englewood Cliffs, N.J.: Prentice-Hall, 1987. [Useful introduction for beginners.]*
5. Fay, Brian. *Contemporary Philosophy of Social Science.* Cambridge: Blackwell's, 1997. [Social sciences in a multicultural world. Moderate.]*
6. Habermas, Jürgen. *On the Logic of the Social Sciences.* Cambridge: Massachusetts Institute of Technology Press, 1988. [Difficult; must read book.]
7. Little, David. *Varieties of Social Explanation.* Boulder, Colo.: Westview Press, 1990 [Excellent, but moderately advanced read.]*
8. Manicas, Peter T. *A History of the Philosophy of the Social Sciences.* Cambridge: Blackwell's, 1987. [An interesting history, with developments in American disciplines included.]*
9. McIntyre, Lee C. *Laws and Explanation in the Social Sciences.* Boulder, Colo.: Westview Press, 1996. [Defends a naturalistic view of behavior.]
10. Outwaithe, William. *New Philosophies of Social Science.* New York: St. Martin's Press, 1987. [Focuses on Realism, Hermeneutics, Critical Theory. Accessible to beginners.]*
11. Popper, Karl. *The Poverty of Historicism.* New York: Harper, 1957. [A classic text.]
12. Root, Michael. *Philosophy of Social Science.* Cambridge: Blackwell's, 1995. [An interesting, accessible discussion, focused on "liberal social science."]*
13. Rosenau, Pauline Marie. *Post-Modernism and the Social Sciences.* Princeton: Princeton University Press, 1992. [Interesting survey, a bit oversimplified.]*
14. Rosenberg, Alexander. *Philosophy of Social Science.* 2nd ed. Boulder, Colo.: Westview Press, 1995. [Zeroes in on Folk Psychology. An important but difficult book.]*
15. Von Wright, G. H. *Explanation and Understanding.* Ithaca, N.Y.: Cornell University Press, 1971. [A classic study of naturalism vs. antinaturalism.]*
16. Winch, Peter. *The Idea of a Social Science and Its Relation to Philosophy.* New York: Routledge, 1958. [A most controversial book on the social sciences. A classic.]

C. Specialized Texts

Anthropology

17. Hollis, Martin, and Lukes, Steven, eds. *Rationality and Relativism.* Oxford: Blackwell's, 1982. [Essential readings. Advanced.]*
18. Wilson, Bryan, ed. *Rationality.* Oxford: Blackwell's, 1970, [Basic essays.]

Economics

19. Hausman, Daniel, ed. *Philosophy of Economics.* Cambridge: Cambridge University Press, 1984. [Advanced.]*

History

20. Iggers, George. *Historiography in the 20th Century: From Scientific Objectivity to Postmodernism.* Hanover, N.H.: Wesleyan University Press, 1997. [Excellent and lucid survey.]*
21. Novick, Peter. *That Noble Dream.* Cambridge: Cambridge University Press, 1988. [Study of American historians' methodological debates.]*

Political Science

22. Ball, Terence, ed. *Idioms of Inquiry.* Albany: State University of New York Press, 1987. [Advanced essays on political science.]*

Psychology

23. Borger, Roger, and Cioffi, Frank, eds. *Explanation and the Behavioral Sciences.* Cambridge: Cambridge University Press, 1984. [Advanced.]
24. Haugeland, John, ed. *Mind Design.* Rev ed. Cambridge: Massachusetts Institute of Technology Press, 1995. [Moderate.]*
25. McDonald, Cynthia, and MacDonald, Graham, eds. *Philosophy of Psychology: Debates on Psychological Explanation.* Cambridge: Cambridge University Press, 1984. [Advanced.]*

Sociology

26. Brown, S. C., ed. *Philosophical Disputes in Social Sciences.* Manchester (UK): Harvester Books, 1980. [Advanced.]*
27. Ryan, Alan, ed. *Philosophy of Social Explanation.* Oxford: Oxford University Press, 1973. [Must read essays; moderate to advanced.]*

Part 3

Explanation and Law

Introduction

Part 1 suggested that science may be distinguished from other endeavors in part because it provides genuine explanations of phenomena. Part 2 pointed out that one important difference between the natural and social sciences has to do with the nature of explanation in the two contexts. It is now time to consider the topic of scientific explanation more explicitly. Just what is a scientific explanation anyway, and how does it differ from explanations in other settings? How do explanations in the natural sciences differ from those of the social sciences? Part 3 also discusses the nature of scientific laws and what role (if any) they have in scientific explanations.

I. SOME EXAMPLES

Science aims at providing explanations of particular events and regularities in nature. We request an explanation whenever we ask why an event occurred or the regularity is present. And it's generally conceded that an explanation must go beyond simply describing the event or regularity in question. Parsing out just what this something extra is has been a major task of philosophy of science. Some examples of scientific explanations at this point clarify just how heterogeneous the class of explanations is, and how difficult it is to identify one or more feature(s) they all hold in common.

Why do the planets undergo circular motion in the heavens? Aristotle explained the motion of the planets by pointing out that planets are part of the perfect and permanent region of the heavens, and circular motion is the only motion appropriate to that region.

Why did the pressure of this gas increase from time 1 to time 2? According to Boyle's law, if the temperature of a gas is constant, a decrease in its volume is invariably accompanied by an inversely proportional increase in its pressure. Since we know the temperature remained constant from time 1 to time 2, we may conclude that the increase in pressure is the direct result of our decreasing the volume of the container.

Why are the planets in the solar system at regular increasing distances from the sun? Kepler, at one point in his curious career, explained the distances of the various planets from the sun by drawing an analogy between the distances and the way the five Platonic solids will fit inside one another. (Platonic solids are solid figures that have faces of identical equilateral plane figures, e.g., a square and a tetrahedron. There are only five such objects.)

Why do we see such diversity among living plants and animals? Creationists explain this diversity with reference to the book of Genesis and God's plan; evolutionists explain this in terms of natural processes such as mutation, heredity, natural selection, and common descent.

Why is the frequency of the dark form of this moth increasing over time in areas downwind of manufacturing centers at the expense of the pale form? H. B. D. Kettlewell argued that in areas downwind of manufacturing centers, pollution covers the resting sites of the moths, making dark moths less visible to birds. Dark moths thus have a survival advantage compared to pale moths in polluted settings. According to the principle of natural selection, whenever one form has a statistical advantage with regard to either survival or reproduction (and the population is not in equilibrium), the favored form will increase in frequency.

Why did I dream I was on a cruise in the Caribbean? According to Sigmund Freud's theory, a dream represents either a wish-fulfillment, a counterwish fulfillment, or something else.

Why did Caesar cross the Rubicon? H. M. Collingwood explained Caesar's decision by pointing out that any rational person who was confronted by the same circumstances and had the same goals would have done the same thing. It is interesting to note that each of the above at one time or another has been proposed as a scientific explanation of a natural phenomenon, but whereas the Aristotelian account endured for a considerable period of time, the Keplerian account was quickly dismissed as quackery. As philosophers of science we need to consider why these explanations are not simply false, but in some deeper sense, misconceived. Likewise, regardless of one's religious views, the creationist/evolutionist debate noted above is at its heart a fundamental dispute about the nature of explanation. Legislators and others cannot hope to think clearly about this issue without first forming a general and competent view about explanation.

Debates about the nature of explanation are also at the heart of longstanding controversies in science. For instance, B. F. Skinner argued that explaining human behavior by appeal to what occurs inside the skin is pointless. Indeed, he went so far as to claim that inquiry along these lines will only lead to confusions and banalities, not real explanations. Much of the conflict between two branches of cognitive psychology, information-processing psychology and behaviorism, can be traced back to this fundamental disagreement about the "proper" way to explain human behavior. The continued use of functional and teleological language in biology and psychology, to cite another example, has been the topic of a long-standing debate: is the use of functional ascriptions "legitimate" or does it reflect the immaturity of biology and psychology as sciences?

Space limitations prevent a discussion of all these issues. The essays in this part focus specifically on the nature of scientific explanation. The Selected Bibliography for Part 3 provides access to some of the vast literature on this topic, including a few references on the exciting topic of teleological and functional explanations in both psychology and biology.

II. HEMPEL'S D-N MODEL OF EXPLANATION

Contemporary work on explanation begins with the seminal work of Carl G. Hempel, who in a series of papers presented a widely accepted view of explanation [1], which for a brief period of time took on the status of a received view. Since Hempel's model identifies the process of explaining events and regularities as a process of subsuming the event or regularity under a general law, it is sometimes referred to as an example of a "covering law" model of explanation.

Hempel initially proposed his D-N, or deductive-nomological, model for singular deterministic events, and shortly thereafter sketched out how the same model could be used to explain deterministic laws. In later papers Hempel proposed similar models for the explanation of statistical or indeterministic events and laws, referred to as his I-S (inductive-statistical) and D-S (deductive-statistical) models of explanation, respectively. Our discussion will focus on his initial, most well-developed and successful, the D-N model, for the explanation of singular events. The rest of this section briefly reviews the main features of this model, but is, of course, no substitute for reading Hempel's paper.

Hempel's D-N model of singular events is developed from the fundamental intuition that explanations are arguments offered to establish that the event-to-be-explained had to occur given the initial conditions and the presence of certain regularities in nature. More technically, a scientific explanation is an argument that has as its premises a set of statements, the *explanans*, that describe the initial conditions $(C_1, C_2, \ldots, C_k)$ and contain at least one scientific law (L), and at least one sentence, the *explanandum* (a statement that describes the event-to-be-explained), that deductively follows as its conclusion (E). Schematically, one can diagram this model as follows:

Explanans:	Statements of initial conditions.	$C_1, C_2, \ldots C_k$
	At least one scientific law.	$L_1, L_2 \ldots L_r$
Explanandum:	A statement describing the	$E.$
	event-to-be-explained.	

Hempel's "Studies in the Logic of Explanation" (reproduced as the first essay in this part) presents and defends a set of requirements genuine scientific explanations must meet on this model:

Logical conditions of adequacy

(R1) The explanandum must be a deductively logical consequence of the explanans.

(R2) The explanans must contain at least one general law necessary for the derivation of the explanandum.

(R3) The explanans must have empirical content, i.e., it must be empirically testable.

Empirical conditions of adequacy

(R4) The sentences comprising the explanans must be true.

John Hospers in an early essay titled "What is Explanation?" provides a simple example of what Hempel had in mind that illustrates these conditions.[1]

(a) All copper conducts electricity.
(b) This wire is made of copper.
(c) Therefore, this wire conducts electricity.

The explanandum (c) follows as a deductive consequence of the explanans (a) and (b); (b) is a law of nature required for the derivation of (c); and the explanans (a) and (b) are each testable. Presuming for the moment the explanans (a) and (b) are true, we have a genuine explanation of why this wire conducts electricity.

In recognition of the fact that not all explanations of events make reference to deterministic laws, Hempel developed the I-S (inductive-statistical) model to handle the explanation of probabilistic or statistical events. An example from Salmon's essay illustrates the formal parallel Hempel draws between D-N and I-S explanations:

Almost all cases of streptococcus infection clear up quickly after the administration of penicillin.
Jones had a streptococcus infection and was treated with penicillin.
——————————————————— [.90 = strength of inductive support]
Jones recovered quickly.

Here the major difference between the two types of explanation is indicated by the double line. Whereas the D-N model required at least one deterministic law that would allow the explanandum to be logically deduced *deductively* from the explanans, the I-S model requires use of a statistical law, from which the explanandum is *inductively* derived from the explanans. The D-N model explains an event by establishing that the event had to occur given the initial conditions and at least one deterministic law. The I-S model similarly explains an event by establishing that the event was likely to occur given the initial conditions and at least one statistical law.

III. Laws of Nature

Laws are central to Hempel's account of explanation and covering law models of explanation in general—they are the "something else" that distinguishes scientific explanations from mere descriptions. But what are laws? We might consider characterizing laws simply as true universal generalizations. But a little reflection should convince us that this won't do. Consider the following two examples: "All pieces of copper expand when heated" and "All of the coins in my pocket are copper." Each is a true universal generalization; yet while the former is clearly a candidate for a law, there is something altogether happenstance about the latter that makes us reluctant to designate it as a law of nature even if it is true. Hempel confronts this problem of distinguishing laws from accidental generalizations by proposing that what distinguishes the two is that whereas the first is spatiotemporally unrestricted (i.e., it applies to copper in all places at all times), the second makes reference to a specific object (my pocket). To account for a few obvious counterexamples to this claim (e.g., Kepler's laws make reference to the sun), Hempel makes a further distinction between fundamental and derivative laws, the idea being that whereas fundamental laws must be spatiotemporally unrestricted in character, derivative laws are legitimate if they are of universal character and can be derived from some more fundamental law(s). The readings by Lambert and Britten and Cartwright both stem from perceived deficiencies in Hempel's account of laws.

IV. Other Problems with Hempel's D-N Model

Salmon's essay briefly reviews the history of work on explanation since the D-N model was first proposed. Many reactions to Hempel's work directly question the sufficiency and/or necessity of one or more of Hempel's four conditions of adequacy (RI)–(R4). To be fair, the Hempel essay anticipates several of these concerns, as, for instance, when he sketches out how his model might be extended to include motivational and teleological explanations in psychology and biology, examples at least superficially quite divorced from illustrations he draws from physics. There are two fundamental problems with Hempel's account that in part motivate the three main alternatives offered by Salmon, van Frassen, and Kitcher: the *asymmetry problem* and the *irrelevance problem.*

1. The Asymmetry Problem

According to Hempel's D-N model, a good explanation is one that makes the occurrence of the event-to-be-explained a certainty. This implies that any good explanation could have been used to predict the event as well, provided the facts and laws stated in the explanans were available prior to the event-to-be-explained's occurrence. Thus logically the only difference between a good explanation and a good prediction has to do with when the argument is stated relative to the event-to-be-explained: if before, it is a prediction; if after, it is an explanation. (Hempel made analogous remarks with respect to his I-S model of explanation—a good statistical explanation establishes that the event-to-be-explained was to be expected [highly probable] and likewise is a potential prediction if provided prior to the event.)

However intuitive this symmetry thesis might seem, it is manifestly false. There are many good predictors of events that no one would regard as explanatory when provided after the event occurred. For instance, barometers serve as reliable predictors of storms, yet no one would explain the storm's occurrence with reference to the barometer's reading. Conversely, something that we regard as a good explanation may nevertheless not serve as a good predictor of the event should it be known prior to the event's occurrence. This is perhaps easiest to see when one considers the explanation of improbable events. What is regarded as a good explanation of why a roulette wheel stops on a particular number doesn't serve as a strong basis for prediction were it known prior to spinning the wheel.

2. The Irrelevance Problem

A second major problem with Hempel's account has to do with the problem of how to prevent spurious derivations from counting as genuine explanations, and more generally how to ensure that all and only relevant information is included in the explanation of events. Two examples provided in Salmon's essay illustrate this problem. First consider an "explanation" of Kepler's laws that has as its explanans a single statement conjoining Kepler's and Boyle's laws. It is trivially true that Kepler's laws deductively follow from the conjunction, yet such a derivation doesn't seem explanatory at all.

A second example reveals the converse problem, namely, how to ensure that all relevant information is included. Consider the Jones example discussed earlier in connection with the I-S model of explanation, this time including the additional information that Jones is infected with a streptococcus bacterium that is penicillin-resistant. This bit of information dramatically changes the probability of Jones's recovery, and thereby compromises the argument as an explanation (at least if we agree with Hempel's insistence that a genuine statistical explanation must make the explanandum event probable). Salmon's essay reviews Hempel's attempts to get around these difficulties.

V. ALTERNATIVES TO HEMPEL'S ACCOUNT

In the years since Hempel first proposed his account, three major alternatives have been proposed by Wesley Salmon, Bas van Frassen, and Philip Kitcher. Each can be seen to arise in part from the perceived shortcomings of Hempel's model.

1. The Role of Causality in Explanation

Sylvain Bromberger is credited as the author of a well-known example that establishes that the conditions Hempel has identified are insufficient to distinguish genuine explanations. Imagine a flagpole that casts a shadow of a definite length during some specified time of the day. It's clear that we can explain the length of the shadow by virtue of the flagpole's height, the sun's particular elevation, and the laws of optics, all of which can easily be formulated in accordance with the D-N model. The problem arises when we consider how the length of the shadow, the sun's elevation, et cetera, could similarly be used to determine the height of the flagpole. Our intuitions are that whereas the height of the flagpole does indeed explain the length of the shadow, the length of the shadow does not explain the flagpole's height.

From Salmon's point of view, the missing element in Hempel's account is the fundamental role of causality in most scientific explanations. The reason why the height of the flagpole explains the length of the shadow but not vice-versa, is that whereas the flagpole is causally responsible for the shadow's length (by blocking the sun's rays it creates the shadow), the same cannot be said for the shadow with respect to the flagpole. Salmon's model is developed in reaction to a number of other problems in Hempel's account, such as its tacit presumption that only highly probable events can be explained (which Salmon traces to an inherent commitment to determinism). Salmon's account also addresses the asymmetry and irrelevance problems noted previously by restricting explanatory elements to those that are (causally) relevant to the event being explained.

2. The Logic of Explanation vs.
the Pragmatics of Explanation

As noted above, the Hempelian account is focused on the *logic of explanation,* i.e., understanding the logical relationships between the sentences comprising an explanation. This contrasts with what may be loosely referred to as the *pragmatics of explanation,* or, consideration of the circumstances under which an explanation is requested or provided. This distinction is quite important to advocates of the Hempelian account, because it forestalls what they might consider facile criticisms of their model.

Two examples illustrate this point: (1) when asked, "Why does this conduct electricity?" one might respond, "It's copper"; and (2) when explaining to a group of fourth graders why the pressure on the walls of a closed pot increased, one might give the kinetic theory and the fact that the temperature had increased. Our intuitions encourage us to believe that the first example is a genuine explanation even if it does not meet Hempel's criteria. On the other hand, the second example, even if it is filled out in splendid Hempelian detail, fails to explain. After all, it is a rare fourth grader who has the slightest idea of what the mean molecular speed squared is.

Hempelians have a response to these worries. Regarding the first example, they can admit we do talk that way when we explain things to someone. Nevertheless, they can maintain that a genuine explanation must meet conditions (R1)–(R4) by pointing out that the example seems explanatory only because it implicitly assumes certain rather obvious statements that if made explicit would satisfy Hempel's requirements. The pragmatics of the situation we are in dictate the extent to which we must state the details required by Hempel's model; but recognition that there are certain situations where we can safely assume our listener has certain knowledge does not detract from their explanatory importance. The second example may be similarly defused. Clearly we need to be cognizant of the background and mental abilities of our audience to judge whether or not an explanation we provide will make sense to them. But recognition that the second example is not appropriate for children is an issue of pragmatics not logic.

The present essay has drawn the distinction between logic and pragmatics in a very rough way. Van Frassen's account develops these considerations more systematically, ultimately questioning whether one can indeed divorce the logic of explanation from the pragmatics of explanation in the manner Hempel has indicated. From van Frassen's point of view, it is only with reference to the pragmatic features of explanation that the asymmetry and irrelevance problems noted above can be circumvented.

3. Explanation as Unification

A third alternative approach emphasizes another dimension of explanation altogether, the idea that understanding in science increases when we are able to reduce the number of independently available hypotheses needed to explain phenomena. Whereas Salmon's account of explanation focuses on particular causal connections at a local level to explain particular facts (bottom up), Kitcher's account identifies explanation as fitting the event to be explained into a more global pattern (top down). Kitcher's essay highlights Newton's and Darwin's theories as examples of how scientific understanding increased as various disparate phenomena were united under single theories. He also briefly sketches how his account overcomes the asymmetry and irrelevance problems that appear to compromise Hempel's account.

VI. THE READINGS IN PART 3

As noted above, Hempel's essay is *the* classic work in philosophy of science. It must be taken account of by all subsequent work on the topic of explanation. The Hempelian view referred to in this introduction is elaborated in Hempel's essay and other works by him [1]. The essay contains some technical concepts from logic but they are worth mastering, as they occur in some of the subsequent articles.

Lambert and Britten challenge Hempel's account of law by means of a series of counterexamples. They argue that genuine laws of nature can be distinguished from accidental generalizations by their ability to support counterfactuals. Cartwright disputes the empirical adequacy requirement (R4) of Hempel's model of explanation, and in particular his emphasis that the laws of a genuine scientific explanation must be true. This, at least on the surface, is an extraordinary claim. You will have to see if it can be justified.

The articles by van Frassen, Kitcher, and Salmon present three major alternatives to the Hempelian approach to scientific explanation. Salmon's article reviews several well-known deficiencies of Hempel's account and provides an alternative causal model of explanation. Van Frassen's account is developed from a consideration of the pragmatic features of explanation. Kitcher's account identifies unification as the important goal of explanation.

D. W. R.

NOTE

1. In A. Flew, ed., *Essays in Conceptual Analysis*, pp. 94–119. London: Macmillan Publishers Ltd., 1956.

11

Studies in the Logic of Explanation[1]

Carl G. Hempel

INTRODUCTION

To explain the phenomena in the world of our experience, to answer the question "why?" rather than only the question "what?" is one of the foremost objectives of empirical science. While there is rather general agreement on this point there exists considerable difference of opinion as to the function and the essential characteristics of scientific explanation. The present essay is an attempt to shed some light on these issues by means of an elementary survey of the basic pattern of scientific explanation and a subsequent more rigorous analysis of the concept of law and the logical structure of explanatory arguments. . . .

I. ELEMENTARY SURVEY OF SCIENTIFIC EXPLANATION

1. *Some Illustrations.* A mercury thermometer is rapidly immersed in hot water; there occurs a temporary drop of the mercury column, which is then followed by a swift rise. How is this phenomenon to be explained? The increase in temperature affects at first only the glass tube of the thermometer; it expands and thus provides a larger space for the mercury inside, whose surface therefore drops. As soon as by heat conduction the rise in temperature reaches the mercury, however, the latter expands, and as its coefficient of expansion is considerably larger than that of glass, a rise of the mercury level results.—This account consists of statements of two kinds. Those of the first kind indicate certain conditions which are realized prior to, or at the same time as, the phenomenon to be explained; we shall refer to them briefly as antecedent conditions. In our illustration, the antecedent conditions include, among others, the fact that the thermometer consists of a glass tube which is

206

partly filled with mercury, and that it is immersed into hot water. The statements of the second kind express certain general laws; in our case, these include the laws of the thermic expansion of mercury and of glass, and a statement about the small thermic conductivity of glass. The two sets of statements, if adequately and completely formulated, explain the phenomenon under consideration: they entail the consequence that the mercury will first drop, then rise. Thus, the event under discussion is explained by subsuming it under general laws, i.e., by showing that it occurred in accordance with those laws, in virtue of the realization of certain specified antecedent conditions.

Consider another illustration. To an observer in a rowboat, that part of an oar which is under water appears to be bent upwards. The phenomenon is explained by means of general laws—mainly the law of refraction and the law that water is an optically denser medium than air—and by reference to certain antecedent conditions—especially the facts that part of the oar is in the water, part in the air, and that the oar is practically a straight piece of wood. Thus, here again, the question *"Why* does the phenomenon occur?" is construed as meaning "according to what general laws, and by virtue of what antecedent conditions does the phenomenon occur?"

So far, we have considered only the explanation of particular events occurring at a certain time and place. But the question "Why?" may be raised also in regard to general laws. Thus, in our last illustration, the question might be asked: Why does the propagation of light conform to the law of refraction? Classical physics answers in terms of the undulatory theory of light, i.e., by stating that the propagation of light is a wave phenomenon of a certain general type, and that all wave phenomena of that type satisfy the law of refraction. Thus, the explanation of a general regularity consists in subsuming it under another, more comprehensive regularity, under a more general law. Similarly, the validity of Galileo's law for the free fall of bodies near the earth's surface can be explained by deducing it from a more comprehensive set of laws, namely Newton's laws of motion and his law of gravitation, together with some statements about particular facts, namely, about the mass and the radius of the earth.

2. *The Basic Pattern of Scientific Explanation.* From the preceding sample cases let us now abstract some general characteristics of scientific explanation. We divide an explanation into two major constituents, the *explanandum* and the *explanans.*[2] By the explanandum, we understand the sentence describing the phenomenon to be explained (not that phenomenon itself); by the explanans, the class of those sentences which are adduced to account for the phenomenon. As was noted before, the explanans falls into two subclasses; one of these contains certain sentences $C_1, C_2 \ldots, C_k$ which state specific antecedent conditions; the other is a set of sentences $L_1, L_2 \ldots, L_r$ which represent general laws.

If a proposed explanation is to be sound, its constituents have to satisfy certain conditions of adequacy, which may be divided into logical and empirical conditions. For the following discussion, it will be sufficient to formulate these requirements in a slightly vague manner. . . .

Logical conditions of adequacy

(R1) The explanandum must be a logical consequence of the explanans; in other words, the explanandum must be logically deducible from the information contained in the explanans; for otherwise, the explanans would not constitute adequate grounds for the explanandum.

logical deduction

(R2) The explanans must contain general laws, and these must actually be required for the derivation of the explanandum. We shall not make it a necessary condition for a sound explanation, however, that the explanans must contain at least one statement which is not a law; for, to mention just one reason, we would surely want to consider as an explanation the derivation of the general regularities governing the motion of double stars from the laws of celestial mechanics, even though all the statements in the explanans are general laws.

(R3) The explanans must have empirical content; i.e., it must be capable, at least in principle, of test by experiment or observation. This condition is implicit in (R1); for since the explanandum is assumed to describe some empirical phenomenon, it follows from (R1) that the explanans entails at least one consequence of empirical character, and this fact confers upon it testability and empirical content. But the point deserves special mention because, as will be seen in §3, certain arguments which have been offered as explanations in the natural and in the social sciences violate this requirement.

Empirical condition of adequacy

(R4) The sentences constituting the explanans must be true. That in a sound explanation, the statements constituting the explanans have to satisfy some condition of factual correctness is obvious. But it might seem more appropriate to stipulate that the explanans has to be highly confirmed by all the relevant evidence available rather than that it should be true. This stipulation, however, leads to awkward consequences. Suppose that a certain phenomenon was explained at an earlier stage of science, by means of an explanans which was well supported by the evidence then at hand, but which has been highly disconfirmed by more recent empirical findings. In such a case, we would have to say that originally the explanatory account was a correct explanation, but that it ceased to be one later, when unfavorable evidence was discovered. This does not appear to accord with sound common usage, which directs us to say that on the basis of the limited initial evidence, the truth of the explanans, and thus the soundness of the explanation, had been quite probable, but that the ampler evidence now available makes it highly probable that the explanans is not true, and hence that the account in question is not—and never has been—a correct explanation.[3] (A similar point will be made and illustrated, with respect to the requirement of truth for laws, [elsewhere].)

Some of the characteristics of an explanation which have been indicated so far may be summarized in the following schema:

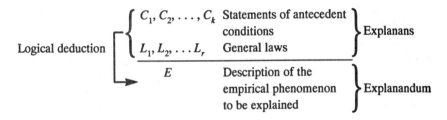

Logical deduction $\begin{cases} C_1, C_2, \ldots, C_k & \text{Statements of antecedent conditions} \\ L_1, L_2, \ldots L_r & \text{General laws} \end{cases}$ Explanans

E — Description of the empirical phenomenon to be explained — Explanandum

Let us note here that the same formal analysis, including the four necessary conditions, applies to scientific prediction as well as to explanation. The difference between the two is of a pragmatic character. If E is given, i.e., if we know that the phenomenon described by E has occurred, and a suitable set of statements $C_1, C_2, \ldots, C_k, L_1, L_2, \ldots, L_r$, is provided afterwards, we speak of an explanation of the phenomenon in question. If the latter statements are given and E is derived prior to the occurrence of the phenomenon it describes, we speak of a prediction. It may be said, therefore, that an explanation of a particular event is not fully adequate unless its explanans, if taken account of in time, could have served as a basis for predicting the event in question. Consequently, whatever will be said in this article concerning the logical characteristics of explanation or prediction will be applicable to either, even if only one of them should be mentioned.

Many explanations which are customarily offered, especially in pre-scientific discourse, lack this potential predictive force, however. Thus, we may be told that a car turned over on the road "because" one of its tires blew out while the car was traveling at high speed. Clearly, on the basis of just this information, the accident could not have been predicted, for the explanans provides no explicit general laws by means of which the prediction might be effected, nor does it state adequately the antecedent conditions which would be needed for the prediction. The same point may be illustrated by reference to W. S. Jevons's view that every explanation consists in pointing out a resemblance between facts, and that in some cases this process may require no reference to laws at all and "may involve nothing more than a single identity, as when we explain the appearance of shooting stars by showing that they are identical with portions of a comet."[4] But clearly, this identity does not provide an explanation of the phenomenon of shooting stars unless we presuppose the laws governing the development of heat and light as the effect of friction. The observation of similarities has explanatory value only if it involves at least tacit reference to general laws.

In some cases, incomplete explanatory arguments of the kind here illustrated suppress parts of the explanans simply as "obvious"; in other cases, they seem to involve the assumption that while the missing parts are not obvious, the incomplete explanans could at least, with appropriate effort, be so supplemented as to make a strict derivation of the explanandum possible. This assumption may be justifiable in some cases, as when we say that a lump of sugar disappeared "because" it was put into hot tea, but it surely is not satisfied in many other cases. Thus, when certain peculiarities in the work of an artist are explained as outgrowths of a specific type of neurosis, this observation may contain significant clues, but in general it does not afford a sufficient basis for a potential prediction of those peculiarities. In cases of this kind, an incomplete explanation may at best be considered as indicating some positive correlation between the antecedent conditions adduced and the type of phenomenon to be explained, and as pointing out a direction in which further research might be carried on in order to complete the explanatory account.

The type of explanation which has been considered here so far is often referred to as causal explanation.[5] If E describes a particular event, then the antecedent circumstances described in the sentences $C_1, C_2, \ldots, C_k$ may be said jointly to "cause" that event, in the sense that there are certain empirical regularities, expressed by the laws $L_1, L_2, \ldots, L_r$, which imply that whenever conditions of the kind indicated by $C_1, C_2, \ldots, C_k$ occur, an event of the kind described in E will take place. Statements such as $L_1, L_2, \ldots, L_r$, which assert general and unexceptional connections between specified characteristics of events, are customarily called causal, or deterministic, laws. They must be distinguished from the so-called statistical laws which assert that in the long run, an explicitly stated percentage of all cases satisfying a given set of conditions are accompanied by an event of a certain specified kind. Certain cases of scientific explanation involve "subsumption" of the explanandum under a set of laws of which at least some are statistical in character. Analysis of the peculiar logical structure of that type of subsumption involves difficult special problems. The present essay will be restricted to an examination of the deductive type of explanation, which has retained its significance in large segments of contemporary science, and even in some areas where a more adequate account calls for reference to statistical laws.[6]

3. *Explanation in the Nonphysical Sciences. Motivational and Teleological Approaches.* Our characterization of scientific explanation is so far based on a study of cases taken from the physical sciences. But the general principles thus obtained apply also outside this area.[7] Thus, various types of behavior in laboratory animals and in human subjects are explained in psychology by subsumption under laws or even general theories of learning or conditioning; and while frequently the regularities invoked cannot be stated with the same generality and precision as in physics or chemistry, it is clear at least that the general character of those explanations conforms to our earlier characterization.

Let us now consider an illustration involving sociological and economic factors. In the fall of 1946, there occurred at the cotton exchanges of the United States a price drop which was so severe that the exchanges in New York, New Orleans, and Chicago had to suspend their activities temporarily. In an attempt to explain this occurrence, newspapers traced it back to a large-scale speculator in New Orleans who had feared his holdings were too large and had therefore begun to liquidate his stocks; smaller speculators had then followed his example in a panic and had thus touched off the critical decline. Without attempting to assess the merits of the argument, let us note that the explanation here suggested again involves statements about antecedent conditions and the assumption of general regularities. The former include the facts that the first speculator had large stocks of cotton, that there were smaller speculators with considerable holdings, that there existed the institution of the cotton exchanges with their specific mode of operation, et cetera. The general regularities referred to are—as often in semipopular explanations—not explicitly mentioned; but there is obviously implied some form of the law of supply and demand to account for the drop in cotton prices in terms of the greatly increased supply under conditions of practically unchanged demand; besides, reliance is necessary on certain regularities in the behavior of individuals who are trying to preserve or improve their economic position. Such laws cannot be formulated at present with satisfactory precision and generality, and therefore, the suggested explanation is surely incomplete, but its intention is unmistakably to account for the phenomenon by integrating it into a general pattern of economic and sociopsychological regularities.

We turn to an explanatory argument taken from the field of linguistics.[8] In northern France, there are in use a large variety of words synonymous with the English "bee," whereas in southern France, essentially only one such word is in existence. For this discrepancy, the explanation has been suggested that in the Latin epoch, the South of France used the word "apicula," the

North the word "apis." The latter, because of a process of phonologic decay in northern France, became the monosyllabic word "e," and monosyllables tend to be eliminated, especially if they contain few consonantic elements, for they are apt to give rise to misunderstandings. Thus, to avoid confusion, other words were selected. But "apicula" which was reduced to "abelho," remained clear enough and was retained, and finally it even entered into the standard language, in the form "abeille." While the explanation here described is incomplete in the sense characterized in the previous section, it clearly exhibits reference to specific antecedent conditions as well as to general laws.[9]

While illustrations of this kind tend to support the view that explanation in biology, psychology, and the social sciences has the same structure as in the physical sciences, the opinion is rather widely held that in many instances, the causal type of explanation is essentially inadequate in fields other than physics and chemistry, and especially in the study of purposive behavior. Let us examine briefly some of the reasons which have been adduced in support of this view. One of the most familiar among them is the idea that events involving the activities of humans singly or in groups have a peculiar uniqueness and irrepeatability which makes them inaccessible to causal explanation because the latter, with its reliance upon uniformities, presupposes repeatability of the phenomena under consideration. This argument which, incidentally, has also been used in support of the contention that the experimental method is inapplicable in psychology and the social sciences, involves a misunderstanding of the logical character of causal explanation. Every individual event, in the physical sciences no less than in psychology or the social sciences, is unique in the sense that it, with all its peculiar characteristics, does not repeat itself. Nevertheless, individual events may conform to, and thus be explainable by means of, general laws of the causal type. For all that a causal law asserts is that any event of a specified kind, i.e., any event having certain specified characteristics, is accompanied by another event which in turn has certain specified characteristics; for example, that in any event involving friction, heat is developed. And all that is needed for the testability and applicability of such laws is the recurrence of events with the antecedent characteristics, i.e., the repetition of those characteristics, but not of their individual instances. Thus, the argument is inconclusive. It gives occasion, however, to emphasize an important point concerning our earlier analysis: When we spoke of the explanation of a single event, the term "event" referred to the occurrence of some more or less complex characteristic in a specific spatio-temporal location or in a certain individual object, and not to *all* the characteristics of that object, or to all that goes on in that space-time region.

A second argument that should be mentioned here[10] contends that the establishment of scientific generalizations—and thus of explanatory principles—for human behavior is impossible because the reactions of an individual in a given situation depend not only upon that situation, but also upon the previous history of the individual. But surely, there is no a priori reason why generalizations should not be attainable which take into account this dependence of behavior on the past history of the agent. That indeed the given argument "proves" too much, and is therefore a non sequitur, is made evident by the existence of certain physical phenomena, such as magnetic hysteresis and elastic fatigue, in which the magnitude of a specific physical effect depends upon the past history of the system involved, and for which nevertheless certain general regularities have been established.

A third argument insists that the explanation of any phenomenon involving purposive behavior calls for reference to motivations and thus for teleological rather than causal analysis. For example, a fuller statement of the suggested explanation for the break in the cotton prices would have to indicate the large-scale speculator's motivations as one of the factors determining the event in question. Thus, we have to refer to goals sought; and this, so the argument runs, introduces a type of explanation alien to the physical sciences. Unquestionably, many of the—frequently incomplete—explanations which are offered for human actions involve reference to goals and motives; but does this make them essentially different from the causal explanations of physics and chemistry? One difference which suggests itself lies in the circumstance that in motivated behavior, the future appears to affect the present in a manner which is not found in the causal explanations of the physical sciences. But clearly, when the action of a person is motivated, say, by the desire to reach a certain objective, then it is not the as yet unrealized future event of attaining that goal which can be said to determine his present behavior, for indeed the goal may never be actually reached; rather—to put it in crude terms—it is (a) his desire, present before the action, to attain that particular objective, and (b) his belief, likewise present before the action, that such and such a course of action is most likely to have the desired effect. The determining motives and beliefs, therefore, have to be classified among the antecedent conditions of a motivational explanation, and there is no formal difference on this account between motivational and causal explanation.

Neither does the fact that motives are not accessible to direct observation by an outside observer constitute an essential difference between the two kinds of explanation; for the determining factors adduced in physical explanations also are very frequently inaccessible to direct observation. This is the case, for instance, when opposite electric charges are adduced in explanation of the mutual attraction of two metal spheres. The presence of those charges, while eluding direct observation, can be ascertained by various kinds of indirect test, and that is sufficient to guarantee the empirical character of the explanatory statement. Similarly, the presence of certain motivations may be ascertainable only by indirect methods, which may include reference to linguistic utterances of the subject in question, slips of pen or tongue, et cetera; but as long as these methods are "operationally determined" with reasonable clarity and precision, there is no essential difference in this respect between motivational explanation and causal explanation in physics.

A potential danger of explanation by motives lies in the fact that the method lends itself to the facile construction of *ex post facto* accounts without predictive force. An action is often explained by attributing it to motives conjectured only after the action has taken place. While this procedure is not in itself objectionable, its soundness requires that (1) the motivational assumptions in question be capable of test, and (2) that suitable general laws be available to lend explanatory power to the assumed motives. Disregard of these requirements frequently deprives alleged motivational explanations of their cognitive significance.

The explanation of an action in terms of the agent's motives is sometimes considered as a special kind of teleological explanation. As was pointed out above, motivational explanation, if adequately formulated, conforms to the conditions for causal explanation, so that the term "teleological" is a misnomer if it is meant to imply either a noncausal character of the explanation or a peculiar determination of the present by the future. If this is borne in mind, however, the term "teleological" may be viewed, in this context, as referring to causal explanations in which some of the antecedent conditions are motives of the agent whose actions are to be explained.[11]

Teleological explanations of this kind have to be distinguished from a much more sweeping type, which has been claimed by certain schools of thought to be indispensable especially in biology. It consists in explaining characteristics of an organism by reference to certain ends or purposes which the characteristics are said to serve. In contradistinction to the cases examined before, the ends are not assumed here to be consciously or subconsciously pursued by the organism in question. Thus, for the phenomenon of mimicry, the explanation is sometimes offered that it serves the purpose of protecting the animals endowed with it from detection by its pursuers and thus tends to preserve the species. Before teleological hypotheses of this kind can be appraised as to their potential explanatory power, their meaning has to be clarified. If they are intended somehow to express the idea that the purposes they refer to are inherent in the design of the universe, then clearly they are not capable of empirical test and thus violate the requirement (R3) stated in §2. In certain cases, however, assertions about the purposes of biological characteristics may be translatable into statements in nonteleological terminology which assert that those characteristics function in a specific manner which is essential to keeping the organism alive or to preserving the species.[12] An attempt to state precisely what is meant by this latter assertion—or by the similar one that without those characteristics, and other things being equal, the organism or the species would not survive—encounters considerable difficulties. But these need not be discussed here. For even if we assume that biological statements in teleological form can be adequately translated into descriptive statements about the life-preserving function of certain biological characteristics, it is clear that (1) the use of the concept of purpose is not essential in these contexts, since the term "purpose" can be completely eliminated from the statements in question, and (2) teleological assumptions, while now endowed with empirical content, cannot serve as explanatory principles in the customary contexts. Thus, e.g., the fact that a given species of butterfly displays a particular kind of coloring cannot be inferred from—and therefore cannot be explained by means of—the statement that this type of coloring has the effect of protecting the butterflies from detection by pursuing birds, nor can the presence of red corpuscles in the human blood be inferred from the statement that those corpuscles have a specific function in assimilating oxygen and that this function is essential for the maintenance of life.

One of the reasons for the perseverance of teleological considerations in biology probably lies in the fruitfulness of the teleological approach as a heuristic device: Biological research which was psychologically motivated by a teleological orientation, by an interest in purposes in nature, has frequently led to important results which can be stated in nonteleological terminology and which increase our knowledge of the causal connections between biological phenomena.

Another aspect that lends appeal to teleological considerations is their anthropomorphic character. A teleological explanation tends to make us feel that we really "understand" the phenomenon in question, because it is accounted for in terms of purposes, with which we are familiar from our own experience of purposive behavior. But it is important to distinguish here understanding in the psychological sense of a feeling of empathic familiarity from understanding in the theoretical, or cognitive, sense of exhibiting the phenomenon to be explained as a special case of some general regularity. The frequent insistence that explanation means the reduction of something unfamiliar to ideas or experiences already familiar to us is indeed misleading. For while some scientific explanations do have this psychological effect, it is by no means universal: The free fall of a physical body may well be said to be a more familiar phenomenon than the law of gravitation, by means of which it can be explained; and surely the basic ideas of the theory of relativity will appear to many to be far less familiar than the phenomena for which the theory accounts.

"Familiarity" of the explanans is not only not necessary for a sound explanation, as has just been noted; it is not sufficient either. This is shown by the many cases in which a proposed explanans sounds suggestively familiar, but upon closer inspection proves to be a mere metaphor, or to lack testability, or to include no general laws and therefore to lack explanatory power. A case in point is the neovitalistic attempt to explain biological phenomena by reference to an entelechy or vital force. The crucial point here is not—as is sometimes said—that entelechies cannot be seen or otherwise directly observed; for that is true also of gravitational fields, and yet, reference to such fields is essential in the explanation of various physical phenomena. The decisive difference between the two cases is that the physical explanation provides (1) methods of testing, albeit indirectly, assertions about gravitational fields, and (2) general laws concerning the strength of gravitational fields, and the behavior of objects moving in them. Explanations by entelechies satisfy the analogue of neither of these two conditions. Failure to satisfy the first condition represents a violation of (R3); it renders all statements about entelechies inaccessible to empirical test and thus devoid of empirical meaning. Failure to comply with the second condition involves a violation of (R2). It deprives the concept of entelechy of all explanatory import; for explanatory power never resides in a concept, but always in the general laws in which it functions. Therefore, notwithstanding the feeling of familiarity it may evoke, the neovitalistic account cannot provide theoretical understanding.

The preceding observations about familiarity and understanding can be applied, in a similar manner, to the view held by some scholars that the explanation, or the understanding, of human actions requires an empathic understanding of the personalities of the agents.[13] This understanding of another person in terms of one's own psychological functioning may prove a useful heuristic device in the search for general psychological principles which might provide a theoretical explanation; but the existence of empathy on the part of the scientist is neither a necessary nor a sufficient condition for the explanation, or the scientific understanding, of any human action. It is not necessary, for the behavior of psychotics or of people belonging to a culture very different from that of the scientist may sometimes be explainable and predictable in terms of general principles even though the scientist who establishes or applies those principles may not be able to understand his subjects empathically. And empathy is not sufficient to guarantee a sound explanation, for a strong feeling of empathy may exist even in cases where we completely misjudge a given personality. Moreover, as Zilsel has pointed out, empathy leads with ease to incompatible results; thus, when the population of a town has long been subjected to heavy bombing attacks, we can understand, in the empathic sense, that its morale should have broken down completely, but we can understand with the same ease also that it should have developed a defiant spirit of resistance. Arguments of this kind often appear quite convincing; but they are of an *ex post facto* character and lack cognitive significance unless they are supplemented by testable explanatory principles in the form of laws or theories.

Familiarity of the explanans, therefore, no matter whether it is achieved through the use of teleological terminology, through neovitalistic metaphors, or through other means, is no indication of the cognitive import and the predictive force of a proposed explanation. Besides, the extent to which an idea will be considered as familiar varies from person to person and from time to time, and a psychological factor of this kind certainly cannot serve as a standard in assessing the worth of a proposed explanation. The decisive requirement for every sound explanation remains that it subsume the explanandum under general laws. . . .

II. LOGICAL ANALYSIS OF LAW AND EXPLANATION

4. *Problems of the Concept of General Law.* From our general survey of the characteristics of scientific explanation, we now turn to a closer examination of its logical structure. The explanation of a phenomenon, we noted, consists in its subsumption under laws or under a theory. But what is a law, what is a theory? While the meaning of these concepts seems intuitively clear, an attempt to construct adequate explicit definitions for them encounters considerable difficulties. In the present section, some basic problems of the concept of law will be described and analyzed; in the next section, we intend to propose, on the basis of the suggestions thus obtained, definitions of law and of explanation for a formalized model language of a simple logical structure [omitted here—Eds.].

The concept of law will be construed here so as to apply to true statements only. The apparently plausible alternative procedure of requiring high confirmation rather than truth of a law seems to be inadequate: It would lead to a relativized concept of law, which would be expressed by the phrase "sentence S is a law relative to the evidence E." This does not accord with the meaning customarily assigned to the concept of law in science and in methodological inquiry. Thus, for example, we would not say that Bode's general formula for the distance of the planets from the sun was a law relative to the astronomical evidence available in the 1770s, when Bode propounded it, and that it ceased to be a law after the discovery of Neptune and the determination of its distance from the sun; rather, we would say that the limited original evidence had given a high probability to the assumption that the formula was a law, whereas more recent additional information reduced that probability so much as to make it practically certain that Bode's formula is not generally true, and hence not a law.[14]

Apart from being true, a law will have to satisfy a number of additional conditions. These can be studied independently of the factual requirement of truth, for they refer, as it were, to all logically possible laws, no matter whether factually true or false. Adopting a term proposed by Goodman,[15] we will say that a sentence is *lawlike* if it has all the characteristics of a general law, with the possible exception of truth. Hence, every law is a lawlike sentence, but not conversely.

Our problem of analyzing the notion of law thus reduces to that of explicating the concept of lawlike sentence. We shall construe the class of lawlike sentences as including analytic general statements, such as "A rose is a rose," as well as the lawlike sentences of empirical science, which have empirical content.[16] It will not be necessary to require that each lawlike sentence permissible in explanatory contexts be of the second kind; rather, our definition of explanation will be so constructed as to guarantee the factual character of the totality of the laws—though not of every single one of them—which function in an explanation of an empirical fact.

What are the characteristics of lawlike sentences? First of all, lawlike sentences are statements of universal form, such as "All robins' eggs are greenish-blue," "All metals are conductors of electricity," "At constant pressure, any gas expands with increasing temperature." As these examples illustrate, a lawlike sentence usually is not only of universal, but also of conditional form; it makes an assertion to the effect that universally, if a certain set of conditions, C, is realized, then another specified set of conditions, E, is realized as well. The standard form for the symbolic expression of a lawlike sentence is therefore the universal conditional. However, since any conditional statement can be transformed into a nonconditional one, conditional form will not be considered as essential for a lawlike sentence, while universal character will be held indispensable.

But the requirement of universal form is not sufficient to characterize lawlike sentences. Suppose, for example, that a given basket, b, contains at a certain time t a number of red apples and nothing else.[17] Then the statement

(S_1) Every apple in basket b at time t is red

is both true and of universal form. Yet the sentence does not qualify as a law; we would refuse, for example, to explain by subsumption under it the fact that a particular apple chosen at random from the basket is red. What distinguishes S_1 from a lawlike sentence? Two points suggest themselves, which will be considered in turn, namely, finite scope and reference to a specified object.

First, the sentence S_1 makes in effect an assertion about a finite number of objects only, and this seems irreconcilable with the notion of law.[18] But are not Kepler's laws considered as lawlike although they refer to a finite set of planets only? And might we not even be willing to consider as lawlike a sentence such as the following?

(S_2) All the sixteen ice cubes in the freezing tray of this refrigerator have
    a temperature of less than 10 degrees centigrade.

This point might well be granted; but there is an essential difference between S_1, on the one hand, and Kepler's laws, as well as S_2, on the other: The latter, while finite in scope, are known to be consequences of more comprehensive laws whose scope is not limited, while for S_1 this is not the case.

Adopting a procedure recently suggested by Reichenbach,[19] we will therefore distinguish between fundamental and derivative laws. A statement will be called a derivative law if it is of universal character and follows from some fundamental laws. The concept of fundamental law requires further clarification; so far, we may say that fundamental laws, and similarly fundamental lawlike sentences, should satisfy a certain condition of nonlimitation of scope.

It would be excessive, however, to deny the status of fundamental lawlike sentence to all statements which, in effect, make an assertion about a finite class of objects only, for that would rule out also a sentence such as "All robins' eggs are greenish-blue," since presumably the class of all robins' eggs—past, present, and future—is finite. But again, there is an essential difference between this sentence and, say, S_1. It requires empirical knowledge to establish the finiteness of the class of robins' eggs, whereas, when the sentence S_1 is construed in a manner which renders it intuitively unlawlike, the terms "basket b" and "apple" are understood so as to imply finiteness of the class of apples in the basket at time t. Thus, so to speak, the meaning of its constitutive terms alone—without additional factual information—entails that S_1 has a finite scope. Fundamental laws, then, will have to be construed so as to satisfy a condition of nonlimited scope; our formulation of that condition however, which refers to what is entailed by "the meaning" of certain expressions, is too vague and will have to be revised later. Let us note in passing that the stipulation here envisaged would bar from the class of fundamental lawlike sentences also such undesirable candidates as "All uranic objects are spherical," where "uranic" means the property of being the planet Uranus; indeed, while this sentence has universal form, it fails to satisfy the condition of nonlimited scope.

In our search for a general characterization of lawlike sentences, we now turn to a second clue which is provided by the sentence S_1. In addition to violating the condition of nonlimited scope, that sentence has the peculiarity of making reference to a particular object, the basket b; and this, too, seems to violate the universal character of a law.[20] The restriction which seems indicated here should again be applied to fundamental lawlike sentences only; for a true general statement about the free fall of physical bodies on the moon, while referring to a particular object, would still constitute a law, albeit a derivative one.

It seems reasonable to stipulate, therefore, that a fundamental lawlike sentence must be of universal form and must contain no essential—i.e., uneliminable—occurrences of designations for particular objects. But this is not sufficient; indeed, just at this point, a particularly serious difficulty presents itself. Consider the sentence.

(S_3) Everything that is either an apple in basket b at time t or a sample of ferric oxide is red.

If we use a special expression, say "x is ferple," as synonymous with "x is either an apple in b at t or a sample of ferric oxide," then the content of S_3 can be expressed in the form.

(S_4) Everything that is ferple is red.

The statement thus obtained is of universal form and contains no designations of particular objects, and it also satisfies the condition of nonlimited scope; yet clearly, S_4 can qualify as a fundamental lawlike sentence no more than can S_3.

As long as "ferple" is a defined term of our language, the difficulty can readily be met by stipulating that after elimination of defined terms, a fundamental lawlike sentence must not contain essential occurrences of designations for particular objects. But this way out is of no avail when "ferple," or another term of its kind, is a primitive predicate of the language under consideration. This reflection indicates that certain restrictions have to be imposed upon those predicates—i.e., terms for properties or relations—which may occur in fundamental lawlike sentences.[21]

More specifically, the idea suggests itself of permitting a predicate in a fundamental lawlike sentence only if it is purely universal, or, as we shall say, purely qualitative, in character; in other words, if a statement of its meaning does not require reference to any one particular object or spatio-temporal location. Thus, the terms "soft," "green," "warmer than," "as long as," "liquid," "electrically charged," "female," "father of," are purely qualitative predicates, while "taller than the Eiffel Tower," "medieval," "lunar," "arctic," "Ming" are not.[22]

Exclusion from fundamental lawlike sentences of predicates which are not purely qualitative would at the same time ensure satisfaction of the condition of nonlimited scope; for the meaning of a purely qualitative predicate does not require a finite extension; and indeed, all the sentences considered above which violate the condition of nonlimited scope make explicit or implicit reference to specific objects.

The stipulation just proposed suffers, however, from the vagueness of the concept of purely qualitative predicate. The question whether indication of the meaning of a given predicate in English does or does not require reference to some specific object does not always permit of an unequivocal answer since English as a natural language does not provide explicit definitions or other clear explications of meaning for its terms. It seems therefore reasonable to attempt definition of the concept of law not with respect to English or any other natural language, but rather with respect to a formalized language—let us call it a model language *L*—which is governed by a well-determined system of logical rules, and in which every term either is characterized as primitive or is introduced by an explicit definition in terms of the primitives.

This reference to a well-determined system is customary in logical research and is indeed quite natural in the context of any attempt to develop precise criteria for certain logical distinctions. But it does not by itself suffice to overcome the specific difficulty under discussion. For while it is now readily possible to characterize as not purely qualitative all those among the defined predicates in *L* whose definiens contains an essential occurrence of some individual name, our problem remains open for the primitives of the language, whose meanings are not determined by definitions with the language, but rather by semantical laws of interpretation. For we want to permit the interpretation of the primitives of *L* by means of such attributes as blue, hard, solid, warmer, but not by the properties of being a descendant of Napoleon, or an arctic animal, or a Greek statue; and the difficulty is precisely that of stating rigorous criteria for the distinction between the permissible and the nonpermissible interpretations. Thus the problem of finding an adequate definition for purely qualitative attributes now arises again; namely for the concepts of the metalanguage in which the semantical interpretation of the primitives is formulated. We may postpone an encounter with the difficulty by presupposing formalization of the semantical metalanguage, the meta-metalanguage, and so forth, but somewhere, we will have to stop at a nonformalized metalanguage; and for it, a characterization of purely qualitative predicates will be needed and will present much the same problems as nonformalized English, with which we began. The characterization of a purely qualitative predicate as one whose meaning can be made explicit without reference to any one particular object points to the intended meaning but does not explicate it precisely, and the problem of an adequate definition of purely qualitative predicates remains open.

There can be little doubt, however, that there exists a large number of predicates which would be rather generally recognized as purely qualitative in the sense here pointed out, and as permissible in the formulation of fundamental lawlike sentences; some examples have beeen given above, and the list could be readily enlarged. When we speak of purely qualitative predicates, we shall henceforth have in mind predicates of this kind. . . .

NOTES

1. This essay grew out of discussions with Dr. Paul Oppenheim; it was published in coauthorship with him and is here reprinted with his permission. Our individual contributions cannot be separated in detail; the present author is responsible . . . for the final formulation of the entire text.

Some of the ideas set forth in part II originated with our common friend, Dr. Kurt Grelling, who suggested them to us in a discussion carried on by correspondence. Grelling and his wife subsequently became victims of the Nazi terror during the Second World War; by including in this essay at least some of Grelling's contributions, which are explicitly identified, we hope to realize his wish that his ideas on this subject might not entirely fall into oblivion.

Paul Oppenheim and I are much indebted to Professors Rudolf Carnap, Herbert Feigl, Nelson Goodman, and W. V. Quine for stimulating discussions and constructive criticism.

2. These two expressions, derived from the Latin *explanare*, were adopted in preference to the perhaps more customary terms "explicandum" and "explicans" in order to reserve the latter for use in the context of explication of meaning, or analysis. . . .

3. (Added in 1964.) Requirement (R4) characterizes what might be called a correct or *true explanation*. In analysis of the logical structure of explanatory arguments, therefore, that requirement may be disregarded. . . .

4. (1924) p. 533.

5. (Added in 1964.) Or rather, causal explanation is one variety of the deductive type of explanation here under discussion.

6. The account given above of the general characteristics of explanation and prediction in science is by no means novel; it merely summarizes and states explicitly some fundamental points which have been recognized by many scientists and methodologists.

Thus, e.g., Mill says: "An individual fact is said to be explained, by pointing out its cause, that is, by stating the law or laws of causation, of which its production is an instance," and "a law or uniformity in nature is said to be explained, when another law or laws are pointed out, of which that law itself is but a case, and from which it could be deduced" (1858, book III, chapter XII, section 1). Similarly, Jevons, whose general characterization of explanation was critically discussed above, stresses that "the most important process of explanation consists in showing that an observed fact is one case of a general law or tendency" (1924, p. 533). Ducasse states the same point as follows: "Explanation essentially consists in the offering of a hypothesis of fact, standing to the fact to be explained as case of antecedent to case of consequent of some already known law of connection" (1925, pp. 150–51). A lucid analysis of the fundamental structure of explanation and prediction was given by Popper in (1935), section 12, and, in an improved version, in his work (1945), especially in chapter 25 and in note 7 for that chapter.—For a recent characterization of explanation as subsumption under general theories, cf., for example, Hull's concise discussion in (1943a), chapter 1. A clear elementary examination of certain aspects of explanation is given in Hospers (1946), and a concise survey of many of the essentials of scientific explanation which are considered in the first two parts of the present study may be found in Feigl (1945), pp. 284 ff.

7. On the subject of explanation in the social sciences, especially in history, cf. also the following publications, which may serve to supplement and amplify the brief discussion to be presented here: Hempel (1942); Popper (1945); White (1943); and the articles *Cause* and *Understanding* in Beard and Hook (1946).

8. The illustration is taken from Bonfante (1946), section 3.

9. While in each of the last two illustrations, certain regularities are unquestionably relied upon in the explanatory argument, it is not possible to argue convincingly that the intended laws, which at present cannot all be stated explicitly, are of a causal rather than a statistical character. It is quite possible that most or all of the regularities which will be discovered as sociology develops will be of a statistical type. Cf., on this point, the suggestive observations in Zilsel (1941), section 3, and (1941a). This issue does not affect, however, the main point we wish to make here, namely that in the social no less than in the physical sciences, subsumption under general regularities is indispensable for the explanation and the theoretical understanding of any phenomenon.

10. Cf., for example, F. H. Knight's presentation of this argument in (1924), pp. 252–52.

11. For a detailed logical analysis of the concept of motivation in psychological theory, see Koch (1941). A stimulating discussion of teleological behavior from the standpoint of contemporary physics and biology is contained in the article (1943) by Rosenblueth, Wiener, and Bigelow. . . .

12. An analysis of teleological statements in biology along these lines may be found in Woodger (1929), especially pp. 432 ff; essentially the same interpretation is advocated by Kaufmann in (1944), chapter 8.

13. For a more detailed discussion of this view on the basis of the general principles outlined above, cf. Zilsel (1941), sections 7 and 8, and Hempel (1942), section 6.

14. The requirement of truth for laws has the consequence that a given empirical statement S can never be definitely known to be a law, for the sentence affirming the truth of S is tantamount to S and is therefore capable only of acquiring a more or less high probability, or degree of confirmation, relative to the experimental evidence available at any given time. On this point, cf. Carnap (1946). For an excellent nontechnical exposition of the semantical concept of truth, which is here invoked, the reader is referred to Tarski (1944).

15. (1947), p. 125.

16. This procedure was suggested by Goodman's approach in (1947). Reichenbach, in a detailed examination of the concept of law, similarly construes his concept of nomological statement as including both analytic and synthetic sentences: cf. (1947), chapter VIII.

17. The difficulty illustrated by this example was stated concisely by Langford (1941), who referred to it as the problem of distinguishing between universals of fact and causal universals. For further discussion and illustration of this point, see also Chisholm (1946), especially pp. 301f. A systematic analysis of the problem was given by Goodman in (1947), especially part III. While not concerned with the specific point under discussion, the detailed examination of counterfactual conditionals and their relation to laws of nature, in chapter VIII of Lewis (1946), contains important observations on several of the issues raised in the present section.

18. The view that laws should be construed as not being limited to a finite domain has been expressed, among others, by Popper (1935), section 13 and by Reichenbach (1947), p. 369.

19. (1947), p. 361. Our terminology as well as the definitions to be proposed later for the two types of law do not coincide with Reichenbach's, however.

20. In physics, the idea that a law should not refer to any particular object has found its expression in the maxim that the general laws of physics should contain no reference to specific space-time points, and that spatio-temporal coordinates should occur in them only in the form of differences or differentials.

21. The point illustrated by the sentences S_3 and S_4 above was made by Goodman, who has also emphasized the need to improve certain restrictions upon the predicates whose occurrence is to be permissible in lawlike sentences. These predicates are essentially the same as those which Goodman calls projectible. Goodman has suggested that the problems of establishing precise criteria for projectibility, of interpreting counterfactual conditionals, and of defining the concept of law are so intimately related as to be virtually aspects of a single problem. Cf. his articles (1946) and (1947). One suggestion for an analysis of projectibility has been made by Carnap in (1947). Goodman's note (1947a) contains critical observations on Carnap's proposals.

22. That laws, in addition to being of universal form, must contain only purely universal predicates was argued by Popper (1935, sections 14, 15). Our alternative expression "purely qualitative predicate" was chosen in analogy to Carnap's term "purely qualitative property" cf. (1947). The above characterization of purely universal predicates seems preferable to a simpler and perhaps more customary one, to the effect that a statement of the meaning of the predicate must require no reference to particular objects. That formulation might be too restrictive since it could be argued that stating the meaning of such purely qualitative terms as "blue" or "hot"

requires illustrative reference to some particular object which has the quality in question. The essential point is that no one specific object has to be chosen; any one in the logically unlimited set of blue or of hot objects will do. In explicating the meaning of "taller than the Eiffel Tower," "being an apple in basket *b* at time *t*," "medieval," et cetera, however, reference has to be made to one specific object or to some one in a limited set of objects.

12

Laws and Conditional Statements

Karel Lambert and Gordon Britten

We can begin with the concept of a law. The first thing to notice is that "law," like "cause" and "explanation," is used in a variety of ways, many of them only vaguely related.[1] In fact, we ourselves have used "law" (or "general law") interchangeably with "generalization," "regularity," and "statement of *universal form.*" Before proceeding much further, these various expressions will have to be sorted out.

Following what is perhaps the most widespread custom, let us say that laws are *statements* that express *regularities.* The properties of laws in which we will be interested will be properties of statements. It is important to keep this in mind, for confusion results when one fails to distinguish between laws as statements and laws as what certain statements express. "Generalization" is ambiguous in exactly the same way: for reasons of clarity, we construe generalizations also as statements and not as what statements of a certain kind express. The question, then, is to decide whether there is a distinction to be made between laws and generalizations, as statements of different types.

One further preliminary. We are going to restrict our attention to the simplest case, where a law has the form of a nonstatistical *generalized conditional* statement. In other words: where a law has the form "For any (every, all) object(s) x, if x is such and such then x is so and so." Such nonstatistical generalized conditionals may also be said to be statements of universal form. For they assert that every object (or event or condition) that has certain properties also has certain other properties. One of the laws we have mentioned previously is a case in point: "For every piece of thread T, there is a threshold weight W, such that if an object is suspended from T and exceeds the threshold weight W, then the thread will break." It should be noticed that mathematically phrased laws merely appear to have form. "$G_S = W/V$." What is suppressed in the original mathematical statement is the reference to all objects, that is, its generality.

Let us assume that laws are statements of universal form.[2] At the same time, it does not seem to be the case that all statements of universal form are laws (for example, "All animals are men"). This claim can be defended by considering additional criteria that statements of universal form must satisfy to be counted as laws.

One criterion is that laws are universal statements that have empirical content. A use to which this criterion is frequently put is to rule out nonsense laws and spurious explanations based on them. For example, one might want to say that the universal statement "All glubbifiers are irascible" is patent nonsense, hence no law, and any "explanation" in which it figured would be unacceptable. It is not at all clear, however, that the line between "sensical" and nonsensical statements coincides with the (for the moment presumed) line between statements that have empirical content and those that do not. It is even to be wondered whether there is *any* effective way of separating sense from nonsense. For many of his sixteenth-century contemporaries, Copernicus's heliocentric hypothesis was the "Copernican Paradox." For one thing, it was virtually part of the meaning of the word "earth" that this planet be the center of the universe.

The criterion of empirical content is also intended to rule out, for example, "All bachelors are unmarried men," as a law, and the following as a genuine explanation of the fact that Jones is an unmarried man: "Jones is a bachelor and all bachelors are unmarried men; therefore, Jones is an unmarried man."

Indeed, we would undoubtedly be very hesitant to accept this as an *explanation* of the fact that Jones is an unmarried man. And it is tempting, as a result, to distinguish between laws and nonlaws in terms of the empirical content of the former. . . . *Analytic* statements (that is, statements true by virtue of the meanings of the words they contain), even of universal form, are not *laws*. There is something circular about the "explanatory" arguments into which they enter. Thus, to consider another example:

1. All paranoids suffer delusions of grandeur.
2. Jones is a paranoid.

Therefore,

3. Jones suffers from delusions of grandeur.

If we assume that the definition of the word "paranoid" includes as part of its meaning the expression "suffers from delusions of grandeur" (that is, [1] is analytic), this "explanation" of Jones's suffering delusions of grandeur because he is paranoid does not appear to advance our understanding.

Despite the intuitive appeal of these remarks, however, we must be very careful. The notion of "empirical content" and the attendant analytic/synthetic distinction are unclear. Admittedly in the cases just considered, the intended analytic/synthetic, empirical/nonempirical contrasts are fairly easy to draw.[3] But cases of this kind are by and large trivial. In more complicated cases of the sort one typically encounters in science, a great many principles are involved in the explanation of particular events. Perhaps one can distinguish between these as having "more" or "less" empirical content; within the context of some specified theory, one can make the distinction sharper still. If the line of argument sketched in the final section of the previous chapter was sound, however, a general distinction between statements that have empirical content and those that do not, for example, between the laws of mathematics and the laws of physics, is at least problematic (once again excepting trivialities like "All bachelors are unmarried men," which are of negligible scientific and systematic importance). We do not have to abandon our earlier requirements that explanations have empirical content. But we should be alert to the possibility that they might need modification.

A second criterion laws might be expected to fulfill is that they be true. Thus, "All swans are white" does not count as a law (assuming provisionally that it satisfies the other criteria) because there are black swans; we cannot explain a particular swan's being white on the basis of it.

This criterion seems even more intuitive than the first. Surely we would not countenance false laws of nature. But there are difficulties with it as well.

The first is that many statements commonly regarded as laws, for example, those of classical mechanics, hold only approximately. They are not, in all strictness, true. If we insist that explanation essentially involve laws and that laws must be true, we are faced with the unpleasant choice of saying either that, for example, Galileo did not *really* explain certain facts about the behavior of objects on inclined planes, or that when he made them his statements were true but now no longer are. Perhaps a less painful move is to relax the requirement that laws be true. There are also reasons . . . for maintaining that at least some laws of nature are neither true nor false. Rather, they serve as general principles, somewhat on the order of moral rules,[4] whose function is to provide a theoretical framework with respect to which research can be carried out and empirical generalizations tested. To put it in a slightly different way, at least some laws of nature seem to have a primarily normative, not descriptive, function.

In any case, there are reasons for leaving it open for now whether laws must be true or not. We can do this by restricting our discussion to *lawlike* statements. Statements are lawlike if they satisfy all the other criteria for laws, independent of whether or not they have a truth value.

The third, and most important, of these criteria can best be taken up by way of example.[5] Consider the two statements:

1. All persons sitting on a certain bench in Boston are Irish.
2. Any body subject to no external forces maintains a constant velocity.

There are many (interrelated) differences between these two statements that might lead us to call the second, but not the first, "lawlike," although both are of universal form. Recall that it is not in virtue of their form that laws are to be distinguished from nonlaws.

To begin with, the first statement seems to be an *accidental* generalization, about what in fact happens to be the case, whereas the second is a law of nature. It does not follow from the first that if someone *were* to sit on a certain bench in Boston he *would* be Irish, nor could we very well explain someone's being Irish by referring to the fact that he was sitting on that bench. But it does follow from the second that if a body *were* subject to no external forces it *would* maintain a constant velocity, and it does (intuitively) explain a body's having a constant velocity to be told that it is subject to no external forces.

Statements of the form "If *A* were (had been) the case, then *B* would be (would have been) the case," where in fact *A* is not (has not been) the case, are called *counterfactual conditionals.* The present suggestion is that lawlike statements support counterfactual conditionals, while nonlawlike statements do not. In turn, the ability to support counterfactuals (as they are called for short) is linked to the predictive, and hence explanatory, force that laws in contrast to nonlaws have.

Conditional statements quite generally allow us to speak about potential (as distinct from actual) events, cases in which should *A* occur, *B* would also. There is thus a sense in which lawlike statements, insofar as they support conditionals, possess a generality that nonlawlike statements do not. The accidental generalization, "All persons sitting on a certain bench in Boston are Irish," is equivalent to the finite conjunction of statements, "Pat is sitting on a certain bench in Boston and is Irish," and so on, whereas the lawlike statement, "Any body subject to no external forces maintains a constant velocity," is unrestrictedly about all objects whatsoever. This is perhaps another way to indicate its explanatory force.

A distinction between laws and generalizations on this basis has recently been given a more than academic prominence in the controversy surrounding cigarette smoking and health. Spokesmen for the tobacco industry are inclined to say that statements expressing a correlation between cigarette smoking and, for example, lung cancer are merely accidental generalizations. A high statistical correlation should not be mistaken for a causal connection; generalizations linking cigarette smoking and lung cancer do not support conditionals to the effect that if someone were, for example, to smoke two packs of cigarettes daily for a specified period of time, the chances of his getting lung cancer would be such and such. The Public Health Service, on the other hand, tends to grant the status of laws to these same generalizations. Which is to say, at least, that they do support the appropriate conditionals. Waiving difficulties with the notion of "confirmation" for the moment, the

P.H.S. might want to stake its case further on the extent to which projections made on the basis of the purported laws were in fact realized. Of course, this still would not force the tobacco industry to say that a "causal" connection had been demonstrated; other criteria of lawlikeness, they could insist, remain to be satisfied.[6]

We take it, then, that it is apparently a necessary (if not also a sufficient) criterion for being a lawlike statement that a statement of universal form support conditionals, in particular counterfactual conditionals. The usefulness of this criterion depends in part on the sureness of our grip on the notions of "support" and "counterfactual conditional."

One way of analyzing such conditionals, in terms of the so-called material conditional of classical logic, will not work. The material conditional— "If P then Q"—is so defined as to be true whenever "P" is false or "Q" is true. Since the antecedents of counterfactual conditionals are always unrealized, that is, "P" is always false, counterfactuals construed as material conditionals would always be true. But this is not acceptable, as can be seen by comparing the two counterfactuals.

1. If any body were subject to no external force, it would maintain a constant velocity.
2. If any body were subject to no external force, it would *not* maintain a constant velocity.

On the model of the material conditional, both counterfactuals are true, although their consequents are incompatible! A related way of rejecting the material conditional analysis is to point up the importance of distinguishing between conditionals that are vacuously true (true simply because their antecedents are unrealized) and those that are not. The material conditional analysis obviously does not allow us to do this.

More generally, many conditionals simply are not truth-functional. We may know the truth values of their component statements and, from this information alone, be unable to determine the truth value of the compound. For example, I may know that "Willie Mays played in the American League" is false, as is "Willie Mays hit four hundred," without knowing whether or not Mays would have hit four hundred had he played in the American League.[7]

A more promising suggestion is that a conditional is like an enthymematic argument, one premise of which is the antecedent of the conditional, the conclusion of which is the consequent. What is omitted, and what serves to connect antecedent with consequent, is simply the law (or laws) that we have taken to support it. A (vastly oversimplified) case in point: "Had the match been struck it would have lighted" is a counterfactual, the consequent of which is derivable from the antecedent with the help of certain laws about matches being struck and lighting.

The proponents of this suggestion maintain that, intuitively, there is a "connection" between maintaining a constant velocity and being acted on by no external forces, whereas there is no such "connection" between sitting on a certain bench in Boston and being Irish. It is this connection that allows for valid predictions, and hence, explanations. But the connection between antecedent and consequent in a true conditional is none other than that which a law provides. From the antecedent and the law, we can derive the consequent. It is in this way that laws serve to support conditionals. Similarly, the *necessity* of a match's lighting when struck is simply a function of the fact that it follows from the law as a logical consequence.

There is a circle here, of course. We have distinguished between laws and generalizations by claiming that only the former supported counterfactual conditionals, and we have analyzed such conditionals in terms of the lawlike connection that exists between their antecedent and consequent. The circle does not undermine the distinction suggested. But it does indicate that we do not understand the distinction very well unless the circle can be broken. One way in which to break the circle is to say that lawlike statements, in contrast to accidental generalizations, are *confirmed by their instances.*[8] . . .

One could also argue, as have some philosophers, that additional criteria of lawlikeness are required if laws are to play the explanatory role assigned to them. Support of counterfactual conditionals is not enough. For example, even if it were a conditional-supporting generalization that every time the sky is red in the morning it rains in the afternoon,[9] we would not explain the fact that it is raining in the afternoon by pointing to this purportedly "lawlike" statement and the fact that the sky was red this morning. Other philosophers reject this sort of distinction between explanatory laws and nonexplanatory generalizations, contending that it is rooted in naive (unexamined) causal and teleological intuitions. For them, the only distinction to be made is that between laws and accidental generalizations, by means of the counterfactual conditional criterion.

But consider these two statements:

1. All pieces of copper wire at $-270°$ C are good conductors.
2. All unicorns are fleet of foot.

Both of these statements are unrestricted universals, and presumably both would support (insofar as they implied) the corresponding counterfactual conditionals. Take it as (nonanalytically) true that if anything were a unicorn it would be fleet of foot. Still, it is unlikely that anyone would regard the statement about unicorns as a law. Nor can the two statements be distinguished by virtue of the fact that the first but not the second has positive instances. We can easily suppose that no copper has been checked out at

−270° C. The difference, rather, seems to lie in the fact that the copper-conducting statement is systematically connected with a larger theoretical framework, whereas the unicorn-fleeting statement is not. To put it in a slightly different way, what enables us to call the first but not the second statement "lawlike" is the *position* of the first in the existing body of knowledge. In particular, whether or not it has any positive instances, it will be granted law status if it *follows from* other laws and theories. That copper at −270° C is a good conductor, for instance, ostensibly is a consequence of the law that all copper is a good conductor.

On the other hand, even well-confirmed unrestricted universals will not count as laws if they are not systematically interconnected with a larger theoretical framework. What leads us to reject the red sky in the morning/rain in the afternoon generalization as a law, or an *explanatory* law, is that it is an isolated assertion having no apparent theoretical ramifications.[10]

This point can be elaborated. It is not enough that a particular universal statement be exceptionless to qualify as a law. It must also be the case that the theoretical framework to which it is connected does not permit exceptions to it. Certainly nothing in current scientific theory precludes, for example, that anyone sitting on a certain bench in Boston is not Irish, nor that unicorns are not fleet of foot. Similarly, whether there are laws linking cigarette smoking to lung cancer depends on how the statistical generalizations correlating them relate to a larger body of scientific theory.[11] Generalizations that are not systematically connected with other laws are merely accidental, they rule out potential instances (for instance, a non-Irishman sitting on that bench in Boston) that scientific theory otherwise permits. . . .

NOTES

1. In the case of laws of nature and laws of the land, this relation seems to be very vague indeed.

2. We said earlier that certain laws are of statistical or probabilistic form. Probabilistic laws raise their own problems, but none of these affect the points to be made in this section.

3. There is a complication in the "paranoid" case. Very plausibly, we could say that suffering delusions of grandeur was a *symptom* of or criterion for paranoia and that although in virtue of this fact there was a kind of "meaning" connection between them this did not preclude "All paranoids suffer delusions of grandeur" from functioning as a law.

4. We do not mean to suggest that they are like moral rules in all respects.

5. A classic contemporary discussion of many of the points that follow can be found in Nelson Goodman's book, *Fact, Fiction, and Forecast* (Indianapolis, Ind.: The Bobbs-Merrill Company, Inc., 1965).

6. The cigarette smoking–lung cancer controversy illustrates in a very sharp way almost all the issues raised in this chapter. The Public Health Service, for example, is explicitly "inductivist" about explanation: it maintains that a high statistical correlation can, when subjected to a variety of tests, be used to assert a casual connection, that is, can be used to explain individual instances of lung cancer, and so on.

Both sides of the controversy are also quite explicitly aware that in many respects the issue is philosophical: How is the concept of cause to be analyzed? See the very interesting discussion in the Report of the Advisory Committee to the Surgeon General of the Public Health Service, *Smoking and Health* (Princeton, N.J.: Van Nostrand Reinhold Co., 1964), pp. 19–21 and passim.

7. The example is Robert Stalnaker's; his analysis of conditional statements will be mentioned shortly.

8. Another extremely interesting analysis of conditions has recently been advanced by Robert Stalnaker. See his paper. "A Theory of Conditionals," in *Studies in Logical Theory,* ed. Nicholas Rescher (American Philosophical Quarterly Monograph Series, 1968). Stalnaker is concerned with the analysis of conditional statements generally, not primarily, as we have been, with counterfactuals. He rejects both of the analyses suggested above, and proposes instead an analysis in terms of possible or hypothetical states of affairs.

Stalnaker's analysis has a great deal to recommended it. It does not, as he himself notes, however, allow us to bypass the circle described in analyzing counterfactual conditionals. We may know the truth-conditions for conditional statements (that is, the conditions with respect to which they are evaluated) without knowing whether particular conditionals are true or false. And presumably those counterfactuals, at any rate, are true which are supported by laws in the way we suggested originally.

9. That is, even if this were a generalization that held without exception and had unrestricted application.

10. This point is emphasized by Ernest Nagel in the course of an extended discussion of laws and counterfactuals, in *The Structure of Science* (New York: Harcourt, Brace, & World, 1961), pp. 47–78.

11. In this case, perhaps most importantly to a molecular-chemical theory.

13

The Truth Doesn't Explain Much

Nancy Cartwright

INTRODUCTION

Scientific theories must tell us both what is true in nature, and how we are to explain it. I shall argue that these are entirely different functions and should be kept distinct. Usually the two are conflated. The second is commonly seen as a by-product of the first. Scientific theories are thought to explain by dint of the descriptions they give of reality. Once the job of describing is done, science can shut down. That is all there is to do. To describe nature—to tell its laws, the values of its fundamental constants, its mass distributions—is *ipso facto* to lay down how we are to explain it.

This is a mistake, I shall argue; a mistake that is fostered by the covering-law model of explanation. The covering law model supposes that all we need to know are the laws of nature—and a little logic, perhaps a little probability theory—and then we know which factors can explain which others. For example, in the simplest deductive-nomological [D-N] version,[1] the covering law model says that one factor explains another just in case the occurrence of the second can be deduced from the occurrence of the first given the laws of nature.

But the D-N model is just an example. In the sense which is relevant to my claims here, most models of explanation offered recently in the philosophy of science are covering law models. This includes not only Hempel's own inductive-statistical model,[2] but also Patrick Suppes's probabilistic model of causation,[3] Wesley Salmon's statistical relevance model,[4] and even Bengt Hanson's contextualistic model.[5] All these accounts rely on the laws of nature, and just the laws of nature, to pick out which factors we can use in explanation.

A good deal of criticism has been aimed at Hempel's original covering-law models. Much of the criticism objects that these models let in too much.

On Hempel's account it seems we can explain Henry's failure to get pregnant by his taking birth control pills, and we can explain the storm by the falling barometer. My objection is quite the opposite. Covering law models let in too little. With a covering law model we can explain hardly anything, even the things of which we are most proud—like the role of DNA in the inheritance of genetic characteristics, or the formation of rainbows when sunlight is refracted through raindrops. We cannot explain these phenomena with a covering law model, I shall argue, because we do not have laws that cover them. Covering laws are scarce.

Many phenomena which have perfectly good scientific explanations are not covered by any laws. No true laws, that is. They are at best covered by *ceteris paribus* generalizations—generalizations that hold only under special conditions, usually ideal conditions. The literal translation is "other things being equal"; but it would be more apt to read "*ceteris paribus*" as "other things being *right.*"

Sometimes we act as if this does not matter. We have in the back of our minds an "understudy" picture of *ceteris paribus* laws: *ceteris paribus* laws are real laws; they can stand in when the laws we would like to see are not available and they can perform all the same functions, only not quite so well. But this will not do. *Ceteris paribus* generalizations, read literally without the "*ceteris paribus*" modifier, are false. They are not only false, but held by us to be false; and there is no ground in the covering law picture for false laws to explain anything. On the other hand, with the modifier the *ceteris paribus* generalizations may be true, but they cover only those few cases where the conditions are right. For most cases, either we have a law that purports to cover, but cannot explain because it is acknowledged to be false, or we have a law that does not cover. Either way, it is bad for the covering-law picture.

1. CETERIS PARIBUS LAWS

When I first started talking about the scarcity of covering laws, I tried to summarize my view by saying "There are no exceptionless generalizations." Then a friend asked, "How about 'All men are mortal'?" She was right. I had been focusing too much on the equations of physics. A more plausible claim would have been that there are no exceptionless quantitative laws in physics. Indeed not only are there no exceptionless laws, but in fact our best candidates are known to fail. This is something like the Popperian thesis that *every theory is born refuted.* Every theory we have proposed in physics, even at the time when it was most firmly entrenched, was known to be deficient in specific and detailed ways. I think this is also true for every precise quantitative law within a physics theory.

But this is not the point I had wanted to make. Some laws are treated, at least for the time being, as if they were exceptionless, whereas others are not, even though they remain "on the books." Snell's law (about the angle of incidence and the angle of refraction for a ray of light) is a good example of this latter kind. In the optics text I use for reference (Miles V. Klein, *Optics*),[6] it first appears on page 21, and without qualification:

Snell's Law: At an interface between dielectric media, there is (also) a *refracted* ray in the second medium, lying in the plane of incidence, making an angle θ_t with the normal, and obeying Snell's law:

$$sin \ \theta/sin \ \theta_t = n_2/n_1$$

where v_1 and v_2 are the velocities of propagation in the two media, and $n_1 = (c/v_1)$, $n_2 = (c/v_2)$ are the indices of refraction.

It is only some 500 pages later, when the law is derived from the "full electromagnetic theory of light," that we learn that Snell's law as stated on page 21 is true only for media whose optical properties are *isotropic*. (In anisotropic media, "there will generally be *two* transmitted waves.") So what is deemed true is not really Snell's law as stated on page 21, but rather a refinement of Snell's law:

Refined Snell's Law: For any two media which are optically isotropic, at an interface between dielectrics there is a refracted ray in the second medium, lying in the plane of incidence, making an angle θ_t with the normal, such that:

$$sin \ \theta/sin \ \theta_t = n_2/n_1$$

The Snell's law of page 21 in Klein's book is an example of a *ceteris paribus* law, a law that holds only in special circumstances—in this case when the media are both isotropic. Klein's statement on page 21 is clearly not to be taken literally. Charitably, we are inclined to put the modifier "*ceteris paribus*" in front to hedge it. But what does this *ceteris paribus* modifier do? With an eye to statistical versions of the covering law model (Hempel's I-S picture, or Salmon's statistical relevance model, or Suppes's probabilistic model of causation) we may suppose that the unrefined Snell's law is not intended to be a universal law, as literally stated, but rather some kind of statistical law: *for the most part*, at an interface between dielectric media there is *a* refracted ray. . . . But this will not do. For *most* media are optically anisotropic, and in an anisotropic medium there are *two* rays. I think there are no more satisfactory alternatives. If *ceteris paribus* laws are to be true laws, there are no statistical laws with which they can generally be identified.

2. WHEN LAWS ARE SCARCE

Why do we keep Snell's law on the books when we both know it to be false and have a more accurate refinement available? There are obvious pedagogic reasons. But are there serious scientific ones? I think there are, and these reasons have to do with the task of explaining. Specifying which factors are explanatorily relevant to which others is a job done by science over and above the job of laying out the laws of nature. Once the laws of nature are known, we still have to decide what kinds of factors can be cited in explanation.

One thing that *ceteris paribus* laws do is to express our explanatory commitments. They tell what kinds of explanations are permitted. We know from the refined Snell's law that in any isotropic medium, the angle of refraction can be explained by the angle of incidence, according to the equation $sin\ \theta/sin\ \theta_t = n_2/n_1$. To leave the unrefined Snell's law on the books is to signal that the same kind of explanation can be given even for some anisotropic media. The pattern of explanation derived from the ideal situation is employed even where the conditions are less then ideal; and we assume that we can understand what happens in *nearly* isotropic media by rehearsing how light rays behave in pure isotropic cases.

This assumption is a delicate one. It fits far better with the simulacrum account of explanation . . . than it does with any covering law model. For the moment I intend only to point out that it *is* an assumption, and an assumption which (prior to the "full electromagnetic theory") goes well beyond our knowledge of the facts of nature. We *know* that in isotropic media, the angle of refraction is due to the angle of incidence under the equation $sin\ \theta/sin\ \theta_t = n_2/n_1$. We *decide* for the two refracted rays in anisotropic media in the same manner. We may have good reasons for the decision; in this case if the media are nearly isotropic, the two rays will be very close together, and close to the angle predicted by Snell's law; or we believe in continuity of physical processes. But still this decision is not forced by our knowledge of the laws of nature.

Obviously this decision could not be taken if we also had on the books a second refinement of Snell's law, implying that in any anisotropic media the angles are quite different from those given by Snell's law. But laws are scarce, and often we have no law at all about what happens in conditions that are less than ideal.

Covering law theorists will tell a different story about the use of *ceteris paribus* laws in explanation. From their point of view, *ceteris paribus* explanations are elliptical for genuine covering law explanations from true laws which we do not yet know. When we use a *ceteris paribus* "law" which we know to be false, the covering law theorist supposes us to be making a bet about what form the true law takes. For example, to retain Snell's unqualified law would be to bet that the (at the time unknown) law for anisotropic media will entail values "close enough" to those derived from the original Snell law.

I have two difficulties with this story. The first arises from an extreme metaphysical possibility, in which I in fact believe. Covering law theorists tend to think that nature is well-regulated; in the extreme, that there is a law to cover every case. I do not. I imagine that natural objects are much like people in societies. Their behavior is constrained by some specific laws and by a handful of general principles, but it is not determined in detail, even statistically. What happens on most occasions is dictated by no law at all. This is not a metaphysical picture that I urge. My claim is that this picture is as plausible as the alternative. God may have written just a few laws and grown tired. We do not know whether we are in a tidy universe or a untidy one. Whichever universe we are in, the ordinary commonplace activity of giving explanations ought to make sense.

The second difficulty for the ellipsis version of the covering-law account is more pedestrian. Elliptical explanations are not explanations: they are at best assurances that explanations are to be had. The law that is supposed to appear in the complete, correct D-N explanation is not a law we have in our theory, not a law that we can state, let alone test. There may be covering law explanations in these cases. But those explanations are not our explanations; and those unknown laws cannot be our grounds for saying of a nearly isotropic medium, "*sin* $\theta_t \approx k \, (n_2/n_1)$ *because sin* $\theta = k$."

What then are our grounds? I assert only what they are not: they are not the laws of nature. The laws of nature that we know at any time are not enough to tell us what kinds of explanations can be given at that time. That requires a decision; and it is just this decision that covering law theorists make when they wager about the existence of unknown laws. We may believe in these unknown laws, but we do so on no ordinary grounds: they have not been tested, nor are they derived from a higher level theory. Our grounds for believing in them are only as good as our reasons for adopting the corresponding explanatory strategy, and no better.

3. When Laws Conflict

I have been maintaining that there are not enough covering laws to go around. Why? The view depends on the picture of science that I mentioned earlier. Science is broken into various distinct domains: hydrodynamics, genetics, laser theory, . . . We have many detailed and sophisticated theories about what happens within the various domains. But we have little theory about what happens in the intersection of domains.

Diagrammatically, we have laws like

ceteris paribus, (x) (S(x) $\hookrightarrow$ I(x))

and

ceteris paribus, (x) (A (x) $\hookrightarrow$ ⌐ I(x)).

For example, (*ceteris paribus*) adding salt to water decreases the cooking time of potatoes; taking the water to higher altitudes increases it. Refining, if we speak more carefully we might say instead, "Adding salt to water while keeping the altitude constant decreases the cooking time; whereas increasing the altitude while keeping the saline content fixed increases it"; or

(x)(S(x) & ⌐ A(x) $\hookrightarrow$ I(x))

and

(x)(A(x) & ⌐ S(x) $\hookrightarrow$ ⌐ I(x))

But neither of these tells what happens when we both add salt to the water and move to higher attitudes.

Here we think that probably there is a precise answer about what would happen, even though it is not part of our common folk wisdom. But this is not always the case. . . . Most real life cases involve some combination of causes; and general laws that describe what happens in these complex cases are not always available. Although both quantum theory and relativity are highly developed, detailed, and sophisticated, there is no satisfactory theory of relativistic quantum mechanics. . . . The general lesson is this: where theories intersect, laws are usually hard to come by.

4. WHEN EXPLANATIONS CAN BE GIVEN ANYWAY

So far, I have only argued that covering laws are scarce, and that *ceteris paribus* laws are not true laws. It remains to argue that, nevertheless, *ceteris paribus* laws have a fundamental explanatory role. But this is easy, for most of our explanations are explanations from *ceteris paribus* laws.

Let me illustrate with a humdrum example. Last year I planted camellias in my garden. I know that camellias like rich soil, so I planted them in composted manure. On the other hand, the manure was still warm, and I also know that camellia roots cannot take high temperatures. So I did not know what to expect. But when many of my camellias died, despite otherwise perfect care, I knew what went wrong. The camellias died because they were planted in hot soil.

This is surely the right explanation to give. Of course, I cannot be absolutely certain that this explanation is the correct one. Some other factor may have been responsible, nitrogen deficiency or some genetic defect in the plants, a factor that I did not notice or may not even have known to be relevant. But this uncertainty is not peculiar to cases of explanation. It is just the uncertainty that besets all of our judgments about matters of fact. We must allow for oversight; still, since I made a reasonable effort to eliminate other menaces to my camellias, we may have some confidence that this is the right explanation.

So we have an explanation for the death of my camellias. But it is not an explanation from any true covering law. There is no law that says that camellias just like mine, planted in soil which is both hot and rich, die. To the contrary, they do not all die. Some thrive; and probably those that do, do so *because* of the richness of the soil they are planted in. We may insist that there must be some differentiating factor which brings the case under a covering law: in soil which is rich and hot, camellias of one kind die; those of another thrive. I will not deny that there may be such a covering law, I merely repeat that our ability to give this humdrum explanation precedes our knowledge of that law. On the Day of Judgment, when all laws are known, these may suffice to explain all phenomena. But in the meantime we do give explanations; and it is the job of science to tell us what kinds of explanations are admissible.

In fact I want to urge a stronger thesis. If, as is possible, the world is not a tidy deterministic system, this job of telling how we are to explain will be a job which is still set when the descriptive task of science is complete. Imagine for example (what I suppose actually to be the case) that the facts about camellias are irreducibly statistical. Then it is possible to know all the general nomological facts about camellias which there are to know—for example, that 62 percent of all camellias in just the circumstances of my camellias die, and 38 percent survive.[7] But one would not thereby know how to explain what happened in my garden. You would still have to look to the *Sunset Garden Book* to learn that the *heat* of the soil explains the perishing, and the *richness* explains the plants that thrive.

5. CONCLUSION

Most scientific explanations use *ceteris paribus* laws. These laws, read literally as descriptive statements, are false, not only false but deemed false even in the context of use. This is no surprise: we want laws that unify; but what happens may well be varied and diverse. We are lucky that we can organize phenomena at all. There is no reason to think that the principles that best organize will be true, nor that the principles that are true will organize much.

NOTES

1. See C. G. Hempel, "Scientific Explanation," in C. G. Hempel (ed.), *Aspects of Scientific Explanation* (New York: Free Press, 1965).

2. See C. G. Hempel, "Scientific Explanation," ibid.

3. See Patrick Suppes, *A Probabilistic Theory of Causality* (Amsterdam: North-Holland Publishing Co., 1970).

4. See Wesley Salmon, "Statistical Explanation," in Wesley Salmon (ed.), *Statistical Explanation and Statistical Relevance* (Pittsburgh: University of Pittsburgh Press. 1971).

5. See Bengt Hanson, "Explanations—Of What?" (mimeograph, Stanford University, 1974).

6. Miles V. Klein, *Optics* (New York: John Wiley & Sons, 1970), p. 21, italics added. θ is the angle of incidence.

7. Various writers, especially Suppes (footnote 3) and Salmon (footnote 4), have urged that knowledge of more sophisticated statistical facts will suffice to determine what factors can be used in explanation. I do not believe that this claim can be carried out. . . .

14

Scientific Explanation:
*How We Got from There to Here**

Wesley Salmon

Is there a *new* consensus in philosophy of science? That is the question we must discuss. But even to pose the question in this way implies that there was an old consensus—and indeed there was, at least with respect to scientific explanation. It is with the old consensus that we should start. In order to understand the present situation we need to see how we got from there to here.[1]

I recall with some amusement a personal experience that occurred in the early 1960s. J. J. C. Smart, a distinguished Australian philosopher, visited Indiana University, where I was teaching at the time. Somehow we got into a conversation about the major unsolved problems in philosophy of science, and he mentioned the problem of scientific explanation. I was utterly astonished—literally, too astonished for words. I considered *that* problem essentially solved by the deductive-nomological (D-N) account that had been promulgated by R. B. Braithwaite (1953), Carl G. Hempel (Hempel and Oppenheim, [1948] 1965), Ernest Nagel (1961), and Karl Popper (1935, 1959), among many others—supplemented, perhaps, by Hempel's then recent account of statistical explanation (1962a). Although this general view had a few rather vocal critics, such as N. R. Hanson (1959) and Michael Scriven (1958, 1959, 1962), it was widely accepted by scientifically minded philosophers; indeed, it qualified handily as *the* received view. What is now amusing about the incident is my naiveté in thinking that a major philosophical problem had actually been solved.

1. THE RECEIVED VIEW

The cornerstone of the old consensus was the *deductive-nomological (D-N) model* of scientific explanation. The fullest and most precise early character-

*The essays mentioned in the text and endnotes of this article refer to other essays in Salmon's (1998) anthology from which this article has been taken.—Eds.

ization of this model was given in the classic article "Studies in the Logic of Explanation" (Hempel and Oppenheim, [1948] 1965). According to that account, a D-N explanation of a particular event is a valid deductive argument whose conclusion states that the event-to-be-explained did occur. Its premises must include essentially at least one general law. The explanation is said to subsume the fact to be explained under these laws. Hence, it is often called the *covering law model.* On the surface this account is beautiful for its clarity and simplicity, but as we shall see in §2, it contains a number of serious hidden difficulties.

Consider one of Hempel's familiar examples. Suppose someone asks why the flame of a particular Bunsen burner turned yellow at a particular moment. This why-question is a request for a scientific explanation. The answer is that a piece of rock salt was placed in the flame, rock salt is a sodium compound, and Bunsen flames always turn yellow when sodium compounds are introduced. The explanation can be laid out formally as follows:

(1) All Bunsen flames turn yellow when sodium compounds are placed in them.
　　All rock salt consists of a sodium compound.
　　A piece of rock salt was placed in this Bunsen flame at a particular time.
　　―――――――――――――――――――――――――――――――――――
　　This Bunsen flame turned yellow at that time.

This explanation is a valid deductive argument with three premises. The first two premises are statements of natural law; the third premise formulates an initial condition in this explanation. The premises constitute the *explanans*— that which does the explaining. The conclusion is the *explanandum*—that which is explained.

From the beginning, however, Hempel and Oppenheim ([1948] 1965, pp. 250–251) recognized that not all scientific explanations are of the D-N variety. Some are probabilistic or statistical. In "Deductive-Nomological vs. Statistical Explanation" (1962a) Hempel offered his first treatment of statistical explanation, and in "Aspects of Scientific Explanation" (1965b) he provided an improved account. This theory includes two types of statistical explanation. The first of these, the inductive-statistical (I-S), explains particular occurrences by subsuming them under statistical laws, much as D-N explanations subsume particular events under universal laws. To cite another of Hempel's famous examples, if we ask why John Jones recovered rapidly from his streptococcus infection, the answer is that he was given a dose of penicillin, and almost all strep infections clear up quickly upon administration of penicillin. More formally,

(2) Almost all cases of streptococcus infection clear up quickly after the administration of penicillin.
Jones had a streptococcus infection.
Jones received treatment with penicillin.

=== $[r]$

Jones recovered quickly.

This explanation is an argument that has three premises (the explanans); the first premise states a statistical regularity—a statistical law—while the other two state initial conditions. The conclusion (the explanandum) states the fact to be explained. There is, however, a crucial difference between explanations (1) and (2): D-N explanations subsume the events to be explained deductively, while I-S explanations subsume them inductively. The single line separating the premises from the conclusion in (1) signifies a relation of deductive entailment between the premises and conclusion. The double line in (2) represents a relationship of inductive support, and the attached variable r stands for the strength of that support. This strength of support may be expressed exactly, as a numerical value of a probability, or vaguely, by means of phrases such as "very probably" or "almost certainly."

An explanation of either of these two kinds can be described as an argument to the effect that *the event to be explained was to be expected by virtue of certain explanatory facts.* In a D-N explanation the event to be explained is deductively certain, given the explanatory facts; in an I-S explanation the event to be explained has high inductive probability relative to the explanatory facts. This feature of expectability is closely related to the *explanation/ prediction symmetry thesis* for explanations of particular facts. According to this thesis—which was advanced for D-N explanation in Hempel-Oppenheim ([1948] 1965, p. 249), and reiterated, with some qualifications, for D-N and I-S explanations in Hempel (1965b, §2.4, §3.5)—any acceptable explanation of a particular fact is an argument, deductive or inductive, that could have been used to predict the fact in question if the facts stated in the explanans had been available prior to its occurrence. As we shall see in §2, this symmetry thesis met with serious opposition.

Hempel was not by any means the only philosopher in the early 1960s to notice that statistical explanations play a highly significant role in modern science. He was, however, the first to present a detailed account of the nature of statistical explanation, and the first to bring out a fundamental problem concerning statistical explanations of particular facts. The case of Jones and the quick recovery can be used as an illustration. It is well known that certain strains of the streptococcus bacterium are penicillin-resistant, and if Jones's infection is of that type, the probability of the quick recovery after treatment with penicillin would be very small. We could, in fact, set up the following inductive argument:

(2′)   Almost no cases of penicillin-resistant streptococcus infection clear up quickly after the administration of penicillin.
Jones had a penicillin-resistant streptococcus infection.
Jones received treatment with penicillin.

$$==[q]$$

Jones did not recover quickly.

The remarkable fact about arguments (2) and (2′) is that their premises are mutually compatible—they could all be true. Nevertheless, their conclusions contradict each other. This is a situation that can never occur with deductive arguments. Given two valid deductions with incompatible conclusions, their premises must also be incompatible. Thus, the problem that has arisen in connection with I-S explanations has no analogue in D-N explanations. Hempel called this *the problem of ambiguity of I-S explanation,* and he sought to resolve it by means of his *requirement of maximal specificity (RMS).*

The source of the problem of ambiguity is a simple and fundamental difference between universal laws and statistical laws. Given the proposition that all *A* are *B,* it follows immediately that all things that are both *A* and *C* are *B.* If all men are mortal, then all men who are over six feet tall are mortal. However, if almost all men who are alive now will be alive five years from now, *it does not follow* that almost all living men with advanced cases of lung cancer will be alive five years hence. There is a parallel fact about arguments. Given a valid deductive argument, the argument will remain valid if additional premises are supplied, as long as none of the original premises is taken away. Given a strong inductive argument—one that supports its conclusion with a very high degree of probability—the addition of one more premise may undermine it completely. Europeans, for example, had for many centuries a great body of inductive evidence to support the proposition that all swans are white, but one true report of a black swan in Australia completely refuted that conclusion.

There is a well-known strategy for dealing with the problem of ambiguity as it applies to inductive arguments per se: it is to impose the *requirement of total evidence.* According to this requirement, one should not rely on the conclusion of an inductive argument—for purposes of making predictions or wagers, for example—unless that argument includes among its premises all available relevant evidence. This approach is entirely unsuitable for the context of scientific explanation because normally, when we seek an explanation for some fact, we already know that it obtains. Thus, knowledge of the fact-to-be-explained is part of our body of available knowledge. We ask why Jones recovered quickly from the strep infection only after we know that the quick recovery occurred. But if we include in the explanans the statement that the quick recovery occurred, the resulting "explanation"

(2″) Almost all cases of streptococcus infection clear up quickly after the administration of penicillin.

Jones had a streptococcus infection.

Jones received treatment with penicillin.

Jones recovered quickly.

Jones recovered quickly.

is trivial and uninteresting. Although the conclusion follows deductively from the augmented set of premises, (2″) does not even qualify as a D-N explanation, for no law is essential for the derivation of the conclusion from the new set of premises. We could eliminate the first three premises and the resulting argument would still be valid.

Hempel (1965b, §3.4) was clearly aware of all of these considerations, and he designed his requirement of maximal specificity (RMS) to circumvent them. The purpose of this requirement is to ensure that all relevant information *of an appropriate sort* is included in any given I-S explanation. Although it is extremely tricky to say just what constitutes *appropriate information*, one could say, very roughly, that it is information that is in principle available prior to the occurrence of the event-to-be-explained.[2] Suppose that we have a putative explanation of the fact that some entity x has the property B. Suppose that this explanation appeals to a statistical law of the form "The probability that an A is a B is equal to r." Suppose, in addition, that we know that this particular x also belongs to a class C that is a subset of A. Then, if the explanation is to satisfy RMS, our body of knowledge must include the knowledge that the probability of a C being a B is equal to q, and q must be equal to r, unless the class C is *logically related* to the property B (or the class of things having the property B) in a certain way. That is, q need not equal r if the statement that the probability of a C being a B is equal to q is a theorem of the mathematical calculus of probability.

In order to clarify this rather complicated requirement, let us refer again to the example of Jones. Consider three separate cases:

(a) Suppose that we know, in addition to the facts stated in the premises of (2), that Jones's penicillin treatment began on Thursday. According to RMS that would have no bearing on the legitimacy of (2) as an explanation, for the day of the week on which the treatment is initiated has no bearing on the efficacy of the treatment. The probability of a rapid recovery after penicillin treatment that began on a Thursday is equal to the probability of rapid recovery after treatment by penicillin (regardless of the day on which it began).

(b) Suppose we were to offer argument (2) as an explanation of Jones's rapid recovery, knowing that the infection was of the penicillin-resistant type. Since we know that the probability of quick recovery from a penicillin-resistant strep infection after treatment by penicillin *is not equal* to the probability of rapid recovery from an unspecified type of strep infection after treatment with penicillin, the explanation would not be legitimate. RMS would outlaw argument (2) as an I-S explanation if the highly relevant information about the penicillin-resistant character of the infection were available.

(c) When asking for an explanation of Jones's quick recovery, we already know that Jones belongs to the class of people with strep infections. Moreover, we know that Jones belongs to the subclass of people with strep infections cured quickly by penicillin, and we know that the probability of anyone in *that* class having a quick recovery is equal to one. This knowledge does not rule out (2) as an explanation. The reason is the "unless" clause of RMS. It is a trivial consequence of mathematical probability theory that the probability of quick recovery among those who experience quick recovery is one. If Y is a proper or improper subclass of X, then the probability of Y, given X, is necessarily equal to one.[3]

Having recognized the problem of ambiguity of I-S explanation, Hempel introduced the requirement of maximal specificity. As can easily be seen, RMS makes explicit reference to our state of knowledge. Whether a given argument qualifies as an I-S explanation depends not only on the objective facts in the world but also on what knowledge the explainer happens to possess. This result led Hempel to enunciate the principle of *essential epistemic relativity of I-S explanation.* D-N explanation, in contrast, does not suffer from any such epistemic relativity. If the premises of argument (1) are true, argument (1) qualifies as a correct D-N explanation of the fact that the Bunsen flame turned yellow. The fact that it is a correct explanation does not depend in any way on our knowledge situation. Of course, whether we *think* that it is a correct explanation will surely depend on our state of knowledge. What is *considered* a correct D-N explanation at one time may be *judged* incorrect at another time because our body of knowledge changes in the meantime. But the objective correctness of the explanation does not change accordingly. By contrast, argument (2) may have true premises and correct inductive logical form, but those features do not guarantee that it is a correct I-S explanation. Relative to one knowledge situation it is legitimate; relative to another it is not. As we shall see in §2, the requirement of maximal specificity and the doctrine of essential epistemic relativization became sources of fundamental difficulty for the received view of explanations of particular facts.

On Hempel's theory it is possible to explain not only particular events but also general regularities. Within the D-N model universal generalizations are explained by deduction from more comprehensive universal generalizations. For example, the law of conservation of linear momentum can be deduced—with the aid of a little mathematics—from Newton's second and third laws of motion. In classical mechanics, consequently, the following argument constitutes an explanation of the law of conservation of linear momentum:

(3) Newton's second law: $F = ma$.

Newton's third law: *For every action there is an equal and opposite reaction.*

Law of conservation of linear momentum: *In every physical interaction, linear momentum is conserved.*

Notice that this explanans contains only statements of law; inasmuch as no particular occurrence is being explained, no statements of particular initial conditions are required.

In the second type of statistical explanation, the deductive-statistical (D-S), statistical regularities are explained by deduction from more comprehensive statistical laws. A famous example comes from the birth of the mathematical theory of probability. A seventeenth-century gentleman, the Chevalier de Méré, wondered whether, in twenty-four tosses of a standard pair of dice, one has a better than fifty-fifty chance of getting double six ("boxcars") at least once, or whether twenty-five tosses are needed. He posed the question to Pascal, who proved that twenty-five is the correct answer. His derivation can be viewed as an explanation of this somewhat surprising fact. It can be set out as follows:

(4) A standard die is a physically homogeneous cube whose six faces are marked with the numbers 1–6.

When a standard die is tossed in the standard manner, each side has an equal probability—namely, one-sixth—of ending uppermost.

When two standard dice are tossed in the standard manner, the outcome on each die is independent of the outcome on the other.

When two standard dice are tossed repeatedly in the standard manner, the result on any given throw is independent of the results on the preceding tosses.

Twenty-five is the smallest number of standard tosses of a standard pair of dice for which the probability of double six occurring at least once is greater than one-half.

A small amount of arithmetic is needed to show that the conclusion of this argument follows deductively from the premises.[4] The first premise is a definition; the three remaining premises are statistical generalizations. The conclusion is also a statistical generalization.

Figure 14.1 shows the four categories of scientific explanations recognized in Hempel (1965b). However, in their explication of D-N explanation in 1948, Hempel and Oppenheim restrict their attention to explanations of particular facts, and do not attempt to provide any explication of explanations of general regularities. The reason for this restriction is given in the notorious footnote 33 (Hempel and Oppenheim, [1948] 1965, p. 273):

Figure 14.1. Hempelian Models of Explanation

Explanada Laws	Particular Facts	General Regularities
Universal Laws	D-N Deductive-Nomological	D-N Deductive-Nomological
Statistical Laws	I-S Inductive-Statistical	D-S Deductive-Statistical

The precise rational reconstruction of explanation as applied to general regularities presents peculiar problems for which we can offer no solution at present.
The core of the difficulty can be indicated by reference to an example: Kepler's
laws, K, may be conjoined with Boyle's law, B, to [form] a stronger law $K.B$;
but derivation of K from the latter would not be considered an explanation of
the regularities stated in Kepler's laws; rather, it would be viewed as representing, in effect, a pointless "explanation" of Kepler's laws by themselves. The
derivation of Kepler's laws from Newton's laws of motion and gravitation, on
the other hand, would be recognized as a genuine explanation in terms of more
comprehensive regularities, or so-called higher-level laws. The problem therefore arises of setting up clear-cut criteria for the distinction of levels of explanation or for a comparison of generalized sentences as to their comprehensiveness. The establishment of adequate criteria for this purpose is as yet an open
problem.

This problem is not resolved in any of Hempel's subsequent writings,
including "Aspects of Scientific Explanation." It was addressed by Michael
Friedman (1974); I shall discuss his seminal article in §4. Since the same
problem obviously applies to D-S explanations, it affects both sectors in the
right-hand column of figure 14. 1. The claim of the received view to a comprehensive theory of scientific explanation thus carries a large promissory
note regarding explanations of laws.

The Hempel-Oppenheim ([1948] 1965) article marks the division
between the prehistory and the history of the modern discussions of scientific
explanation. Although Aristotle, John Stuart Mill (1843), and Karl Popper
(1935), among many others, had previously expressed similar views about the
nature of deductive explanation, the Hempel-Oppenheim essay spells out the
D-N model with far greater precision and clarity. Hempel's 1965 "Aspects"
article is *the* central document in the hegemony (with respect to scientific
explanation) of logical empiricism. I shall refer to the account given there as
the received view. According to the received view, *every legitimate scientific
explanation* fits into one of the four compartments in figure 14.1.

2. ATTACKS ON THE RECEIVED VIEW

The hegemony of logical empiricism regarding scientific explanation did not endure for very long. Assaults came from many directions; most of them can be presented in terms of old familiar counterexamples. Some of the counterexamples are cases that satisfy all of the criteria set forth in the received view but which clearly are not admissible as scientific explanations. These are designed to show that the requirements imposed by the received view are not *sufficient* to determine what constitutes a correct scientific explanation. Other counterexamples are cases that appear intuitively to be satisfactory scientific explanations but that fail to satisfy the criteria of the received view. They are designed to show that these requirements are not *necessary* either.

(1) One of the best known is Sylvain Bromberger's flagpole example.[5] A certain flagpole casts a shadow of a certain length at some particular time. Given the height of the flagpole, its opacity, the elevation of the sun in the sky, and the rectilinear propagation of light, it is possible to deduce the length of the shadow and, ipso facto, to provide a D-N explanation of its length. There is no puzzle about this. But given the length of the shadow, the position and opacity of the flagpole, the elevation of the sun, and the rectilinear propagation of light, we can deduce the height of the flagpole. Yet hardly anyone would allow that the length of the shadow explains the height of the flagpole.[6]

(2) It has often been noted that, given a sudden drop in the reading of a barometer, we can reliably infer the occurrence of a storm. It does not follow that the barometric reading explains the storm; rather, a drop in atmospheric pressure explains both the barometric reading and the storm.[7]

Examples (1) and (2) show something important about causality and explanation. The first shows that we explain effects in terms of their causes; we do not explain causes in terms of their effects. See "Explanatory Asymmetry" (essay 10) for a deeper analysis of this problem. The second shows that we do not explain one effect of a common cause in terms of another effect of that same cause. Our common sense has told us for a long time that to explain an event is, in many cases, to find and identify its cause. One important weakness of the received account is its failure to make explicit reference to causality—indeed, Hempel has explicitly denied that explanations must always involve causes (1965b, pp. 352–353).

(3) Many years ago Scriven (1959) noticed that we can explain the occurrence of paresis in terms of the fact that the patient had latent syphilis untreated by penicillin. However, given someone with latent untreated syphilis, the chance that he or she will develop paresis is about one-fourth, and there is no known way to separate those who will develop paresis from those who won't.

(4) My favorite example is the case of the man who regularly takes his wife's birth control pills for an entire year, and who explains the fact that he did not become pregnant during the year on the basis of his consumption of oral contraceptives (Salmon, 1971, p. 34).

Examples (3) and (4) have to do with expectability, and consequently with the explanation/prediction symmetry thesis. Scriven (1959) had offered example (3) in order to show that we can have explanations of events that are improbable, and hence are not to be expected; indeed, he argued that evolutionary biology is a science containing many explanations but virtually no predictions. Example (4) shows that an argument that fully qualifies as a D-N explanation, and consequently provides expectability, can fail to be a bona fide explanation. Peter Railton has pointed out that Hempel's view can be characterized in terms of the *nomic expectability* of the event to be explained. He argues—quite correctly, I believe—that nomicity may well be a bona fide requirement for scientific explanation, but that expectability cannot be demanded.

My own particular break with the received doctrine occurred in 1963, very shortly after the aforementioned conversation with Smart. At the 1963 meeting of the AAAS, I argued that Hempel's I-S model, with its high probability requirement and its demand for expectability, is fundamentally mistaken.[8] Statistical relevance instead of high probability, I argued, is the key concept in statistical explanation.

In support of this contention I offered the following example (which, because of serious questions about the efficacy of psychotherapy, happens to have some medical importance). Suppose that Jones, instead of being afflicted with a strep infection, has a troublesome neurotic symptom. Under psychotherapy this symptom disappears. Can we explain the recovery in terms of the treatment? We could set out the following inductive argument, in analogy with argument (2):

(5) Most people who have a neurotic symptom of type N and who undergo psychotherapy experience relief from that symptom.
Jones had a symptom of type N and underwent psychotherapy.
$$\overline{\qquad\qquad\qquad\qquad\qquad\qquad\qquad\qquad}\; [r]$$
Jones experienced relief from this symptom.

Before attempting to evaluate this proffered explanation, we should take account of the fact that there is a fairly high spontaneous remission rate—that is, many people who suffer from that sort of symptom get better regardless of treatment. No matter how large the number r, if the rate of recovery for people who undergo psychotherapy is no larger than the spontaneous remission rate, it would be a mistake to consider argument (5) a legitimate explanation. A high probability is not sufficient for a correct explanation. If, however, the number r is not very large, but is greater than the spontaneous remission rate, the fact that the patient underwent psychotherapy has at least some degree of explanatory force. A high probability is not necessary for a sound explanation.[9]

Examples (3) and (4) both pertain to the issue of relevance. In example (3) we have a factor (syphilis not treated with penicillin) that is highly relevant to the explanandum (contracting paresis) even though no high probabilities are involved. This example exhibits the explanatory force of relevance. In example (4) we have an obviously defective "explanation" because of the patent irrelevance of the consumption of birth control pills to the nonpregnancy of a man. Furthermore, in my example of psychotherapy and relief from a neurotic symptom, the issue of whether the explanation is legitimate or not hinges on the question whether the psychotherapy was, indeed, relevant to the disappearance of the symptom. Henry Kyburg, who commented on my AAAS paper, pointed out—through an example similar in principle to example (4)—that the same sort of criticism could be leveled against the D-N model. It, too, needs to be guarded against the introduction of irrelevancies into putative explanations.

While my initial criticism of the received view centered on issues of high probability versus relevancy, other philosophers attacked the requirement of maximal specificity and the associated doctrine of essential epistemic relativization of I-S explanation. On this front the sharpest critic was J. Alberto Coffa (1974), who challenged the very intelligibility of an epistemically relativized notion of inductive explanation. His argument ran roughly as follows. Suppose someone offers a D-N explanation of some particular fact, such as we had in argument (1) above. If the premises are true and the logical form correct, then (1) is a *true* D-N explanation. In our present epistemic state we may not know for sure that (1) is a true explanation; for instance, I might not be sure that placing a sodium compound in a Bunsen flame always turns the flame yellow. Given this uncertainty, I might consult chemical textbooks, ask chemists of my acquaintance, or actually perform experiments to satisfy myself that the first premise of (1) is true. If I have doubts about any other premises, there are steps I could take to satisfy myself that they are true. In the end, although I cannot claim to be *absolutely certain* of the truth of the premises of (1), I can conclude that they are well confirmed. If I am equally confident of the logical correctness of the argument, I can then claim to have good reasons to believe that (1) is a true D-N explanation. Crucial to this conclusion is the fact that I know what sort of thing a true D-N explanation is.

The situation regarding D-N explanations is analogous to that for more commonplace entities. Suppose I see a bird in a bush but I am not sure what kind of bird it is. If I approach it more closely, listen to its song, look at it through binoculars, and perhaps ask an ornithologist, I can establish that it is a hermit thrush. I can have good reason to believe that it is a hermit thrush.

It is a well-confirmed hermit thrush. But all of this makes sense only because we have objective nonepistemically relativized criteria for what an actual hermit thrush is. Without that, the concept of a well-confirmed hermit thrush would make no sense because there would be literally nothing we could have good reason to believe we are seeing.

When we turn to I-S explanations, a serious complication develops. If we ask, prior to an inquiry about a particular I-S explanation, what sort of thing constitutes a true I-S explanation, Hempel must reply that he does not know. All he can tell us about are epistemically relativized I-S explanations. He can tell us the criteria for determining that an I-S explanation is acceptable in a given knowledge situation. It appears that he is telling us the grounds for justifiably believing, in such a knowledge situation, that we have a genuine I-S explanation. But what can this mean? Since, according to Hempel's 1965 view, there is no such thing as a bona fide I-S explanation, unrelativized to any knowledge situation, what is it that we have good reason to believe that we have? We can have good reason to believe that we have a true D-N explanation because we know what sort of thing a true D-N explanation is. The main burden of Hempel and Oppenheim, ([1948] 1965) was to spell out just that. According to Hempel, it is impossible in principle to spell out any such thing for I-S explanations.

On the basis of a careful analysis of Hempel's doctrine of essential epistemic relativity, it is possible to conclude that Hempel has offered us not an independent stand-alone conception of inductive explanation of particular facts, but rather a conception of inductive explanation that is completely parasitic on D-N explanation. One is strongly tempted to draw the conclusion that an I-S explanation is essentially an enthymeme—an incomplete deductive argument. Faced with an enthymeme, we may try to improve it by supplying missing premises, and in so doing we may be more or less successful. But the moment we achieve complete success by supplying all of the missing premises, we no longer have an enthymeme—instead we have a valid deductive argument. Similarly, it appears, given an epistemically relativized I-S explanation, we may try to improve our epistemic situation by increasing our body of knowledge. With more knowledge we may be able to furnish more complete explanations. But when we finally succeed in accumulating all of the relevant knowledge and incorporating it into our explanation, we will find that we no longer have an inductive explanation—instead we have a D-N explanation.

A doctrine of inductive explanations that construes them as incomplete deductive explanations seems strongly to suggest determinism. According to the determinist every fact of nature is amenable, in principle, to complete deductive explanation. We make do with inductive explanations only because of the incompleteness of our knowledge. We appeal to probabilities only as a reflection of our ignorance. An ideal intelligence, such as Laplace's famous demon, would have no use for probabilities or inductive explanations (see Salmon, 1974a). Although Hempel has explicitly denied any commitment to determinism, his theory of I-S explanation fits all too neatly into the determinist's scheme of things. Eventually Hempel (1977) retracted his doctrine of *essential* epistemic relativization.

Careful consideration of the various difficulties in Hempel's I-S model led to the development of the statistical-relevance (S-R) model. Described concisely, an S-R explanation is an assemblage of all and only those factors relevant to the fact-to-be-explained. For instance, to explain why Albert, an American teenager, committed an act of delinquency, we cite such relevant factors as his sex, the socioeconomic status of his family, his religious background, his place of residence (urban versus suburban or rural), ethnic background, et cetera (see Greeno, [1970] 1971). It would clearly be a mistake to mention such factors as the day of the week on which he was born or whether his social security number is odd or even, for they are statistically irrelevant to the commission of delinquent acts.

It should be pointed out emphatically that an assemblage of relevant factors—along with an appropriate set of probability values—is not an argument of any sort, deductive or inductive. Acceptance of the S-R model thus requires abandonment of what I called the third dogma of empiricism, namely, the general thesis that every bona fide scientific explanation is an argument (see essay 6). It was Richard Jeffrey ([1969] 1971) who first explicitly challenged that dogma.

The S-R model could not long endure as an independent conception of scientific explanation, for it embodied only statistical correlations, without appeal to causal relations. Reacting to Hempel's I-S model, I thought that statistical relevance, rather than high inductive probability, has genuine explanatory import. I no longer think so. Statistical-relevance relations are important to scientific explanation for a different reason, namely, because they constitute important evidence of causal relations. Causality, rather than statistical relevance, is what has explanatory import.

It may seem strange that the received view excised causal conceptions from its characterization of scientific explanation. Have we not known since Aristotle that explanations involve causes? It would be reasonable to think so. But putting the "cause" back into "because" is no simple matter, for Hume's searching analysis strongly suggested that to embrace physical causality might involve a rejection of empiricism. Those philosophers who have strongly insisted on the causal character of explanation—e.g., Scriven —have simply evaded Hume's critique. My own view is that the "cause" cannot be put back into "because" without a serious analysis of causality. The essays in part III of this book offer some suggestions as to how such an analysis might go (see also Salmon, 1984b, chaps. 5–7).

One of the major motivations for the received view was, I believe, the hope that scientific explanations could be characterized in a completely formal manner. (Note that the title of the Hempel-Oppenheim article is "Studies in the *Logic* of Explanation.") This makes it natural to think of explanations as arguments, for, as Carnap showed in his major treatise on probability (1950), both deductive logic and inductive logic can be explicated within a single semantic framework. Hempel and Oppenheim ([1948] 1965) offer a semantical analysis of lawlike statements.[10] This makes it possible to characterize a *potential explanation* as an argument of correct form containing at least one lawlike statement among its premises. A true explanation fulfills in addition the empirical condition that its premises and conclusion be true. A correct I-S explanation must satisfy still another condition, namely, Hempel's requirement of maximal specificity. This relevance requirement is also formulated in logical terms.

The upshot for the received view is that there are two models (three if D-S is kept separate from D-N) of scientific explanation, and that every legitimate explanation conforms to one or the other. Accordingly, any phenomenon in our universe, even in domains in which we do not yet have any scientific knowledge, must be either amenable to explanation by one of these models or else not susceptible to any sort of scientific explanation. The same would hold, it seems, for scientific explanations in any possible world.

Such universalistic ambitions strike me as misplaced. In our world, for example, we impose the demand that events be explained by their temporal antecedents, not by events that come later. But the structure of time itself is closely connected with entropic processes in our universe, and these depend on de facto conditions in our universe. In another universe the situation might be quite different—for example, time might be symmetric rather than asymmetric. In the macrocosm of our world, causal influence is apparently propagated continuously; action-at-a-distance does not seem to occur. In the microcosm of our world, what Einstein called "spooky action-at-a-distance" seems to occur. What counts as acceptable scientific explanation depends crucially on the causal and temporal structure of the world, and these are matters of fact rather than matters of logic. The moral I would draw is just this: we should not hope for formal models of scientific explanation that are universally applicable. We do best by looking at explanations in various domains of science, and by attempting adequately to characterize their structures. If it turns out—as I think it does—that very broad stretches of science employ common explanatory structures, that is an extremely interesting fact about our world.

3. THE PRAGMATICS OF EXPLANATION

From the beginning Hempel and the other proponents of the received view recognized the obvious fact that scientific monographs, textbooks, articles, lectures, and conversation do not present scientific explanations that conform precisely to their models. They also realized that to do so would be otiose. Therefore, in the writing and speech of scientists we find partial explanations, explanation sketches, and elliptically formulated explanations. What sort of presentation is suitable is determined by factors such as the knowledge and interests of those who do the explaining and of their audiences. These are pragmatic factors.

Hempel devoted two sections of "Aspects" (1965b, §4–5) to the pragmatics of explanation, but the discussion was rather narrow. In 1965 (and a fortiori in 1948) formal pragmatics was not well developed, especially in those aspects that bear on explanation. Bromberger's path-breaking article "Why-Questions" appeared in 1966, but it dealt only with D-N explanation; the most prominent subsequent treatment of the pragmatics of explanation was provided in van Fraassen's *Scientific Image* (1980). A rather different pragmatic approach can be found in Peter Achinstein's *Nature of Explanation* (1983).

Van Fraassen adopts a straightforward conception of explanations (scientific and other) as answers to why-questions. Why-questions are posed in various contexts, and they have presuppositions. If the presuppositions are not fulfilled, the question does not arise; in such cases the question should be rejected rather than answered. If the question does arise, then the context heavily determines what constitutes an appropriate answer. Now, van Fraassen is not offering an "anything goes as long as it satisfies the questioner" view of explanation, for there are objective criteria for the evaluation of answers. But there is a deep problem.

Van Fraassen characterizes a why-question as an ordered triple $\langle P_k, X, R \rangle$. P_k is the topic (what Hempel and most others call the "explanandum"). X is the contrast class, a set of alternatives with respect to which P_k is to be explained. In the Bunsen flame example, the contrast class might be:

P_1 = the flame turned orange;
P_2 = the flame turned green;

.
.
.

P_k = the flame turned yellow

.
.
.

P_s = the flame did not change color.

A satisfactory answer to the why-question is an explanation of the fact that P_k rather than any other member of the contrast class is true. R is the relevance relation; it relates the answer to the topic and contrast class. In the Bunsen flame example we can construe R as a causal relation; putting the rock salt into the Bunsen flame is what causes it to turn yellow. The problem is that van Fraassen places no restrictions on what sort of relation R may be. That is presumably freely chosen by the questioner. An answer A is relevant if A bears relation R to the topic P_k. In "Van Fraassen on Explanation" (essay 11) Philip Kitcher and I have shown that, without some restrictions on the relation R, any answer A can be the explanation of any topic P_k. Thus, van Fraassen needs to provide a list of types of relations that qualify as bona fide relevance relations. This is precisely the problem that philosophers who have not emphasized pragmatic aspects of explanation have been concerned to resolve. Even acknowledging this serious problem, van Fraassen and others have clearly demonstrated the importance of pragmatic features of explanation; what they have not shown is that pragmatics is the whole story.

One of the most important works on scientific explanation since Hempel's "Aspects" is Peter Railton's doctoral dissertation, "Explaining Explanation" (1980). In this work he introduces a valuable pair of concepts: *ideal explanatory texts* and *explanatory information*.[11] An ideal explanatory text for any fact to be explained is an extremely extensive and detailed account of everything that contributed to that fact—everything that is causally or lawfully relevant to it. Such texts are ideal entities; they are virtually never written out in full. To understand the fact being explained, we do not have to have the whole ideal text; what is required is that we be able to fill in the needed parts of it. Explanatory information is any information that illuminates any portion of the ideal text. Once we are clear on just what it is that we are trying to explain, the ideal explanatory text is fully objective; its correctness is determined by the objective causal and nomic features of the world. It has no pragmatic dimensions.

Pragmatic considerations arise when we decide which portions of the ideal explanatory text are to be illuminated in any given situation. This is the contextual aspect. When a why-question is posed, various aspects of the context—including the interests, knowledge, and training of the questioner—determine what explanatory information is salient. The resulting explanation must reflect objective relevance relations, but it must also honor the salience of the information it includes. Seen in this perspective, van Fraassen's account of the pragmatics of explanation fits admirably into the overall picture by furnishing guidelines that determine what sort of explanatory information is appropriate in the context.

4. THE MORAL OF THE STORY

What have we learned from all of this? Several lessons, I believe. First, we must put the "cause" back into "because." Even if some types of explanation turn out not to be causal, many explanations do appeal essentially to causes. We must build into our theory of explanation the condition that causes can explain effects but effects do not explain causes. By the same token, we must take account of temporal asymmetries; we can explain later events in terms of earlier events, but not vice versa. Temporal asymmetry is closely related to causal asymmetry (see "Explanatory Asymmetry" (essay 10)).

Second, the high probability or expectedness requirement of the received view is not acceptable. High probability is neither necessary nor sufficient for scientific explanation, as examples (3) and (4) respectively show.

Third, we can dispense—as Hempel himself did (1977)—with his doctrine of *essential epistemic relativity* of I-S explanation. Many authors have found this aspect of the received view unpalatable. Coffa (1974), Fetzer (1974b), and Railton (1978) employ a propensity conception of probability to characterize types of statistical explanation that are not epistemically relativized. In my (1984b, chap. 3) I try to avoid such relativization by means of objectively homogeneous reference classes. Railton's notion of the ideal explanatory text provides the basis for a fully objective concept of statistical explanation.

Fourth, our theory of scientific explanation should have a place for a robust treatment of the pragmatics of explanation. Considerations of salience arise whenever we attempt to express or convey scientific explanations.

Fifth, we can relinquish the search for one or a small number of formal models of scientific explanation that are supposed to have universal applicability. This point has been argued with considerable care by Achinstein (1983).

Do we have the basis for a new consensus? Not quite yet, I fear. It would, of course, be silly to expect unanimous agreement among philosophers on any major topic. But leaving that impossible dream aside, there are serious issues on which fundamental disagreements exist. One of these concerns the nature of laws. Is there an objective distinction between true lawlike generalizations and generalizations that just happen to be true? Or is the distinction merely epistemic or pragmatic? The problem of laws remains unsolved, I believe, and—given the enormous influence of the covering law conception of explanation—of fundamental importance.

Another major issue concerns the question whether there are bona fide statistical explanations of particular events. Hempel's I-S model, with its high probability requirement and its essential epistemic relativization, has encountered too many difficulties. It is not likely to be resuscitated. The S-R model gives rise to results that seem strongly counterintuitive to many. For instance, on that model it is possible that factors negatively relevant to an occurrence help to explain it. Even worse, suppose (as contemporary physical theory strongly suggests) that our world is indeterministic. Under circumstances of a specified type C, an event of a given type E sometimes occurs and sometimes does not. There is, in principle, no way to explain why, on a given occasion, E *rather than* non-E occurs. Moreover, if on one occasion C explains why E occurs, then on another occasion the same kind of circumstances explain why E fails to occur. Although I do not find this consequence intolerable, I suspect that the majority of philosophers do.

One frequent response to this situation is to claim that all explanations are deductive. Where statistical explanations are concerned, they are of the kind classified by Hempel as D-S. Thus, we do not have statistical explanations of particular events; all statistical explanations are explanations of statistical generalizations. We can explain why the vast majority of tritium atoms now in existence will very probably decay within the next fifty years, for the half-life of tritium is about $12\frac{1}{4}$ years. Perhaps we can explain why a particular tritium atom has a probability of just over $^{15}/_{16}$ of decaying within the next fifty years. But we cannot, according to this line of thought, explain why a given tritium atom decayed within a given half-century. The consequence of this view is that, insofar as indeterminism holds, we cannot explain what happens in the world. If we understand the stochastic mechanisms that indeterministically produce all of the various facts, we may claim to be able to explain *how the world works*. That is not the same as being able to explain *what happens*. To explain why an event has a high probability of occurring is *not* the same as explaining why it occurred. Moreover, we can explain why some event that did not occur—such as the disintegration of an atom that did not disintegrate—had a certain probability of occurring. But we cannot explain an event that did not happen.

Let me mention a third point of profound disagreement. Kitcher (1985) has suggested that there are two widely different approaches to explanation; he characterizes them as *bottom up* and *top down*. They could be described, respectively, as *local* and *global*. Both Hempel's approach and mine fall into the bottom-up or local variety. We look first to the particular causal connections or narrow empirical generalizations. We believe that there can be local explanations of particular facts. We try to work up from there to more fundamental causal mechanisms or more comprehensive theories.

Kitcher favors a top-down approach. Although many scientists and philosophers had remarked on the value of unifying our scientific knowledge, the first philosopher to provide a detailed account of explanation as unification is Friedman (1974). On his view, we increase our understanding of the world to the extent that we are able to reduce the number of independently acceptable hypotheses needed to account for the phenomena in the world. Both Kitcher (1976) and I (1989, 1990b) have found problems in the technical details of Friedman's theory; nevertheless, we both agree that Friedman's basic conception has fundamental importance. The main idea of the top-down approach is that one looks first to the most comprehensive theories and to the unification of our knowledge that they provide. To explain something is to fit it into a global pattern. What qualifies as a law or a causal relation is determined by its place in the simplest and most comprehensive theories. In his (1981) Kitcher began the development of an approach to explanatory unification along rather different lines from that of Friedman; in his (1989 and 1993) he works his proposals out in far greater detail.

Let us return, finally, to the fundamental question of this essay: Is there a new consensus concerning scientific explanation? At present, quite obviously, there is not. I do not know whether one will emerge in the foreseeable future, though I have recently come to see a basis for some hope in that direction (Salmon, 1989, 1990b, §5). However that may be, I am convinced that we have learned a great deal about this subject in the years since the publication of Hempel's magisterial "Aspects" essay. To my mind, that signifies important progress.[12]

APPENDIX

The preceding essay is a summary of material treated at much greater length in *Four Decades of Scientific Explanation* (Salmon, 1990b). Near the end I wrote:

> We have arrived, finally, at the conclusion of the saga of four decades. It has been more the story of a personal odyssey than an unbiased history. Inasmuch as I was a graduate student in philosophy in 1948 [the beginning of the first decade], my career as a philosopher spans the entire period. . . . My specific research on scientific explanation began in 1963, and I have been an active participant in the discussions and debates during the past quarter-century. Full objectivity can hardly be expected
>
> . . . I know that there are . . . important pieces of work . . . that have not been mentioned. . . . My decisions about what to discuss and what to omit are, without a doubt, idiosyncratic, and I apologize to the authors of such works for my neglect. (p. 180)

One philosopher to whom such an apology was due is Adolf Grunbaum. "Explanatory Asymmetry" (essay 10) embodies my attempt to make amends.

James H. Fetzer is another worker who deserves a major apology. At the close of §3.3, which is devoted to a discussion of Alberto Coffa's dispositional theory of inductive explanation, I wrote:

> Another of the many partisans of propensities in the third decade is James H. Fetzer. Along with Coffa, he deserves mention because of the central place he accords that concept in the theory of scientific explanation. Beginning in 1971, he published a series of papers dealing with the so-called propensity interpretation of probability and its bearing on problems of scientific explanation (Fetzer 1971, 1974a, 1974b, 1975, 1976, 1977). However, because the mature version

of his work on these issues is contained in his 1981 book, *Scientific Knowledge,* we shall deal with his views in the fourth decade. (p. 89)

Although these remarks are true, they fall far short of telling the whole truth. In my praise for Coffa, I said:

> In his doctoral dissertation Coffa (1973, chap. IV) argues that an appeal to the propensity interpretation of probability enables us to develop a theory of inductive explanation that is a straightforward generalization of deductive-nomological explanation, and that avoids both epistemic relativization and the reference class problem. This ingenious approach has, unfortunately, received no attention, for it was never extracted from his dissertation for publication elsewhere. (p. 83)

Without retracting my positive comments about Coffa, I must now point out that Fetzer's paper "A Single Case Propensity Theory of Explanation," published in 1974, contains a systematically developed theory of statistical explanation that has the same virtues I claimed for Coffa's approach. The issue here is not one of priority but one of complementarity. Coffa and Fetzer approach the problems in very different ways; both authors are highly deserving of our attention.

With complete justice, Fetzer has articulated his dissatisfaction in two articles (Fetzer, 1991, 1992). Those who want a more balanced view than I gave should refer to these writings as well.

I should emphasize, however, that although I find the approach to explanation via propensities valuable, I cannot agree that propensities furnish an admissible interpretation of the probability calculus. As Paul Humphreys argues cogently in his (1985), the probability calculus requires probabilities that propensities cannot furnish. For my view on the matter see Salmon (1979b).

NOTES

This essay resulted from an NEH institute titled "Is There a New Consensus in Philosophy of Science?" held at the Center for Philosophy of Science, University of Minnesota. The section on scientific explanation was held in the fall term, 1985. Kitcher and Salmon (1989) contains results of this portion of the institute.

1. For a much more complete and detailed account of this development, see Salmon (1989, 1990b).
2. Neither Hempel nor I would accept this as a precise formulation, but I think it is an intuitively clear way of indicating what is at issue here.
3. Y is a *proper subclass* of X if and only if every Y is an X but some X are not Y; Y is an *improper subclass* of X if and only if X is identical to Y.
4. It goes as follows. Since the probability of 6 on each die is $\frac{1}{6}$, and the outcomes are

independent, the probability of double-6 is $\frac{1}{36}$. Consequently, the probability of not getting double-6 on any given toss is $\frac{35}{36}$. Since the successive tosses are independent, the probability of not getting double-6 on n successive tosses is $(\frac{35}{36})^n$. The probability of getting double-6 at least once in n successive tosses is $1 - (\frac{35}{36})^n$. That quantity exceeds $\frac{1}{2}$ if and only if $n > 24$.

5. As far as I know, Bromberger never published this example, though he offers a similar one in his (1966).

6. In his stimulating book *The Scientific Image* (1980), Bas van Fraassen offers a charming philosophy-of-science-fiction story in which he maintains that, in the context, the length of a shadow does explain the height of a tower. Most commentators, I believe, remain skeptical on this point, See "Van Fraassen on Explanation" (essay 11).

7. This example is so old, and has been cited by so many philosophers, that I am reluctant to attribute it to any individual.

8. This essay is based on a presentation at the 1986 meeting of the American Association for the Advancement of Science.

9. I also offered another example. Around that time Linus Pauling's claims about the value of massive doses of vitamin C in the prevention of common colds was receiving a great deal of attention. To ascertain the efficacy of vitamin C in preventing colds, I suggested, it is *not* sufficient to establish that people who take large doses of vitamin C avoid colds. What is required is a double-blind controlled experiment in which the rate of avoidance for those who take vitamin C is compared with the rate of avoidance for those who receive only a placebo. If there is a significant difference in the probability of avoidance for those who take vitamin C and for those who do not, then we may conclude that vitamin C has some degree of causal efficacy in preventing colds. If, however, there is no difference between the two groups, then it would be a mistake to try to explain a person's avoidance of colds by constructing an argument analogous to (2) in which that result is attributed to treatment with vitamin C.

10. It is semantical rather than syntactical because it involves not only the characterization of a formal language but also the intended interpretation of that language.

11. These concepts are discussed more briefly and more accessibly in Railton (1981).

12. Suggestions regarding rapproachment between the unification theory and the causal theory are offered in "Scientific Explanation: Causation *and* Unification" (essay 4).

REFERENCES

Achinstein, Peter. 1983. *The nature of explanation.* New York: Oxford University Press.

Braithwaite, R. B. 1953. *Scientific explanation.* Cambridge: Cambridge University Press.

Bromberger, Sylvain. 1966. Why-Questions. In *Mind and cosmos,* edited by Robert G. Colodny, 86–111. Pittsburgh, Penn.: University of Pittsburgh Press.

Carnap, Rudolf. 1950. *Logical foundations of probability.* Chicago: University of Chicago Press.

Coffa, J. Alberto. 1973. Foundations of inductive explanation. Ph.D. diss. University of Pittsburgh.

———. 1974. Hempel's ambiguity. *Synthese* 28: 141–64.

Fetzer, James H. 1971. Dispositional probabilities. In *PSA 1970,* edited by Roger C. Buck and Robert S. Cohen, 473–82. Dordrecht: D. Reidel.

———. 1974a. Statistical explanations. In *PSA 1972,* edited by Kenneth Schaffner and Robert S. Cohen, 337–47. Dordrecht: D. Reidel.

———. 1974b. A single case propensity theory of explanation. *Synthese* 28: 171–98.

———. 1975. On the historical explanation of unique events. *Theory and Decision* 6: 87–97.

———. 1976. The likeness of lawlikeness. In *PSA 1974,* edited by Robert S. Cohen, 377–91. Dordrecht: D. Reidel.

————. 1977. A world of dispositions. *Synthese* 34: 397–421.

————. 1991. *Scientific knowledge.* Dordrecht: D. Reidel.

————. 1992. What's wrong with Salmon's history: The third decade. *Philosophy of Science* 59: 246–62.

Friedman, Michael. 1974. Explanation and scientific understanding. *Journal of Philosophy* 71: 5–19.

Greeno, James G. [1970] 1971. Explanation and information. In Salmon, 1971, pp. 89–104.

Hanson, N. R. 1959. On the symmetry between explanation and prediction. *Philosophical Review* 68: 349–58.

Hempel, Carl G. 1962. Deductive-nomological vs. statistical explanation. In *Scientific explanation, space, and time.* Minnesota Studies in the Philosophy of Science, vol. 3. Edited by Herbert Feigl and Grover Maxwell, 98–169.

————. 1965a. *Aspects of scientific explanation and other essays in the philosophy of science.* New York: Free Press.

————. 1965b. Aspects of scientific explanation. In Hempel, 1965a, pp. 331–496.

————. 1977. Nachwort 1976: Neuere ideen zsu den problemen der statistischen erklärung. In *Aspekte wissenschaftlicher erklärung,* by Carl G. Hempel, 98–123. Berlin: Walter de Gruyter.

Hempel, Carl G., and Paul Oppenheim. [1948] 1965. Studies in the logic of explanation. In Hempel 1965a, pp. 245–96.

Humphreys, Paul W. 1985. Why propensities cannot be probabilities. *Philosophical Review* 94: 557–70.

Jeffrey, R. C. [1969] 1971. Statistical explanation vs. statistical inference. In Salmon, 1971, pp. 19–28. Originally published in Rescher, ed. 1969.

Kitcher, Philip. 1976. Explanation, conjunction, and unification. *Journal of Philosophy* 73: 207–12.

————. 1981. Explanatory unification. *Philosophy of Science* 48: 507–31.

————. 1985. Two approaches to explanation. *Journal of Philosophy* 82: 632–39.

————. 1989. Explanatory unification and the causal structure of the world. In Kitcher and Salmon, eds., 1989, pp. 410–505.

————. 1993. *The advancement of science.* New York: Oxford University Press.

Kitcher, Philip, and Wesley C. Salmon. 1989. *Scientific explanation.* Minnesota Studies in the Philosophy of Science, vol. 13. Minneapolis: University of Minnesota Press.

Mill, John Stuart. 1843. *A system of logic.* London: John W. Parker.

Nagel, Ernst. 1961. *The structure of science: Problems in the logic of scientific explanation.* New York: Harcourt, Brace, and World.

Popper, Karl R. 1935. *Logik der forschung.* Vienna: Springer.

————. 1959. *The logic of scientific discovery.* New York: Basic Books. Translation with added appendices, of Popper, 1935.

Railton, P. 1978. A deductive-nomological model of probabilistic explanation. *Philosophy of Science* 45: 206–26.

————. 1980. Explaining explanation: A realist account of scientific explanation and understanding. Ph.D. diss. Princeton University, Princeton, N.J.

————. 1981. Probability, explanation, and information. *Synthese* 48: 233–56.

Rescher, Nicholas, ed. 1969. *Essays in honor of Carl G. Hempel.* Dordrecht: D. Reidel.

Salmon, Wesley C. 1971. *Statistical explanation and statistical relevance.* With contributions by J. G. Greeno and R. C. Jeffrey. Pittsburgh, Penn.: University of Pittsburgh Press.

————. 1974. Comments on "Hempel's ambiguity" by J. A. Coffa, *Synthese* 28: 165–69.

————. 1979. Propensities: A discussion review. *Erkenntnis* 14: 183–216.

————. 1984. *Scientific explanation and the causal structure of the world.* Princeton, N.J.: Princeton University Press.

—————. 1989. Four decades of scientific explanation. In Kitcher and Salmon, eds. 1989, pp. 3–219.

—————. 1990. *Four decades of scientific explanation.* Minneapolis: University of Minnesota Press. Reprinted from Kitcher and Salmon, eds. 1989, pp. 3–219.

Scriven, Michael. 1958. Definitions, explanations, and theories. In *Concepts, theories, and the mind-body problem,* Minnesota Studies in the Philosophy of Science, vol. 2. Edited by Herbert Feigl, Grover Maxwell, and Michael Scriven, 99–195. Minneapolis: University of Minnesota Press.

—————. 1959. Explanation and prediction in evolutionary theory. *Science* 130: 477–82.

—————. 1962. Explanations, predictions, and laws. In *Scientific explanation, space, and time,* Minnesota Studies in the Philosophy of Science, vol. 3. Edited by Herbert Feigl and Grover Maxwell, 170–230. Minneapolis: University of Minnesota Press.

15

The Pragmatics of Explanation

Bas C. van Fraassen

There are two problems about scientific explanation. The first is to describe it: when is something explained? The second is to show why (or in what sense) explanation is a virtue. Presumably we have no explanation unless we have a good theory; one which is independently worthy of acceptance. But what virtue is there in explanation over and above this? I believe that philosophical concern with the first problem has been led thoroughly astray by mistaken views on the second.

I. FALSE IDEALS

To begin I wish to dispute three ideas about explanation that seem to have a subliminal influence on the discussion. The first is that explanation is a relation simply between a theory or hypothesis and the phenomena or facts, just like truth for example. The second is that explanatory power cannot be logically separated from certain other virtues of a theory, notably truth or acceptability. And the third is that explanation is the overriding virtue, the end of scientific inquiry.

When is something explained? As a foil to the above three ideas, let me propose the simple answer: *when we have a theory which explains*. Note first that "have" is not "have on the books"; I cannot claim to have such a theory without implying that this theory is acceptable all told. Note also that both "have" and "explains" are tensed; and that I have allowed that we can have a theory which does not explain, or "have on the books" an unacceptable one that does. Newton's theory explained the tides but not the advance in the perihelion of mercury; we used to have an acceptable theory, provided by Newton, which bore (or bears timelessly?) the explanation relationship to some facts but not to all. My answer also implies that we can intelligibly say

264

that the theory explains, and not merely that people can explain by means of the theory. But this consequence is not very restrictive, because the former could be an ellipsis for the latter.

There are questions of usage here. I am happy to report that the history of science allows systematic use of both idioms. In Huygens and Young the typical phrasing seemed to be that phenomenon may be explained *by means of* principles, laws and hypotheses, or *according to* a view.[1] On the other hand, Fresnel writes to Arago in 1815 "Tous ces phénomènes . . . sont réunis et expliqués par la même théorie des vibrations," and Lavoisier says that the oxygen hypothesis he proposes *explains* the phenomena of combustion.[2] Darwin also speaks in the latter idiom: "In scientific investigations it is permitted to invent any hypothesis, and if it explains various large and independent classes of facts it rises to the rank of a well-grounded theory"; though elsewhere he says that the facts of geographical distribution are *explicable on* the theory of migration.[3]

My answer did separate acceptance of the theory from its explanatory power. Of course, the second can be a reason for the first; but *that* requires their separation. Various philosophers have held that explanation logically requires true (or acceptable) theories as premises. Otherwise, they hold, we can at most mistakenly believe that we have an explanation.

This is also a question of usage, and again usage is quite clear. Lavoisier said of the phlogiston hypothesis that it is too vague and consequently "s'adapte a toutes les explications dans lesquelles on veut le faire entrer."[4] Darwin explicitly allows explanations by false theories when he says "It can hardly be supposed that a false theory would explain, in so satisfactory a manner as does the theory of natural selection, the several large classes of facts above specified."[5] More recently, Gilbert Harman has argued similarly: that a theory explains certain phenomena is part of the evidence that leads us to accept it. But that means that the explanation-relation is visible beforehand. Finally, we criticize theories selectively: a discussion of celestial mechanics around the turn of the century would surely contain the assertion that Newton's theory does explain many planetary phenomena, though not the advance in the perihelion of Mercury.

There is a third false ideal, which I consider worst: that explanation is the *summum bonum* and exact aim of science. A virtue could be overriding in one of two ways. The first is that it is a minimal criterion of acceptability. Such is consistency with the facts in the domain of application (though not necessarily with all data, if these are dubitable!). Explanation is not like that, or else a theory would not be acceptable at all unless it explained all facts in its domain. The second way in which a virtue may be overriding is that of being required when it can be had. This would mean that if two theories pass other tests (empirical adequacy, simplicity) equally well, then the one which explains more must be accepted. As I have argued elsewhere,[6] and as we shall see in connection with Salmon's views below, a precise formulation of this demand requires hidden variables for indeterministic theories. But of course, hidden variables are rejected in scientific practice as so much "metaphysical baggage" when they make no difference in empirical predictions.

II. A BIASED HISTORY

I will outline the attempts to characterize explanation of the past three decades, with no pretense of objectivity. On the contrary, the selection is meant to illustrate the diagnosis, and point to the solution, of the next section.

1. Hempel

In 1966, Hempel summarized his views by listing two main criteria for explanation. The first is the criterion of *explanatory relevance*: "the explanatory information adduced affords good grounds for believing that the phenomenon to be explained did, or does, indeed occur."[7] That information has two components, one supplied by the scientific theory, the other consisting of auxiliary factual information. The relationship of providing good grounds is explicated as (a) implying (D-N case), or (b) conferring a high probability (I-S case), which is not lowered by the addition of other (available) evidence.

As Hempel points out, this criterion is not a sufficient condition for explanation: the red shift gives us good grounds for believing that distant galaxies are receding from us, but does not explain why they do. The classic case is the *barometer example*: the storm will come exactly if the barometers fall, which they do exactly if the atmospheric conditions are of the correct sort; yet only the last factor explains. Nor is the criterion a necessary condition; for this the classic case is the *paresis example*. We explain why the mayor, alone among the townsfolk, contracted paresis by his history of latent, contracted syphilis; yet such histories are followed by paresis in only a small percentage of cases.

The second criterion is the requirement of *testability*; but since all serious candidates for the role of scientific theory meet this, it cannot help to remove the noted defects.

2. Beckner, Putnam, and Salmon

The criterion of explanatory relevance was revised in one direction, informally by Beckner and Putnam and precisely by Salmon. Morton Beckner, in his discussion of evolution theory, pointed out that this often explains a phenomenon only by showing how it could have happened, given certain possible conditions.[8] Evolutionists do this by constructing models of processes which utilize only genetic and natural selection mechanisms, in which the outcome agrees with the actual phenomenon. Parallel conclusions were drawn by Hilary Putnam about the way in which celestial phenomena are explained by Newton's theory of gravity: celestial motions could indeed be as they are, given a certain possible (though not known) distribution of masses in the universe.[9]

We may take the paresis example to be explained similarly. Mere consistency with the theory is of course much too weak, since that is implied by logical irrelevance. Hence Wesley Salmon made this precise as follows: to explain is to exhibit (the) statistically relevant factors.[10] (I shall leave till later the qualifications about "screening off.") Since this sort of explication discards the talk about modeling and mechanisms of Beckner and Putnam, it may not capture enough. And indeed, I am not satisfied with Salmon's arguments that his criterion provides a sufficient condition. He gives the example of an equal mixture of Uranium 238 atoms and Polonium 214 atoms, which makes the Geiger counter click in interval $(t, t + m)$. This means that one of the atoms disintegrated. Why did it? The correct answer will be: because it was a Uranium 238 atom, if that is so—although the probability of its disintegration is much higher relative to the previous knowledge that the atom belonged to the described mixture.[11] The problem with this argument is that, on Salmon's criterion, we can explain not only why there was a disintegration, but also why *that* atom disintegrated *just then.* And surely that is exactly one of those facts which atomic physics leaves unexplained?

But there is a more serious general criticism. Whatever the phenomenon is, we can amass the statistically relevant factors, as long as the theory does not rule out the phenomenon altogether. "What more could one ask of an explanation?" Salmon inquires.[12] But in that case, as soon as we have an empirically adequate theory, we have an explanation of every fact in its domain. We may claim an explanation as soon as we have shown that the phenomenon can be embedded in some model allowed by the theory—that is, does not throw doubt on the theory's empirical adequacy.[13] But surely that is too sanguine?

3. Global Properties

Explanatory power cannot be identified with empirical adequacy; but it may still reside in the performance of the theory as a whole. This view is accompanied by the conviction that science does not explain individual facts but general regularities and was developed in different ways by Michael Friedman and James Greeno. Friedman says explicitly that in his view, "the kind of understanding provided by science is global rather than local" and consists in the simplification and unification imposed on our world picture.[14] That S_1 explains S_2 is a conjunction of two facts: S_1 implies S_1 relative to our background knowledge (and/or belief) K, and S_1 unifies and simplifies the set of its consequences relative to K. Friedman will no doubt wish to weaken the first condition in view of Salmon's work.

The precise explication Friedman gives of the second condition does not work, and is not likely to have a near variant that does.[15] But here we may look at Greeno's proposal.[16] His abstract and closing statement subscribe to the same general view as Friedman. But he takes as his model of a theory one which specifics a single probability space Q as the correct one, plus two partitions (or random variables) of which one is designated *explanandum* and the other *explanans*. An example: sociology cannot explain why Albert, who lives in San Francisco and whose father has a high income, steals a car. Nor is it meant to. But it does explain delinquency in terms of such other factors as residence and parental income. The degree of explanatory power is measured by an ingeniously devised quantity which measures the information I the theory provides of the explanandum variable M on the basis of explanans S. This measure takes its maximum value if all conditional probabilities $P(M_1/S_1)$ are zero or one (D-N case), and its minimum value zero if S and M are statistically independent.

Unfortunately, this way of measuring the unification imposed on our data abandons Friedman's insight that scientific understanding cannot be identified as a function of grounds for rational expectation. For if we let S and M describe the behavior of the barometer and coming storms, with P (barometer falls) = P (storm. comes) = 0.2, P (storm comes/barometer falls) = 1, and P (storm comes/barometer does not fall) = 0, then the quantity I takes its maximum value. Indeed, it does so whether we designate M or S as explanans.

It would seem that such asymmetries as exhibited by the red shift and barometer examples must necessarily remain recalcitrant for any attempt to strengthen Hempel's or Salmon's criteria by global restraints on theories alone.

4. The Major Difficulties

There are two main difficulties, illustrated by the old paresis and barometer examples, which none of the examined positions can handle. The first is that there are cases, clearly in a theory's domain, where the request for explanation is nevertheless rejected. We can explain why John, rather than his brothers contracted paresis, for he had syphilis; but not why he, among all those syphilitics, got paresis. Medical science is incomplete, and hopes to find the answer some day. But the example of the uranium atom disintegrating just then rather than later, is formally similar and we believe the theory to be complete. We also reject such questions as the Aristotelians asked the Galileans: why does a body free of impressed forces retain its velocity? The importance of this sort of case, and its pervasive character, has been repeatedly discussed by Adolf Grünbaum.

The second difficulty is the asymmetry revealed by the barometer: even if the theory implies that one condition obtains when and only when another does, it may be that it explains the one in terms of the other and not vice versa. An example which combines both the first and second difficulty is this: according to atomic physics, each chemical element has a characteristic atomic structure and a characteristic spectrum (of light emitted upon excitation). Yet the spectrum is explained by the atomic structure, and the question why a substance has that structure does not arise at all (except in the trivial sense that the questioner may need to have the terms explained to him).

5. Causality

Why are there no longer any Tasmanian natives? Well, they were a nuisance, so the white settlers just kept shooting them till there were none left. The request was not for population statistics, but for the story; though in some truncated way, the statistics "tell" the story.

In a later paper Salmon gives a primary place to causal mechanisms in explanation.[17] Events are bound into causal chains by two relations: spatio-temporal continuity and statistical relevance. Explanation requires the exhibition of such chains. Salmon's point of departure is Reichenbach's *principle of the common cause*: every relation of statistical relevance ought to be explained by one of causal relevance. This means that a correlation of simultaneous values must be explained by a prior common cause. Salmon gives two statistical conditions that must be met by a common cause C of events A and B:

(a) $P(A \& B/C) = P(A/C)P(B/C)$
(b) $P(A/B \& C) = P(A/C)$ "*C screens off B from A.*"

If $P(B/C) \neq 0$ these are equivalent, and symmetric in A and B.

Suppose that explanation is typically the demand for a common cause. Then we still have the problem: when does this arise? Atmospheric conditions explain the correlation between barometer and storm, say; but are still prior causes required to explain the correlation between atmospheric conditions and falling barometers?

In the quantum domain, Salmon says, causality is violated because "causal influence is not transmitted with spatio-temporal continuity." But the situation is worse. To assume Reichenbach's principle to be satisfiable, continuity aside, is to rule out all genuinely indeterministic theories. As example, let a theory say that C is invariably followed by one of the incompatible events A, B, or D, each with probability $\frac{1}{3}$. Let us suppose the theory complete, and its probabilities irreducible, with C the complete specification of state. Then we will find a correlation for which only C could be the common cause, but it is not. Assuming that A, B, D are always preceded by C and that they have low but equal prior probabilities, there is a statistical correlation between $\phi = (A \text{ or } D)$ and $\psi = (B \text{ or } D)$, for $P(\phi/\psi) = P(\phi/\psi) = \frac{1}{2} \neq P(\phi)$. But C, the only available candidate, does not screen off ϕ from ψ: $P(\phi/C \ \& \ \psi) = P(\phi/\psi) = \frac{1}{2} \neq P(\phi/C)$ which is $\frac{2}{3}$. Although this may sound complicated, the construction is so general that almost any irreducibly probabilistic situation will give a similar example. Thus Reichenbach's *principle of the common cause* is in fact a demand for hidden variables.

Yet we retain the feeling that Salmon has given an essential clue to the asymmetries of explanation. For surely the crucial point about the barometer is that the atmospheric conditions screen off the barometer fall from the storm? The general point that the asymmetries are totally bound up with causality was argued in a provocative article by B.A. Brody.[18] Aristotle certainly discussed examples of asymmetries: the planets do not twinkle because they are near, yet they are near if and only if they do not twinkle (*Posterior Analytics*, I, 13). Not all explanations are causal, says Brody, but the others use a second Aristotelian notion, that of essence. The spectrum angle is a clear case: sodium has that spectrum because it has this atomic structure, which is its essence.

Brody's account has the further advantage that he can say when questions do not arise: other properties are explained in terms of essence, but the request for an explanation of the essence does not arise. However, I do not see how he would distinguish between the questions why the uranium atom disintegrated and why it disintegrated just then. In addition there is the problem that modern science is not formulated in terms of causes and essences, and it seems doubtful that these concepts can be redefined in terms which do occur there.

6. Why-Questions

A why-question is a request for explanation. Sylvain Bromberger called P the *presupposition* of the question Why-P? and restated the problem of explanation as that of giving the conditions under which proposition Q is a correct answer to a why-question with presupposition P.[19] However, Bengt Hannson has pointed out that "Why was it John who ate the apple?" and "Why was it the apple which John ate?: are different why-questions, although the comprised proposition is the same.[20] The difference can be indicated by such phrasing, or by emphasis ("Why did *John* ... ?") or by an auxiliary clause ("Why did John rather than ... ?"). Hannson says that an explanation is requested, not of a proposition or fact, but of an *aspect* of a proposition.

As is at least suggested by Hannson, we can cover all these cases by saying that we wish an explanation of why *P* is true in contrast to other members of a set *X* or propositions. This explains the tension in our reaction to the paresis-example. The question why the mayor, in contrast to other townfolk generally, contracted paresis *has* a true correct answer: because of his latent syphilis. But the question why he did in contrast to the other syphilitics in his country club, has no true correct answer. Intuitively we may say: *Q* is a correct answer to *Why P in contrast to X?* only if *Q* gives reasons to expect that *P*, in contrast to the other members of *X*. Hannson's proposal for a precise criterion is: the probability of *P* given *Q* is higher than the average of the probabilities of *R* given *Q*, for members *R* of *X*.

Hannson points out that the set *X* of alternatives is often left tacit; the two questions about paresis might well be expressed by the same sentence in different contexts. The important point is that explanations are not requested of propositions, and consequently a distinction can be drawn between answered and rejected requests in a clear way. However, Hannson makes *Q* a correct answer to *Why P in contrast to X?* when *Q* is statistically irrelevant, when *P* is already more likely than the rest; or when *Q* implies *P* but not the others. I do not see how he can handle the barometer (or red shift, or spectrum) asymmetries. On his precise criterion, that the barometer fell is a correct answer to why it will storm as opposed to be calm. The difficulty is very deep: if *P* and *R* are necessarily equivalent, according to our accepted theories, how can *Why P in contrast to X?* be distinguished from *Why R in contrast to X?*

III. THE SOLUTION

1. Prejudices

Two convictions have prejudiced the discussion of explanation, one methodological and one substantive.

The first is that a philosophical account must aim to produce necessary and sufficient conditions for theory *T* explaining phenomenon *E*. A similar prejudice plagued the discussion of counterfactuals for twenty years, requiring the exact conditions under which, if *A* were the case, *B* would be. Stalnaker's liberating insight was that these conditions are largely determined by context and speaker's interest. This brings the central question to light: what *form* can these conditions take?

The second conviction is that explanatory power is a virtue of theories by themselves, or of their relation to the world, like simplicity, predictive strength, truth, empirical adequacy. There is again an analogy with counterfactuals: it used to be thought that science contains, or directly implies, counterfactuals. In all but limiting cases, however, the proposition expressed is highly context-dependent, and the implication is there at most relative to the determining contextual factors, such as speakers' interest.

2. Diagnosis

The earlier accounts lead us to the format: C explains E relative to theory T exactly if (a) T has certain global virtues, and (b) T implies a certain proposition $\phi(C, E)$ expressible in the language of logic and probability theory. Different accounts directed themselves to the specification of what should go into (a) and (b). We may add, following Beckner and Putnam, that T explains E exactly if there is a proposition C consistent with T (and presumably, background beliefs) such that C explains E relative to T.

The significant modifications were proposed by Hannson and Brody. The former pointed out that the explanandum E cannot be reified as a proposition: we request the explanation of something F in contrast to its alternatives X (the latter generally tacitly specified by context). This modification is absolutely necessary to handle some of our puzzles It requires that in (b) above we replace "$\phi(C, E)$" by the formula form "$\psi(C, F, X)$." But the problem of asymmetries remains recalcitrant, because if T implies the necessary equivalence of F and F' (say, atomic structure and characteristic spectrum), then T will also imply $\psi(C, F', X)$ if and only if it implies $\psi(C, F, X)$.

The only account we have seen which grapples at all successfully with this is Brody's. For Brody points out that even properties which we believe to be constantly conjoined in all possible circumstances, can be divided into essences and accidents, or related as cause and effect. In this sense, the asymmetries were no problem for Aristotle.

3. The Logical Problem

We have now seen exactly what logical problem is posed by the asymmetries. To put it in current terms: how can we distinguish propositions which are true in exactly the same possible worlds?

There are several known approaches that use impossible worlds. David Lewis, in his discussion of causality, suggests that we should look not only to the worlds theory T allows as possible, but also to those it rules out as impossible, and speaks of counterfactuals which are counterlegal. Relevant logic and entailment draw distinctions between logically equivalent sentences and their semantics devised by Routley and Meyer use both inconsistent and incomplete worlds. I believe such approaches to be totally inappropriate for the problem of explanation, for when we look at actual explanations of phenomena by theories, we do not see any detours through circumstances or events ruled out as impossible by the theory.

A further approach, developed by Rolf Schock, Romane Clark, and myself distinguishes sentences by the facts that make them true. The idea is simple. That it rains, that it does not rain, that it snows, and that it does not snow, are four distinct facts. The disjunction that it rains or does not rain is made true equally by the first and second, and not by the third or fourth, which distinguishes it from the logically equivalent disjunction that it snows or does not snow.[21] The distinction remains even if there is also a fact of its raining or not raining, distinct or identical with that of its snowing or not snowing.

This approach can work for the asymmetries of explanation. Such asymmetries are possible because, for example, the distinct facts that light is emitted with wavelengths λ, μ, . . . conjointly make up the characteristic spectrum, while quite different facts conjoin to make up the atomic structure. So we have shown how such asymmetries *can* arise, in the way that Stalnaker showed how failures of transitivity in counterfactuals can arise. But while we have the distinct facts to classify asymmetrically, we still have the nonlogical problem: whence comes the classification? The only suggestion so far is that it comes from Aristotle's concepts of cause and essence; but if so, modern science will not supply it.

4. The Aristotelian Sieve

I believe that we should return to Aristotle more thoroughly, and in two ways. To begin, I will state without argument how I understand Aristotle's theory of science. Scientific activity is divided into two parts, *demonstration* and *explanation*, the former treated mainly by the *Posterior Analytics* and the latter mainly by book II of the *Physics*. Illustrations in the former are mainly examples of explanations in which the results of demonstration are *applied*; this is why the examples contain premises and conclusions which are not necessary and universal principles, although demonstration is only to and from such principles. Thus the division corresponds to our pure versus applied science. There is no reason to think that principles and demonstrations have such words as "cause" and "essence" in them, although looking at pure science from outside, Aristotle could say that its principles state causes and essences. In applications, the principles may be filtered through a conceptual sieve originating outside science.

The doctrine of the four "causes" (*aitiai*) allows for the systematic ambiguity or context-dependence of why-questions.[22] Aristotle's example (*Physics* II, 3; 195a) is of a lantern. In a modern example, the question why the porch light is on may be answered "because I flipped the switch" or "because we are expecting company," and the context determines which is appropriate. Probabilistic relations cannot distinguish these. Which factors are explanatory is decided not by features of the scientific theory but by concerns brought from outside. This is true even if we ask specifically for an "efficient cause," for how far back in the chain should we look, and which factors are merely auxiliary contributors?

Aristotle would not have agreed that essence is context-dependent. The essence is what the thing is, hence, its sum of classificatory properties. Realism has always asserted that ontological distinctions determine the "natural" classification. But which property is counted as explanatory and which as explained seems to me clearly context dependent. For consider Bromberger's flagpole example: the shadow is so long because the pole has this height, and not conversely. At first sight, no contextual factor could reverse this asymmetry, because the pole's height is a property it has in and by itself, and its shadow is a very accidental feature. The general principle linking the two is that its shadow is a function $f(x, t)$ of its height x and the time t (the latter determining the sun's elevation). But imagine the pole is the pointer on a giant sundial. Then the values of f have desired properties for each time t, and we appeal to these to explain why it is (had to be) such a tall pole.

We may again draw a parallel to counterfactuals. Professor Geach drew my attention to the following spurious argument: If John asked his father for money, then they would not have quarreled (because John is too proud to ask after a quarrel). Also if John asked and they hadn't quarreled, he would receive. By the usual logic of counterfactuals, it follows that if John asked his father for money, he would receive. But we know that he would not, because they have in fact quarreled. The fallacy is of equivocation, because "what was kept constant" changed in the middle of the monologue. (Or if you like, the aspects by which worlds are graded as more or less similar to this one.) Because science cannot dictate what speakers decide to "keep constant" it contains no counterfactuals. By exact parallel, *science contains no explanations*.

5. The Logic of Why-Questions

What remains of the problem of explanation is to study its logic, which is the logic of why-questions. This can be put to some extent, but not totally, in the general form developed by Harrah and Belnap and others.[23]

A question admits of three classes of response: *direct answers, corrections,* and *comments*. A *presupposition*, it has been held, is any proposition implied by all direct answers, or equivalently, denied by a correction. I believe we must add that the question "Why P, in contrast to X?" also presupposes that (a) P is a member of X, (b) P is true and the majority of X are not. This opens the door to the possibility that a question may not be uniquely determined by its set of direct answers. The question itself should decompose into factors which determine that set: the *topic P*, the *alternatives X*, and a *request specification* (of which the doctrine of the four "causes" is perhaps the first description).

We have seen that the propositions involved in question and answer must be individuated by something more than the set of possible worlds. I propose that we use the facts that make them true (see footnote 21). The context will determine an asymmetric relation among these facts, of *explanatory relevance*; it will also determine the theory or beliefs which determine which worlds are *possible*, and what is *probable* relative to what.

We must now determine what direct answers are and how they are evaluated. They must be made true by facts (and only by facts forcing such) which are explanatorily relevant to those which make the topic true. Moreover, these facts must be statistically relevant, telling for the topic in contrast to the alternatives generally; this part I believe to be explicable by probabilities, combining Salmon's and Hannson's account. How strongly the answers count for the topic should be part of their evaluation as better or worse answers.

The main difference from such simple questions as "Which cat is on the mat?" lies in the relation of a why-question to its presuppositions. A why-question may fail to arise because it is ill-posed (*P* is false, or most of *X* is true), or because only question-begging answers tell probabilistically for *P* in contrast to *X* generally, or because none of the factors that do tell for *P* are explanatorily relevant in the question-context. Scientific theory enters mainly in the evaluation of possibilities and probabilities, which is only part of the process, and which it has in common with other applications such as prediction and control.

IV. SIMPLE PLEASURES

There are no explanations in science. How did philosophers come to mislocate explanation among semantic rather than pragmatic relations? This was certainly in part because the positivists tended to identify the pragmatic with subjective psychological features. They looked for measures by which to evaluate theories. Truth and empirical adequacy are such, but they are weak, being preserved when a theory is watered down. Some measure of "goodness of fit" was also needed, which did not reduce to a purely internal criterion such as simplicity, but concerned the theory's relation to the world. The studies of explanation have gone some way toward giving us such a measure, but it was a mistake to call this explanatory power. The fact that seemed to confirm this error was that we do not say that we *have* an explanation unless we have a theory which is acceptable, and victorious in its competition with alternatives, whereby we can explain. Theories are applied in explanation, but the peculiar and puzzling features of explanation are supplied by other factors involved. I shall now redescribe several familiar subjects from this point of view.

When a scientist campaigns on behalf of an advocated theory, he will point out how our situation will change if we accept it. Hitherto unsuspected factors become relevant, known relations are revealed to be strands of an intricate web, some terribly puzzling questions are laid to rest as not arising at all. We shall be in a much better position to explain. But equally, we shall be in a much better position to predict and control. The features of the theory that will make this possible are its empirical adequacy and logical strength, not special "explanatory power" and "control power." On the other hand, it is also a mistake to say explanatory power is nothing but those other features, for then we are defeated by asymmetries having no "objective" basis in science.

Why are *new* predictions so much more to the credit of a theory than agreement with the old? Because they tend to bring to light new phenomena which the older theories cannot explain. But of course, in doing so, they throw doubt on the empirical adequacy of the older theory: they show that a precondition for explanation is not met. As Boltzmann said of the radiometer, "the theories based on older hydrodynamic experience can never describe" these phenomena.[24] The failure in explanation is a by-product.

Scientific inference is inference to the best explanation. That does not rule at all for the supremacy of explanation among the virtues of theories. For we evaluate how good an explanation is given by how good a theory is used to give it, how close it fits to the empirical facts, how internally simple and coherent the explanation. There is a further evaluation in terms of a prior judgment of which kinds of factors are explanatorily relevant. If this further evaluation took precedence, overriding other considerations, explanation would be the peculiar virtue sought above all. But this is not so: instead, science schools our imagination so as to revise just those prior judgments of what satisfies and eliminates wonder.

Explanatory power is something we value and desire. But we are as ready, for the sake of scientific progress, to dismiss questions as not really arising at all. Explanation is indeed a virtue; but still, less a virtue than an anthropocentric pleasure.[25]

NOTES

1. I owe these and following references to my student Mr. Paul Thagard. For instance see C. Huygens, *Treatise on Light,* tr. by S. P. Thompson (New York, 1962), pp. 19, 20, 22, 63; Thomas Young, *Miscellaneous Works,* ed. by George Peacock (London, 1855), Vol. I, pp. 168, 170.

2. Augustin Fresnel, *Oeuvres Complètes* (Paris, 1866), Vol. 1, p. 36 (see also pp. 254, 355); Antoine Lavoisier, *Oeuvres* (Paris, 1862), Vol. II, p. 233.

3. Charles Darwin, *The Variation of Animals and Plants* (London, 1868), Vol. I, p. 9; *On the Origin of Species* (Facs. of first edition, Cambridge, Mass., 1964), p. 408.

4. Antoine Lavoisier, *op. cit.,* p. 640.

5. *Origin* (sixth ed., New York, 1962), p. 476.

6. "Wilfrid Sellars on Scientific Realism," *Dialogue,* Vol. 14 (1975), pp. 606–16.

7. C. G. Hempel, *Philosophy of Natural Science* (Englewood Cliffs, N.J., 1966), p. 48.

8. *The Biological Way of Thought* (Berkeley, Calif., 1968), p. 176; this was first published in 1959.

9. In a paper of which a summary is found in Frederick Suppe (ed.), *The Structure of Scientific Theories* (Urbana, Ill., 1974).

10. "Statistical Explanation," pp. 173–231 in R. G. Colodny (ed.), *The Nature and Function of Scientific Theories* (Pittsburgh, Penn., 1970); reprinted also in Salmon's book cited below.

11. Ibid., pp. 207–209. Nancy Cartwright has further, unpublished, counterexamples to the necessity and sufficiency of Salmon's criterion.

12. Ibid., p. 222.

13. These concepts are discussed in my "To Save the Phenomena," *The Journal of Philosophy* 73 (1976), forthcoming.

14. "Explanation and Scientific Understanding," *The Journal of Philosophy* 71 (1974): 5–19.

15. See Philip Kitcher, "Explanation, Conjunction, and Unification," *The Journal of Philosophy* 73 (1976): 207–12.

16. "Explanation and Information," pp. 89–103 in Wesley Salmon (ed.), *Statistical Explanation and Statistical Relevance* (Pittsburgh, Penn., 1971). This paper was originally published with a different title in *Philosophy of Science* 37 (1970): 279–93.

17. "Theoretical Explanation," pp. 118–45 in Stephan Körner (ed.), *Explanation* (Oxford, 1975).

18. "Towards an Aristotelian Theory of Scientific Explanation," *Philosophy of Science* 39 (1972): 20–31.

19. "Why-Questions," pp. 86–108 in R. G. Colodny (ed.), *Mind and Cosmos* (Pittsburgh, Penn., 1966).

20. "Explanations—Of What?" (mimeographed: Stanford University, 1974)

21. Cf. my "Facts and Tautological Entailments," *The Journal of Philosophy* 66 (1969): 477–87 and in A. R. Anderson et al. (eds.), *Entailment* (Princeton, 1975); and "Extension, Intension, and Comprehension" in Milton Munitz (ed.), *Logic and Ontology* (New York, 1973).

22. Cf. Julius Moravcik, "Aristotle on Adequate Explanations," *Synthese* 28 (1974): 3–18

23. Cf. N. D. Belnap Jr., "Questions: Their Presuppositions, and How They Can Fail to Arise," *The Logical Way of Doing Things,* ed. by Karel Lambert (New Haven, 1969), pp. 23–39.

24 Ludwig Boltzmann, *Lectures on Gas Theory,* tr. by S. G. Brush (Berkeley, Calif., 1964), p. 25

25. The author wishes to acknowledge helpful discussions and correspondence with Professors N. Cartwright, B. Hannson, K. Lambert, and W. Salmon, and the financial support of the Canada Council.

16

Explanatory Unification

Philip Kitcher

1. THE DECLINE AND FALL OF THE COVERING LAW MODEL

One of the great apparent triumphs of logical empiricism was its official theory of explanation. In a series of lucid studies (Hempel 1965, chapters 9, 10, 12; Hempel 1962; Hempel 1966), C. G. Hempel showed how to articulate precisely an idea which had received a hazy formulation from traditional empiricists such as Hume and Mill. The picture of explanation which Hempel presented, the *covering law model,* begins with the idea that explanation is derivation. When a scientist explains a phenomenon, he derives (deductively or inductively) a sentence describing that phenomenon (the *explanandum* sentence) from a set or sentences (the *explanans*) which must contain at least one general law.

Today the model has fallen on hard times. Yet it was never the empiricists' whole story about explanation. Behind the official model stood an unofficial model, a view of explanation which was not treated precisely, but which sometimes emerged in discussions of theoretical explanation. In contrasting scientific explanation with the idea or reducing unfamiliar phenomena to familiar phenomena, Hempel suggests this unofficial view: "What scientific explanation, especially theoretical explanation, aims at is not an intuitive and highly subjective kind of understanding, but an objective kind of insight that is achieved by a systematic unification, by exhibiting the phenomena as manifestations of common, underlying structures and processes that conform to specific, testable, basic principles" (Hempel 1966, p. 83; see also Hempel 1965, pp. 345, 444). Herbert Feigl makes a similar point: "The aim of scientific explanation throughout the ages has been *unification,* that is, the comprehending of a maximum of facts and regularities in terms of a minimum of theoretical concepts and assumptions" (Feigl 1970, p. 12).

This unofficial view, which regards explanation as unification, is, I think, more promising than the official view. My aim in this paper is to develop the view and to present its virtues. Since the picture of explanation which results is rather complex, my exposition will be programmatic, but I shall try to show that the unofficial view can avoid some prominent shortcomings of the covering law model.

Why should we want an account of scientific explanation? Two reasons present themselves. Firstly, we would like to understand and to evaluate the popular claim that the natural sciences do not merely pile up unrelated items of knowledge of more or less practical significance, but that they increase our understanding of the world. A theory of explanation should show us *how* scientific explanation advances our understanding. (Michael Friedman cogently presents this demand in his (1974)). Secondly, an account of explanation ought to enable us to comprehend and to arbitrate disputes in past and present science. Embryonic theories are often defended by appeal to their explanatory power. A theory of explanation should enable us to judge the adequacy of the defense.

The covering law model satisfies neither of these *desiderata*. Its difficulties stem from the fact that, when it is viewed as providing a set of necessary and *sufficient* conditions for explanation, it is far too liberal. Many derivations which are intuitively nonexplanatory meet the conditions of the model. Unable to make relatively gross distinctions, the model is quite powerless to adjudicate the more subtle considerations about explanatory adequacy which are the focus of scientific debate. Moreover, our ability to derive a description of a phenomenon from a set of premises *containing a law* seems quite tangential to our understanding of the phenomenon. Why should it be that exactly those derivations which employ laws advance our understanding?

The unofficial theory appears to do better. As Friedman points out, we can easily connect the notion of unification with that of understanding. (However, as I have argued in my (1976), Friedman's analysis of unification is faulty; the account or unification offered below is indirectly defended by diagnosis of the problems for his approach.) Furthermore, as we shall see below, the acceptance of some major programs of scientific research—such as the Newtonian program of eighteenth-century physics and chemistry and the Darwinian program of nineteenth-century biology—depended on recognizing promises for unifying, and thereby explaining, the phenomena. Reasonable skepticism may protest at this point that the attractions of the unofficial view stem from its unclarity. Let us see.

2. EXPLANATION: SOME PRAGMATIC ISSUES

Our first task is to formulate the problem of scientific explanation clearly, filtering out a host of issues which need not concern us here. The most obvious way in which to categorize explanation is to view it as an activity. In this activity we answer the actual or anticipated questions of an actual or anticipated audience. We do so by presenting reasons. We draw on the beliefs we hold, frequently using or adapting arguments furnished to us by the sciences.

Recognizing the connection between explanations and arguments, proponents of the covering law model (and other writers on explanation) have identified explanations as special types of arguments. But although I shall follow the covering law model in employing the notion of argument to characterize that of explanation, I shall not adopt the ontological thesis that explanations are arguments. Following Peter Achinstein's thorough discussion of ontological issues concerning explanation in his (1977), I shall suppose that an explanation is an ordered pair consisting of a proposition and an act type.[1] The relevance of arguments to explanation resides in the fact that what makes an ordered pair (p, explaining q) an explanation is that a sentence expressing p bears an appropriate relation to a particular argument. (Achinstein shows how the central idea of the covering law model can be viewed in this way.) So I am supposing that there are acts of explanation which draw on arguments supplied by science, reformulating the traditional problem of explanation as the question: What features should a scientific argument have if it is to serve as the basis for an act of explanation?[2]

The complex relation between scientific explanation and scientific argument may be illuminated by a simple example. Imagine a mythical Galileo confronted by a mythical fusilier who wants to know why his gun attains maximum range when it is mounted on a flat plain, if the barrel is elevated at 45° to the horizontal. Galileo reformulates this question as the question of why an ideal projectile, projected with fixed velocity from a perfectly smooth horizontal plane and subject only to gravitational acceleration, attains maximum range when the angle of elevation of the projection is 45°. He defends this reformulation by arguing that the effects of air resistance in the case of the actual projectile, the cannonball, are insignificant, and that the curvature of the earth and the unevenness of the ground can be neglected. He then selects a kinematical argument which shows that, for fixed velocity, an ideal projectile attains maximum range when the angle of elevation is 45°. He adapts this argument by explaining to the fusilier some unfamiliar terms ("uniform acceleration," let us say), motivating some problematic principles (such as the law of composition of velocities), and by omitting some obvious computational steps. Both Galileo and the fusilier depart satisfied.

The most general problem of scientific explanation is to determine the conditions which must be met if science is to be used in answering an explanation-seeking question Q. I shall restrict my attention to explanation-seeking why-questions, and I shall attempt to determine the conditions under which an argument whose conclusion is S can be used to answer the question "Why is it the case that S?" More colloquially, my project will be that of deciding when an argument explains why its conclusion is true.[3]

We leave on one side a number of interesting, and difficult issues. So, for example, I shall not discuss the general relation between explanation-seeking questions and the arguments which can be used to answer them, nor the pragmatic conditions governing the idealization of questions and the adaptation of scientific arguments to the needs of the audience. (For illuminating discussions of some of these issues, see Bromberger 1962.) Given that so much is dismissed, does anything remain?

In a provocative article (van Fraassen 1977), Bas van Fraassen denies, in effect, that there are any issues about scientific explanation other than the pragmatic questions I have just banished. After a survey of attempts to provide a theory of explanation he appears to conclude that the idea that explanatory power is a special virtue of theories is a myth. We accept scientific theories on the basis of their empirical adequacy and simplicity, and, having done so, we use the arguments with which they supply us to give explanations. This activity of applying scientific arguments in explanation accords with extra-scientific, "pragmatic," conditions. Moreover, our views about these extra-scientific factors are revised in the light of our acceptance of new theories: ". . . science schools our imagination so as to revise just those prior judgments or what satisfies and eliminates wonder" (van Fraassen 1977, p. 150). Thus there are no context-independent conditions beyond those of simplicity and empirical adequacy which distinguish arguments for use in explanation.

Van Fraassen's approach does not fit well with some examples from the history of science—such as the acceptance of Newtonian theory of matter and Darwin's theory of evolution—examples in which the explanatory promise of a theory was appreciated in advance of the articulation of a theory with predictive power. (See pp. 170–172.) Moreover, the account I shall offer provides an answer to skepticism that no "global constraints" (van Fraassen 1977, p. 146) on explanation can avoid the familiar problems of asymmetry and irrelevance, problems which bedevil the covering law model. I shall try to respond to van Fraassen's challenge by showing that there are certain context-independent features of arguments which distinguish them for application in response to explanation-seeking why-questions, and that we can assess theories (including embryonic theories) by their ability to provide us with such arguments. Hence I think that it is possible to defend the thesis that historical appeals to the explanatory power of theories involve recognition of a virtue over and beyond considerations of simplicity and predictive power.

Resuming our main theme, we can use the example of Galileo and the fusilier to achieve a further refinement of our problem. Galileo selects and adapts an argument from his new kinematics—that is, he draws an argument from a set of arguments available for explanatory purposes, a set which I shall call the *explanatory store*. We may think of the sciences not as providing us with many unrelated individual arguments which can be used in individual acts of explanation, but as offering a reserve of explanatory arguments, which we may tap as need arises. Approaching the issue in this way, we shall be led to present our problem as that of specifying the conditions which must be met by the explanatory store.

The set of arguments which science supplies for adaptation acts of explanation will change with our changing beliefs. Therefore the appropriate *analysandum* is the notion of the store of arguments relative to a set of accepted sentences. Suppose that, at the point in the history of inquiry which interests us, the set of accepted sentences is K. (I shall assume, for simplicity's sake, that K is consistent. Should our beliefs be inconsistent then it is more appropriate to regard K as some tidied version of our beliefs.) The general problem I have set is that of specifying $E(K)$, the *explanatory store over K*, which is the set of arguments acceptable as the basis for acts of explanation by those whose beliefs are exactly the members of K. (For the purposes of this paper I shall assume that, for each K there is exactly one $E[K]$.)

The unofficial view answers the problem: for each K, $E(K)$ is the set of arguments which best unifies K. My task is to articulate the answer. I begin by looking at two historical episodes in which the desire for unification played a crucial role. In both cases, we find three important features: (i) prior to the articulation of a theory with high predictive power, certain proposals for theory construction are favored on grounds of their explanatory promise; (ii) the explanatory power of embryonic theories is explicably tied to the notion of unification; (iii) particular features of the theories are taken to support their claims to unification. Recognition of (i) and (ii) will illustrate points that have already been made while (iii) will point toward an analysis of the concept of unification.

3. A NEWTONIAN PROGRAM

Newton's achievements in dynamics, astronomy, and optics inspired some of his successors to undertake an ambitious program which I shall call "dynamic corpuscularianism."[4] *Principia* had shown how to obtain the motions of bodies from a knowledge of the forces acting on them, and had also demonstrated the possibility of dealing with gravitational systems in a unified way. The next step would be to isolate a few basic force laws, akin to the law of universal gravitation, so that, applying the basic laws to specifications of the dispositions of the ultimate parts of bodies, all of the phenomena of nature could be derived. Chemical reactions, for example, might be understood in terms of the rearrangement of ultimate parts under the action of cohesive and repulsive forces. The phenomena of reflection, refraction and diffraction of light might be viewed as resulting from a special force of attraction between light corpuscles and ordinary matter. These speculations encouraged eighteenth-century Newtonians to construct very general hypotheses about inter-atomic forces-even in the absence of any confirming evidence for the existence of such forces.

In the preface to *Principia,* Newton had already indicated that he took dynamic corpuscularianism to be a program deserving the attention of the scientific community:

> I wish we could derive the rest of the phenomena of Nature by the same kind of reasoning from mechanical principles, for I am induced by many reasons to suspect that they may all depend upon certain forces by which the particles of bodies, by some causes hitherto unknown, are either mutually impelled towards one another, and cohere in regular figures, or are repelled and recede from one another (Newton 1962, p. xviii. See also Newton 1952, pp. 401–402).

This, and other influential passages, inspired Newton's successors to try to complete the unification of science by finding further force laws analogous to the law of universal gravitation. Dynamic corpuscularianism remained popular so long as there was promise of significant unification. Its appeal began to fade only when repeated attempts to specify force laws were found to invoke so many different (apparently incompatible) attractive and repulsive forces that the goal of unification appeared unlikely. Yet that goal could still motivate renewed efforts to implement the program. In the second half of the eighteenth-century Boscovich revived dynamic corpuscularian hopes by claiming that the whole of natural philosophy can be reduced to "one law of forces existing in nature.[5]

The passage I have quoted from Newton suggests the nature of the unification that was being sought. *Principia* had exhibited how one style of argument, one "kind of reasoning from mechanical principles," could be used in the derivation of descriptions of many, diverse phenomena. The unifying power of Newton's work consisted in its demonstration that one *pattern* of argument could be used again and again in the derivation of a wide range of accepted sentences. (I shall give a representation of the Newtonian pattern in section 3.) In searching for force laws analogous to the law of universal gravitation, Newton's successors were trying to generalize the pattern or argument presented in *Principia,* so that one "kind of reasoning" would suffice to derive all phenomena of motion. If, furthermore, the facts studied by chemistry, optics, physiology, and so forth, could be related to facts about particle motion, then one general pattern of argument would be used in the derivation of all phenomena. I suggest that this is the ideal of unification at which Newton's immediate successors aimed, which came to seem less likely to be attained as the eighteenth century wore on, and which Boscovich's work endeavored, with some success, to reinstate.

4. THE RECEPTION OF DARWIN'S EVOLUTIONARY THEORY

The picture of unification which emerges from the last section may be summarized quite simply: a theory unifies our beliefs when it provides one (or more generally, a few) pattern(s) of argument which can be used in the derivation of a large number of sentences which we accept. I shall try to develop this idea more precisely in later sections. But first I want to show how a different example suggests the same view of unification.

In several places, Darwin claims that his conclusion that species evolve through natural selection should be accepted because of its explanatory power, that . . . the doctrine must sink or swim according as it groups and explains phenomena" (F. Darwin 1887; vol. 2, p. 153, quoted in Hull 1974, p. 292). Yet, as he often laments, he is unable to provide any complete derivation of any biological phenomenon—our ignorance of the appropriate facts and regularities is "profound." How, then, can he contend that the primary virtue of the new theory is its explanatory power?

The answer lies in the fact that Darwin's evolutionary theory promises to unify a host of biological phenomena (C. Darwin 1964, pp. 243–44). The eventual unification would consist in derivations of descriptions of these phenomena which would instantiate a common pattern. When Darwin expounds his doctrine what he offers us is the pattern. Instead of detailed explanations of the presence of some particular trait in some particular species, Darwin presents two "imaginary examples" (C. Darwin 1964, pp. 90–96) and a diagram which shows, in a general way, the evolution of species *represented by schematic letters* (1964, pp. 116–26). In doing so, he exhibits a pattern of argument, which he maintains can be instantiated, *in principle,* by a complete and rigorous derivation of descriptions of the characteristics of any current species. The derivation would employ the principle of natural selection—as well as premises describing ancestral forms and the nature of their environment and the (unknown) laws of variation and inheritance. In place of detailed evolutionary stories, Darwin offers *explanation-sketches.* By showing how a particular characteristic would be advantageous to a particular species, he indicates an explanation of the emergence of that characteristic in the species, suggesting the outline of an argument instantiating the general pattern.

From this perspective, much of Darwin's argumentation in the *Origin* (and in other works) becomes readily comprehensible. Darwin attempts to show how his pattern can be applied to a host of biological phenomena. He claims that, by using arguments which instantiate the pattern, we can account for analogous variations in kindred species, for the greater variability of specific (as opposed to generic) characteristics, for the facts about geographical distribution, and so forth. But he is also required to resist challenges that the pattern cannot be applied in some cases, that premises for arguments instantiating the pattern will not be forthcoming. So, for example, Darwin must show how evolutionary stories, fashioned after his pattern, can be told to account for the emergence of complex organs. In both aspects of his argument, he is responding to those who would limit the application of his pattern or whether he is campaigning for its use within a realm of biological phenomena. Darwin has the same goal. He aims to show that his theory should be accepted because it unifies and explains.

5. ARGUMENT PATTERNS

Our two historical examples[6] have led us to the conclusion that the notion of an argument pattern is central to that of explanatory unification. Quite different considerations could easily have pointed us in the same direction. If someone were to distinguish between the explanatory worth of two arguments instantiating a common pattern, then we would regard that person as an explanatory deviant. To grasp the concept of explanation is to see that if one accepts an argument as explanatory, one is thereby committed to accepting as explanatory other arguments which instantiate the same pattern.

To say that members of a set of arguments instantiate a common pattern is to recognize that the arguments in the set are similar in some interesting way. With different interests, people may fasten on different similarities, and may arrive at different notions of argument pattern. Our enterprise is to characterize the concept of argument pattern which plays a role in the explanatory activity of scientists.

Formal logic, ancient and modern, is concerned in one obvious sense with patterns of argument. The logician proceeds by isolating a small set of expressions (the logical vocabulary), considers the schemata formed from sentences by replacing with dummy letters all expressions which do not belong to this set, and tries to specify which sequences of these schemata are valid patterns of argument. The pattern of argument which is taught to students of Newtonian dynamics is not a pattern of the kind which interests logicians. It has instantiations with different logical structures. (A rigorous derivation of the equations of motion of different dynamical systems would have a logical structure depending on the number of bodies involved and the mathematical details of the integration.) Moreover, an argument can only instantiate the Newtonian pattern if particular *non*logical terms, "force," "mass," and "acceleration," occur in it in particular ways. However, the logician's approach can help us to isolate the notion of argument pattern which we require.

Let us say that a *schematic sentence* is an expression obtained by replacing some, but not necessarily all, the nonlogical expressions occurring in a sentence with dummy letters. A set of *filling instructions* for a schematic sentence is a set of directions for replacing the dummy letters of the schematic sentence, such that, for each dummy letter, there is a direction which tells us how it should be replaced. A *schematic argument* is a sequence of schematic sentences. A *classification* for a schematic argument is a set of sentences which describe the inferential characteristics of the schematic argument: its function is to tell us which terms in the sequence are to be regarded as premises, which are to be inferred from which, what rules of inference are to be used, and so forth.

We can use these ideas to define the concept of a *general argument pattern.* A general argument pattern is a triple consisting of a schematic argument, a set of sets of filling instructions containing one set of filling instructions for each term of the schematic argument, and a classification for the schematic argument. A sequence of sentences instantiates the general argument pattern just in case it meets the following conditions:

(i)   The sequence has the same number or terms as the schematic argument of the general argument pattern.

(ii) Each sentence in the sequence is obtained from the corresponding schematic sentence in accordance with the appropriate set of filling instructions.

(iii) It is possible to construct a chain of reasoning which assigns to each sentence the status accorded to the corresponding schematic sentence by the classification.

We can make these definitions more intuitive by considering the way in which they apply to the Newtonian example. Restricting ourselves to the basic pattern used in treating systems which contain one body (such as the pendulum and the projectile) we may represent the schematic argument as follows:

(1) The force on α is β.
(2) The acceleration of α is γ.
(3) Force = mass $\cdot$ acceleration.
(4) (Mass of α) $\cdot$ (γ) = (β)
(5) $\delta = \theta$

The filling instructions tell us that all occurrences of "α" are to be replaced by an expression referring to the body under investigation; occurrences of "β" are to be replaced by an algebraic expression referring to a function of the variable coordinates and of time; "γ" is to be replaced by an expression which gives the acceleration of the body as a function of its coordinates and their time-derivatives (thus, in the case of a one-dimensional motion along the x-axis of a Cartesian coordinate system, γ would be relaced by the expression "d^2x/dt^2"); "δ" is to be replaced by an expression referring to the variable coordinates of the body, and "θ" is to be replaced by an explicit function of time (thus the sentences which instantiate (5) reveal the dependence of the variable coordinates on time, and so provide specifications of the positions of the body in question throughout the motion). The classification of the argument tells us that (1)–(3) have the status of premises, that (4) is obtained from them by substituting identicals, and that (5) follows from (4) using algebraic manipulation and the techniques of the calculus.

Although the argument patterns which interest logicians are general argument patterns in the sense just defined, our example exhibits clearly the features which distinguish the kinds of patterns which scientists are trained to use. Whereas logicians are concerned to display all the schematic premises which are employed and to specify exactly which rules of inference are used, our example allows for the use or premises (mathematical assumptions) which do not occur as terms of the schematic argument, and it does not give a complete description of the way in which the route from (1) to (5) is to go. Moreover, our pattern does not replace all nonlogical expressions by dummy letters. Because some nonlogical expressions remain, the pattern imposes special demands on arguments which instantiate it. In a different way, restrictions are set by the instructions for replacing dummy letters. The patterns or logicians are very liberal in both these latter respects. The conditions for replacing dummy letters in Aristotelian syllogisms, or first-order schemata, require only that some letters be relaced with predicates, others with names.

Arguments may be similar either in terms of their logical structure or in terms of the nonlogical vocabulary they employ at corresponding places. I think that the notion of similarity (and the corresponding notion of pattern) which is central to the explanatory activity of scientists results from a compromise in demanding these two kinds of asimilarity. I propose that scientists are interested in *stringent* patterns of argument, patterns which contain some nonlogical expressions and which are fairly similar in terms of logical structure. The Newtonian pattern cited above furnishes a good example. Although arguments instantiating this pattern do not have exactly the same logical structure, the classification imposes conditions which ensure that there will be similarities in logical structure among such arguments. Moreover, the presence of the nonlogical terms sets strict requirements on the instantiations and so ensures a different type of kinship among them. Thus, without trying to provide an exact analysis of the notion of stringency, we may suppose that the stringency of a pattern is determined by two different constraints: (1) the conditions on the substitution of expressions for dummy letters, jointly imposed by the presence of nonlogical expressions in the pattern and by the filling instructions, and (2) the conditions on the logical structure, imposed by the classification. If both conditions are relaxed completely then the notion of pattern degenerates so as to admit *any* argument. If both conditions are simultaneously made as strict as possible, then we obtain another degenerate case, a "pattern" which is its own unique instantiation. If condition (2) is tightened at the total expense of (1), we produce the logician's notion of pattern. The use of condition (1) requires that arguments instantiating a common pattern draw on a common nonlogical vocabulary. We can glimpse here that ideal of unification through the use of a few theoretical concepts which the remarks of Hempel and Feigl suggest.

Ideally, we should develop a precise account of how these two kinds of similarity are weighted against one another. The best strategy for obtaining such an account is to see how claims about stringency occur in scientific discussions. But scientists do not make explicit assessments of the stringency of argument patterns. Instead they evaluate the ability of a theory to explain and to unify. The way to a refined account of stringency lies through the notions of explanation and unification.

6. EXPLANATION AS UNIFICATION

As I have posed it, the problem of explanation is to specify which set of arguments we ought to accept for explanatory purposes given that we hold certain sentences to be true. Obviously this formulation can encourage confusion: we must not think of a scientific community as *first* deciding what sentences it will accept and *then* adopting the appropriate set of arguments. The Newtonian and Darwinian examples should convince us that the promise of explanatory power enters into the modification of our beliefs. So, in proposing that $E(K)$ is a function of K, I do not mean to suggest that the acceptance of K must be temporally prior to the adoption of $E(K)$.

$E(K)$ is to be that set of arguments which best unifies K. There are, of course, usually many ways of deriving some sentences in K from others. Let us call a set of arguments which derives some members of K from other members of K a *systematization* of K. We may then think of $E(K)$ as the best systematization of K.

Let us begin by making explicit an idealization which I have just made tacitly. A set of arguments will be said to be *acceptable relative* to K just in case every argument in the set consists of a sequence of steps which accord with elementary valid rules of inference (deductive or inductive) and if every premise of every argument in the set belongs to K. When we are considering ways of systematizing K we restrict our attention to those sets of arguments which are acceptable relative to K. This is an idealization because we sometimes use as the basis of acts of explanation arguments furnished by theories whose principles we no longer believe. I shall not investigate this practice nor the considerations which justify us in engaging in it. The most obvious way to extend my idealized picture to accommodate it is to regard the explanatory store over K, as I characterize it here, as being supplemented with an extra class of arguments meeting the following conditions: (a) from the perspective of K, the premises of these arguments are approximately true; (b) these arguments can be viewed as approximating the structure of (parts of) arguments in $E(K)$; (c) the arguments are simpler than the corresponding arguments in $E(K)$. Plainly, to spell out these conditions precisely would lead into issues which are tangential to my main goal in this paper.

The moral of the Newtonian and Darwinian examples is that unification is achieved by using similar arguments in the derivation of many accepted sentences. When we confront the set of possible systematizations of K we should therefore attend to the *patterns* of argument which are employed in each systematization. Let us introduce the notion of a *generating set*: if Σ is a set of arguments then a generating set for Σ is a set of argument patterns Π such that each argument in Σ is an instantiation of some pattern in Π. A generating set for Σ will be said to be *complete with respect to K* if and only if every argument which is acceptable relative to K and which instantiates a pattern in Π belongs to Σ. In determining the explanatory store $E(K)$ we first narrow our choice to those sets of arguments which are acceptable relative to K, the systematizations of K. Then we consider, for each such set of arguments, the various generating sets of argument patterns which are complete with respect to K. (The importance of the requirement of completeness is to debar explanatory deviants who use patterns selectively.) Among these latter sets we select that set with the greatest unifying power (according to criteria shortly to be indicated) and we call the selected set the *basis* of the set of arguments in question. The explanatory store over K is that systematization whose basis does best by the criteria of unifying power.

This complicated picture can be made clearer, perhaps, with the help of a diagram.

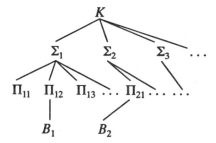

Systematizations. sets of arguments acceptable relative to K.

Complete generating sets. Π_{ij}, is a generating set for Σ_i which is complete with respect to K.

Bases. B_i is the basis for Σ_i and is selected as the best of the Π_{ij} on the basis of unifying power.

If B_k is the basis with the greatest unifying power then $E(K) = \Sigma_k$.

The task which confronts us is now formulated as that of specifying the factors which determine the unifying power of a set of argument patterns. Our Newtonian and Darwinian examples inspire an obvious suggestion: unifying power is achieved by generating a large number of accepted sentences as the conclusions of acceptable arguments which instantiate a few, stringent patterns. With this in mind, we define the *conclusion set* of a set of arguments Σ, $C(\Sigma)$, to be the set of sentences which occur as conclusions of some argument in Σ. So we might propose that the unifying power of a basis B_i with respect to K varies directly with the size of $C(\Sigma_i)$, varies directly with the stringency of the patterns which belong to B_i, and varies inversely with the number of members of B_i. This proposal is along the right lines, but it is, unfortunately, too simple.

The pattern of argument which derives a specification of the positions of bodies as explicit functions of time from a specification of the forces acting on those bodies is, indeed, central to Newtonian explanations. But not every argument used in Newtonian explanations instantiates this pattern. Some Newtonian derivations consist of an argument instantiating the pattern followed by further derivations from the conclusion. Thus, for example, when we explain why a pendulum has the period it does we may draw on an argument which *first* derives the equation of motion of the pendulum and *then* continues by deriving the period. Similarly, in explaining why projectiles projected with fixed velocity obtain maximum range when projected at 45° to the horizontal, we first show how the values of the horizontal and vertical coordinates can be found as functions of time and the angle of elevation, use our results to compute the horizontal distance traveled by the time the projectile returns to the horizontal, and then show how this distance is a maximum when the angle of elevation of projection is 45°. In both cases we take further steps beyond the computation of the explicit equations of motion—and the further steps in each case are different.

If we consider the entire range of arguments which Newtonian dynamics supplies for explanatory purposes, we find that these arguments instantiate a number of different patterns. Yet these patterns are not entirely distinct, for all of them proceed by using the computation of explicit equations of motion as a prelude to further derivation. It is natural to suggest that the pattern of computing equations of motion is the *core* pattern provided by Newtonian theory, and that the theory also shows how conclusions generated by arguments instantiating the core pattern can be used to derive further conclusions. In some Newtonian explanations, the core pattern is supplemented by a *problem-reducing pattern*, a pattern of argument which shows how to obtain a further type of conclusion from explicit equations of motion.

This suggests that our conditions on unifying power should be modified, so that, instead of merely counting the number of different patterns in a basis, we pay attention to similarities among them. All the patterns in the basis may contain a common core pattern, that is, each of them may contain some pattern as a subpattern. The unifying power of a basis is obviously increased if some (or all) of the patterns it contains share a common core pattern.

As I mentioned at the beginning of this chapter, the account of explanation as unification is complicated. The explanatory store is determined on the basis of criteria which pull in different directions, and I shall make no attempt here to specify precisely the ways in which these criteria are to be balanced against one another. Instead, I shall show that some traditional problems of scientific explanation can be solved without more detailed specification of the conditions on unifying power. For the account I have indicated has two important corollaries.

(A) Let Σ, Σ' be sets of arguments which are acceptable relative to K and which meet the following conditions:

    (i)   the basis of Σ' is as good as the basis of Σ in terms of the criteria of stringency of patterns, paucity of patterns, presence of core patterns, and so forth.

    (ii) $C(\Sigma)$ is a proper subset of $C(\Sigma')$.

Then $\Sigma \neq E(K)$.

(B) Let Σ, Σ' be sets of arguments which are acceptable relative to K and which meet the following conditions:

    (i) $C(\Sigma) = C(\Sigma')$

    (ii)   the basis of Σ' is a proper subset of the basis of Σ.

Then $\Sigma \neq E(K)$.

(A) and (B) tell us that sets of arguments which do equally well in terms of some of our conditions are to be ranked according to their relative ability to satisfy the rest. I shall try to show that (A) and (B) have interesting consequences.

7. ASYMMETRY, IRRELEVANCE, AND ACCIDENTAL GENERALIZATION

Some familiar difficulties beset the covering law model. The *asymmetry problem* arises because some scientific laws have the logical form of equivalences. Such laws can be used "in either direction." Thus a law asserting that the satisfaction of a condition C_1 is equivalent to the satisfaction of a condition C_2 can be used in two different kinds of argument. From a premise asserting that an object meets C_1, we can use the law to infer that it meets C_2; conversely, from a premise asserting that an object meets C_2, we can use the law to infer that it meets C_1. The asymmetry problem is generated by noting that in many such cases one of these derivations can be used in giving explanations while the other cannot.

Consider a hoary example, (For further examples, see Bromberger 1966.) We can explain why a simple pendulum has the period it does by deriving a specification of the period from a specification of the length and the law which relates length and period. But we cannot explain the length of the pendulum by deriving a specification of the length from a specification of the period and the same law. What accounts for our different assessment of these two arguments? Why does it seem that one is explanatory while the other "gets things backwards"? The covering law model fails to distinguish the two, and thus fails to provide answers.

The *irrelevance problem* is equally vexing. The problem arises because we can sometimes find a lawlike connection between an accidental and irrelevant occurrence and an event or state which would have come about independently of that occurrence. Imagine that Milo the magician waves his hands over a sample of table salt, thereby "hexing" it. It is true (and I shall suppose, lawlike) that all hexed samples of table salt dissolve when placed in water. Hence we can construct a derivation of the dissolving of Milo's hexed sample of salt by citing the circumstances of the hexing. Although this derivation fits the covering law model, it is, by our ordinary lights, nonexplanatory. (This example is given by Wesley Salmon in his (1970); Salmon attributes it to Henry Kyburg. For more examples, see Achinstein 1971.)

The covering law model explicitly debars a further type of derivation which any account of explanation ought to exclude. Arguments whose premises contain no laws, but which make essential use of accidental generalizations are intuitively nonexplanatory. Thus, if we derive the conclusion that Horace is bald from premises stating that Horace is a member of the Greenbury School Board and that all members of the Greenbury School Board are bald, we do not thereby explain why Horace is bald. (See Hempel 1965, p. 339.) We shall have to show that our account does not admit as explanatory derivations of this kind.

I want to show that the account of explanation I have sketched contains sufficient resources to solve these problems.[7] In each case we shall pursue a common strategy. Faced with an argument we want to exclude from the explanatory store we endeavor to show that any set of arguments containing the unwanted argument could not provide the best unification of our beliefs. Specifically, we shall try to show either that any such set of arguments will be more limited than some other set within an equally satisfactory basis, or that the basis of the set must fare worse according to the criterion of using the smallest number of most stringent patterns. That is, we shall appeal to the corollaries (A) and (B) given above. In actual practice, this strategy for exclusion is less complicated than one might fear, and, as we shall see, its applications to the examples just discussed brings out what is intuitively wrong with the derivations we reject.

Consider first the irrelevance problem. Suppose that we were to accept as explanatory the argument which derives a description of the dissolving of the salt from a description of Milo's act of hexing. What will be our policy for explaining the dissolving of samples of salt which have not been hexed? If we offer the usual chemical arguments in these latter cases then we shall commit ourselves to an inflated basis for the set of arguments we accept as explanatory. For, unlike the person who explains *all* cases of dissolving of samples of salt by using the standard chemical pattern of argument, we shall be committed to the use of two different patterns of argument in covering such cases. Nor is the use of the extra pattern of argument offset by its applicability in explaining other phenomena. Our policy employs one extra pattern of argument without extending the range of things we can derive from our favored set of arguments. Conversely, if we eschew the standard chemical pattern of argument (just using the pattern which appeals to the hexing) we shall find ourselves unable to apply our favored pattern to cases in which the sample of salt dissolved has not been hexed. Moreover, the pattern we use will not fall under the more general patterns we employ to explain chemical phenomena such as solution, precipitation, and so forth. Hence the unifying power of the basis for our preferred set of arguments will be less than that of the basis for the set of arguments we normally accept as explanatory.[8]

If we explain the dissolving of the sample of salt which Milo has hexed by appealing to the hexing then we are faced with the problems of explaining the dissolving of unhexed samples of salt. We have two options: (a) to adopt two patterns of argument corresponding to the two kinds of case or (b) to adopt one pattern of argument whose instantiations apply just to the cases of hexed salt. If we choose (a) then we shall be in conflict with (B), whereas choice of (b) will be ruled out by (A). The general moral is that appeals to hexing fasten on a local and accidental feature of the cases of solution. By contrast our standard arguments instantiate a pattern which can be generally applied.[9]

A similar strategy succeeds with the asymmetry problem. We have general ways of explaining why bodies have the dimensions they do. Our practice is to describe the circumstances leading to the formation or the object in question and then to show how it has since been modified. Let us call explanations of this kind "origin and development derivations." (In some cases, the details of the original formation of the object are more important; with other objects, features of its subsequent modification are crucial.) Suppose now that we admit as explanatory a derivation of the length of a simple pendulum from a specification of the period. Then we shall either have to explain the lengths of *non*swinging bodies by employing quite a different style of explanation (an origin and development derivation) or we shall have to forego explaining the lengths of such bodies. The situation is exactly parallel to that of the irrelevance problem. Admitting the argument which is intuitively nonexplanatory saddles us with a set of arguments which is less good at unifying our beliefs than the set we normally choose for explanatory purposes.

Our approach also solves a more refined version of the pendulum problem (given by Paul Teller in his [1974]). Many bodies which are not currently executing pendulum motion *could* be making small oscillations, and, were they to do so, the period of their motion would be functionally related to their dimensions. For such bodies we can specify the *dispositional period* as the period which the body would have if it were to execute small oscillations. Someone may now suggest that we can construct derivations of the dimensions of bodies from specifications of their dispositional periods, thereby generating an argument pattern which can be applied as generally as that instantiated in origin and development explanations. This suggestion is mistaken. There are some objects—such as the earth and the Crab Nebula—which *could not* be pendulums, and for which the notion of a dispositional period makes of no sense. Hence the argument pattern proposed cannot entirely supplant our origin and development derivations, and, in consequence, acceptance of it would fail to achieve the best unification of our beliefs.

The problem posed by accidental generalizations can be handled in parallel fashion. We have a general pattern of argument, using principles of physiology, which we apply to explain cases of baldness. This pattern is generally applicable, whereas that which derives ascriptions of baldness using the principle that all members of the Greenbury School Board are bald is not. Hence, as in the other cases, sets which contain the unwanted derivation would be ruled out by one of the conditions (A), (B).

Of course, this does not show that an account of explanation along the lines I have suggested would sanction only derivations which satisfy the conditions imposed by the covering law model. For I have not argued that an explanatory derivation need contain *any* sentence or universal form. What *does* seem to follow from the account of explanation as unification is that explanatory arguments must not use accidental generalization, and, in this respect, the new account appears to underscore and generalize an important insight of the covering law model. Moreover, our success with the problems of asymmetry and irrelevance indicates that, even in the absence or a detailed account of the notion of stringency and of the way in which generality of the consequence set is weighed against paucity and stringency of the patterns in the basis, the view of explanation as unification has the resources to solve some traditional difficulties for theories of explanation.

8. SPURIOUS UNIFICATION

Unfortunately there is a fly in the ointment. One of the most aggravating problems for the covering law model has been its failure to exclude certain types of self-explanation. (For a classic source of difficulties see Eberle, Kaplan, and Montague 1961.) As it stands, the account of explanation as unification seems to be even more vulnerable on this score. The problem derives from a phenomenon which I shall call *spurious unification.*

Consider, first, a difficulty which Hempel and Oppenheim noted in a seminal article (Hempel 1965, chapter 10). Suppose that we conjoin two laws. Then we can derive one of the laws from the conjunction, and the derivation conforms to the covering law model (unless, of course, the model is restricted to cover only the explanation of singular sentences: Hempel and Oppenheim do, in fact, make this restriction). To quote Hempel and Oppenheim:

> The core of the difficulty can be indicated briefly by reference to an example: Kepler's laws, K, may be conjoined with Boyle's law, B, to a stronger law $K \cdot B$; but derivation of K from the latter would not be considered as an explanation of the regularities stated in Kepler's laws; rather it would be viewed as representing, in effect, a pointless "explanation" of Kepler's laws by themselves. (Hempel 1965, p. 273, fn. 33.)

This problem is magnified for our account. For why may we nor unify our beliefs *completely* by deriving all of them using arguments which instantiate the one pattern?

$$\frac{\alpha \text{ and } \beta}{\alpha} \text{ ["}\alpha\text{" is to be replaced by any sentence we accept.]}$$

Or, to make matters even more simple, why should we not unify our beliefs by using the most official pattern of self-derivation?

α ["α" is to be replaced by any sentence we accept.]

There is an obvious reply. The patterns just cited may succeed admirably in satisfying our criteria of using a few patterns of argument to generate many beliefs, but they fail dismally when judged by the criterion of stringency. Recall that the stringency of a pattern is assessed by adopting a compromise between two constraints: stringent patterns are not only to have instantiations with similar logical structures; their instantiations are also to contain similar nonlogical vocabulary at similar places. Now both of the above argument patterns are very lax in allowing any vocabulary whatever to appear in the place of "α." Hence we can argue that, according to our intuitive concept of stringency, they should be excluded as nonstringent.

Although this reply is promising, it does not entirely quash the objection. A defender of the unwanted argument patterns may *artificially* introduce restrictions on the pattern to make it more stringent. So, for example, if we suppose that one of *our* favorite patterns (such as the Newtonian pattern displayed above) is applied to generate conclusions meeting a particular condition *C*, the defender of the patterns just cited may propose that "α" is to be replaced, not by any sentence, but by a sentence which meets *C*. He may then legitimately point out that his newly contrived pattern is as stringent as our favored pattern. Inspired by this partial success, he may adopt a general strategy. Wherever we use an argument pattern to generate a particular type of conclusion, he may use some argument pattern which involves self-derivation, placing an appropriate restriction on the sentences to be substituted for the dummy letters. In this way, he will mimic whatever unification we achieve. His "unification" is obviously spurious. How do we debar it?

The answer comes from recognizing the way in which the stringency of the unwanted patterns was produced. Any condition on the substitution of sentences for dummy letters would have done equally well, provided only that it imposed constraints comparable to those imposed by acceptable patterns. Thus the stringency of the restricted pattern seems accidental to it. This accidental quality is exposed when we notice that we can vary the filling instructions, while retaining the same syntactic structure, to obtain a host of other argument patterns with equally many instantiations. By contrast, the constraints imposed on the substitution of nonlogical vocabulary in the Newtonian pattern (for example) cannot be amended without destroying the stringency of the pattern or without depriving it of its ability to furnish us with many instantiations. Thus the constraints imposed in the Newtonian pattern are essential to its functioning; those imposed in the unwanted pattern are not.

Let us formulate this idea as an explicit requirement. If the filling instructions associated with a pattern *P* could be replaced by different filling instructions, allowing for the substitution of a class of expressions or the same syntactic category, to yield a pattern *P′* and if *P′* would allow the derivation *any* sentence, then the unification achieved by *P* is spurious. Consider, in this light, any of the patterns which we have been trying to debar. In each case, we can vary the filling instructions to produce an even more "successful" pattern. So, for example, given the pattern:

α ["α" is to be replaced by a sentence meeting condition *C*]

is we can generalize the filling instructions to obtain

α ["α" is to be replaced by any sentence].

Thus, under our new requirement, the unification achieved by the original pattern is spurious.

In a moment I shall try to show how this requirement can be motivated, both by appealing to the intuition which underlies the view or explanation as unification and by recognizing the role that something like my requirement has played in the history or science. Before I do so, I want to examine a slightly different kind of example which initially appears to threaten my account. Imagine that a group of religious fanatics decides to argue for the explanatory power of some theological doctrines by claiming that these doctrines unify their beliefs about the world. They suggest that their beliefs can be systematized by using the following pattern:

God wants it to be the case that α.
What God wants to be the case is the case. ["α" is to be replaced by any accepted sentence describing the physical world]

$$\alpha$$

The new requirement will also identify as spurious the pattern just presented, and will thus block the claim that the theological doctrines that God exists and has the power to actualize his wishes have explanatory power. For it is easy to see that we can modify the filling instructions to obtain a pattern that will yield any sentence whatsoever.

Why should patterns whose filling instructions can be modified to accommodate any sentence be suspect? The answer is that, in such patterns, the nonlogical vocabulary which remains is idling. The presence of that nonlogical vocabulary imposes no constraints on the expressions we can substitute for the dummy symbols, so that, beyond the specification that a place be filled by expressions of a particular syntactic category, the structure we impose by means of filling instructions is quite incidental. Thus the patterns in question do not genuinely reflect the contents of our beliefs. The explanatory store should present the order of natural phenomena which is exposed by what we think we know. To do so, it must exhibit connections among our beliefs beyond those which could be found among any beliefs. Patterns of self-derivation and the type of pattern exemplified in the example of the theological community merely provide trivial, omnipresent connections, and, in consequence, the unification they offer is spurious.

My requirement obviously has some kinship with the requirement that the principles put forward in giving explanations be testable. As previous writers have insisted that genuine explanatory theories should not be able to cater to all possible evidence. I am demanding that genuinely unifying patterns should not be able to accommodate all conclusions. The requirement that I have proposed accords well with some of the issues which scientists have addressed in discussing the explanatory merits of particular theories. Thus several of Darwin's opponents complain that the explanatory benefits claimed for the embryonic theory of evolution are illusory, on the grounds that the style of reasoning suggested could be adapted to any conclusion. (For a particularly acute statement of the complaint, see the review by Fleeming Jenkin, printed in Hull 1974, especially p. 342.) Similarly, Lavoisier denied that the explanatory power or the phlogiston theory was genuine, accusing that theory of using a type of reasoning, which could adapt itself to any conclusion (Lavoisier 1862, vol. II, p. 233). Hence I suggest that some problems of spurious unification can be solved in the way I have indicated, and that the solution conforms both to our intuitions about explanatory unification and to the considerations which are used in scientific debate.

However, I do not wish to claim that my requirement will debar all types of spurious unification. It may be possible to find other unwanted patterns which circumvent my requirement. A full characterization of the notion of a stringent argument pattern should provide a criterion for excluding the unwanted patterns. My claim in this section is that it will do so by counting as spurious the unification achieved by patterns which adapt themselves to any conclusion and by patterns which accidentally restrict such universally hospitable patterns. I have also tried to show how this claim can be developed to block the most obvious cases of spurious unification.

9. CONCLUSIONS

I have sketched an account of explanation as unification, attempting to show that such an account has the resources to provide insight into episodes in the history of science and to overcome some traditional problems for the covering law model. In conclusion, let me indicate very briefly how my view of explanation as unification suggests how scientific explanation yields understanding. By using a few patterns of argument in the derivation of many beliefs we minimize the number of *types* of premises we must take as underived. That is, we reduce, insofar as possible, the number of types of facts we must accept as brute. Hence we can endorse something close to Friedman's view of the merits of explanatory unification (Friedman 1974, pp. 18–19).

Quite evidently, I have only *sketched* an account of explanation. To provide precise analyses of the notions I have introduced, the basic approach to explanation offered here must be refined against concrete examples of scientific practice. What needs to be done is to look closely at the argument patterns favored by scientists and attempt to understand what characteristics they share. If I am right, the scientific search for explanation is governed by a maxim, once formulated succinctly by E. M. Forster. Only connect.

NOTES

A distant ancestor of this paper was read to the Dartmouth College Philosophy Colloquium in the Spring of 1977. I would like to thank those who participated, especially Merrie Bergmann and Jim Moor, for their helpful suggestions. I am also grateful to two anonymous referees for

Philosophy of Science whose extremely constructive criticisms have led to substantial improvements. Finally, I want to acknowledge the amount I have learned from the writing and teaching of Peter Hempel. The present essay is a token payment on an enormous debt.

1. Strictly speaking, this is one of two views which emerge from Achinstein's discussion and which he regards as equally satisfactory. As Achinstein goes on to point out, either of these ontological theses can be developed to capture the central idea of the covering law model.

2. To pose the problem in this way we may still invite the change that *arguments* should not be viewed as the bases for acts of explanation. Many of the criticisms leveled against the covering law model by Wesley Salmon in his seminal paper on statistical explanation (Salmon 1970) can be reformulated to support this charge. My discussion in section 7 will show how some of the difficulties raised by Salmon for the covering law model do not bedevil my account. However, I shall not respond directly to the points about statistical explanation and statistical inference advanced by Salmon and by Richard Jeffrey in his (1970). I believe that Peter Railton has shown how these specific difficulties concerning statistical explanation can be accommodated by an approach which takes explanations to be (or be based on) arguments (see Railton 1978), and that the account offered in section 4 of his paper can be adapted to complement my own.

3. Of course, in restricting my attention to why-questions I am following the tradition of philosophical discussion or scientific explanation: as Bromberger notes in section IV of his (1966) not all explanations are directed at why-questions, but attempts to characterize explanatory responses to why-questions have a special interest for the philosophy of science because of the connection to a range of methodological issues. I believe that the account of explanation offered in the present paper could be extended to cover explanatory answers to some other kinds of questions (such as how-questions). But I do want to disavow the claim that unification is relevant to all types of explanation. If one believes that explanations are sometimes offered in response to what-questions (for example), so that it is correct to talk of someone explaining what a gene is, then one should allow that some types of explanation can be characterized independently of the notions of unification or of argument. I ignore these kinds of explanation in part because they lack the methodological significance of explanations directed at why-questions and in part because the problem of characterizing explanatory answers to what-questions seems so much less recalcitrant than that of characterizing explanatory answers to why-questions (for a similar assessment, see Belnap and Steel 1976, pp. 86–87). Thus I regard a full account of explanation as a heterogeneous affair, because the conditions required of adequate answers to different types of questions are rather different, and I intend the present essay to make a proposal about how part of this account (the most interesting part) should be developed.

4. For illuminating accounts of Newton's influence on eighteenth-century research see Cohen (1956) and Schofield (1969). I have simplified the discussion by considering only *one* of the programs which eighteenth-century scientists derived from Newton's work. A more extended treatment would reveal the existence of several different approaches aimed at unifying science, and I believe that the theory of explanation proposed in this paper may help in the historical task of understanding the diverse aspirations of different Newtonians. (For the problems involved in this enterprise, see Heimann and McGuire 1971).

5. See Boscovich (1966) part III, especially p. 134. For an introduction to Boscovich's work, see the essays by L L. Whyte and Z. Markovic in Whyte (1961). For the influence of Boscovich on British science, see the essays of Pearce Williams and Schofield in the same volume, and Schofeild (1969).

6. The examples could easily be multiplied. I think it is possible to understand the structure and explanatory power of such theories as modern evolutionary theory, transmission genetics, plate tectonics, and sociobiology in the terms I develop here.

7. More exactly, I shall try to show that my account can solve some of the principal ver-

sions of these difficulties which have been used to discredit the covering law model. I believe that it can also overcome more refined versions of the problems than I consider here, but to demonstrate that would require a more lengthy exposition.

8. There is an objection this line of reasoning. Can't we view the arguments $\langle (x) \ (Sx$ and $Hx) \rightarrow Dx)$, Sa and $Ha, Da\rangle$, $\langle (x)((Sx$ and $\sim Hx) \rightarrow Dx)$, Sb and $\sim Hb, Db\rangle$ as instantiating a common pattern? I reply that, insofar as we can view these arguments as instantiating a common pattern, the standard pair of comparable (low-level) derivations—$\langle (x) \ (Sx \rightarrow Dx)$, Sa, $Da\rangle$, $\langle (x) \ (Sx \rightarrow Dx)$, Sb, $Db\rangle$—share a more stringent common pattern. Hence incorporating the deviant derivations in the explanatory store would give us an inferior basis. We can justify the claim that the pattern instantiated by the standard pair of derivations is more stringent than that shared by the deviant derivations, by noting that representation of the deviant pattern would compel us to broaden our conception of schematic sentence, and, even were we to do so, the deviant pattern would contain a "degree of freedom" which the standard pattern lacks. For a representation of the deviant "pattern" would take the form $\langle (x) \ ((Sx$ and $xHx) \rightarrow Dx)$, Sa and $\alpha Ha, Da\rangle$, where "α" is to be replaced uniformly either with the null symbol or with "$\sim$." Even if we waive my requirement that, in schematic sentences, we substitute for *non*logical vocabulary, it is evident that this "pattern" is more accommodating than the standard pattern.

9. However, the strategy I have recommended will not avail a different type of case. Suppose that a deviant wants to explain the dissolving of the salt by appealing to some property which holds universally. That is, the "explanatory" arguments are to begin from some premise such as "$(x)((x$ is a sample at salt and x does not violate conservation or energy) $\rightarrow x$ dissolves in water)" or "$(x)((x$ is a sample of salt and $x = x) \rightarrow x$ dissolves in water)." I would handle these cases somewhat differently. If the deviant's explanatory store were to be as unified as our own, then it would contain arguments corresponding to ours in which a redundant conjunct systematically occurred, and I think it would be plausible to invoke a criterion of simplicity to advocate dropping that conjunct.

REFERENCES

Achinstein, P. *Law and Explanation.* Oxford University Press, 1971.

———. "What is an Explanation?" *American Philosophical Quarterly* 14 (1977): 1–13.

Belnap, N., and T. B. Steel. *The Logic of Questions and Answers.* New Haven, Conn.: Yale University Press, 1976.

Boscovich, R. J. *A Theory of Natural Philosophy* (trans. J. M. Child). Cambridge, Mass.: MIT Press, 1966.

Bromberger, S. "An Approach to Explanation." In *Analytical Philosophy* (First Series), edited by R. J. Butler. Oxford: Blackwell, 1962.

———. "Why-Questions." In *Mind and Cosmos,* edited by R. Colodny. Pittsburgh, Penn.: University of Pittsburgh Press, 1966.

Cohen, I. B. *Franklin and Newton.* Philadelphia: American Philosophical Society, 1956.

Darwin, C. *On the Origin of Species.* Facsimile of the First Edition, edited by E. Mayr. Cambridge, Mass.: Harvard University Press, 1964.

Darwin, F. *The Life and Letters of Charles Darwin.* London: John Murray, 1987.

Eberle, R., D. Kaplan, and R. Montague. "Hempel and Oppenheim on Explanation." *Philosophy of Science* 28 (1961): 418–28.

Feigl, H. "The 'Orthodox' View of Theories: Remarks in Defense as Well as Critique." In *Minnesota Studies in the Philosophy of Science,* Vol. 4, edited by M. Radner and S. Winokur. Minneapolis: University of Minnesota Press, 1970.

Friedman, M. "Explanation and Scientific Understanding." *Journal of Philosophy* 71 (1974): 5–19.

Heimann, F., and J. E. McGuire. "Newtonian Forces and Lockean Powers." *Historical Studies in the Physical Sciences* 3 (1971): 233–306.

Hempel, C. G. *Aspects of Scientific Explanation.* New York: The Free Press, 1965.

———. "Deductive-Nonlogical vs. Statistical Explanation." In *Minnesota Studies in the Philosophy of Science,* Vol. 3, edited by H. Feigl and G. Maxwell. Minneapolis: University of Minnesota Press, 1962.

———. *Philosophy of Natural Science.* Englewood Cliffs, N.J.: Prentice-Hall, 1966.

Hull, D. (ed.). *Darwin and His Critics.* Cambridge, Mass.: Harvard University Press, 1974.

Jeffrey, R. "Statistical Explanation vs. Statistical Inference." In *Essays in Honor of Carl G. Hempel,* edited by N. Rescher. Dordrecht: D. Reidel, 1970.

Kitcher, P. S. "Explanation, Conjunction, and Unification." *Journal of Philosophy* 73 (1976): 207–12.

Lavoisier, A. *Oeuvres.* Paris, n.p., 1862.

Newton, I. *The Mathematical Principles of Natural Philosophy,* translated by A. Motte and F. Cajori. Berkeley: University of California Press, 1962.

———. *Opticks.* New York: Dover, 1952.

Railkon, P. "A Deductive-Nomological Model of Probabilistic Explanation." *Philosophy of Science* 45 (1978): 206–26.

Salmon, W. "Statistical Explanation." In *The Nature and Function of Scientific Theories,* edited by R. Colodny. Pittsburgh, Penn.: University of Pittsburgh Press, 1970.

Schofield, R. E. *Mechanism and Materialism.* Princeton, N.J.: Princeton University Press, 1969.

Teller, F. "On Why-Questions," *Nous* 8 (1974): 371–80.

van Fraassen, B. "The Pragmatics of Explanation." *American Philosophical Quarterly* 14 (1977): 143–50.

Whyte, L. L. (ed.). *Roger Joseph Boscovich.* London: Allen and Unwin, 1961.

CASE STUDY FOR PART 3

It is sometimes argued that since certain specific properties of the universe are necessary for the origin of intelligent life in the universe, the fact that we can pose questions at all in some sense explains the origin/existence of those special properties. One way of stating this observation, known as the *Anthropic principle,* is as follows:

> The conditions necessary for intelligent life can only be met under the most unusual of circumstances. The universe we live in is clearly not representative of a purely random set of initial conditions, but rather the product of a unique set of circumstances favorable for the evolution of intelligent life. Our ability to observe and ask questions presupposes these circumstances, and thus explains their presence.

1. Does the Anthropic principle qualify as a scientific law on Hempel's, Lambert and Britten's, or Cartwright's accounts?

2. Is the Anthropic principle an a posteriori truth, i.e., something whose truth is discovered by experience? Or is it an a priori truth, i.e., something whose truth can be known prior to experience?

3. Does the Anthropic principle embody a causal regularity? Explain your answer.

4. Do Hempel's, Lambert and Britten's, or Cartwright's accounts of scientific laws allow for the possibility of a priori truths counting as scientific laws?

5. Is the Anthropic principle explanatory? If so, in what sense?

6. Assuming for the moment that the Anthropic principle is explanatory, can Hempel, Salmon, van Fraassen, or Kitcher account for the explanatory element captured by the Anthropic principle on their respective models of explanation?

STUDY QUESTIONS FOR PART 3

1. Is explaining an event merely a matter of pointing out how it arises from processes and/or events that are familiar? Can you think of any familiar events or processes whose explanation involves recourse to unfamiliar events or processes?

2. What reasons does Hempel provide for believing explanations are arguments?

3. Are any of Hempel's conditions of adequacy sufficient for a genuine scientific explanation? Can you think of counterexamples for each one?

4. Are any of Hempel's conditions of adequacy necessary for a genuine scientific explanation? Can you think of counterexamples for each one?

5. What is Hempel's account of functional/teleological explanation according to the essay you read? Is this an adequate account of such explanations? Defend your answer.

6. Hempel mentions he is offering an account of what are often referred to as causal explanations. To what extent do Hempelian explanations rely on causality and how does Salmon's account differ in this regard?

7. What is Hempel's conception of law? Why is it deficient according to Lambert and Britten?

8. What is Lambert and Britten's conception of scientific explanation and how does it differ from Hempel's model?

9. Should all true counterfactual conditionals be regarded as laws?

10. How does conceiving of laws as counterfactual conditionals improve Hempel's concept of laws and his model of explanation according to Lambert and Britten?

11. After reading the articles in this part, a student remarks that whereas Lambert and Britten's article establishes that at least one of Hempel's conditions is not sufficient, Cartwright's article points out that one of the conditions is not necessary. Which conditions does this student have in mind? Is the student correct in her assessment? Why or why not?

12. Nancy Cartwright argues that scientific theories have two functions that should be kept distinct. What are these two functions?

13. What reasons does Cartwright provide for believing that covering law models of scientific explanation conflate the two functions identified above? Why does Cartwright believe they should be kept distinct?

14. What are *ceteris paribus* laws? Can you think of any examples outside of the domain of physics?

15. What does Cartwright mean in her assertion that "the truth doesn't explain much"? Do you agree with her assessment? Why or why not?

16. What are scientific explanations according to Cartwright's essay?

17. Compare Salmon's, van Fraassen's, and Kitcher's solutions to the asymmetry problem. Do any of them succeed? Explain your answer.

18. Compare Salmon's, van Fraassen's, and Kitcher's solutions to the irrelevancy problem. Do any of them succeed? Explain your answer.

19. Why does Salmon insist that explanatory factors must be statistically relevant? What does van Fraassen say on this point? Who presents the better argument and why?

20. Leaving genuinely indeterministic phenomena to one side for the moment, is causality necessary for an adequate explanation of phenomena? What reasons does Salmon offer for this thesis? Do you find them compelling? Why or why not?

21. Is there any way that Salmon's model of explanation can make sense of the "explanation as unification" intuition discussed in Kitcher's article?

22. What reasons does van Fraassen provide for considering pragmatic

aspects of explanation as central? How might Hempel, Salmon, and Kitcher reply to this challenge?

23. What reasons does Kitcher provide for identifying explanation with unification? What does van Fraassen say on this point? Who presents the better argument and why?

24. Compare and contrast van Fraassen's and Kitcher's use of historical examples of explanations to make their respective points. Is this just a matter of the diversity of history providing evidence for any claim one likes? Or is there something more going on in these examples?

25. Are Salmon's, van Fraassen's, and Kitcher's proposals mutually exclusive alternatives to understanding the nature of explanation? Or is there some way in which they can be considered as complementary approaches? Explain your answer.

SELECTED BIBLIOGRAPHY

(* indicates contains an extensive bibliography. For additional references on explanation in the social sciences, see the Selected Bibliography of part 2.)

A. *General Works on Explanation*

[1] Hempel, C. *Aspects of Scientific Explanation.* New York: The Free Press, 1965. [A collection of Hempel's classic papers on explanation, including one on functional explanation (see below).]*

[2] Kitcher, P. and Salmon, W., eds. *Scientific Explanation.* Minneapolis, Minn.: University of Minnesota Press, 1989. [A comprehensive and historical survey of four decades of work on explanations.]*

[3] Pitt, J., ed. *Theories of Explanation.* New York: Oxford University Press, 1988. [A collection of classic papers on explanation.]

[4] Salmon, W. *Scientific Explanation and the Causal Structure of the World.* Princeton, N.J.: Princeton University Press, 1984. [Systematic non-Hempelian account of explanation.]

[5] Van Fraassen, B. *The Scientific Image.* Oxford: Oxford University (Clarendon) Press, 1980. [Presents a fuller treatment of his theory of explanation.]

B. *Works on Scientific Laws and Their Connection to Explanation*

[6] Achinstein, P. *Law and Explanation.* Oxford: Clarendon Press, 1971. [Does not sever the logical from the pragmatic aspects of explanation.]

[7] Armstrong, D. *What is a Law of Nature?* Cambridge: Cambridge University Press, 1983. [Nonregularity account of laws.]*

[8] Dilworth, C., ed. *The Metaphysics of Science.* Boston Studies in the Philosophy of Science, vol. 173, Kluwer Academic Publishers, 1996. [Includes sections on empirical laws, and how laws are used in scientific explanations.]*

[9] Kline, A. D., and C. Matheson. "How the Laws of Physics Don't Even Fib." In *P.S.A.*

1986. Vol. 1, edited by A. Fine and P. Machamer. East Lansing, Mich.: Philosophy of Science Association, 1986. [Critique of Cartwright.]

C. Functional/Teleological Explanation

[10] Brandon, R. *Adaptation and Environment.* Princeton, N.J.: Princeton University Press, 1990, esp. chap 5. [Presents an account of natural-selectionist explanations of adaptations that includes insights from both Salmon's and Kitcher's views.]

[11] Cummins, R. *The Nature of Psychological Explanation.* Cambridge, Mass.: MIT Press, 1983. [Presents a view of functional explanations influential among both philosophers of biology and philosophers of mind.]

[12] Schaffner, K. *Discovery and Explanation in Biology and Medicine.* Chicago: University of Chicago Press, 1993, esp. chapters 6–8. [Discusses the nature of explanation in a broad biomedical context.]

[13] Wright, L. *Teleological Explanation.* Berkeley: University of California Press, 1976. [Elaborates an etiological view of teleological explanations favored by many philosophers of evolutionary biology.]

Part 4

Theory and Observation

Introduction

I. THE PROBLEM

Our student newspaper recently reported two episodes. In one, a biology student was reported to have said, in defense of some religious view of his, that the theory of evolution was just that—a theory. His professor supposedly shouted back, "It's not a *theory*!" In another story, a mathematician used, as an example of one of man's greatest cognitive achievements, Einstein's *theory* of special relativity.

Though the student and biologist have some disagreements, they do not appear to be over what a theory is. But the mathematician seems to have a different concept of what it is to call something a theory. For the student and biologist to call something a theory is to reduce its cognitive authority. The mathematician thinks that at least one theory is without cognitive superiors. Someone (perhaps all three) is confused.

In this part our goal is to begin to form a coherent and informed view on the nature of theories. Before attempting to characterize the structure of a theory, let's consider a pair of examples. The explanans of two of the examples in part 3, when considered along with the wider set of laws and assumptions in which they are embedded, provide splendid examples of theories.

Aristotle's explanation of the circular movements of the planets is just a part of his wider astronomical theory. His theory "tells" us much more about the universe than why planets move in circular orbits (e.g., it tells us that the stationary earth is at the center of the universe, it gives the spatial order of the planets and their relation to the stars, and it provides for predictions of the movements of the planets).

There is an evolutionary *explanation* of certain increases in the "complexity" of life-forms as evidenced in the fossil record. But the wider *theory* of evolution provides explanations and predictions of a plethora of phenomena. For example, the theory allows us to explain why only certain species of moths in Birmingham, England, survived the pollution of that city brought on by industrialization. The theory provides us with insights into how we should go about creating more rust-resistant grains. The theory provides the causal mechanism for understanding why the descendants of certain mutants in a population come to dominate the population. And we could go on and on.

Our philosophical problem can be simply stated: What is a scientific theory? We shall see that the answer is considerably more complex.

II. THE STRUCTURE OF THEORIES

The topic of theories is perhaps the most difficult one in the philosophy of science. It is certainly the most difficult of those found in this anthology. Like the mind-body problem and the problem of perception, the topic of theories is really a constellation of intertwined problems. Furthermore, as in the area of explanation, there is a historically important view. Though this view is widely recognized as problematic, the reader of current literature often must have a substantial acquaintance with it. But unlike the Hempelian view on explanation, which can be presented to a student with minimal background, the standard view on theories cannot. Even elementary presentations require training in symbolic logic and the philosophy of language.

Fortunately, much about theories can be discussed without taking on all the details of the standard account. A place to begin is with a rough and tentative characterization of theories.

Structurally, theories are sets of statements, some of which state laws, others of which are singular factual or existential claims (e.g., that electrons exist and that electrons have a charge of minus one). Furthermore, theories contain some terms that refer to unobservable entities or properties. The statements of a theory are interrelated in such a way as to embody certain virtues: generality or comprehensiveness of explanatory and predictive power, ability to unify diverse phenomena and laws, depth of explanatory power. Theories explain not a particular law or phenomenon but whole ranges of each—this range is typically called the domain of the theory. Theories show how apparently diverse or unrelated phenomena and laws really are related, at least for explanatory purposes. Theories aim at a deep understanding of phenomena (i.e., theories often describe the causal mechanism behind regularities or appeal to microstructure in accounting for the macroscopic properties of objects).

We can consider in a little detail an example—the kinetic theory of matter and heat. In the mid-nineteenth century, J. P. Joule synthesized the work of Rumford, Bernoulli, and others to form the kinetic theory. This theory exercised the talents of many of the great nineteenth-century physicists.

Structurally, the theory contains (a) a set of laws—in particular, what we now call the laws of classical mechanics, the most central of which are Newton's laws of motion; (b) a set of singular existential and factual statements. We now refer to this set as the model of a gas. It includes the following claims: gases consist of molecules; the size of molecules is negligible; the number of molecules is very large; molecules are in random motion; molecular collisions are perfectly elastic. Notice that some of the terms in these statements refer to unobservable entities and properties.

It was claimed earlier that theories have certain virtues. The generality of the explanatory power of the kinetic theory is witnessed by the fact that it explains the gas laws (laws that relate to the pressure, temperature and volume of a gas), laws about the specific heats of gases, laws about the rates of diffusion of one gas in another, and so on. The theory unifies apparently diverse phenomena. It shows, for example, that the fact that an odiferous substance released, say, in the corner of a room spreads at the rate it does throughout the room and the fact that the earth has not lost its atmosphere are, from the point of view of the theory, rather similar phenomena. To use the gas-diffusion example again, the theory does, of course, explain this well-known phenomenon, but it also illuminates the underlying mechanism by which it occurs. The theory also takes common notions such as temperature and pressure and shows one what they "really" are. It gives the microstructure that "underlies" these properties.

III. OBSERVATION

Most philosophers agree that an appeal to unobservables is an important mark of scientific theories. But here the agreement ends. When one considers the details of this rather vague criterion, a plethora of issues are forced upon us. Exactly how are observables and unobservables to be distinguished? Is what is observable subject to changes in historical contingencies—the invention of a new instrument, for example? Is what is observable subject to theoretical contingencies; that is, given two theories, T and T' at time t, can an entity or property be observable relative to T and unobservable relative to T'?

Consider the following view that has three parts: (i) At any time, t, all the nonlogical terms of science (terms other than "all," "and," "or," "a," and so on) can be uniquely parsed into one of two classes: (a) those that refer to observable entities or properties (e.g., "red," "circular," "cup," "cat"; these are called observation terms). (b) those that refer to unobservable entites or properties, e.g., "valence," "charge," "electron." (These are called theoretical terms). (ii) If a term is correctly classified as an observation or theoretical term it is permanently so classified. (iii) An entity or property is an observable if and only if normal observers in standard conditions can ascertain the presence of the entity or property by direct observation. Otherwise the entity or property is unobservable.

Something like the view above is embedded in the standard account of theories. Though, in fairness, it must be admitted to be an extreme and perhaps unsympathetic reconstruction of the standard view. Nevertheless, it will serve for anchoring our discussion. We need to be asking whether the observational-theoretical distinction can be drawn in a way that illuminates scientific practice or certain philosophical goals or both.

IV. STATUS OF UNOBSERVABLES

Supposing that a satisfactory way has been found to make the observable/unobservable distinction, another issue beckons us. Are unobservables real? Do they actually exist? This may strike one as an odd question. It is the case from a grammatical point of view that theoretical terms appear to refer and describe just as observation terms do. But as we shall see, philosophers have thought that the actual function of theoretical talk is quite different.

A question closely related to whether unobservables exist is whether theoretical statements are true. Theoretical statements are statements that contain only theoretical and logical terms (e.g., "electrons have a charge of minus one"). Analogously, observation statements are statements that contain only observation and logical terms (e.g., "that table is red").

There are several distinct views on the questions we have been asking. It will be helpful to chart a taxonomy of the more important views. The chart can be generated by asking three questions: (Q1) Do unobservables exist? (Q2) Are theoretical statements true? (Q3) Are observation statements true? Do not be misled by (Q2) and (Q3). The question is not: Are all the theoretical or observational sentences of theories true or are some of them true; or do we know that some of them are true? The question is: Are they candidates for being true? They will not be candidates if there is some general philosophical reason for suspecting the truth of them as a class. The chart of views follows:

	Superrealism	Realism	Instrumentalism	Descriptivism
(Q1) Do unobservables exist?	Yes	Yes	No	No
(Q2) Are theoretical statements true?	Yes	Yes	No	Yes
(Q3) Are observation statements true?	No	Yes	Yes	Yes

[Handwritten annotations: "Maxwell" above Realism; "Toulmin" above Instrumentalism; "Stace" above Descriptivism. Under Realism Q1: "Don't Play"; under Instrumentalism Q1: "not necessarily". Under Realism Q2: "useful". Under Realism Q3: "useful". Under Instrumentalism Q3: "Gross physical world"; "Materialist"; "5 senses". Under Descriptivism Q3: "reflections of our consciousness".]

There are other possibilities, of course. The trinity, No, No, No, is an obvious one. This possibility has been taken seriously by some philosophers. But the arguments for it are not usually produced from a close look at science. They are often generated from a global attack on the possibility of knowledge. We could construct a more elaborate and finer-grained taxonomy. But since those above are the most discussed possibilities and since our purposes are merely to initiate the discussion, the above major possibilities and grain will do.

Realists are fond of regarding their view as common sense. Whether they are right or not it is true that the dialectic of the issue is such that it is tacitly assumed that the burden of proof rests with the antirealist. We can say from personal experience that scientists often talk the instrumentalist's line and that a little philosophy often jars students from a realist stance. Whether a little philosophy is a dangerous thing or not you will have to decide.

We do not wish to review in detail the reasons for the four views—the readings do that. We do want to say enough about the views so that they are at least on the surface distinguishable.

Sir Arthur Eddington (1881–1944), the famous English astronomer, held the view I have dubbed superrealism. Eddington begins the presentation of his view in a now famous passage that both excites and teases the reader.

> I have settled down to the task of writing these lectures and have drawn up my chairs to my two tables. Two tables! Yes; there are duplicates of every object about me—two tables, two chairs, two pens.[1]

The first table, the table of common sense, is extended, permanent, colored, and substantial. The second table, the table of physics, is mostly emptiness. It is a vast number of electric charges speeding about.

For our purposes, what is interesting about Eddington is that he denies that the first table exists, much less that there is a colored table there. In his words, "modern physics has by delicate test and remorseless logic assured me that my second table is the only one which is really there—whatever 'there' may be."[2]

Exactly how physics teaches us, if it does, that observation sentences are false is a very complicated matter. If you are interested in pursuing superrealism, [3] and [5] of the bibliography may help.

A descriptivist claims that we have absolutely no reason to believe that the theoretical "entities" and "properties" of a theory exist. This leads him to propose a new function for theoretical talk. Theoretical statements supposedly provide us with "shorthand formulae." These formulae prove convenient for organizing observables and making predictions about observables, but they should not be taken as referring to entities or properties over and above observables.

A simple example will help clarify the point. The caterer of a wedding might find it convenient to plan in terms of the "average wedding-goer." Such a concept will be helpful in summarizing certain data from past weddings and predicting features of the wedding to be. But the caterer does not think average wedding-goers are real or exist. After all, not only does the average wedding-goer drink 3.2 whiskey sours and eat 6 cocktail knishes, but he has 2.1 children.

In the same way that the average wedding-goers do not exist over and above John, Mildred, and so on, and their actual and dispositional behavior, electrons do not exist over and above the complex of actual and possible data they stand in for.

Notice that, for the descriptivist, theoretical statements are true. This, at the outset, might appear strange since theoretical entities do not exist. But given Stace's view, we see that theoretical statements are just very complex, though briefly stated, observation statements. Since observation statements are true, so are theoretical statements.

Though descriptivism and instrumentalism are often confused they are distinct views. The "no" answer for the instrumentalist to (Q2) may be misleading—misleading in the same way as your answering "no" if a child asked you if you ever sprained your kidney. A better answer would be to point out that kidneys are not the kinds of things that can be sprained or for that matter not sprained. Similarly, for the instrumentalist, theoretical statements are not the kinds of things that are true or false. Instrumentalists often refer to theoretical statements as rules, or inference tickets, or predictive instruments. Just as the rules of chess (e.g., the rule that pawns can only move forward) are neither true nor false but useful for certain purposes, playing an interesting game, so the theoretical sentences of science are neither true nor false but useful for certain purposes, making predictions. (Of course, "'pawns can only move forward' is a rule of chess" is true or false.)

V. THE READINGS IN PART 4

The essay by Rudolf Carnap is crucial in setting the context for the remaining essays. His essay is a nontechnical presentation of the traditional conception of scientific theories. He discusses the observational-theoretical distinction and how the theoretical components of a theory are linked to the observational. The remaining essays in this part can be seen as criticizing the Carnapian conception of theories or elaborating the philosophical status of theories.

Hilary Putnam and N. R. Hanson both attack the observational-theoretical dichotomy. Putnam wants to show that the distinction cannot be coherently drawn. It turns out that some terms end up in the wrong bin from where they preanalytically belong. Hanson's charge is of a different sort. He wants to show that there is no sort of observation important to science that is independent of theory. All significant observation is connected with theoretical claims.

Three of the main positions on the status of unobservables, descriptivism, instrumentation, and realism, are covered by the essays of Stace, Toulmin, and Maxwell respectively. Stace's argument for descriptivism relies very heavily on a strict observational-theoretical distinction. At the end of his piece he makes an extraordinary claim which deserves examination: "If the views which I have been expressing are followed out, they will lead to the conclusion that, strictly speaking, *nothing exists except sensations* (and the minds which perceive them)."

Toulmin's argument probes the meaning of existence claims. He points out that not every existence claim may be of the same piece. What we are really asking when we worry about the existence of theoretical entities might be quite different from what we are asking when we worry about the existence of dodos.

Maxwell's essay contains a statement and critique of both descriptivism and instrumentalism on the way to defending realism. Maxwell tries to show that the observational-theoretical distinction is not sharp. It is more like a continuous change than an abrupt one. Hence we ought not read ontological significance into it.

The essay by Matheson and Kline is on a different level. It looks at what motivates the sharp observation theory distinction and the prospects for realizing those motives. The paper argues that, at least to some extent, where you come down on the status of unobservables is a function of commitments to theses on the *structure* of theories.

A. D. K.

NOTES

1. A. Eddington, *The Nature of the Physical World,* New York: Cambridge University Press, 1929, p. ix.
2. Ibid., p. xii.

17

The Nature of Theories

Rudolf Carnap

1. Theories and Nonobservables

One of the most important distinctions between two types of laws in science is the distinction between what may be called (there is no generally accepted terminology for them) empirical laws and theoretical laws. Empirical laws are laws that can be confirmed directly by empirical observations. The term "observable" is often used for any phenomenon that can be directly observed, so it can be said that empirical laws are laws about observables.

Here, a warning must be issued. Philosophers and scientists have quite different ways of using the terms "observable" and "nonobservable." To a philosopher, "observable" has a very narrow meaning. It applies to such properties as "blue," "hard," "hot." These are properties directly perceived by the senses. To the physicist, the word has a much broader meaning. It includes any quantitative magnitude that can be measured in a relatively simple, direct way. A philosopher would not consider a temperature of, perhaps, 80 degrees centigrade, or a weight of 93½ pounds, an observable because there is no direct sensory perception of such magnitudes. To a physicist, both are observables because they can be measured in an extremely simple way. The object to be weighed is placed on a balance scale. The temperature is measured with a thermometer. The physicist would not say that the mass of a molecule, let alone the mass of an electron, is something observable, because here the procedures of measurement are much more complicated and indirect. But magnitudes that can be established by simple procedures—length with a ruler, time with a clock, or frequency of light waves with a spectrometer—are called observables.

A philosopher might object that the intensity of an electric current is not really observed. Only a pointer position was observed. An ammeter was attached to the circuit and it was noted that the pointer pointed to a mark

316

labeled 5.3. Certainly the current's intensity was not observed. It was *inferred* from what was observed.

The physicist would reply that this was true enough, but the inference was not very complicated. The procedure of measurement is so simple, so well established, that it could not be doubted that the ammeter would give an accurate measurement of current intensity. Therefore, it is included among what are called observables.

There is no question here of who is using the term "observable" in a right or proper way. There is a continuum which starts with direct sensory observations and proceeds to enormously complex, indirect methods of observation. Obviously no sharp line can be drawn across this continuum; it is a matter of degree. A philosopher is sure that the sound of his wife's voice, coming from across the room, is an observable. But suppose he listens to her on the telephone. Is her voice an observable or isn't it? A physicist would certainly say that when he looks at something through an ordinary microscope, he is observing it directly. Is this also the case when he looks into an electron microscope? Does he observe the path of a particle when he sees the track it makes in a bubble chamber? In general, the physicist speaks of observables in a very wide sense compared with the narrow sense of the philosopher, but, in both cases, the line separating observable from nonobservable is highly arbitrary. It is well to keep this in mind whenever these terms are encountered in a book by a philosopher or scientist. Individual authors will draw the line where it is most convenient, depending on their points of view, and there is no reason why they should not have this privilege.

Empirical laws, in my terminology, are laws containing terms either directly observable by the senses or measurable by relatively simple techniques. Sometimes such laws are called empirical generalizations, as a reminder that they have been obtained by generalizing results found by observations and measurements. They include not only simple qualitative laws (such as, "All ravens are black") but also quantitative laws that arise from simple measurements. The laws relating pressure, volume, and temperature of gases are of this type. Ohm's law, connecting the electric potential difference, resistance, and intensity of current, is another familiar example. The scientist makes repeated measurements, finds certain regularities, and expresses them in a law. These are the empirical laws. As indicated in earlier chapters, they are used for explaining observed facts and for predicting future observable events.

There is no commonly accepted term for the second kind of laws, which I call *theoretical laws*. Sometimes they are called abstract or hypothetical laws. "Hypothetical" is perhaps not suitable because it suggests that the distinction between the two types of laws is based on the degree to which the laws are confirmed. But an empirical law, if it is a tentative hypothesis, confirmed only to a low degree, would still be an empirical law although it might be said that it was rather hypothetical. A theoretical law is not to be distinguished from an empirical law by the fact that it is not well established, but by the fact that it contains terms of a different kind. The terms of a theoretical law do not refer to observables even when the physicist's wide meaning for what can be observed is adopted. They are laws about such entities as molecules, atoms, electrons, protons, electromagnetic fields, and others that cannot be measured in simple, direct ways. . . .

It is true, as shown earlier, that the concepts "observable" and "nonobservable" cannot be sharply defined because they lie on a continuum. In actual practice, however, the difference is usually great enough so there is not likely to be debate. All physicists would agree that the laws relating pressure, volume, and temperature of a gas, for example, are empirical laws. Here the amount of gas is large enough so that the magnitudes to be measured remain constant over a sufficiently large volume of space and period of time to permit direct, simple measurements which can then be generalized into laws. All physicists would agree that laws about the behavior of single molecules are theoretical. Such laws concern a microprocess about which generalizations cannot be based on simple, direct measurements.

Theoretical laws are, of course, more general than empirical laws. It is important to understand, however, that theoretical laws cannot be arrived at simply by taking the empirical laws, then generalizing a few steps further. How does a physicist arrive at an empirical law? He observes certain events in nature. He notices a certain regularity. He describes this regularity by making an inductive generalization. It might be supposed that he could now put together a group of empirical laws, observe some sort of pattern, make a wider inductive generalization, and arrive at a theoretical law. Such is not the case.

To make this clear, suppose it has been observed that a certain iron bar expands when heated. After the experiment has been repeated many times, always with the same result, the regularity is generalized by saying that this bar expands when heated. An empirical law has been stated, even though it has a narrow range and applies only to one particular iron bar. Now further tests are made of other iron objects with the ensuing discovery that every time an iron object is heated it expands. This permits a more general law to be formulated, namely that all bodies of iron expand when heated. In similar fashion, the still more general laws "All metals . . . ," then "All solid bodies . . . ," are developed. These are all simple generalizations, each a bit more general than the previous one, but they are all empirical laws. Why? Because in each case, the objects dealt with are observable (iron, copper, metal, solid bodies); in each case the increases in temperature and length are measurable by simple, direct techniques.

In contrast, a theoretical law relating to this process would refer to the behavior of molecules in the iron bar. In what way is the behavior of the molecules connected with the expansion of the bar when heated? You see at once

that we are now speaking of nonobservables. We must introduce a theory—
the atomic theory of matter—and we are quickly plunged into atomic laws
involving concepts radically different from those we had before. It is true that
these theoretical concepts differ from concepts of length and temperature
only in the degree to which they are directly or indirectly observable, but the
difference is so great that there is no debate about the radically different
nature of the laws that must be formulated.

Theoretical laws are related to empirical laws in a way somewhat analogous to the way empirical laws are related to single facts. An empirical law
helps to explain a fact that has been observed and to predict a fact not yet
observed. In similar fashion, the theoretical law helps to explain empirical
laws already formulated, and to permit the derivation of new empirical laws.
Just as the single, separate facts fall into place in an orderly pattern when they
are generalized in an empirical law, the single and separate empirical laws fit
into the orderly pattern of a theoretical law. This raises one of the main problems in the methodology of science. How can the kind of knowledge that will
justify the assertion of a theoretical law be obtained? An empirical law may
be justified by making observations of single facts. But to justify a theoretical law, comparable observations cannot be made because the entities
referred to in theoretical laws are nonobservables. . . .

How can theoretical laws be discovered? We cannot say: "Let's just collect more and more data, then generalize beyond the empirical laws until we
reach theoretical ones." No theoretical law was ever found that way. We
observe stories and trees and flowers, noting various regularities and
describing them by empirical laws. But no matter how long or how carefully
we observe such things, we never reach a point at which we observe a molecule. The term "molecule" never arises as a result of observations. For this
reason, no amount of generalization from observations will ever produce a
theory of molecular processes. Such a theory must arise in another way. It is
stated not as a generalization of facts but as a hypothesis. The hypothesis is
then tested in a manner analogous in certain ways to the testing of an empirical law. From the hypothesis, certain empirical laws are derived, and these
empirical laws are tested in turn by observation of facts. Perhaps the empirical laws derived from the theory are already known and well confirmed.
(Such laws may even have motivated the formulation of the theoretical law.)
Regardless of whether the derived empirical laws are known and confirmed,
or whether they are new laws confirmed by new observations, the confirmation of such derived laws provides indirect confirmation of the theoretical law.

The point to be made clear is this. A scientist does not start with one
empirical law, perhaps Boyle's law for gases, and then seek a theory about
molecules from which this law can be derived. The scientist tries to formulate a much more general theory from which a variety of empirical laws can
be derived. The more such laws, the greater their variety and apparent lack

of connection with one another, the stronger will be the theory that explains them. Some of these derived laws may have been known before, but the theory may also make it possible to derive new empirical laws which can be confirmed by new tests. If this is the case, it can be said that the theory made it possible to predict new empirical laws. The prediction is understood in a hypothetical way. If the theory holds, certain empirical laws will also hold. The predicted empirical law speaks about relations between observables, so it is now possible to make experiments to see if the empirical law holds. If the empirical law is confirmed, it provides indirect confirmation of the theory. Every confirmation of a law, empirical or theoretical, is, of course, only partial, never complete and absolute. But in the case of empirical laws, it is a more direct confirmation. The confirmation of a theoretical law is indirect, because it takes place only through the confirmation of empirical laws derived from the theory.

The supreme value of a new theory is its power to predict new empirical laws. It is true that it also has value in explaining known empirical laws, but this is a minor value. If a scientist proposes a new theoretical system, from which no new laws can be derived, then it is logically equivalent to the set of all known empirical laws. The theory may have a certain elegance, and it may simplify to some degree the set of all known laws, although it is not likely that there would be an essential simplification. On the other hand, every new theory in physics that has led to a great leap forward has been a theory from which new empirical laws could be derived. If Einstein had done no more than propose his theory of relativity as an elegant new theory that would embrace certain known laws—perhaps also simplify them to a certain degree—then his theory would not have had such a revolutionary effect.

Of course it was quite otherwise. The theory of relativity led to new empirical laws which explained for the first time such phenomena as the movement of the perihelion of Mercury, and the bending of light rays in the neighborhood of the sun. These predictions showed that relativity theory was more than just a new way of expressing the old laws. Indeed, it was a theory of great predictive power. The consequences that can be derived from Einstein's theory are far from being exhausted. These are consequences that could not have been derived from earlier theories. Usually a theory of such power does have an elegance, and a unifying effect on known laws. It is simpler than the total collection of known laws. But the great value of the theory lies in its power to suggest new laws that can be confirmed by empirical means.

II. CORRESPONDENCE RULES

An important qualification must now be added to the discussion of theoretical laws and terms given in section I. The statement that empirical laws are derived from theoretical laws is an oversimplification. It is not possible to derive them directly because a theoretical law contains theoretical terms, whereas an empirical law contains only observable terms. This prevents any direct deduction of an empirical law from a theoretical one.

To understand this, imagine that we are back in the nineteenth century, preparing to state for the first time some theoretical laws about molecules in a gas. These laws are to describe the number of molecules per unit volume of the gas, the molecular velocities, and so forth. To simplify matters, we assume that all the molecules have the same velocity. (This was indeed the original assumption; later it was abandoned in favor of a certain probability distribution of velocities.) Further assumptions must be made about what happens when molecules collide. We do not know the exact shape of molecules, so let us suppose that they are tiny spheres. How do spheres collide? There are laws about colliding spheres, but they concern large bodies. Since we cannot directly observe molecules, we assume their collisions are analogous to those of large bodies; perhaps they behave like perfect billiard balls on a frictionless table. These are, of course, only assumptions; guesses suggested by analogies with known macrolaws.

But now we come up against a difficult problem. Our theoretical laws deal exclusively with the behavior of molecules, which cannot be seen. How, therefore, can we deduce from such laws a law about observable properties such as the pressure or temperature of a gas or properties of sound waves that pass through the gas? The theoretical laws contain only theoretical terms. What we seek are empirical laws containing observable terms. Obviously, such laws cannot be derived without having something else given in addition to the theoretical laws.

The something else that must be given is this: a set of rules connecting the theoretical terms with the observable terms. Scientists and philosophers of science have long recognized the need for such a set of rules, and their nature has often been discussed. An example of such a rule is: "If there is an electromagnetic oscillation of a specified frequency, then there is a visible greenish-blue color of a certain hue." Here something observable is connected with a nonobservable microprocess.

Another example is: "The temperature (measured by a thermometer and, therefore, an observable in the wider sense explained earlier) of a gas is proportional to the mean kinetic energy of its molecules." This rule connects a nonobservable in molecular theory, the kinetic energy of molecules, with an observable, the temperature of the gas. If statements of this kind did not exist, there would be no way of deriving empirical laws about observables from theoretical laws about nonobservables. Different writers have different names for these rules. I call them "correspondence rules." P. W. Bridgman calls them operational rules. Norman R. Campbell speaks of them as the "Dictionary."[1] Since the rules connect a term in one terminology with a term in another terminology, the use of the rules is analogous to the use of a French-English dictionary. What does the French word *cheval* mean? You look it up in the dictionary and find that it means "horse." It is not really that simple when a set of rules is used for connecting nonobservables with observables; nevertheless, there is an analogy here that makes Campbell's "Dictionary" a suggestive name for the set of rules.

There is a temptation at times to think that the set of rules provides a means for defining theoretical terms, whereas just the opposite is really true. A theoretical term can never be explicitly defined on the basis of observable terms, although sometimes an observable can be defined in theoretical terms. For example, "iron" can be defined as a substance consisting of small crystalline parts, each having a certain arrangement of atoms and each atom being a configuration of particles of a certain type. In theoretical terms then, it is possible to express what is meant by the observable term "iron," but the reverse is not true.

There is no answer to the question: "Exactly what is an electron?" Later we shall come back to this question, because it is the kind that philosophers are always asking scientists. They want the physicist to tell them just what he means by "electricity," "magnetism," "gravity," "a molecule." If the physicist explains them in theoretical terms, the philosopher may be disappointed. "That is not what I meant at all," he will say. "I want you to tell me, in ordinary language, what those terms mean." Sometimes the philosopher writes a book in which he talks about the great mysteries of nature. "No one," he writes, "has been able so far, and perhaps no one ever will be able, to give us a straightforward answer to the question: 'What is electricity?' And so electricity remains forever one of the great, unfathomable mysteries of the universe."

There is no special mystery here. There is only an improperly phrased question. Definitions that cannot, in the nature of the case, be given, should not be demanded. If a child does not know what an elephant is, we can tell him it is a huge animal with big ears and a long trunk. We can show him a picture of an elephant. It serves admirably to define an elephant in observable terms that a child can understand. By analogy, there is a temptation to believe that, when a scientist introduces theoretical terms, he should also be able to define them in familiar terms. But this is not possible. There is no way a physicist can show us a picture of electricity in the way he can show his child a picture of an elephant. Even the cell of an organism, although it cannot be seen with the unaided eye, can be represented by a picture because the cell can be seen when it is viewed through a microscope. But we do not possess a picture of the electron. We cannot say how it looks or how it feels, because it cannot be seen or touched. The best we can do is to say that it is an extremely small body that behaves in a certain manner. This may seem to be analogous to our description of an elephant. We can describe an elephant as a large animal that behaves in a certain manner. Why not do the same with an electron?

The answer is that a physicist can describe the behavior of an electron only by stating theoretical laws, and these laws contain only theoretical terms. They describe the field produced by an electron, the reaction of an electron to a field, and so on. If an electron is in an electrostatic field, its velocity will accelerate in a certain way. Unfortunately, the electron's acceleration is an unobservable. It is not like the acceleration of a billiard ball, which can be studied by direct observation. There is no way that a theoretical concept can be defined in terms of observables. We must, therefore, resign ourselves to the fact that definitions of the kind that can be supplied for observable terms cannot be formulated for theoretical terms.

It is true that some authors, including Bridgman, have spoken of the rules as "operational definitions." Bridgman had a certain justification, because he used his rules in a somewhat different way, I believe, than most physicists use them. He was a great physicist and was certainly aware of his departure from the usual use of rules, but he was willing to accept certain forms of speech that are not customary, and this explains his departure. It was pointed out . . . that Bridgman preferred to say that there is not just one concept of intensity of electric current, but a dozen concepts. Each procedure by which a magnitude can be measured provides an operational definition for that magnitude. Since there are different procedures for measuring current, there are different concepts. For the sake of convenience, the physicist speaks of just one concept of current. Strictly speaking, Bridgman believed, he should recognize many different concepts, each defined by a different operational procedure of measurement.

We are faced here with a choice between two different physical languages. If the customary procedure among physicists is followed, the various concepts of current will be replaced by one concept. This means, however, that you place the concept in your theoretical laws, because the operational rules are just correspondence rules, as I call them, which connect the theoretical terms with the empirical ones. Any claim to possessing a definition— that is, an operational definition—of the theoretical concept must be given up. Bridgman could speak of having operational definitions for his theoretical terms only because he was not speaking of a general concept. He was speaking of partial concepts, each defined by a different empirical procedure.

Even in Bridgman's terminology, the question of whether his partial concepts can be adequately defined by operational rules is problematic. Reichenbach speaks often of what he calls "correlative definitions." . . . Perhaps correlation is a better term than definition for what Bridgman's rules actually do. In geometry, for instance, Reichenbach points out that the axiom system of geometry, as developed by David Hilbert, for example, is an uninterpreted axiom system. The basic concepts of point, line, and plane could just as well be called "class alpha," "class beta," and "class gamma." We must not be seduced by the sound of familiar words, such as "point" and "line," into thinking they must be taken in their ordinary meaning. In the axiom system, they are uninterpreted terms. But when geometry is applied to physics, these terms must be connected with something in the physical world. We can say, for example, that the lines of the geometry are exemplified by rays of light in a vacuum or by stretched cords. In order to connect the uninterpreted terms with observable physical phenomena, we must have rules for establishing the connection.

What we call these rules is, of course, only a terminological question; we should be cautious and not speak of them as definitions. They are not definitions in any strict sense. We cannot give a really adequate definition of the geometrical concept of "line" by referring to anything in nature. Light rays, stretched strings, and so on are only approximately straight; moreover, they are not lines, but only segments of lines. In geometry, a line is infinite in length and absolutely straight. Neither property is exhibited by any phenomenon in nature. For that reason, it is not possible to give an operational definition, in the strict sense of the word, of concepts in theoretical geometry. The same is true of all the other theoretical concepts of physics. Strictly speaking, there are no "definitions" of such concepts. I prefer not to speak of "operational definitions" or even to use Reichenbach's term "correlative definitions." In my publications (only in recent years have I written about this question), I have called them "rules of correspondence" or, more simply, "correspondence rules."

Campbell and other authors often speak of the entities in theoretical physics as mathematical entities. They mean by this that the entities are related to each other in ways that can be expressed by mathematical functions. But they are not mathematical entities of the sort that can be defined in pure mathematics. In pure mathematics, it is possible to define various kinds of numbers, the function of logarithm, the exponential function, and so forth. It is not possible, however, to define such terms as "electron" and "temperature" by pure mathematics. Physical terms can be introduced only with the help of nonlogical constants, based on observations of the actual world. Here we have an essential difference between an axiomatic system in mathematics and an axiomatic system in physics.

If we wish to give an interpretation of a term in a mathematical axiom system, we can do it by giving a definition in logic. Consider, for example, the term "number" as it is used in Peano's axiom system. We can define it in logical terms, by the Frege-Russell method, for example. In this way the concept of "number" acquires a complete, explicit definition on the basis of pure logic. There is no need to establish a connection between the number 5 and such observables as "blue" and "hot." The terms have only a logical interpretation; no connection with the actual world is needed. Sometimes an axiom system in mathematics is called a theory. Mathematicians speak of set theory, group theory, matrix theory, probability theory. Here the word "theory" is used in a purely analytic way. It denotes a deductive system that makes no reference to the actual world. We must always bear in mind that such a use of the word "theory" is entirely different from its use in reference to empirical theories such as relativity theory, quantum theory, psychoanalytical theory, and Keynesian economic theory.

A postulate system in physics cannot have, as mathematical theories have, a splendid isolation from the world. Its axiomatic terms "electron," "field," and so on—must be interpreted by correspondence rules that connect the terms with observable phenomena. This interpretation is necessarily incomplete. Because it is always incomplete, the system is left open to make it possible to add new rules of correspondence. Indeed, this is what continually happens in the history of physics. I am not thinking now of a revolution in physics, in which an entirely new theory is developed, but of less radical changes that modify existing theories. Nineteenth-century physics provides a good example, because classical mechanics and electromagnetics had been established, and, for many decades, there was relatively little change in fundamental laws. The basic theories of physics remained unchanged. There was, however, a steady addition of new correspondence rules, because new procedures were continually being developed for measuring this or that magnitude.

Of course, physicists always face the danger that they may develop correspondence rules that will be incompatible with each other or with the theoretical laws. As long as such incompatibility does not occur, however, they are free to add new correspondence rules. The procedure is never-ending. There is always the possibility of adding new rules, thereby increasing the amount of interpretation specified for the theoretical terms; but no matter how much this is increased, the interpretation is never final. In a mathematical system, it is otherwise. There a logical interpretation of an axiomatic term is complete. Here we find another reason for reluctance in speaking of theoretical terms as "defined" by correspondence rules. It tends to blur the important distinction between the nature of an axiom system in pure mathematics and one in theoretical physics.

Is it not possible to interpret a theoretical term by correspondence rules so completely that no further interpretation would be possible? Perhaps the actual world is limited in its structure and laws. Eventually a point may be reached beyond which there will be no room for strengthening the interpretation of a term by new correspondence rules. Would not the rules then provide a final, explicit definition for the term? Yes, but then the term would no longer be theoretical. It would become part of the observation language. The history of physics has not yet indicated that physics will become complete; there has been only a steady addition of new correspondence rules and a continual modification in the interpretations of theoretical terms. There is no way of knowing whether this is an infinite process or whether it will eventually come to some sort of end.

It may be looked at this way. There is no prohibition in physics against making the correspondence rules for a term so strong that the term becomes explicitly defined and therefore ceases to be theoretical. Neither is there any basis for assuming that it will always be possible to add new correspondence rules. Because the history of physics has shown such a steady, unceasing modification of theoretical concepts, most physicists would advise against correspondence rules so strong that a theoretical term becomes explicitly defined. Moreover, it is a wholly unnecessary procedure. Nothing is gained by it. It may even have the adverse effect of blocking progress.

Of course, here again we must recognize that the distinction between observables and nonobservables is a matter of degree. We might give an explicit definition, by empirical procedures, to a concept such as length, because it is so easily and directly measured, and is unlikely to be modified by new observations. But it would be rash to seek such strong correspondence rules that "electron" would be explicitly defined. The concept "electron" is so far removed from simple, direct observations that it is best to keep it theoretical, open to modifications by new observations.

III. How New Empirical Laws Are Derived from Theoretical Laws

... The [previous] discussion concerned the ways in which correspondence rules are used for linking the nonobservable terms of a theory with the observable terms of empirical laws. This can be made clearer by a few examples of the manner in which empirical laws have actually been derived from the laws of a theory.

The first example concerns the kinetic theory of gases. Its model, or schematic picture, is one of small particles called molecules, all in constant agitation. In its original form, the theory regarded these particles as little balls, all having the same mass and, when the temperature of the gas is constant, the same constant velocity. Later it was discovered that the gas would not be in a stable state if each particle had the same velocity; it was necessary to find a certain probability distribution of velocities that would remain stable. This was called the Boltzmann-Maxwell distribution. According to this distribution, there was a certain probability that any molecule would be within a certain range on the velocity scale.

When the kinetic theory was first developed, many of the magnitudes occurring in the laws of the theory were not known. No one knew the mass of a molecule, or how many molecules a cubic centimeter of gas at a certain temperature and pressure would contain. These magnitudes were expressed by certain parameters written into the laws. After the equations were formulated, a dictionary of correspondence rules was prepared. These correspondence rules connected the theoretical terms with observable phenomena in a way that made it possible to determine indirectly the values of the parameters in the equations. This, in turn, made it possible to derive empirical laws. One correspondence rule states that the temperature of the gas corresponds to the mean kinetic energy of the molecules. Another correspondence rule connects the pressure of the gas with the impact of molecules on the confining wall of a vessel. Although this is a continuous process involving discrete molecules, the total effect can be regarded as a constant force pressing on the wall. Thus, by means of correspondence rules, the pressure that is measured macroscopically by a manometer (pressure gauge) can be expressed in terms of the statistical mechanics of molecules.

What is the density of the gas? Density is mass per unit volume, but how do we measure the mass of a molecule? Again our dictionary—a very simple dictionary—supplies the correspondence rule. The total mass M of the gas is the sum of the masses m of the molecules. M is observable (we simply weigh the gas), but m is theoretical. The dictionary of correspondence rules gives the connection between the two concepts. With the aid of this dictionary, empirical tests of various laws derived from our theory are possible. On the basis of the theory, it is possible to calculate what will happen to the pressure of the gas when its volume remains constant and its temperature is increased. We can calculate what will happen to a sound wave produced by striking the side of the vessel, and what will happen if only part of the gas is heated. These theoretical laws are worked out in terms of various parameters that occur within the equations of the theory. The dictionary of correspondence rules enables us to express these equations as empirical laws, in which concepts are measurable, so that empirical procedures can supply values for the parameters. If the empirical laws can be confirmed, this provides indirect confirmation of the theory. Many of the empirical laws for gases were known, of course, before the kinetic theory was developed. For these laws, the theory provided an explanation. In addition, the theory led to previously unknown empirical laws.

The power of a theory to predict new empirical laws is strikingly exemplified by the theory of electromagnetism, which was developed about 1860 by two great physicists, Michael Faraday and James Clerk Maxwell. (Faraday did most of the experimental work, and Maxwell did most of the mathematical work.) The theory dealt with electric charges and how they behaved in electrical and magnetic fields. The concept of the electron—a tiny particle with an elementary electric charge—was not formulated until the very end of the century. Maxwell's famous set of differential equations, for describing electromagnetic fields, presupposed only small discrete bodies of unknown nature, capable of carrying an electric charge or a magnetic pole. What happens when a current moves along a copper wire? The theory's dictionary made this observable phenomenon correspond to the actual movement along the wire of little charged bodies. From Maxwell's theoretical model, it became possible (with the help of correspondence rules, of course) to derive many of the known laws of electricity and magnetism.

The model did much more than this. There was a certain parameter c in Maxwell's equations. According to his model, a disturbance in an electromagnetic field would be propagated by waves having the velocity c. Electrical experiments showed the value of c to be approximately 3×10^{10} centimeters per second. This was the same as the known value for the speed of light, and it seemed unlikely that it was an accident. Is it possible, physicists asked themselves, that light is simply a special case of the propagation of an electromagnetic oscillation? It was not long before Maxwell's equations were providing explanations for all sorts of optical laws, including refraction, the velocity of light in different media, and many others.

Physicists would have been pleased enough to find that Maxwell's model explained known electrical and magnetic laws; but they received a double bounty. The theory also explained optical laws! Finally, the great strength of the new model was revealed in its power to predict, to formulate empirical laws that had not been previously known.

The first instance was provided by Heinrich Hertz, the German physicist. About 1890, he began his famous experiments to see whether electromagnetic waves of low frequency could be produced and detected in the laboratory. Light is an electromagnetic oscillation and propagation of waves at very high frequency. But Maxwell's laws made it possible for such waves to have *any* frequency. Hertz's experiments resulted in his discovery of what at first were called Hertz waves. They are now called radio waves. At first, Hertz was able to transmit these waves from one oscillator to another over only a small distance—first a few centimeters, then a meter or more. Today a radio broadcasting station sends its waves many thousands of miles.

The discovery of radio waves was only the beginning of the derivation of new laws from Maxwell's theoretical model. X-rays were discovered and were thought to be particles of enormous velocity and penetrative power. Then it occurred to physicists that, like light and radio waves, these might be electromagnetic waves, but of extremely high frequency, much higher than the frequency of visible light. This also was later confirmed, and laws dealing with X-rays were derived from Maxwell's fundamental field equations. X-rays proved to be waves of a certain frequency range within the much broader frequency band of gamma rays. The X-rays used today in medicine are simply gamma rays of certain frequency. All this was largely predictable on the basis of Maxwell's model. His theoretical laws, together with the correspondence rules, led to an enormous variety of new empirical laws.

The great variety of fields in which experimental confirmation was found contributed especially to the strong overall confirmation of Maxwell's theory. The various branches of physics had originally developed for practical reasons; in most cases, the divisions were based on our different sense organs. Because the eyes perceive light and color, we call such phenomena optics; because our ears hear sounds, we call a branch of physics acoustics; and because our bodies feel heat, we have a theory of heat. We find it useful to construct simple machines based on the movements of bodies, and we call it mechanics. Other phenomena, such as electricity and magnetism, cannot be directly perceived, but their consequences can be observed.

In the history of physics, it is always a big step forward when one branch of physics can be explained by another. Acoustics, for instance, was found to be only a part of mechanics, because sound waves are simply elasticity waves in solids, liquids, and gases. We have already spoken of how the laws of gases were explained by the mechanics of moving molecules. Maxwell's theory was another great leap forward toward the unification of physics. Optics was found to be a part of electromagnetic theory. Slowly the notion grew that the whole of physics might some day be unified by one great theory. At present there is an enormous gap between electromagnetism on the one side and gravitation on the other. Einstein made several attempts to develop a unified field theory that might close this gap; more recently, Heisenberg and others have made similar attempts. So far, however, no theory has been devised that is entirely satisfactory or that provides new empirical laws capable of being confirmed.

Physics originally began as a descriptive macrophysics, containing an enormous number of empirical laws with no apparent connections. In the beginning of a science, scientists may be very proud to have discovered hundreds of laws. But, as the laws proliferate, they become unhappy with this state of affairs; they begin to search for unifying principles. In the nineteenth century, there was considerable controversy over the question of underlying principles. Some felt that science must find such principles, because otherwise it would be no more than a description of nature, not a real explanation. Others thought that that was the wrong approach, that underlying principles belong only to metaphysics. They felt that the scientist's task is merely to describe, to find out *how* natural phenomena occur, not *why*.

Today we smile a bit about the great controversy over description versus explanation. We can see that there was something to be said for both sides, but that their way of debating the question was futile. There is no real opposition between explanation and description. Of course, if description is taken in the narrowest sense, as merely describing what a certain scientist did on a certain day with certain materials, then the opponents of mere description were quite right in asking for more, for a real explanation. But today we see that description in the broader sense, that of placing phenomena in the context of more general laws, provides the only type of explanation that can be given for phenomena. Similarly, if the proponents of explanation mean a metaphysical explanation, not grounded in empirical procedures, then their opponents were correct in insisting that science should be concerned only with description. Each side had a valid point. Both description and explanation, rightly understood, are essential aspects of science.

The first efforts at explanation, those of the Ionian natural philosophers, were certainly partly metaphysical; the world is all fire, or all water, or all change. Those early efforts at scientific explanation can be viewed in two different ways. We can say: "This is not science, but pure metaphysics. There is no possibility of confirmation, no correspondence rules for connecting the theory with observable phenomena." On the other hand, we can say: "These Ionian theories are certainly not scientific, but at least they are pictorial visions of theories. They are the first primitive beginnings of science."

It must not be forgotten that, both in the history of science and in the psychological history of a creative scientist, a theory has often first appeared as a kind of visualization, a vision that comes as an inspiration to a scientist long before he has discovered correspondence rules that may help in confirming his theory. When Democritus said that everything consists of atoms, he certainly had not the slightest confirmation for this theory. Nevertheless, it was a stroke of genius, a profound insight, because two thousand years later his vision was confirmed. We should not, therefore, reject too rashly any anticipatory vision of a theory, provided it is one that may be tested at some future time. We are on solid ground, however, if we issue the warning that no hypothesis can claim to be scientific unless there is the *possibility* that it can be tested. It does not have to be confirmed to be a hypothesis, but there must be correspondence rules that will permit, in principle, a means of confirming or disconfirming the theory. It may be enormously difficult to think of experiments that can test the theory; this is the case today with various unified field theories that have been proposed. But if such tests are possible in principle, the theory can be called a scientific one. When a theory is first proposed, we should not demand more than this.

The development of science from early philosophy was a gradual, step-by-step process. The Ionian philosophers had only the most primitive theories. In contrast, the thinking of Aristotle was much clearer and on more solid scientific ground. He made experiments, and he knew the importance of experiments, although in other respects he was an apriorist. This was the beginning of science. But it was not until the time of Galileo Galilei, about 1600, that a really great emphasis was placed on the experimental method in preference to aprioristic reasoning about nature. Even though many of Galileo's concepts had previously been stated as theoretical concepts, he was the first to place theoretical physics on a solid empirical foundation. Certainly Newton's physics (about 1670) exhibits the first comprehensive, systematic theory, containing unobservables as theoretical concepts: the universal force of gravitation, a general concept of mass, theoretical properties of light rays, and so on. His theory of gravity was one of great generality.

Between any two particles, small or large, there is a force proportional to the square of the distance between them. Before Newton advanced this theory, science provided no explanation that applied to both the fall of a stone and the movements of planets around the sun.

It is very easy for us today to remark how strange it was that it never occurred to anyone before Newton that the same force might cause the apple to drop and the moon to go around the earth. In fact, this was not a thought likely to occur to anyone. It is not that the *answer* was so difficult to give; it is that nobody had asked the *question*. This is a vital point. No one had asked: "What is the relation between the force that heavenly bodies exert upon each other and terrestrial forces that cause objects to fall to the ground?" Even to speak in such terms as "terrestrial" and "heavenly" is to make a bipartition, to cut nature into two fundamentally different regions. It was Newton's great insight to break away from this division, to assert that there is no such fundamental cleavage. There is one nature, one world. Newton's universal law of gravitation was the theoretical law that explained for the first time both the fall of an apple and Kepler's laws for the movements of planets. In Newton's day, it was a psychologically difficult, extremely daring adventure to think in such general terms.

Later, of course, by means of correspondence rules, scientists discovered how to determine the masses of astronomical bodies. Newton's theory also said that two apples, side by side on a table, attract each other. They do not move toward each other because the attracting force is extremely small and the friction on the table very large. Physicists eventually succeeded in actually measuring the gravitational forces between two bodies in the laboratory. They used a torsion balance consisting of a bar with a metal ball on each end, suspended at its center by a long wire attached to a high ceiling. (The longer and thinner the wire, the more easily the bar would turn.) Actually, the bar never came to an absolute rest but always oscillated a bit. But the mean point of the bar's oscillation could be established. After the exact position of the mean point was determined, a large pile of lead bricks was constructed near the bar. (Lead was used because of its great specific gravity. Gold had an even higher specific gravity, but gold bricks are expensive.) It was found that the mean of the oscillating bar had shifted a tiny amount to bring one of the balls on the end of the bar nearer to the lead pile. The shift was only a fraction of a millimeter, but it was enough to provide first observation of a gravitational effect between two bodies in a laboratory—an effect that had been predicted by Newton's theory of gravitation.

It had been known before Newton that apples fall to the ground and that the moon moves around the earth. Nobody before Newton could have predicted the outcome of the experiment with the torsion balance. It is a classic instance of the power of a theory to predict a new phenomenon not previously observed.

NOTE

1. See Percy W. Bridgman, *The Logic of Modern Physics* (New York: Macmillan, 1927), and Norman R. Campbell, *Physics: The Elements* (Cambridge: Cambridge University Press, 1920). Rules of correspondence are discussed by Ernest Nagel, *The Structure of Science* (New York: Harcourt, Brace, & World, 1961), pp. 97–105.

18

What Theories Are Not

Hilary Putnam

The announced topic for this symposium was the role of models in empirical science; however, in preparing for this symposium, I soon discovered that I had first to deal with a different topic, and this different topic is the one to which this paper actually will be devoted. The topic I mean is the role of *theories* in empirical science; and what I do in this paper is attack what may be called the "received view" on the role of theories—that theories are to be thought of as "partially interpreted calculi" in which only the "observation terms" are "directly interpreted" (the theoretical terms being only "partially interpreted," or, some people even say, "partially understood").

To begin, let us review this received view. The view divides the nonlogical vocabulary of science into two parts:

OBSERVATION TERMS
such terms as
"red,"
"touches,"
"stick," etc.

THEORETICAL TERMS
such terms as
"electron,"
"dream,"
"gene," etc.

The basis for the division appears to be as follows: the observation terms apply to what may be called publicly observable things and signify observable qualities of these things, while the theoretical terms correspond to the remaining unobservable qualities and things.

This division of terms into two classes is then allowed to generate a division of statements into two[1] classes as follows:

OBSERVATIONAL STATEMENTS
statements containing only observation terms and logical vocabulary

THEORETICAL STATEMENTS
statements containing
theoretical terms

Lastly, a scientific theory is conceived of as an axiomatic system which may be thought of as initially uninterpreted, and which gains "empirical meaning" as a result of a specification of meaning *for the observation terms alone.* A kind of partial meaning is then thought of as drawn up to the theoretical terms, by osmosis, as it were.

One can think of many distinctions that are crying out to be made ("new" terms vs. "old" terms, technical terms vs. nontechnical ones, terms more-or-less peculiar to one science vs. terms common to many, for a start). My contention here is simply:

(1) The *problem* for which this dichotomy was invented ("how is it possible to interpret theoretical terms?") does not exist.

(2) A basic reason some people have given for introducing the dichotomy is false: namely, justification in science does *not* proceed "down" in the direction of observation terms. In fact, justification in science proceeds in any direction that may be handy—more observational assertions sometimes being justified with the aid of more theoretical ones, and vice versa. Moreover, as we shall see, while the notion of an *observation report* has some importance in the philosophy of science, such reports cannot be identified on the basis of the vocabulary they do or do not contain.

(3) In any case, whether the reasons for introducing the dichotomy were good ones or bad ones, the double distinction (observation terms–theoretical terms, observation statements–theoretical statements) presented above is, in fact, completely broken-backed. This I shall try to establish now.

In the first place, it should be noted that the dichotomy under discussion was intended as an explicative and not merely a stipulative one. That is, the words "observational" and "theoretical" are not having arbitrary new meanings bestowed upon them; rather, preexisting uses of these words (especially in the philosophy of science) are presumably being sharpened and made clear. And, in the second place, it should be recalled that we are dealing with a double, not just a single, distinction. That is to say, part of the contention I am criticizing is that, once the distinction between observational and theoretical *terms* has been drawn as above, the distinction between theoretical statements and observational reports or assertions (in something like the sense usual in methodological discussions) can be drawn in terms of it. What I mean when I say that the dichotomy is "completely broken-backed" is this:

(A) If an "observation term" is one that cannot apply to an unobservable, then there are no observation terms.[2]

(B) Many terms that refer primarily to what Carnap would class as "unobservables" are not theoretical terms; and at least some theoretical terms refer primarily to observables.

(C) Observational reports can and frequently do contain theoretical terms.

(D) A scientific theory, properly so-called, may refer only to observables. (Darwin's theory of evolution, as originally put forward, is one example.)

To start with the notion of an "observation term": Carnap's formulation in *Testability and Meaning* [1] was that for a term to be an observation term not only must it correspond to an observable quality, but the determination whether the quality is present or not must be able to be made by the observer in a relatively short time, and with a high degree of confirmation. In his most recent authoritative publication [2], Carnap is rather brief. He writes, "the terms of V_0 [the 'observation vocabulary'—H.P.] are predicates designating observable properties of events or things (e.g., 'blue', 'hot', 'large', etc.) or observable relations between them (e.g., 'x is warmer than y', 'x is contiguous to y', etc.)" [2, p. 41]. The only other clarifying remarks I could find are the following: "The name 'observation language' may be understood in a narrower or in a wider sense; the observation language in the wider sense includes the disposition terms. In this article I take the observation language L_0 in the narrower sense" [2, p. 63]. "An observable property may be regarded as a simple special case of a testable disposition: for example, the operation for finding out whether a thing is blue or hissing or cold, consists simply in looking or listening or touching the thing, respectively. Nevertheless, *in the reconstruction of the language* [italics mine—H.P.] it seems convenient to take some properties for which the test procedure is extremely simple (as in the examples given) as directly observable, and use them as primitives in L_0" [2, p. 63].

These paragraphs reveal that Carnap, at least, thinks of observation terms as corresponding to qualities that can be detected without the aid of instruments. But always so detected? Or can an observation term refer sometimes to an observable thing and sometimes to an unobservable? While I have not been able to find any explicit statement on this point, it seems to me that writers like Carnap must be *neglecting* the fact that all terms—including the 'observation terms'—have at least the possibility of applying to unobservables. Thus their problem has sometimes been formulated in quasi-historical terms—"How could theoretical terms have been introduced into the language?" And the usual discussion strongly suggests that the following puzzle is meant: if we imagine a time at which people could only talk about observables (had not available any theoretical terms), how did they ever manage to *start* talking about unobservables?

It is possible that I am here doing Carnap and his followers an injustice. However, polemics aside, the following points must be emphasized:

(1) Terms referring to unobservables are *invariably* explained, in the actual history of science, with the aid of already present locutions referring to unobservables. There never was a stage of language at which it was impossible to talk about unobservables. Even a three-year-old child can understand a story about "people too little to see"[3] and not a single "theoretical term" occurs in this phrase.

(2) There is not even a single *term* of which it is true to say that it *could*

not (without changing or extending its meaning) be used to refer to unobservables. "Red," for example, was so used by Newton when he postulated that red light consists of *red corpuscles*.[4]

In short: if an "observation term" is a term which *can*, in principle, only be used to refer to observable things, then *there are no observation terms*. If, on the other hand, it is granted that locutions consisting of just observation terms can refer to unobservables, there is no longer any reason to maintain *either* that theories and speculations about the unobservable parts of the world must contain "theoretical (= nonobservation) terms" *or* that there is any general problem as to how one can introduce terms referring to unobservables. Those philosophers who find a difficulty in how we understand theoretical terms should find an equal difficulty in how we understand "red" and "smaller than."

So much for the notion of an "observation term." Of course, one may recognize the point just made—that the "observation terms" also apply, in some contexts, to unobservables—and retain the class (with a suitable warning as to how the label "observation term" is to be understood). But can we agree that the complementary class—what should be called the "nonobservation terms"—is to be labelled "theoretical terms"? No, for the identification of "theoretical terms" with "term (other than the 'disposition terms,' which are given a special place in Carnap's scheme) designating an unobservable quality" is unnatural and misleading. On the one hand, it is clearly an enormous (and, I believe, insufficiently motivated) extension of common usage to classify such terms as "angry," "loves," and so forth, as "theoretical terms" simply because they allegedly do not refer to public observables. A theoretical term, properly so-called, is one which comes from a scientific *theory* (and the almost untouched problem, in thirty years of writing about "theoretical terms" is what is *really* distinctive about such terms). In this sense (and I think it the sense important for discussions of science) "satellite" is, for example, a theoretical term (although the things it refers to are quite observable[5]) and "dislikes" clearly is not.

Our criticisms so far might be met by relabeling the first dichotomy (the dichotomy of terms) "observation vs. nonobservation," and suitably "hedging" the notion of "observation." But more serious difficulties are connected with the identification upon which the second dichotomy is based—the identification of "theoretical statements" with statements containing nonobservation ("theoretical") terms, and "observation statements" with "statements in the observational vocabulary."

That observation statements may contain theoretical terms is easy to establish. For example, it is easy to imagine a situation in which the following sentence might occur: "We also *observed* the creation of two electron-positron pairs."

This objection is sometimes dealt with by proposing to "relativize" the observation-theoretical dichotomy to the context. (Carnap, however, rejects this way out in the article we have been citing.) This proposal to "relativize" the dichotomy does not seem to me to be very helpful. In the first place, one can easily imagine a context in which "electron" would occur, in the same text, in *both* observational reports and in theoretical conclusions from those reports. (So that one would have distortions if one tried to put the term in either the "observational term" box or in the "theoretical term" box.) In the second place, for what philosophical problem or point does one require even the relativized dichotomy?

The usual answer is that sometimes a statement A (observational) is offered in support of a statement B (theoretical). Then, in order to explain why A is not itself questioned in the context, we need to be able to say that A is functioning, in that context, as an observation report. But this misses the point I have been making! I do not deny the need for some such notion as "observation report." What I deny is that the distinction between observation reports and, among other things, theoretical statements, can or should be drawn on the basis of vocabulary. In addition, a relativized dichotomy will not serve Carnap's purposes. One can hardly maintain that theoretical terms are only partially interpreted, whereas observational terms are completely interpreted, if no sharp line exists between the two classes. (Recall that Carnap takes his problem to be "reconstruction of the language," not of some isolated scientific context.) . . .

NOTES

1. Sometimes a *tripartite* division is used: observation statements, theoretical statements (containing *only* theoretical terms). and "mixed" statements (containing both kinds of terms). This refinement is not considered here, because it avoids none of the objections presented below.

2. I neglect the possibility of trivially constructing terms that refer only to observables: namely, by conjoining "and is an observable thing" to terms that would otherwise apply to some unobservables. "Being an observable thing" is, in a sense, highly theoretical and yet applies only to observables!

3. Von Wright has suggested (in conversation) that this is an *extended* use of language (because we first learn words like "people" in connection with people we *can* see). This argument from "The way we learn to use the word" appears to be unsound however (cf. [4]).

4. Some authors (although not Carnap) explain the intelligibility of such discourse in terms of logically possible submicroscopic observers. But (a) such observers could not see single photons (or light corpuscles) even on Newton's theory; and (b) once such physically impossible (though logically possible) "observers" are introduced, why not go further and have observers with sense organs for electric charge, or the curvature of space, et cetera! Presumably because *we* can see *red*, but not *charge*. But then, this just makes the point that we *understand* "red" even when applied outside our normal "range," even though we learn it ostensively, without *explaining* that fact. (The explanation lies in this: that understanding any term—even "red"—involves at least two elements: internalizing the syntax of a natural language, and

acquiring a background of ideas. Overemphasis on the way "red" is *taught* has led some philosophers to misunderstand how it is *learned.*)

5. Carnap might exclude "satellite" as an observation term, on the ground that it takes a comparatively long time to verify that something is a satellite with the naked eye, even if the satellite is close to the parent body (although this could be debated). However, "satellite" cannot be excluded on the quite different ground that many satellites are too far away to see (which is the ground that first comes to mind) since the same is true of the huge majority of all *red* things.

REFERENCES

[1] Carnap, R. "Testability and Meaning." In *Reading in the Philosophy of Science,* edited by H. Feigl and M. Brodbeck, 47–92. New York, Appleton-Century-Crofts, 1955, x + 517 pp. Reprinted from *Philosophy of Science* 3 (1936) and 4 (1937).

[2] Carnap, R. "The Methodological Character of Theoretical Concepts." In *Minnesota Studies in the Philosophy of Science.* Vol. 1, edited by H. Feigi et al., 1–74. Minneapolis: University of Minnesota Press, 1956, x + 517 pp.

[3] Carnap, R. *The Foundations of Logic and Mathematics.* Vol. 4, no. 3 of *International Encyclopedia of United Science.* Chicago: University of Chicago Press, 1939, 75 pp.

[4] Fodor, J. "Do Words Have Uses?" Submitted to *Inquiry.*

[5] Putnam, H. "Mathematics and the Existence of Abstract Entities," *Philosophical Studies* 7 (1957): 81–88.

[6] Quine, W. V. O. "The Scope and Language of Science." *British Journal for the Philosophy of Science* 8 (1957): 1–17.

19

Observation

N. R. Hanson

Were the eye not attuned to the Sun,
The sun could never be seen by it.
                    GOETHE[1]

A

Consider two microbiologists. They look at a prepared slide; when asked what they see, they may give different answers. One sees in the cell before him a cluster of foreign matter: it is an artifact, a coagulum resulting from inadequate staining techniques. This clot has not more to do with the cell, in *vivo*, than the scars left on it by the archaeologist's spade have to do with the original shape of some Grecian urn. The other biologist identifies the clot as a cell organ, a "Golgi body." As for techniques, he argues: "The standard way of detecting a cell organ is by fixing and staining. Why single out this one technique as producing artifacts, while others disclose genuine organs?"

The controversy continues.[2] It involves the whole theory of microscopical technique; nor is it an obviously experimental issue. Yet it affects what scientists say they see. Perhaps there is a sense in which two such observers do not see the same thing, do not begin from the same data, though their eyesight is normal and they are visually aware of the same object.

Imagine these two observing a Protozoan—*Amoeba*. One sees a one-celled animal, the other a noncelled animal. The first sees *Amoeba* in all its analogies with different types of single cells: liver cells, nerve cells, epithelium cells. These have a wall, nucleus, cytoplasm, et cetera. Within this class *Amoeba* is distinguished only by its independence. The other, however, sees *Amoeba*'s homology not with single cells, but with whole animals. Like all animals *Amoeba* ingests its food, digests and assimilates it. It excretes, reproduces, and is mobile—more like a complete animal than an individual tissue cell.

This is not an experimental issue, yet it can affect experiment. What either man regards as significant questions or relevant data can be determined by whether he stresses the first or the last term in "unicellular animal."[3]

Some philosophers have a formula ready for such situations: "Of course they see the same thing. They make the same observation since they begin from the same visual data. But they interpret what they see differently. They construe the evidence in different ways."[4] The task is then to show how these data are molded by different theories or interpretations or intellectual constructions.

Considerable philosophers have wrestled with this task. But in fact the formula they start from is too simple to allow a grasp of the nature of observation within physics. Perhaps the scientists cited above do not begin their inquiries from the same data, do not make the same observations, do not even see the same thing? Here many concepts run together. We must proceed carefully, for wherever it makes sense to say that two scientists looking at x do not see the same thing, there must always be a prior sense in which they do see the same thing. The issue is, then, "Which of these senses is most illuminating for the understanding of observational physics?"

These biological examples are too complex. Let us consider Johannes Kepler: imagine him on a hill watching the dawn. With him is Tycho Brahe. Kepler regarded the sun as fixed: it was the earth that moved. But Tycho followed Ptolemy and Aristotle in this much at least: the earth was fixed and all other celestial bodies moved around it. *Do Kepler and Tycho see the same thing in the east at dawn?*

We might think this an experimental or observational question, unlike the questions "Are there Golgi bodies?" and "Are Protozoa one-celled or non-celled?" Not so in the sixteenth and seventeenth centuries. Thus Galileo said to the Ptolemaist ". . . neither Aristotle nor you can prove that the earth is de facto the centre of the universe. . . ."[5] "Do Kepler and Tycho see the same thing in the east at dawn?" is perhaps not a de facto question either, but rather the beginning of an examination of the concepts of seeing and observation.

The resultant discussion might run
"Yes, they do."
"No, they don't."
"Yes, they do!"
"No, they don't!' . . .

That this is possible suggests that there may be reasons for both contentions.[6] Let us consider some points in support of the affirmative answer.

The physical processes involved when Kepler and Tycho watch the dawn are worth noting. Identical photons are emitted from the sun; these traverse solar space, and our atmosphere. The two astronomers have normal vision; hence these photons pass through the cornea, aqueous humor, iris, lens, and vitreous body of their eyes in the same way. Finally their retinas are affected. Similar electrochemical changes occur in their selenium cells. The same configuration is et ceterahed on Kepler's retina as on Tycho's. So they see the same thing.

Locke sometimes spoke of seeing in this way: a man sees the sun if his is a normally-formed retinal picture of the sun. Dr. Sir W. Russell Brain speaks of our retinal sensations as indicators and signals. Everything taking place behind the retina is, as he says, "an intellectual operation based largely on non-visual experience. . . ."[7] What we see are the changes in the *tunica retina*. Dr. Ida Mann regards the macula of the eye as itself "seeing details in bright light," and the rods as "seeing approaching motor-cars." Dr. Agnes Arber speaks of the eye as itself seeing.[8] Often, talk of seeing can direct attention to the retina. Normal people are distinguished from those for whom no retinal pictures can form: we may say of the former that they can see whilst the latter cannot see. Reporting when a certain red dot can be seen may supply the oculist with direct information about the condition of one's retina.[9]

This need not be pursued, however. These writers speak carelessly: seeing the sun is not seeing retinal pictures of the sun. The retinal images which Kepler and Tycho have are four in number, inverted and quite tiny.[10] Astronomers cannot be referring to these when they say they see the sun. If they are hypnotized, drugged, drunk, or distracted they may not see the sun, even though their retinas register its image in exactly the same way as usual.

Seeing is an experience. A retinal reaction is only a physical state—a photochemical excitation. Physiologists have not always appreciated the differences between experiences and physical states.[11] People, not their eyes, see. Cameras, and eyeballs, are blind. Attempts to locate within the organs of sight (or within the neurological reticulum behind the eyes) some nameable called "seeing" may be dismissed. That Kepler and Tycho do, or do not, see the same thing cannot be supported by reference to the physical states of their retinas, optic nerves, or visual cortices: there is more to seeing than meets the eyeball.

Naturally, Tycho and Kepler see the same physical object. They are both visually aware of the sun. If they are put into a dark room and asked to report when they see something—anything at all—they may both report the same object at the same time. Suppose that the only object to be seen is a certain lead cylinder. Both men see the same thing: namely this object—whatever it is. It is just here, however, that the difficulty arises, for while Tycho sees a mere pipe, Kepler will see a telescope, the instrument about which Galileo has written to him.

Unless both are visually aware of the same object there can be nothing of philosophical interest in the question whether or not they see the same thing. Unless they both see the sun in this prior sense our question cannot even strike a spark.

Nonetheless, both Tycho and Kepler have a common visual experience of some sort. This experience perhaps constitutes their seeing the same thing.

Indeed, this may be a seeing logically more basic than anything expressed in the pronouncement "I see the sun" (where each means something different by "sun"). If what they meant by the word "sun" were the only clue, then Tycho and Kepler could not be seeing the same thing, even though they were gazing at the same object.

If, however, we ask, not "Do they see the same thing?" but rather "What is it that they both see?," an unambiguous answer may be forthcoming. Tycho and Kepler are both aware of a brilliant yellow-white disc in a blue expanse over a green one. Such a "sense-datum" picture is single and uninverted. To be unaware of it is not to have it. Either it dominates one's visual attention completely or it does not exist.

If Tycho and Kepler are aware of anything visual, it must be of some pattern of colors. What else could it be? We do not touch or hear with our eyes, we only take in light.[12] This private pattern is the same for both observers. Surely if asked to sket ceterah the contents of their visual fields they would both draw a kind of semicircle on a horizon-line.[13] They say they see the sun. But they do not see every side of the sun at once; so what they really see is discoid to begin with. It is but a visual aspect of the sun. In any single observation the sun is a brilliantly luminescent disc, a penny painted with radium.

So something about their visual experiences at dawn is the same for both: a brilliant yellow-white disc centered between green and blue color patches. Sket ceterahes of what they both see could be identical—congruent. In this sense Tycho and Kepler see the same thing at dawn. The sun appears to them in the same way. The same view, or scene, is presented to them both.

In fact, we often speak in this way. Thus the account of a recent solar eclipse:[14] "Only a thin crescent remains; white light is now completely obscured; the sky appears a deep blue, almost purple, and the landscape is a monochromatic green . . . there are the flashes of light on the disc's circumference and now the brilliant crescent to the left. . . ." Newton writes in a similar way in the *Opticks*: "These Arcs at their first appearance were of a violet and blue Colour, and between them were white Arcs of Circles, which . . . became a little tinged in their inward Limbs with red and yellow. . . ."[15] Every physicist employs the language of lines, color patches, appearances, shadows. Insofar as two normal observers use this language of the same event, they begin from the same data: they are making the same observation. Differences between them must arise in the interpretations they put on these data.

Thus, to summarize, saying that Kepler and Tycho see the same thing at dawn just because their eyes are similarly affected is an elementary mistake. There is a difference between a physical state and a visual experience. Suppose, however, that it is argued as above—that they see the same thing because they have the same sense-datum experience. Disparities in their accounts arise in ex post facto interpretations of what is seen, not in the fundamental visual data. If this is argued, further difficulties soon obtrude.

B

Normal retinas and cameras are impressed similarly by fig. 1.[16] Our visual sense-data will be the same too. If asked to draw what we see, most of us will set out a configuration like fig. 1.

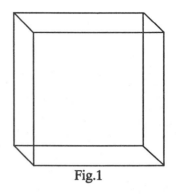

Fig.1

Do we all see the same thing?[17] Some will see a perspex cube viewed from below. Others will see it from above. Still others will see it as a kind of polygonally-cut gem. Some people see only criss-crossed lines in a plane. It may be seen as a block of ice, an aquarium, a wire frame for a kite—or any of a number of other things.

Do we, then, all see the same thing? If we do, how can these differences be accounted for?

Here the "formula" re-enters: "These are different *interpretations* of what all observers see in common. Retinal reactions to fig. 1 are virtually identical; so too are our visual sense-data, since our drawings of what we see will have the same content. There is no place in the seeing for these differences, so they must lie in the interpretations put on what we see."

This sounds as if I do two things, not one, when I see boxes and bicycles. Do I put different interpretations on fig. 1 when I see it now as a box from below, and now as a cube from above? I am aware of no such thing. I mean no such thing when I report that the box's perspective has snapped back into the page.[18] If I do not mean this, then the concept of seeing which is natural in this connection does not designate two diaphanous components, one optical the other interpretative. Fig. 1 is simply seen now as a box from below, now as a cube from above; one does not first soak up an optical pattern and then clamp an interpretation on it. Kepler and Tycho just see the sun. That is all. That is the way the concept of seeing works in this connection.

"But," you say, "seeing fig. 1 first as a box from below, then as a cube from above, involves interpreting the lines differently in each case." Then for you and me to have a different interpretation of fig. 1 just *is* for us to see something different. This does not mean we see the same thing and then interpret it differently. When I suddenly exclaim "Eureka—a box from above," I do not refer simply to a different interpretation. (Again, there is a logically prior sense in which seeing fig. 1 as from above and then as from below is seeing the same thing differently, i.e. being aware of the same diagram in different ways. We can refer just to this, but we need not. In this case we do not.)

Besides, the word "interpretation" is occasionally useful. We know where it applies and where it does not. Thucydides presented the facts objectively; Herodotus put an interpretation on them. The word does not apply to everything—it has a meaning. Can interpreting always be going on when we see? Sometimes, perhaps, as when the hazy outline of an agricultural machine looms up on a foggy morning and, with effort, we finally identify it. Is this the "interpretation" which is active when bicycles and boxes are clearly seen? Is it active when the perspective of fig. 1 snaps into reverse? There was a time when Herodotus was half-through with his interpretation of the Greco-Persian wars. Could there be a time when one is half-through interpreting fig. 1 as a box from above, or as anything else?

"But the interpretation takes very little time—it is instantaneous." Instantaneous interpretation hails from the Limbo that produced unsensed sensibilia, unconscious inference, incorrigible statements, negative facts, and *Objektive*. These are ideas which philosophers force on the world to preserve some pet epistemological or metaphysical theory.

Only in contrast to "Eureka" situations (like perspective reversals, where one cannot interpret the data) is it clear what is meant by saying that though Thucydides could have put an interpretation on history, he did not. Moreover, whether or not an historian is advancing an interpretation is an empirical question: we know what would count as evidence one way or the other. But whether we are employing an interpretation when we see fig. 1 in a certain way is not empirical. What could count as evidence? In no ordinary sense of "interpret" do I interpret fig. 1 differently when its perspective reverses for me. If there is some extraordinary sense of that word it is not clear, either in ordinary language, or in extraordinary (philosophical) language. To insist that different reactions to fig. 1 *must* lie in the interpretations put on a common visual experience is just to reiterate (without reasons) that the seeing of *x must* be the same for all observers looking at *x*.

"But 'I see the figure as a box' means: I am having a particular visual experience which I always have when I interpret the figure as a box, or when I look at a box. . . ." ". . . if I meant this, I ought to know it. I ought to be able to refer to the experience directly and not only indirectly. . . ."[19]

Ordinary accounts of the experiences appropriate to fig. 1 do not require visual grist going into an intellectual mill: theories and interpretations are "there" in the seeing from the outset. How can interpretations "be there" in the seeing? How is it possible to see an object according to an interpretation? "The question represents it as a queer fact; as if something were being forced into a form it did not really fit. But no squeezing, no forcing took place here."[20]

Consider now the reversible perspective figures which appear in textbooks on Gestalt psychology: the tea-tray, the shifting (Schröder) staircase, the tunnel. Each of these can be seen as concave, as convex, or as a flat drawing.[21] Do I really see something different each time, or do I only interpret what I see in a different way? To interpret is to think, to do something;

seeing is an experiential state.[22] The different ways in which these figures are seen are not due to different thoughts lying behind the visual reactions. What could "spontaneous" mean if these reactions are not spontaneous? When the staircase "goes into reverse" it does so spontaneously. One does not think of anything special; one does not think at all. Nor does one interpret. One just sees, now a staircase as from above, now a staircase as from below.

The sun, however, is not an entity with such variable perspective. What has all this to do with suggesting that Tycho and Kepler may see different things in the east at dawn? Certainly the cases are different. But these reversible perspective figures are examples of different things being seen in the same configuration, where this difference is due neither to differing visual pictures, nor to any "interpretation" superimposed on the sensation. . . .

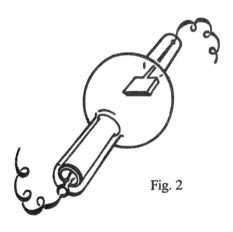

Fig. 2

A trained physicist could see one thing in fig 2: an X-ray tube viewed from the cathode. Would Sir Lawrence Bragg and an Eskimo baby see the same thing when looking at an X-ray tube? Yes, and no. Yes—they are visually aware of the same object. No—the *ways* in which they are visually aware are profoundly different. Seeing is not only the having of a visual experience; it is also the way in which the visual experience is had.

At school the physicist had gazed at this glass-and-metal instrument. Returning now, after years in university and research, his eye lights upon the same object once again. Does he see the same thing now as he did then? Now he sees the instrument in terms of electrical circuit theory, thermodynamic theory, the theories of metal and glass structure, thermionic emission, optical transmission, refraction, diffraction, atomic theory, quantum theory, and special relativity.

Contrast the freshman's view of college with that of his ancient tutor. Compare a man's first glance at the motor of his car with a similar glance ten exasperating years later.

"Granted, one learns all these things," it may be countered, "but it all figures in the interpretation the physicist puts on what he sees. Though the layman sees exactly what the physicist sees, he cannot interpret it in the same way because he has not learned so much."

Is the physicist doing more than just seeing? No, he does nothing over and above what the layman does when he sees an X-ray tube. What are you doing over and above reading these words? Are you interpreting marks on a page? When would this ever be a natural way of speaking? Would an infant see what you see here, when you see words and sentences and he sees but marks and lines? One does nothing beyond looking and seeing when one dodges bicycles, glances at a friend, or notices a cat in the garden.

"The physicist and the layman see the same thing," it is objected, "but they do not make the same thing of it." The layman can make nothing of it. Nor is that just a figure of speech. I can make nothing of the Arab word for *cat*, though my purely visual impressions may be distinguishable from those of the Arab who can. I must learn Arabic before I can see what he sees. The layman must learn physics before he can see what the physicist sees.

If one must find a paradigm case of seeing it would be better to regard as such not the visual apprehension of color patches but things like seeing what time it is, seeing what key a piece of music is written in, and seeing whether a wound is septic.[23]

Pierre Duhem writes:

Enter a laboratory; approach the table crowded with an assortment of apparatus, an electric cell, silk-covered copper wire, small cups of mercury, spools, a mirror mounted on an iron bar; the experimenter is inserting into small openings the metal ends of ebony-headed pins; the iron oscillates, and the mirror attached to it throws a luminous band upon a celluloid scale; the forward-backward motion of this spot enables the physicist to observe the minute oscillations of the iron bar. But ask him what he is doing. Will he answer "I am studying the oscillations of an iron bar which carries a mirror?" No, he will say that he is measuring the electric resistance of the spools. If you are astonished, if you ask him what his words mean, what relation they have with the phenomena he has been observing and which you have noted at the same time as he, he will answer that your question requires a long explanation and that you should take a course in electricity.[24]

The visitor must learn some physics before he can see what the physicist sees. Only then will the context throw into relief those features of the objects before him which the physicist sees as indicating resistance.

This obtains in all seeing. Attention is rarely directed to the space between the leaves of a tree, save when a Keats brings it to our notice.[25] (Consider also what was involved in Crusoe's seeing a vacant space in the sand as a footprint.) Our attention most naturally rests on objects and events which dominate the visual field. What a blooming, buzzing, undifferentiated confusion visual life would be if we all arose tomorrow without attention capable of dwelling only on what had heretofore been overlooked.[26]

The infant and the layman can see: they are not blind. But they cannot see what the physicist sees; they are blind to what he sees.[27] We may not hear that the oboe is out of tune, though this will be painfully obvious to the trained musician. (Who, incidentally, will not hear the tones and *interpret* them as being out of tune, but will simply hear the oboe to be out of tune.[28] We simply see what time it is; the surgeon simply sees a wound to be septic; the physicist sees the X-ray tube's anode overheating.) The elements of the visitor's visual field, though identical with those of the physicist, are not organized for him as for the physicist; the same lines, colors, shapes are apprehended by both, but not in the same way. There are indefinitely many ways in which a constellation of lines, shapes, patches, may be seen. *Why* a visual pattern is seen differently is a question for psychology, but *that* it may be seen differently is important in any examination of the concepts of seeing and observation. Here, as Wittgenstein might have said, the psychological is a symbol of the logical.

You see a bird, I see an antelope; the physicist sees an X-ray tube, the child a complicated lamp bulb; the microscopist sees coelenterate mesoglea, his new student sees only a gooey, formless stuff. Tycho and Simplicius see a mobile sun, Kepler and Galileo see a static sun.[29]

It may be objected, "Everyone, whatever his state of knowledge, will see fig. 1 as a box or cube, viewed as from above or as from below." True; almost everyone, child, layman, physicist, will see the figure as box-like one way or another. But could such observations be made by people ignorant of the construction of box-like objects? No. This objection only shows that most of us —the blind, babies, and dimwits excluded—have learned enough to be able to see this figure as a three-dimensional box. This reveals something about the sense in which Simplicius and Galileo do see the same thing (which I have never denied)—they both see a brilliant heavenly body. The schoolboy and the physicist both see that the X-ray tube will smash if dropped. Examining how observers see different things in x marks something important about their seeing the same thing when looking at x. If seeing different things involves having different knowledge and theories about x, then perhaps the sense in which they see the same thing involves their sharing knowledge and theories about x. Bragg and the baby share no knowledge of X-ray tubes. They see the same thing only in that if they are looking at x they are both having some visual experience of it. Kepler and Tycho agree on more: they see the same thing in a stronger sense. Their visual fields are organized in much the same way. Neither sees the sun about to break out in a grin, or about to crack into ice cubes. (The baby is not "set" even against these eventualities.) Most people today see the same thing at dawn in an even stronger sense: we share much knowledge of the sun. Hence Tycho and Kepler see different things, and yet they see the same thing. That these things can be said *in a different way* depends on their knowledge, experience, and theories.

... The elements of their experiences are identical; but their conceptual organization is vastly different. Can their visual fields have a different organization? Then they can see different things in the east at dawn.

It is the sense in which Tycho and Kepler do not observe the same thing which must be grasped if one is to understand disagreements within micro-physics. Fundamental physics is primarily a search for intelligibility—it is philosophy of matter. Only secondarily is it a search for objects and facts (though the two endeavors are as hand and glove). Microphysicists seek new modes of conceptual organization. If that can be done the finding of new enti-ties will follow. Gold is rarely discovered by one who has not got the lay of the land.

To say that Tycho and Kepler, Simplicius and Galileo, Hooke and Newton, Priestley and Lavoisier, Soddy and Einstein, De Broglie and Born, Heisenberg and Bohm all make the same observations but use them differ-ently is too easy.[30] It does not explain controversy in research science. Were there no sense in which they were different observations they could not be used differently. This may perplex some: that researchers sometimes do not appreciate data in the same way is a serious matter. It is important to realize, however, that sorting out differences about data, evidence, observation, may require more than simply gesturing at observable objects. It may require a comprehensive reappraisal of one's subject matter. This may be difficult, but it should not obscure the fact that nothing less than this may do. . . .

NOTES

1.   Wär' nicht das Auge sonnenhaft,
                      Die Sonne könnt' es nie erblicken;
        Goethe, *Zahme Xenien* (Werke, Weimar, 1887–1918), bk. 3, 1805.

2. Cf. the papers by Baker and Gatonby in *Nature*, 1949–present.

3. This is not a merely conceptual matter, of course. Cf. Wittgenstein, *Philosophical Investigations* (Blackwell, Oxford, 1953), p. 196.

4. (1) G. Berkeley, *Essay Towards a New Theory of Vision* (in *Works*, vol. I [London, T. Nelson, 1948–56]), pp. 51 ff.

(2) James Mill, *Analysis of the Phenomena of the Human Mind* (Longmans, London, 1869), vol. 1, p. 97.

(3) J. Sully, *Outlines of Psychology* (Appleton, New York, 1885).

(4) William James, *The Principles of Psychology* (Holt, New York, 1890–1905), vol. 2, pp. 4, 78, 80 and 81; vol. 1, p. 221.

(5) A. Schopenhauer, *Satz vom Grunde* (in *Sämmtliche Werke*, Leipzig, 1888), ch. 4.

(6) H. Spencer, *The Principles of Psychology* (Appleton, New York, 1897), vol. 4, chs. 9, 10.

(7) E. von Hartmann, *Philosophy of the Unconscious* (K. Paul, London, 1931), B, chs. 7, 8.

(8) W. M. Wundt, *Vorlesungen über die Menschen und Thierseele* (Voss, Hamburg, 1892), 4, 13.

(9) H. L. F. von Helmholtz, *Handbuch der Physiologischen Optik* (Leipzig, 1867), pp. 430, 447.

(10) A. Binet, *La psychologie du raisonnement, recherches expérimentales par l'hyp-notisme* (Alcan, Paris, 1886), chs. 3, 5.

(11) J. Grote, *Exploratio Philosophica* (Cambridge, 1900), vol. 2, pp. 201 ff.

(12) B. Russell, in *Mind* (1913), p. 76. *Mysticism and Logic* (Longmans, New York, 1918), p. 209. *The Problems of Philosophy* (Holt, New York, 1912), pp. 73, 92, 179, 203.

(13) Dawes Hicks, *Arist. Soc. Sup.* vol. 2 (1919), pp. 176–8.

(14) G. F. Stout, *A Manual of Psychology* (Clive, London, 1907, 2nd ed.), vol. 2, 1 and 2, pp. 324, 561–4.

(15) A. C. Ewing, *Fundamental Questions of Philosophy* (New York, 1951), pp. 45 ff.

(16) G. W. Cunningham, *Problems of Philosophy* (Holt, New York, 1924), pp. 96–7.

5. Galileo, *Dialogue Concerning the Two Chief World Systems* (California, 1953), "The First Day," p. 33.

6. "'Das ist doch kein Sehen!'—'Das ist doch ein Sehen!' Beide müssen sich begrifflich rechtfertigen lassen" (Wittgenstein, *Phil Inv.* p. 203).

7. Brain, *Recent Advances in Neurology* (with Strauss) (London, 1929), p. 88. Compare Helmholtz: "The sensations are signs to our consciousness, and it is the task of our intelligence to learn to understand their meaning" (*Handbuch der Physiologischen Optik* (Leipzig, 1867), vol. 3, p. 433).

See also Husserl, "Ideen zu einer Reinen Phaenomenologie," in *Jahrbuch für Philosophie*, vol. 1 (1913), pp. 75, 79, and Wagner's *Handwörterbuch der physiologie*, vol. 3, section 1 (1846), p. 183.

8. Mann, *The Science of Seeing* (London, 1949), pp. 48–9. Arber, *The Mind and the Eye* (Cambridge, 1954). Compare Muller: "In any field of vision, the retina sees only itself in its spatial extension during a state of affection. It perceives itself as . . . et cetera." (*Zur vergleichenden Physiologie des Gesichtesinnes des Menschen und der Thiere* (Leipzig, 1826), p. 54).

9. Kolin: "An astigmatic eye when looking at millimeter paper can accommodate to see sharply either the vertical lines or the horizontal lines" (*Physics* (New York, 1950), pp. 570 ff.).

10. Cf. Whewell, *Philosophy of Discovery* (London, 1860), "The Paradoxes of Vision."

11. Cf. e.g., J. Z. Young, *Doubt and Certainty in Science* (Oxford, 1951, The Reith Lectures), and Gray Walter's article in *Aspects of Form*, ed. by L. L. Whyte (London, 1953). Compare Newton: "Do not the Rays of Light in falling upon the bottom of the Eye excite Vibrations in the Tunica Retina? Which Vibrations, being propagated along the solid Fibres of the Nerves into the Brain, cause the Sense of seeing" (*Opticks* (London, 1769), bk. 3, part 1).

12. "Rot und grün kann ich nur sehen, aber nicht hören" (Wittgenstein, *Phil. Inv.* p. 209).

13. Cf. "An appearance is the same whenever the same eye is affected in the same way" (Lambert, *Photometria* (Berlin, 1760)); "We are justified, when different perceptions offer themselves to us, to infer that the underlying real conditions are different" (Helmholtz, *Wissenschaftliche Abhandlungen* (Leipzig, 1882), vol. 2, p. 656), and Hertz: "We form for ourselves images or symbols of the external objects; the manner in which we form them is such that the logically necessary (*denknotwendigen*) consequences of the images in thought are invariably the images of materially necessary (*naturnotwendigen*) consequences of the corresponding objects" (*Principles of Mechanics* (London, 1889). p. 1).

Broad and Price make depth a feature of the private visual pattern. However, Weyl (*Philosophy of Mathematics and Natural Science* (Princeton, 1949), p. 125) notes that a single eye perceives qualities spread out in a *two*-dimensional field, since the latter is dissected by any one-dimensional line running through it. But our conceptual difficulties remain even when Kepler and Tycho keep one eye closed.

Whether or not two observers are having the same visual sense-data reduces directly to the question of whether accurate pictures of the contents of their visual fields are identical, or differ in some detail. We can then discuss the publicly observable pictures which Tycho and Kepler draw of what they see, instead of those private, mysterious entities locked in their visual consciousness. The accurate picture and the sense-datum must be identical; how could they differ?

14. From the B.B.C. report, 30 June 1954.

15. Newton, *Opticks*, bk. 2, part 1. The writings of Claudius Ptolemy sometimes read like a phenomenalist's textbook. Cf. e.g., *The Almagest* (Venice, 1515), 6, section 2, "On the Directions in the Eclipses," "When it touches the shadow's circle from within," "When the circles touch each other from without." Cf. also 7 and 8, 9 (section 4). Ptolemy continually seeks to chart and predict "the appearances"—the points of light on the celestial globe. *The Almagest* abandons any attempt to explain the machinery behind these appearances.

Cf. Pappus: "The (circle) dividing the milk-white portion which owes its colour to the sun, and the portion which has the ashen colour natural to the moon itself is indistinguishable from a great circle" (*Mathematical Collection* (Hultsch, Berlin and Leipzig, 1864), pp. 554–60).

16. This famous illusion dates from 1832, when L. A. Necker, the Swiss naturalist, wrote a letter to Sir David Brewster describing how when certain rhomboidal crystals were viewed on end the perspective could shift in the way now familiar to us. Cf. *Phil. Mag.* 3, no. 1 (1832), 329–37, especially p. 336. It is important to the present argument to note that this observational phenomenon began life not as a psychologist's trick, but at the very frontiers of observational science.

17. Wittgenstein answers: "Denn wir sehen eben wirklich zwei verschiedene Tatsachen" (*Tractatus* 5. 5423).

18. "Auf welche Vorgänge spiele ich an?" (Wittgenstein, *Phil. Inv.*, p. 214).

19. Ibid., p. 194 (top).

20. Ibid., p. 200.

21. This is not due to eye movements, or to local retinal fatigue. Cf. Flugel, *Brit. J. Psychol.* 6 (1913), 60; *Brit. Psychol.* 5 (1913), 357. Cf. Donahue and Griffiths, *Amer. J. Psychol.* (1931), and Luckiesch, *Visual Illusions and their Applications* (London, 1922). Cf. also Peirce, *Collected Papers* (Harvard, 1931), 5, 183. References to psychology should not be misunderstood; but as one's acquaintance with the psychology of perception deepens, the character of the conceptual problems one regards as significant will deepen accordingly. Cf. Wittgenstein, *Phil. Inv.* p. 206 (top). Again, p. 193: "Its causes are of interest to psychologists. We are interested in the concept and its place among the concepts of experience."

22. Wittgenstein, *Phil. Inv.*, p. 212.

23. Often "What do you see?" only poses the question "Can you identify the object before you?" This is calculated more to test one's knowledge than one's eyesight.

24. Duhem, *La theorie physique* (Paris, 1914), p. 218.

25. Chinese poets felt the significance of "negative features" like the hollow of a clay vessel or the central vacancy of the hub of a wheel (cf. Waley, *Three Ways of Thought in Ancient China* (London, 1939), p. 155).

26. Infants are indiscriminate; they take in spaces, relations, objects, and events as being of equal value. They still must learn to organize their visual attention. The camera-clarity of their visual reactions is not by itself sufficient to differentiate elements in their visual fields. Contrast Mr. W. H. Auden who recently said of the poet that he is "bombarded by a stream of varied sensations which would drive him mad if he took them all in. It is impossible to guess how much energy we have to spend every day in not-seeing, not-hearing, not-smelling, not-reacting."

27. "He was blind to the *expression* of a face. Would his eyesight on that account be defective?" (Wittgenstein, *Phil. Inv.*, p. 210) and "Because they seeing see not; and hearing they hear not, neither do they understand" (Matt. 13: 10–13).

28. "Es hort doch jeder nur, was er versteht" (Goethe, *Maxims* (Werke, Weimar, 1887–1918)).

29. Against this Professor H. H. Price has argued: "Surely it appears to both of them to be rising, to be moving upwards, across the horizon . . . they both see a moving sun: they both see a round bright body which appears to be rising." Philip Frank retorts: "Our sense observation shows only that in the morning the distance between horizon and sun is increasing, but it does

not tell us whether the sun is ascending or the horizon is descending . . ." (*Modern Science and Its Philosophy* (Harvard, 1949), p. 231). Precisely. For Galileo and Kepler the horizon drops; for Simplicius and Tycho the sun rises. This is the difference Price misses, and which is central to this essay.

30. This parallels the too-easy epistemological doctrine that all normal observers see the same things in x, but interpret them differently.

20

Science and the Physical World

W. T. Stace

So far as I know scientists still talk about electrons, protons, neutrons, and so on. We never directly perceive these, hence if we ask how we know of their existence the only possible answer seems to be that they are an inference from what we do directly perceive. What sort of an inference? Apparently a causal inference. The atomic entities in some way impinge upon the sense of the animal organism and cause that organism to perceive the familiar world of tables, chairs, and the rest.

But is it not clear that such a concept of causation, however interpreted, is invalid? The only reason we have for believing in the law of causation is that we *observe* certain regularities or sequences. We observe that, in certain conditions, A is always followed by B. We call A the cause, B the effect. And the sequence A-B becomes a causal law. It follows that all *observed* causal sequences are between sensed objects in the familiar world of perception, and that all known causal laws apply solely to the world of sense and not to anything beyond or behind it. And this in turn means that we have not got, and never could have, one jot of evidence for believing that the law of causation can be applied *outside* the realm of perception, or that that realm can have any causes (such as the supposed physical objects) which are not themselves perceived.

Put the same thing in another way. Suppose there is an observed sequence A-B-C, represented by the vertical lines in the diagram below.

The observer X sees, and can see, nothing except things in the familiar world of perception. What *right* has he, and what *reason* has he, to assert causes of A, B, and C, such as a', b', c', which he can never observe, behind the perceived world? He has no *right*, because the law of causation on which he is relying has never been observed to operate outside the series of perceptions, and he can have, therefore, no evidence that it does so. And he has no *reason* because the phenomenon C is *sufficiently* accounted for by the cause *B, B* by *A,* and so on. It is unnecessary and super- fluous to introduce a *second* cause *b'* for *B, c'* for *C,* and so forth. To give two causes for each phenomenon, one in one world and one in another, is unnecessary, and perhaps even self-contradictory.

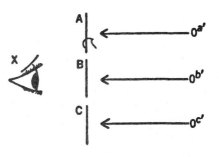

Is it denied, then, it will be asked, that the star causes light waves, that the waves cause retinal changes, that these cause changes in the optic nerve, which in turn causes movement in the brain cells, and so on? No, it is not denied. But the observed causes and effects are all in the world of perception. And no sequences of sense-data can possibly justify going outside that world. If you admit that we never observe anything except sensed objects and their relations, regularities, and sequences, then it is obvious that we are completely shut in by our sen- sations and can never get outside them. Not only causal relations, but all other observed relations, upon which *any* kind of inferences might be founded, will lead only to further sensible objects and their relations. No inference, therefore, can pass from what is sensible to what is not sensible.

The fact is that atoms are *not* inferences from sensations. No one denies, of course, that a vast amount of perfectly valid inferential reasoning takes place in the physical theory of the atom. But it will not be found to be in any strict logical sense inference from *sense-data to atoms.* An *hypothesis* is set up, and the inferential processes are concerned with the application of the hypothesis, that is, with the prediction by its aid of further possible sensations and with its own internal consistency.

That atoms are not inferences from sensations means, of course, that from the existence of sensations we cannot validly infer the existence of atoms. And this means that we cannot have any reason at all to believe that they exist. And that is why I propose to argue that they do not exist—or at any rate that no one could know it if they did, and that we have absolutely no evidence of their existence.

What status have they, then? Is it meant that they are false and worthless, merely untrue? Certainly not. No one supposes that the entries in the nautical almanac "exist" anywhere except on the pages of that book and in the brains of its compilers and readers. Yet they are "true," inasmuch as they enable us to predict certain sensations, namely, the positions and times of certain per- ceived objects which we call the stars. And so the formulae of the atomic theory are true in the same sense, and perform a similar function.

I suggest that they are nothing but shorthand formulae, ingeniously worked out by the human mind, to enable it to predict its experience, i.e., to predict what sensations will be given to it. By "predict" here I do not mean to refer solely to the future. To calculate that there was an eclipse of the sun visible in Asia Minor in the year 585 B.C.E. is, in the sense in which I am using the term, to predict.

In order to see more clearly what is meant, let us apply the same idea to another case, that of gravitation. Newton formulated a law of gravitation in terms of "forces." It was supposed that this law—which was nothing but a mathematical formula—governed the operation of these existent forces. Nowadays it is no longer believed that these forces exist at all. And yet the law can be applied just as well without them to the prediction of astronomical phenomena. It is a matter of no importance to the scientific man whether the forces exist or not. That may be said to be a purely philosophical question. And I think the philosopher should pronounce them fictions. But that would not make the law useless or untrue. If it could still be used to predict phenomena, it would be just as true as it was.

It is true that fault is now found with Newton's law, and that another law, that of Einstein, has been substituted for it. And it is sometimes supposed that the reason for this is that forces are no longer believed in. But this is not the case. Whether forces exist or not simply does not matter. What matters is the discovery that Newton's law does *not* enable us accurately to predict certain astronomical facts such as the exact position of the planet Mercury. Therefore another formula, that of Einstein, has been substituted for it which permits correct predictions. This new law, as it happens, is a formula in terms of geometry. It is pure mathematics and nothing else. It does not contain anything about forces. In its pure form it does not even contain, so I am informed, anything about "humps and hills in space-time." And it does not matter whether any such humps and hills exist. It is truer than Newton's law, not because it substitutes humps and hills for forces, but solely because it is a more accurate formula of prediction.

Not only may it be said that forces do not exist. It may with equal truth be said that "gravitation" does not exist. Gravitation is not a "thing," but a mathematical formula, which exists only in the heads of mathematicians. And as a mathematical formula cannot cause a body to fall, so gravitation cannot cause a body to fall. Ordinary language misleads us here. We speak of the law "of" gravitation, and suppose that this law "applies to" the heavenly bodies. We are thereby misled into supposing that there are *two* things, namely, the gravitation and the heavenly bodies, and that one of these things, the gravitation, causes changes in the other. In reality nothing exists except the moving bodies. And neither Newton's law nor Einstein's law is, strictly speaking, a law of gravitation. They are both laws of moving bodies, that is to say, formulae which tell us how these bodies will move.

Now, just as in the past "forces" were foisted into Newton's law (by himself, be it said), so now certain popularizers of relativity foisted "humps and hills in space-time" into Einstein's law. We hear that the reason why the planets move in curved courses is that they cannot go through these humps and hills, but have to go round them! The planets just get "shoved about," not by forces, but by the humps and hills! But these humps and hills are pure metaphors. And anyone who takes them for "existences" gets asked awkward questions as to what "curved space" is curved "in."

It is not irrelevant to our topic to consider *why* human beings invent these metaphysical monsters of forces and bumps in space-time. The reason is that they have never emancipated themselves from the absurd idea that science "explains" things. They were not content to have laws which merely told them *that* the planets will, as a matter of fact, move in such and such ways. They wanted to know "why" the planets move in those ways. So Newton replied, "Forces." "Oh," said humanity, "that explains it. We understand forces. We feel them every time someone pushes or pulls us." Thus the movements were supposed to be "explained" by entities familiar because analogous to the muscular sensations which human beings feel. The humps and hills were introduced for exactly the same reason. They seem so familiar. If there is a bump in the billiard table, the rolling billiard ball is diverted from a straight to a curved course. Just the same with the planets. "Oh, I see!" says humanity, "that's quite simple. That *explains* everything."

But scientific laws, properly formulated, never "explain" anything. They simply state, in an abbreviated and generalized form, *what happens*. No scientist, and in my opinion no philosopher, knows *why* anything happens, or can "explain" anything. Scientific laws do nothing except state the brute fact that "when *A* happens, *B* always happens too." And laws of this kind obviously enable us to predict. If certain scientists substituted humps and hills for forces, then they have just substituted one superstition for another. For my part I do not believe that *science* has done this, though some *scientists* may have. For scientists, after all, are human beings with the same craving for "explanations" as other people.

I think that atoms are in exactly the same position as forces and the humps and hills of space-time. In reality the mathematical formulae which are the scientific ways of stating the atomic theory are simply formulae for calculating what sensations will appear in given conditions. But just as the weakness of the human mind demanded that there should correspond to the formula of gravitation a real "thing" which could be called "gravitation itself" or "force," so the same weakness demands that there should be a real thing corresponding to the atomic formulae, and this real thing is called the atom. In reality the atoms no more cause sensations than gravitation causes apples to fall. The only causes of sensations are other sensations. And the relation of atoms to sensations to be felt is not the relation of cause to effect, but the relation of a mathematical formula to the facts and happenings which it enables the mathematician to calculate.

Some writers have said that the physical world has no color, no sound, no taste, no smell. It has no spatiality. Probably it has not even number. We must not suppose that it is in any way like our world, or that we can understand it by attributing to it the characters of our world. Why not carry this progress to its logical conclusion? Why not give up the idea that it has even the character of "existence" which our familiar world has? We have given up smell, color, taste. We have given up even space and shape. We have given up number. Surely, after all that, mere existence is but a little thing to give up. No? Then is it that the idea of existence conveys "a sort of halo"? I suspect so. The "existence" of atoms is but the expiring ghost of the pellet and billiard-ball atoms of our forefathers. They, of course, had size, shape, weight, hardness. These have gone. But thinkers still cling to their existence, just as their fathers clung to the existence of forces, and for the same reason. Their reason is not in the slightest that science has any use for the existent atom. But the *imagination* has. It seems somehow to explain things, to make them homely and familiar.

It will not be out of place to give one more example to show how common fictitious existences are in science, and how little it matters whether they really exist or not. This example has no strange and annoying talk of "bent spaces" about it. One of the foundations of physics is, or used to be, the law of the conservation of energy. I do not know how far, if at all, this has been affected by the theory that matter sometimes turns into energy. But that does not affect the lesson it has for us. The law states, or used to state, that the amount of energy in the universe is always constant, that energy is never either created or destroyed. This was highly convenient, but it seemed to have obvious exceptions. If you throw a stone up into the air, you are told that it exerts in its fall the same amount of energy which it took to throw it up. But suppose it does not fall. Suppose it lodges on the roof of your house and stays there. What has happened to the energy which you can nowhere perceive as being exerted? It seems to have disappeared out of the universe. No, says the scientist, it still exists as *potential* energy. Now what does this blessed word "potential"—which is thus brought in to save the situation—mean as applied to energy? It means, of course, that the energy does not exist in any of its regular "forms," heat, light, electricity, et cetera. But this is merely negative. What positive meaning has the term? Strictly speaking, none whatever. Either the energy exists or it does not exist. There is no realm of the "potential" half-way between existence and nonexistence. And the existence of energy can only consist in its being exerted. If the energy is not being exerted, then it is not energy and does not exist. Energy can no more exist without energizing than heat can exist without being hot. The "potential" existence of the energy is, then, a fiction. The actual empirically verifiable facts are that if a certain quantity of energy e exists in the universe and then disappears out of the universe (as happens when the stone lodges on the roof), the same amount of energy *e* will always reappear, begin to exist again, in certain known conditions. That is the fact which the law of the conservation of energy actually expresses. And the fiction of potential energy is introduced simply because it is convenient and makes the equations easier to work. They could be worked quite well without it, but would be slightly more complicated. In either case the function of the law is the same. Its object is to apprise us that if in certain conditions we have certain perceptions (throwing up the stone), then in certain other conditions we shall get certain other perceptions (heat, light, stone hitting skull, or other such). But there will always be a temptation to hypostatize the potential energy as an "existence," and to believe that it is a "cause" which "explains" the phenomena.

If the views which I have been expressing are followed out, they will lead to the conclusion that, strictly speaking, *nothing exists except sensations* (and the minds which perceive them). The rest is mental construction or fiction. But this does not mean that the conception of a star or the conception of an electron are worthless or untrue. Their truth and value consist in their capacity for helping us to organize our experience and predict our sensations.

21

Do Sub-Microscopic Entities Exist?

Stephen Toulmin

Nonscientists are often puzzled to know whether the electrons, genes, and other entities scientists talk about are to be thought of as really existing or not. Scientists themselves also have some difficulty in saying exactly where they stand on this issue. Some are inclined to insist that all these things are just as real, and exist in the same sense as tables and chairs and omnibuses. But others feel a certain embarrassment about them, and hesitate to go so far; they notice the differences between establishing the existence of electrons from a study of electrical phenomena, inferring the existence of savages from depressions in the sand, and inferring the existence of an inflamed appendix from a patient's signs and symptoms; and it may even occur to them that to talk about an electromagnet in terms of "electrons" is a bit like talking of Pyrexia of Unknown Origin when the patient has an unaccountable temperature. Yet the theory of electrons does *explain* electrical phenomena in a way in which no mere translation into jargon, like "pyrexia" can explain a sick man's temperature; and how, we may ask, could the electron theory work at all if, after all, electrons did not really exist?

Stated in this way, the problem is confused: let us therefore scrutinize the question itself a little more carefully. For when we compare Robinson Crusoe's discovery with the physicist's one, it is not only the sorts of discovery which are different in the two cases. To talk of existence in both cases involves quite as much of a shift, and by passing too swiftly from one use of the word to the other we may make the problem unnecessarily hard for ourselves.

Notice, therefore, what different ideas we may have in mind when we talk about things "existing." If we ask whether dodos exist or not, i.e., whether there are any dodos left nowadays, we are asking whether the species has survived or is extinct. But when we ask whether electrons exist or not, we certainly do not have in mind the possibility that they may have become extinct: in whatever sense we ask this question, it is not one in which "exists" is opposed to "does not exist any more." Again, if we ask whether Ruritania exists, i.e., whether there is such a country as Ruritania, we are asking whether there really is such a country as Ruritania or whether it is an imaginary, and so a nonexistent country. But we are not interested in asking of electrons whether they are genuine instances of a familiar sort of thing or nonexistent ones: the way in which we are using the term "exist" is not one in which it is opposed to "are nonexistent." In each case, the word "exist" is used to make a slightly different point, and to mark a slightly different distinction. As one moves from Man Friday to dodos, and on from them to Ruritania, and again to electrons, the change in the nature of the cases brings other changes with it: notably in the way one has to understand sentences containing the word "exist."

What, then, of the question, "Do electrons exist?" How is this to be understood? A more revealing analogy than dodos or Ruritania is to be found in the question, "Do contours exist?" A child who had read that the equator was "an imaginary line drawn round the center of the earth" might be struck by the contours, parallels of latitude, and the rest, which appear on maps along with the towns, mountains, and rivers, and ask of them whether *they* existed. How should we reply? If he asked his question in bare words, "Do contours exist?" one could hardly answer him immediately: clearly the only answer one can give to this question is "Yes and No." They "exist" all right, but do they *exist*? It all depends on your manner of speaking. So he might be persuaded to restate his question, asking now, "Is there really a line on the ground whose height is constant?"; and again the answer would have to be "Yes and No," for there is (so to say) a "line," but then again not what you might call a *line*. . . . And so the cross-purposes would continue until it was made clear that the real question was: "Is there anything to show for contours—anything visible on the terrain, like the white lines on a tennis court? Or are they only cartographical devices, having no geographical counterparts?" Only then would the question be posed in anything like an unambiguous manner. The sense of "exists" in which a child might naturally ask whether contours existed is accordingly one in which "exists" is opposed not to "does not exist any more" or to "is nonexistent," but to "is only a (cartographical) fiction."

This is very much the sense in which the term "exists" is used of atoms, genes, electrons, fields, and other theoretical entities in the physical sciences. There, too, the question "Do they exist?" has in practice the force of "Is there anything to show for them, or are they only theoretical fictions?" To a working physicist, the question "Do neutrinos exist?" acts as an invitation to "produce a neutrino," preferably by making it *visible*. If one could do this one would indeed have something to show for the term "neutrino," and the difficulty of doing it is what explains the peculiar difficulty of the problem. For the problem arises acutely only when we start asking about the existence of *submicroscopic* entities, i.e., things which by all normal standards are invisible. In the nature of the case, to produce a neutrino must be a more sophisticated business than producing a dodo or a nine-foot man. Our problem is accordingly complicated by the need to decide what is to count as "producing" a neutrino, a field, or a gene. It is not obvious what sorts of thing ought to count: certain things are, however, generally regarded by scientists as acceptable—for instance, cloud-chamber pictures of α-ray tracks, electron microscope photographs or, as a second-best, audible clicks from a Geiger counter. They would regard such striking demonstrations as these as sufficiently like being shown a live dodo on the lawn to qualify as evidence of the existence of the entities concerned. And certainly, if we reject these as insufficient, it is hard to see what more we can reasonably ask for: if the term "exists" is to have any application to such things, must not this be it?

What if no such demonstration were possible? If one could not show, visibly, that neutrinos existed, would that necessarily be the end of them? Not at all; and it is worth noticing what happens when a demonstration of the preferred type is not possible, for then the difference between talking about the existence of electrons or genes, and talking about the existence of dodos, unicorns or nine-foot men becomes all-important. If, for instance, I talk plausibly about unicorns or nine-foot men and have nothing to show for them, so that I am utterly unable to say, when challenged, under what circumstances a specimen might be, or might have been seen, the conclusion may reasonably be drawn that my nine-foot men are imaginary and my unicorns a myth. In either case, the things I am talking about may be presumed to be nonexistent, i.e., are discredited and can be written off. But in the case of atoms, genes, and the like, things are different: the failure to bring about or describe circumstances in which one might point and say, "There's one!" need not, as with unicorns, be taken as discrediting them.

Not all those theoretical entities which cannot be shown to exist need be held to be nonexistent: there is for them a middle way. Certainly we should hesitate to assert that any theoretical entity really existed until a photograph or other demonstration had been given. But, even if we had reason to believe that no such demonstration ever could be given, it would be too much to conclude that the entity was nonexistent; for this conclusion would give the impression of discrediting something that, as a fertile explanatory concept, did not necessarily deserve to be discredited. To do so would be like refusing to take any notice of contour lines because there were no visible marks corresponding to them for us to point to on the ground. The conclusion that the notion must be dropped would be justified only if, like "phlogiston," "caloric fluid," and the "ether," it had also lost all explanatory fertility. No doubt scientists would be happy if they could refer in their explanations only to entities which could be shown to exist, but at many stages in the development of science it would have been crippling to have insisted on this condition too rigorously. A scientific theory is often accepted and in circulation for a long time, and may have to advance for quite a long way, before the question of the real existence of the entities appearing in it can even be posed.

The history of science provides one particularly striking example of this. The whole of theoretical physics and chemistry in the nineteenth century was developed round the notions of atoms and molecules: both the kinetic theory of matter, whose contribution to physics was spectacular, and the theory of chemical combinations and reactions, which turned chemistry into an exact science, made use of these notions, and could hardly have been expounded except in terms of them. Yet not until 1905 was it definitively shown by Einstein that the phenomenon of Brownian motion could be regarded as a demonstration that atoms and molecules really existed. Until that time, no such demonstration had ever been recognized, and even a Nobel prize-winner like Ostwald, for whose work as a chemist the concepts "atom" and "molecule" must have been indispensable, could be skeptical until then about the reality of atoms. Moreover by 1905 the atomic theory had ceased to be the last word in physics: some of its foundations were being severely attacked, and the work of Niels Bohr and J. J. Thomson was beginning to alter the physicist's whole picture of the constitution of matter. So, paradoxically, one finds that the major triumphs of the atomic theory were achieved at a time when even the greatest scientist could regard the idea of atoms as hardly more than a useful fiction, and that atoms were definitely shown to exist only at a time when the classical atomic theory was beginning to lose its position as the basic picture of the constitution of matter.

Evidently, then, it is a mistake to put questions about the reality or existence of theoretical entities too much in the center of the picture. In accepting a theory scientists need not, to begin with, answer these questions either way: certainly they do not, as Kneale suggests, commit themselves thereby to a belief in the existence of all the things in terms of which the theory is expressed. To suppose this is a variant of the Man Friday fallacy. In fact, the question whether the entities spoken of in a theory exist or not is one to which we may not even be able to give a meaning until the theory has some accepted position. The situation is rather like that we encountered earlier in connection with the notion of light traveling. It may seem natural to suppose that a physicist who talks of light as traveling must make some assumptions about what it is that is traveling: on investigation, however, this turns out not to be so, for the question, what it is that is traveling, is one which cannot even be asked without going beyond the phenomena which the notion is originally used to explain. Likewise, when a scientist adopts a new theory, in which novel concepts are introduced (waves, electrons, or genes), it may seem natural to suppose that he is committed to a belief in the existence of the things in terms of which his explanations are expressed. But again, the question whether genes, say, really exist takes us beyond the original phenomena explained in terms of "genes." To the scientist, the real existence of his theoretical entities is contrasted with their being only useful theoretical fictions: the fact of an initial explanatory success may therefore leave the question of existence open.

There is a converse to this form of the Man Friday fallacy. Having noticed that a theory may be accepted long before visual demonstrations can be produced of the existence of the entities involved, we may be tempted to conclude that such things as cloud-chamber photographs are rather overrated: in fact, that they only seem to bring us nearer to the things of which the physicist speaks as a result of mere illusion. This is a conclusion which Kneale has advanced, on the ground that physical theories do not stand or fall by the results obtained from cloud-chambers and the like rather then by the results of any other physical experiments. But this is still to confuse two different questions, which may be totally independent: the question of the acceptability of the theories and the question of the reality of the theoretical entities. To regard cloud-chamber photographs as showing us that electrons and α-particles really exist need not mean giving the cloud-chamber a preferential status among our grounds for accepting current theories of atomic structure. These theories were developed and accepted before the cloud-chamber was, or indeed could have been invented. Nevertheless, it was the cloud-chamber which first showed in a really striking manner just how far nuclei, electrons, α-particles, and the rest could safely be thought of as real things; that is to say, as more than explanatory fictions.

22

The Ontological Status of Theoretical Entities

Grover Maxwell

That anyone today should seriously contend that the entities referred to by scientific theories are only convenient fictions, or that talk about such entities is translatable without remainder into talk about sense contents or everyday physical objects, or that such talk should be regarded as belonging to a mere calculating device and, thus, without cognitive content—such contentions strike me as so incongruous with the scientific and rational attitude and practice that I feel this paper *should* turn out to be a demolition of straw men. But the instrumentalist views of outstanding physicists such as Bohr and Heisenberg are too well known to be cited, and in a recent book of great competence, Professor Ernest Nagel concludes that "the opposition between [the realist and the instrumentalist] views [of theories] is a conflict over preferred modes of speech" and "the question as to which of them is the 'correct position' has only terminological interest."[1] The phoenix, it seems, will not be laid to rest.

The literature on the subject is, of course, voluminous, and a comprehensive treatment of the problem is far beyond the scope of one essay. I shall limit myself to a small number of constructive arguments (for a radically realistic interpretation of theories) and to a critical examination of some of the more crucial assumptions (sometimes tacit, sometimes explicit) that seem to have generated most of the problems in this area.[2]

THE PROBLEM

Although this essay is not comprehensive, it aspires to be fairly self-contained. Let me, therefore, give a pseudohistorical introduction to the problem with a piece of science fiction (or fictional science).

In the days before the advent of microscopes, there lived a Pasteur-like scientist whom, following the usual custom, I shall call Jones. Reflecting on the fact that certain diseases seemed to be transmitted from one person to another by means of bodily contact or by contact with articles handled previously by an afflicted person, Jones began to speculate about the mechanism of the transmission. As a "heuristic crutch," he recalled that there is an obvious *observable* mechanism for transmission of certain afflictions (such as body lice), and he postulated that all, or most, infectious diseases were spread in a similar manner but that in most cases the corresponding "bugs" were too small to be seen and, possibly, that some of them lived inside the bodies of their hosts. Jones proceeded to develop his theory and to examine its testable consequences. Some of these seemed to be of great importance for preventing the spread of disease.

After years of struggle with incredulous recalcitrance, Jones managed to get some of his preventative measures adopted. Contact with or proximity to diseased persons was avoided when possible, and articles which they handled were "disinfected" (a word coined by Jones) either by means of high temperatures or by treating them with certain toxic preparations which Jones termed "disinfectants." The results were spectacular: within ten years the death rate had declined 40 percent. Jones and his theory received their well-deserved recognition.

However, the "crobes" (the theoretical term coined by Jones to refer to the disease-producing organisms) aroused considerable anxiety among many of the philosophers and philosophically inclined scientists of the day. The expression of this anxiety usually began something like this: "In order to account for the facts, Jones must assume that his crobes are too small to be seen. Thus the very postulates of his theory preclude their being observed; they are *unobservable in principle*." (Recall that no one had envisaged such a thing as a microscope.) This common prefatory remark was then followed by a number of different "analyses" and "interpretations" of Jones's theory. According to one of these, the tiny organisms were merely convenient fictions—*façons de parler*—extremely useful as heuristic devices for facilitating (in the "context of discovery") the thinking of scientists but not to be taken seriously in the sphere of cognitive knowledge (in the "context of justification"). A closely related view was that Jones's theory was merely an instrument, useful for organizing observation statements and (thus) for producing desired results, and that, therefore, it made no more sense to ask what was the nature of the entities to which it referred than it did to ask what was the nature of the entities to which a hammer or any other tool referred.[3] "Yes," a philosopher might have said, "Jones's theoretical expressions are just meaningless sounds or marks on paper which, when correlated with observation sentences by appropriate syntactical rules, enable us to predict successfully and otherwise organize data in a convenient fashion." These philosophers called themselves "instrumentalists."

According to another view (which, however, soon became unfashionable), although expressions containing Jones's theoretical terms were genuine sentences, they were translatable without remainder into a set (perhaps infinite) of observation sentences. For example, "There are crobes of disease X on this article" was said to translate into something like this: 'If a person handles this article without taking certain precautions, he will (probably) contact disease X; and if this article is first raised to a high temperature, then if a person handles it at any time afterward, before it comes into contact with another person with disease X, he will (probably) not contract disease X; and . . ."

Now virtually all who held any of the views so far noted granted, even insisted, that theories played a useful and legitimate role in the scientific enterprise. Their concern was the elimination of "pseudoproblems" which might arise, say, when one began wondering about the "reality of supraempirical entities," et cetera. However, there was also a school of thought, founded by a psychologist named Pelter, which differed in an interesting manner from such positions as these. Its members held that while Jones's crobes might very well exist and enjoy "full-blown reality," they should not be the concern of medical research at all. They insisted that if Jones had employed the correct methodology, he would have discovered, even sooner and with much less effort, all of the observation laws related to disease contraction, transmission, et cetera without introducing superfluous links (the crobes) into the causal chain.

Now, lest any reader find himself waxing impatient, let me hasten to emphasize that this crude parody is not intended to convince anyone, or even to cast serious doubt upon sophisticated varieties of any of the reductionistic positions caricatured (some of them not too severely, I would contend) above. I am well aware that there are theoretical entities and theoretical entities, some of whose conceptual and theoretical statuses differ in important respects from Jones's crobes. (I shall discuss some of these later.) Allow me, then, to bring the Jonesean prelude to our examination of observability to a hasty conclusion.

Now Jones had the good fortune to live to see the invention of the compound microscope. His crobes were "observed" in great detail, and it became possible to identify the specific kind of *microbe* (for so they began to be called) which was responsible for each different disease. Some philosophers freely admitted error and were converted to realist positions concerning theories. Others resorted to subjective idealism or to a thoroughgoing phenomenalism, of which there were two principal varieties. According to one, the one "legitimate" observation language had for its descriptive terms only those which referred to sense data. The other maintained the stronger thesis that *all* "factual" statements were *translatable* without remainder into the sense-datum language. In either case, any two nonsense data (e.g., a theoretical entity and what would ordinarily be called an "observable physical object") had virtually the same status. Others contrived means of modifying their views much less drastically. One group maintained that Jones's crobes actually never had been unobservable in principle, for, they said, the theory did not imply the impossibility of finding a means (e.g., the microscope) of observing them. A more radical contention was that the crobes were not observed at all; it was argued that what was seen by means of the microscope was just a shadow or an image rather than a corporeal organism.

THE OBSERVATIONAL-THEORETICAL DICHOTOMY

Let us turn from these fictional philosophical positions and consider some of the actual ones to which they roughly correspond. Taking the last one first, it is interesting to note the following passage from Bergmann: "But it is only fair to point out that if this . . . methodological and terminological analysis [for the thesis that there are no atoms] . . . is strictly adhered to, even stars and microscopic objects are not physical things in a literal sense, but merely by courtesy of language and pictorial imagination. This might seem awkward. But when I look through a microscope, all I see is a patch of color which creeps through the field like a shadow over a wall. And a shadow, though real, is certainly not a physical thing."[4]

I should like to point out that it is also the case that if this analysis is strictly adhered to, we cannot observe physical things through opera glasses, or even through ordinary spectacles, and one begins to wonder about the status of what we see through an ordinary windowpane. And what about distortions due to temperature gradients—however small and, thus, always present—in the ambient air? It really *does* "seem awkward" to say that when people who wear glasses describe what they see they are talking about shadows, while those who employ unaided vision talk about physical things—or that when we look through a windowpane, we can only *infer* that it is raining, while if we raise the window, we may "observe directly" that it is. The point I am making is that there is, in principle, a continuous series beginning with looking through a vacuum and containing these as members: looking through a windowpane, looking through glasses, looking through binoculars, looking through a low-power microscope, looking through a high-power microscope, et cetera, in the order given. The important consequence is that, so far, we are left without criteria which would enable us to draw a nonarbitrary line between "observation" and "theory." Certainly, we will often find it convenient to draw such a to-some-extent-arbitrary line; but its position will vary widely from context to context. (For example, if we are determining the resolving characteristics of a certain microscope, we would certainly draw the line beyond ordinary spectacles, probably beyond simple magnifying glasses, and possibly beyond another microscope with a lower power of resolution.) But what ontological ice does a mere methodologically convenient observational-theoretical dichotomy cut? Does an entity attain physical thinghood and/or "real existence" in one context only to lose it in another? Or, we may ask, recalling the continuity from observable to unobservable, is what is seen through spectacles a "little bit less real" or does it "exist to a slightly less extent" than what is observed by unaided vision?[5]

However, it might be argued that things seen through spectacles and binoculars look like ordinary physical objects, while those seen through microscopes and telescopes look like shadows and patches of light. I can only reply that this does not seem to me to be the case, particularly when looking at the moon, or even Saturn, through a telescope or when looking at a small, though "directly observable," physical object through a low-power microscope. Thus, again, a continuity appears.

"But," it might be objected, "theory tells us that what we see by means of a microscope is a real image, which is certainly distinct from the object on the stage." Now first of all, it should be remarked that it seems odd that one who is espousing an austere empiricism which requires a sharp observational-language/theoretical-language distinction (and one in which the former language has a privileged status) should need a theory in order to tell him what is observable. But, letting this pass, what is to prevent us from saying that we still observe the object on the stage, even though a "real image" may be involved? Otherwise, we shall be strongly tempted by phenomenalistic demons, and at this point we are considering a physical-object observation language rather than a sense-datum one. (Compare the traditional puzzles: Do I see one physical object or two when I punch my eyeball? Does one object split into two? Or do I see one object and one image? Etc.)

Another argument for the continuous transition from the observable to the unobservable (theoretical) may be adduced from theoretical considerations themselves. For example, contemporary valency theory tells us that there is a virtually continuous transition from very small molecules (such as those of hydrogen) through "medium-sized" ones (such as those of the fatty acids, polypeptides, proteins, and viruses) to extremely large ones (such as crystals of the salts, diamonds, and lumps of polymeric plastic). The molecules in the last-mentioned group are macro, "directly observable" physical objects but are, nevertheless, genuine, single molecules; on the other hand, those in the first mentioned group have the same perplexing properties as subatomic particles (de Broglie waves, Heisenberg indeterminacy, etc.). Are we to say that a large protein molecule (e.g., a virus) which can be "seen" only with an electron microscope is a little less real or exists to somewhat less an extent than does a molecule of a polymer which can be seen with an optical microscope? And does a hydrogen molecule partake of only an infinitesimal portion of existence or reality? Although there certainly *is* a continuous transition from observability to unobservability, any talk of such a continuity from full-blown existence to nonexistence is, clearly, nonsense.

Let us now consider the next to last modified position which was adopted by our fictional philosophers. According to them, it is only those entities which are *in principle* impossible to observe that present special problems. What kind of impossibility is meant here? Without going into a detailed discussion of the various types of impossibility, about which there is abundant literature with which the reader is no doubt familiar, I shall assume what usually seems to be granted by most philosophers who talk of entities which are unobservable in principle—i.e., that the theory(s) itself (coupled with a physiological theory of perception, I would add) entails that such entities are unobservable.

We should immediately note that if this analysis of the notion of unobservability (and, hence, of observability) is accepted, then its use as a means of delimiting the observation language seems to be precluded for those philosophers who regard theoretical expressions as elements of a calculating device—as meaningless strings of symbols. For suppose they wished to determine whether or not "electron" was a theoretical term. First, they must see whether the theory entails the sentence "Electrons are unobservable." So far, so good, for their calculating devices are said to be able to select genuine sentences, provided they contain no theoretical terms. But what about the selected "sentence" itself? Suppose that "electron" is an observation term. It follows that the expression is a genuine sentence and asserts that electrons are unobservable. But this entails that "electron" is *not* an observation term. Thus if "electron" is an observation term, then it is *not* an observation term. Therefore it is not an observation term. But then it follows that "Electrons are unobservable" is not a genuine sentence and does not assert that electrons are unobservable, since it is a meaningless string of marks and does not assert anything whatever. Of course, it could be stipulated that when a theory "selects" a meaningless expression of the form "*X*s are unobservable," then "*X*" is to be taken as a theoretical term. But this seems rather arbitrary.

But, assuming that well-formed theoretical expressions are genuine sentences, what shall we say about unobservability in principle? I shall begin by putting my head on the block and argue that the present-day status of, say, electrons is in many ways similar to that of Jones's crobes before microscopes were invented. I am well aware of the numerous theoretical arguments for the impossibility of observing electrons. But suppose new entities are discovered which interact with electrons in such a mild manner that if an electron is, say, in an eigenstate of position, then, in certain circumstances, the interaction does not disturb it. Suppose also that a drug is discovered which vastly alters the human perceptual apparatus—perhaps even activates latent capacities so that a new sense modality emerges. Finally, suppose that in our altered state we are able to perceive (not necessarily visually) by means of these new entities in a manner roughly analogous to that by which we now see by means of photons. To make this a little more plausible, suppose that the energy eigenstates of the electrons in some of the compounds present in the relevant perceptual organ are such that even the weak interaction with the new entities alters them and also that the cross sections, relative to the new entities, of the electrons and other particles of the gases of the air are so small that the chance of any interaction here is negligible. Then we might be able to "observe directly" the position and possibly the approximate diameter and other properties of some electrons. It would follow, of course, that quantum theory would have to be altered in some respects, since the new entities do not conform to all its principles. But however improbable this may be, it does not, I maintain, involve any logical or conceptual absurdity. Furthermore, the modification necessary for the inclusion of the new entities would not necessarily change the meaning of the term "electron."[6]

Consider a somewhat less fantastic example, and one which does not involve any change in physical theory. Suppose a human mutant is born who is able to "observe" ultraviolet radiation, or even X-rays, in the same way we "observe" visible light.

Now I think that it is extremely improbable that we will ever observe electrons directly (i.e., that it will ever be reasonable to assert that we have so observed them). But this is neither here nor there; it is not the purpose of this essay to predict the future development of scientific theories, and, hence, it is not its business to decide what actually is observable or what will become observable (in the more or less intuitive sense of "observable" with which we are now working). After all, we are operating, here, under the assumption that it is theory, and thus science itself, which tells us what is or is not, in this sense, observable (the "in principle" seems to have become superfluous). And this is the heart of the matter; for it follows that, at least for this sense of "observable," there are no a priori or philosophical criteria for separating the observable from the unobservable. By trying to show that we can talk about the *possibility* of observing electrons without committing logical or conceptual blunders, I have been trying to support the thesis that any (nonlogical) term is a possible candidate for an observation term.

There is another line which may be taken in regard to delimitation of the observation language. According to it, the proper term with which to work is not "'observable" but, rather "observed." There immediately comes to mind the tradition beginning with Locke and Hume (No idea without a preceding impression!), running through logical atomism and the principle of acquaintance, and ending (perhaps) in contemporary positivism. Since the numerous facets of this tradition have been extensively examined and criticized in the literature, I shall limit myself here to a few summary remarks.

Again, let us consider at this point only observation languages which contain ordinary physical-object terms (along with observation predicates, etc., of course). Now, according to this view, all descriptive terms of the observation language must refer to that which has been observed. How is this to be interpreted? Not too narrowly, presumably, otherwise each language user would have a different observation language. The name of my Aunt Mamie, of California, whom I have never seen, would not be in my observation language, nor would "snow" be an observation term for many Floridians. One could, of course, set off the observation language by means of this awkward restriction, but then, obviously, not being the referent of an observation term would have no bearing on the ontological status of Aunt Mamie or that of snow.

Perhaps it is intended that the referents of observation terms must be members of a *kind* some of whose members have been observed or instances of a *property* some of whose instances have been observed. But there are familiar difficulties here. For example, given any entity, we can always find a kind whose only member is the entity in question; and surely expressions such as "men over fourteen feet tall" should be counted as observational even though no instances of the "property" of being a man over fourteen feet tall have been observed. It would seem that this approach must soon fall back upon some notion of simples or determinables vs. determinates. But is it thereby saved? If it is held that only those terms which refer to observed simples or observed determinates are observation terms, we need only remind ourselves of such instances as Hume's notorious missing shade of blue. And if it is contended that in order to be an observation term an expression must at least refer to an observed determinable, then we can always find such a determinable which is broad enough in scope to embrace any entity whatever. But even if these difficulties can be circumvented, we see (as we knew all along) that this approach leads inevitably into phenomenalism, which is a view with which we have not been concerning ourselves.

Now it is not the purpose of this essay to give a detailed critique of phenomenalism. For the most part, I simply assume that it is untenable, at least in any of its translatability varieties.[7] However, if there are any unreconstructed phenomenalists among the readers, my purpose, insofar as they are concerned, will have been largely achieved if they will grant what I suppose most of them would stoutly maintain anyway, i.e., that theoretical entities are no worse off than so-called observable physical objects.

Nevertheless, a few considerations concerning phenomenalism and related matters may cast some light upon the observational-theoretical dichotomy and, perhaps, upon the nature of the "observation language." As a preface, allow me some overdue remarks on the latter. Although I have contended that the line between the observable and the unobservable is diffuse, that it shifts from one scientific problem to another, and that it is constantly being pushed toward the "unobservable" end of the spectrum as we develop better means of observation—better instruments—it would, nevertheless, be fatuous to minimize the importance of the observation base, for it is absolutely necessary as a confirmation base for statements which do refer to entities which are unobservable at a given time. But we should take as its basis and its unit not the "observational term" but, rather, the quickly decidable sentence. (I am indebted to Feyerabend, *loc. cit.*, for this terminology.) A quickly decidable sentence (in the technical sense employed here) may be defined as a singular, nonanalytic sentence such that a reliable, reasonably sophisticated language user can very quickly decide[8] whether to assert it or deny it when he is reporting on an occurrent situation. "Observation term" may now be defined as a "descriptive (nonlogical) term which may occur in a quickly decidable sentence," and "observation sentence" as a "sentence whose only descriptive terms are observation terms."

Returning to phenomenalism, let me emphasize that I am not among those philosophers who hold that there are no such things as sense contents (even sense data), nor do I believe that they play no important role in our perception of "reality." But the fact remains that the referents of most (not all) of the statements of the linguistic framework used in everyday life and in science are *not* sense contents but, rather, physical objects and other publicly observable entities. Except for pains, odors, "inner states," et cetera, *we do not usually observe sense contents*; and although there is good reason to believe that they play an indispensable role in observation, *we are usually not aware of them when we visually* (or tactilely) observe physical objects. For example, when I observe a distorted, obliquely reflected image in a mirror, I may seem to be seeing a baby elephant standing on its head; later I discover it is an image of Uncle Charles taking a nap with his mouth open and his hand in a peculiar position. Or, passing my neighbor's home at a high rate of speed, I observe that he is washing a car. If asked to report these observations I could quickly and easily report a baby elephant and a washing of a car; I probably would not, without subsequent observations, be able to report what colors, shapes, et cetera (i.e., what sense data) were involved.

Two questions naturally arise at this point. How is it that we can (sometimes) quickly decide the truth or falsity of a pertinent observation sentence? and, What role do sense contents play in the appropriate tokening of such sentences? The heart of the matter is that these are primarily scientific-theoretical questions rather than "purely logical," "purely conceptual," or "purely epistemological." If theoretical physics, psychology, neurophysiology, etc., were sufficiently advanced, we could give satisfactory answers to these questions, using, in all likelihood, the physical-thing language as our observation language and *treating sensations, sense contents, sense data, and "inner states" as theoretical* (yes, theoretical!) *entities.*[9]

It is interesting and important to note that, even before we give completely satisfactory answers to the two questions considered above, we can, with due effort and reflection, train ourselves to "observe directly" what were once theoretical entities—the sense contents (color sensations, etc.)—involved in our perception of physical things. As has been pointed out before, we can also come to observe other kinds of entities which were once theoretical. Those which most readily come to mind involve the use of instruments as aids to observation. Indeed, using our painfully acquired theoretical knowledge of the world, we come to see that we "directly observe" many kinds of so-called theoretical things. After listening to a dull speech while sitting on a hard bench, we begin to become poignantly aware of the presence of a considerably strong gravitational field, and as Professor Feyerabend is fond of pointing out, if we were carrying a heavy suitcase in a changing gravitational field, we could observe the changes of the $G\mu\nu$, of the metric tensor.

I conclude that our drawing of the observational-theoretical line at any given point is an accident and a function of our physiological makeup, our current state of knowledge, and the instruments we happen to have available and, therefore, that it has no ontological significance whatever. . . .

NOTES

1. E. Nagel, *The Structure of Science* (New York: Harcourt, Brace, and World, 1961), ch. 6.

2. For the genesis and part of the content of some of the ideas expressed herein, I am indebted to a number of sources; some of the more influential are H. Feigl, "Existential Hypotheses," *Philosophy of Science* 17: 35–62 (1950); P. K. Feyerabend, "An Attempt at a Realistic Interpretation of Experience," *Proceedings of the Aristotelian Society* 58: 144–170 (1958); N. R. Hanson, *Patterns of Discovery* (Cambridge: Cambridge University Press, 1958); E. Nagel, loc. cit; Karl Popper, *The Logic of Scientific Discovery* (London: Hutchinson, 1959); M. Scriven, "Definitions, Explanations, and Theories," in *Minnesota Studies in the Philosophy of Science,* vol. 2, ed. H. Feigl, M. Scriven, and G. Maxwell (Minneapolis: University of Minnesota Press, 1958); Wilfrid Sellars, "Empiricism and the Philosophy of Mind," in *Minnesota Studies in the Philosophy of Science,* vol. 1, ed. H. Feigl and M. Scriven (Minneapolis: University of Minnesota Press, 1956); and "The Language of Theories," in *Current Issues in the Philosophy of Science,* ed. H. Feigl and G. Maxwell (New York: Holt, Rinehart, and Winston, 1961).

3. I have borrowed the hammer analogy from E. Nagel, "Science and [Feigl's] Semantic Realism," *Philosophy of Science* 17: 174–181 (1950), but it should be pointed out that Professor Nagel makes it clear that he does not necessarily subscribe to the view which he is explaining.

4. G. Bergmann, "Outline of an Empiricist Philosophy of Physics," *American Journal of Physics,* 2 (1943): 248–258, 335–342, reprinted in *Readings in the Philosophy of Science,* ed. H. Feigl and M. Brodbeck (New York: Appleton-Century-Crofts, 1953), pp. 262–287.

5. I am not attributing to Professor Bergmann the absurd views suggested by these questions. He seems to take a sense-datum language as his observation language (the base of what he called "the empirical hierarchy"), and, in some ways, such a position is more difficult to refute than one which purports to take an "observable-physical-object" view. However, I believe that demolishing the straw men with which I am now dealing amounts to desirable preliminary "therapy." Some nonrealist interpretations of theories which embody the presupposi-

tion that the observable-theoretical distinction is sharp and ontologically crucial seem to me to entail positions which correspond to such straw men rather closely.

6. For arguments that it is possible to alter a theory without altering the meanings of its terms, see my "Meaning Postulates in Scientific Theories" in *Current Issues in the Philosophy of Science,* ed. Feigl and Maxwell.

7. The reader is no doubt familiar with the abundant literature concerned with this issue. See, for example, Sellars's "Empiricism and the Philosophy of Mind," which also contains references to other pertinent works.

8. We may say "noninferentially" decide, provided this is interpreted liberally enough to avoid starting the entire controversy about observability all over again.

9. Cf. Sellars, "Empiricism and the Philosophy of Mind." As Professor Sellars points out, this is the crux of the "other-minds" problem. Sensations and inner states (relative to an intersubjective observation language, I would add) are theoretical entities (and they "really exist") and *not* merely actual and/or possible behavior. Surely it is the unwillingness to countenance theoretical entities—the hope that every sentence is translatable not only into some observation language but into the physical-thing language—which is responsible for the "logical behaviorism" of the neo-Wittgensteinians.

23

Is There a Significant Observational-Theoretical Distinction?

Carl A. Matheson and A. David Kline

Introduction

Twentieth-century philosophy of science has been dominated by efforts to develop and criticize the two-tier conception of scientific theories.[1] Given the most optimistic prognosis the view was dead by 1970 and seriously ill for at least a decade prior to that. Nevertheless the view still has profound effects. Much of the now fashionable debates over scientific progress and realism/antirealism get their wind from the critique of the two-tier conception and the lack of a well-developed, coherent, alternative story.

At the quick of the old view is an observational-theoretical distinction. This essay will be concerned with two very important questions: (1) What led to the demise of the observational-theoretical distinction? Our concern will not be so much with detailed criticisms but rather structural defects. We want to show how the distinction did not realize the goals that motivated it. (2) How does the distinction spawn the realism/antirealism controversy? Furthermore given its demise what is left of the controversy? More particularly, section I gives a standard way of making the distinction plus an account of what motivates it. In section II it is shown how the distinction plus the framework that motivates it leads to what we call the "old" realism issue. After the critique of the distinction in section III, we show in section IV how the "new" realism issue is born. In the final section some general remarks are made on what is left of the distinction. Does it retain any philosophical importance?

I

Defenders of the two-tier conception begin with an intuitive or preanalytical belief that terms such as "desk" and "brown" are importantly different from

374

"electron" and "valence." Surprisingly little effort has been spent on developing a careful criterion for making the distinction. The following by Carl Hempel is an instance of a common and very influential way of making the distinction:

> In regard to an observational term it is possible, under suitable circumstances, to decide by means of direct observation whether the term does or does not apply to a given situation. . . . Theoretical terms, on the other hand, usually purport to refer to not directly observable entities and their characteristics.[2]

What are "suitable" circumstances? How are we to understand "direct observation"? It is not plausible to think that the logical empiricists were unaware of these difficulties. Their lack of effort in sharpening the criterion can perhaps be attributed to the belief that, no matter how the distinction finally gets put, there will be the issue of how observational and theoretical terms are related. Furthermore this latter issue is more suitably investigated by the methods of the philosophers.[3]

Rather than rehearse ways of formulating the observational-theoretical distinction, we shall consider the more fruitful and fundamental issues that motivate the distinction. There are in fact three interrelated motivations. They concern issues pertaining to meaning, justification, and theory comparison.

The explanans of adequate explanations must have empirical significance. Consider the following example:

> Why has the water cooled after being placed over a flame then removed? "The water had initially been deprived of its glubbification through action of the flame (which generally has this effect); removed from the flame, it has become increasingly glubbified again and hence cooler, in accordance with the general principle that glubbification varies inversely with warmth."[4]

One suspects that the problem with the above is deeper than that some of the claims are merely false. They lack meaning or significance. This sort of case would not stump anyone but, more interestingly, violations of meaning occur in serious science. Bishop Berkeley, an eighteenth-century divine and philosopher, had a number of worries about the physics of his day.

> In works on motion by more recent and sober thinkers of our age, not a few terms of somewhat abstract and obscure signification are used, such as *solicitation of gravity, urge, dead forces,* etc., terms which darken writings in other respects very learned, and beget opinions at variance with truth and the common sense of men. These terms must be examined with great care. . . .[5]

Without being so arrogant as to actually doubt the significance of some of the claims of contemporary physics or psychology, we might ask what provides for the significance of talk about charm and attenuators. More generally what provides for the meaningfulness of theoretical discourse?

Logical empiricism provides a general answer.

> It is a basic principle of contemporary empiricism that a sentence makes a cognitively significant assertion, and thus can be said to be true or false, if and only if either (1) it is analytic or contradictory—in which case it is said to have purely logical meaning or significance—or else (2) it is capable, at least potentially, of test by experiential evidence—in which case it is said to have empirical meaning or significances.[6]

Notice that the empiricist account of meaningfulness connects issues of meaning with issues of knowledge or justification. We can now see clearly what motivates a sharp observational-theoretical distinction. In contrast to what is problematic, i.e., theoretical discourse, observation reports or statements appear to be transparently meaningful in virtue of their obvious openness to testing.[7] The project for the philosopher will be to connect the observational components of a theory, by some to be specified relation, with the theoretical components in order to insure the testability and hence meaningfulness of the latter. This is precisely the project of the two-tier conception of science. It all rests on exploiting the supposed virtues of the observational domain.

Another motivation concerns providing a framework in which theory comparison is possible. If T and T^* are to be compared, we must find some implication of T, say, I, for which T^* implies I or the negation of I. It must be the case that (i) I has the same meaning for proponents of T and T^*, otherwise I and its negation would not be contradictory, and (ii) the truth or falsity of I must be determinable independently of T and T^*, otherwise the testing procedure would be circular. Furthermore, we want the proponents of T and T^* to be able to come to agreement about the truth or falsity of I and hence the truth or falsity of T and T^*. A sharp observational-theoretical dichotomy would solve all these problems. The truth of observation sentences is easily determinable independently of any theory.

II

As we saw in section I a goal of the traditional conception of theories is to specify the meaning of the theoretical in terms of the observational vocabulary. In other words, we must specify a relation, R, between the terms comprising the observational and theoretical vocabularies. The history of efforts to specify R is very rich. Two importantly different formulations are crucial for setting the context for the "old" realism issue.

Consider a theoretical term, *T*, e.g., "fragile." According to the *explicit definition* account of *R*, necessary and sufficient conditions should be given for the application of *T* in observation terms. Those conditions, to continue the example, may be in terms of "being sharply struck" and "breaking." More precisely consider the following definition:

> Object, *O*, is fragile.
> If and only if
> *O* is sharply struck, then *O* will break

Given this account of *R*, what should we say about the reality of theoretical entities? Let us begin with the view that seems to be embedded in common sense:

REALISM: (i) Theoretical sentences have truth values, i.e., they are either true or false and (ii) theoretical terms can refer to unobservable properties and entities.

If we ask whether theories that require explicit definitions are to be interpreted realistically, the answer is no. But we need to be careful here. Since theoretical talk summarizes observational talk or functions as a shorthand for observational talk, theoretical sentences like the observational sentences they summarize, have a truth value. But theoretical terms do not refer to nonobservables. They in fact refer to observables or relations among observables. To use another example, we might try to define "electric current" in terms of the meter readings of a galvanometer. There is no unobservable entity or property, electric current, but rather talk of electric current is a way to simplify a good deal of talk about observables. Let us call the interpretation of theories generated from the insistence upon explicit definitions, *reductionism*. Reductionism is a species of antirealism.

There are two fundamental difficulties with explicit definition as an account of *R*. First consider the sample definition for "fragile." On standard interpretations of logic the if-then or conditional sentence is true in those cases where the antecedent is false. That is, if *O* is *not* sharply struck the whole conditional is true. The disastrous consequence of this is that if *O* is not struck, *O* is fragile given our definition of fragile. Crowbars and properly cooked linguine, as long as they are not struck, will be fragile.

Secondly, if additional explicit definitions are given for "fragile" we will have to regard them, given that they provide necessary and sufficient conditions, as defining new concepts. Since this will hold for all theoretical terms and since many theoretical terms are given multiple definitions, concepts will proliferate at a dizzying rate. Put another way, a description of scientific practice does not reveal a proliferation of concepts for each observationally different way of indicating the application of a term; therefore, the explicit definition procedure does not reflect scientific practice. Even putting scientific practice aside, the methodological virtue of simplicity seems to rule out explicit definitions as a general procedure.

Given the above criticisms, an effort was made to provide an improved specification of R. The requirement of giving a necessary condition for the application of a given theoretical term was dropped. Consider the form of the new *partial definitions* on our old example of fragility:

> If object O is sharply struck,
> Then
> O breaks if and only if O is fragile.

Partial definitions immediately repair the defects of explicit definitions: no longer does every definition define a new concept and it is not the case that every unstruck object is fragile.

What commitment to unobserved entities does this account encourage? The commonly held view is that partial interpretation is consistent both with a realist and an instrumentalist interpretation.[8]

> METAPHYSICAL INSTRUMENTALISM: (i) Theoretical sentences do not have a truth value, (ii) theoretical terms cannot refer to unobservable properties or entities, and (iii) theories are sets of rules or algorithms for making predictions.

On an instrumentalist interpretation, one should speak of theories being adequate or inadequate, more or less useful, but not true or false. Instrumentalism is a species of antirealism.

Instrumentalism, it is thought, does not follow merely from the partial interpretation specification of R. Some additional argument is needed: for example, the fact that most of the theoretical concepts of science are highly idealized or that without blushing, incompatible theories are employed for the same subject. A liquid may be regarded as a set of discrete particles for one purpose and as a continuous medium for another. We are not insisting that these are good arguments. They are simple examples of considerations that have pushed people to instrumentalism.[9]

Despite the received wisdom here, we believe that partial interpretation by itself encourages instrumentalism. After all, explicit definitions are historically and perhaps logically the ideal fulfillment of the empiricist program. Once one buys into specifying the meaning of terms by test conditions, can one really pull back from demanding explicit definitions? It does not seem that just giving sufficient conditions for the application of a term is sufficient for determining the reference of the term. But if the terms have no reference, how can the sentences containing them have a truth value? They appear to be meaningless. Hence instrumentalism. Even if we granted that the meaning of a term is partially specified by a partial definition, this would allow for experimental conditions in which there is no truth of the matter as to whether the term applies or not—again, albeit a modified, instrumentalism.

There are difficulties with the partial interpretation specification of *R*, but to consider them would take us too far afield. The task of this section was to indicate how the "old" realism/antirealism issue is a consequence of the specification of *R*. Recall that the need to specify *R* was a result of the strict observational-theoretical distinction. Theoretical talk, unlike its counterpart observational, is suspect. There is an opening to raise questions about the reality of the entities that seem to be mentioned by theoretical discourse.

III

We want to turn to efforts to criticize the observational-theoretical dichotomy. The most popular form of criticism has been to challenge particular ways of *drawing* the distinction. Challenges have been of two importantly different species. An example of one species is Peter Achinstein's critique of Carnap's formulation according to which observation terms designate observable properties and relations whereas theoretical terms designate unobservable properties and relations.[10] Achinstein points out that according to the ordinary use of "observe," scientists can observe electrons passing through a cloud chamber. Let us call this species of objection one from linguistic usage. There are a number of such objections in the literature, none of which are particularly persuasive. The reason is that the logical empiricists are not trying to formulate a distinction that necessarily fits ordinary or scientific usage. They are willing to deviate from ordinary usage to achieve a reconstruction of science that is philosophically sound.

The second type of challenge takes a particular way of drawing the distinction and shows that in its own terms it does not parse the terms of science in the way the proponent of the distinction had in mind. In fairness to Achinstein there is a way to read his criticism along the second line. "Look, you haven't told me what you mean by 'observe'. All I've got to go on is the ordinary usage. On that usage electrons turn out to be observable—a result that you do not want. If you have some technical sense of 'observe' let's hear it. Then we can see if it parses terms like you want." This is clearly a more powerful objection. It takes a proponent's distinction and shows that in its own terms it does not achieve the desired goals.

There is a considerable body of literature challenging given ways of drawing the distinction.[11] The account mentioned in section I in terms of "direct observation" has been the subject of much criticism. Unfortunately, even when of the second species, these critiques fail to end the matter, for one always wonders whether with a little work the distinction could be put more successfully.

Rather than repeat specific challenges, we want to turn to two different sorts of critiques. They are aimed at undermining the motivation for the distinction. One of those motivations concerned the epistemological transparency of observational discourse. This motivation is doubly important in that the meaningfulness of discourse was thought to be parasitic upon its epistemological status. Is the observational epistemologically privileged and in what way? There is a philosophical program that made strong claims for the observational—it provides a class of sentences whose truth or falsity could be determined with certainty.

It should be immediately clear that what we would normally classify as an observational matter fails to meet the condition of certainty. According to our everyday use of the term "observable," a barn is observable: that I see a barn counts as an observational truth. However, as Descartes and many others have shown, I *could* be wrong about any observational report. What I see could be no more than a barn facade. Worse, I could be hallucinating or dreaming. There is no certainty to be had here. We must look elsewhere for the observational basis. With the concept of a sense-datum, the basis is located inside the mind of the observer.

According to proponents of sense-datum theory, a sense-datum is what one is really aware of when one makes an observation. In a sense, the man who sees real pink elephants at the animal decoration center and the drunkard who merely hallucinates them, are not seeing the same thing. The drunkard is not seeing anything in the world. He is merely falsely believing himself to do so. But in another sense, what they see is the same: both of them have the same mental images and it is that similar pair of mental images that leads to their similar utterances of "I see pink elephants." A sense-datum statement, then is just a report of that internal mental state.

The proponents of sense-datum theory go on to say that, while we can be wrong in our perceptual reports of the outside world, we simply cannot be wrong in our sense-datum reports. If I think I see pink elephants, there may be no pink elephants out there in the world, but I must have pink elephant sense-data. Supposedly, sense-datum reports are infallible. They are known with certainty. As such, they and they alone provide the proper realm of the observational.

Sense-datum reports, of course, will not provide one wing of an observational-theoretical distinction that mimics anything in common talk or scientific practice. But they do seem to help secure a class of sentences that could accomplish the philosophical purposes of empiricism. The reports claim epistemic privilege in a strong and clear way.

Unfortunately matters have not gone well for the sense-datum theory. An immediate problem concerns getting from sense-datum reports to claims about ordinary objects. We do want to talk about dogs and chairs. How is such talk related to sense-datum talk? Some relation not unlike our previous *R*, will have to be specified to bridge the gap. It has not proved to be the case that statements about ordinary objects can be captured by (translated into) sense-datum statements alone.

But there is a much less technical difficulty. Recall that the reason we are interested in sense-datum reports is that they are supposedly knowable with certainty: if I think that I am having a pink square sense-datum, then I must be having one. However, an examination of the nature of sense data might cause us to doubt this claim. For instance suppose that the sense-datum in question is the mental image. Then, the sense-datum report will be a sentence (or perhaps belief) about that image. As such, the report presupposes an *interpretation* of the datum. Suppose I have a mental image consisting of fifteen green spots, but due to an unfortunate inability to count properly, I acquire the belief that I have a fourteen-spotted sense-datum. My mistake came through my interpretation. Whenever there is an interpretation, there is a possibility of mistake.

The observational-theoretical distinction is supposed to capture an important epistemological difference. Sense-datum theory fails as an effort to show what is privileged about the observational. Interestingly there is no serious alternative to sense-datum theory as an account of what is privileged about the observational. If there is an important epistemological difference we need a convincing story about it. (See section V.)

Let us examine the final motivation for a strict observational-theoretical distinction, viz., theory comparison. If we are to compare *T* and *T** we must find some implications of the theories whose truth or falsity can be determined *independently* of the truth of the theories. This seems to be part of what is meant by a fair, impartial, or objective test. We do not ask Nancy Reagan for an assessment of some of her husband's actions in order to test President Reagan's "theory" of the Iran arms deal. We suspect that Nancy and Ron hold the same theory and assess events in terms of it. The truth of the assessment is not independent of the theory, or so we think.

Observation is alleged to provide a theory-neutral framework for testing theories. The most penetrating critique of this idea comes from Norwood Hanson.[12] Hanson begins by asking whether the famous observational astronomer Tycho Brahe and Johannes Kepler saw the *same thing* when they looked east at dawn. (They held very different theories, the Ptolemaic and Copernican views, respectively.) According to a popular account, the sensory core theory, they do see the same thing. Seeing involves (i) having a sense-datum and (ii) interpreting the sense-datum. So seeing an apple is having a reddish, heart-shaped, pushed-in-on-the-bottom datum and interpreting it as an apple. Brahe and Kepler both see the sun in the east but they may have different interpretations—the sun rising or the earth rotating.

Hanson has a number of criticisms of the sensory core theory. He appeals to psychological evidence to question seeing as a two-stage process. In seeing one does not experience the two stages. One simply sees. Interpretation takes time. It makes sense to talk about being half done. But seeing is, at least as experienced, instantaneous.

Yet these sorts of considerations and others like them are not the quick of Hanson's objection. Hanson confronts the sensory core theory with a dilemma: If seeing the same thing comes down to having the same sense-datum, seeing in that sense is irrelevant to science. If seeing is such that it is relevant to science, it is theory-laden. But then theory-laden seeing is not the neutral observation that the proponents of the observational-theoretical distinction wanted. It is not capable of solving the issue of theory comparison as they conceive it.

Let's consider the horns of the dilemma in more detail. To be relevant to science in this context, we mean being relevant to the testing of a scientific theory. Theories, ontologically, are sets of statements. It is statements that can come into logical relationships with theories. Now Hanson can admit that Brahe and Kepler have a number of similar physiological states, e.g., they have similar patterns of retinal stimulation. But, of course, having these patterns is not seeing in any interesting sense even though they may be necessary for seeing. Physiological states strictly speaking are irrelevant to theories. Physiological states and theories are from different incomparable categories. It will not help to bring in sense-data or any other kind of entity. To get something that will bear on scientific theories it must have a propositional form.

Turning to the second horn, an example of observation or seeing relevant to science is the following: John *sees that* the paper turned blue. "That the paper turned blue" is potentially relevant to testing theories. But Hanson is quick to point out that what one can "see that" is dependent upon one's theories. What true propositions are produced in seeing depends upon what one knows. One person might see that something turned red. Another person might see that the driver in front is putting on his brakes. Walking into a physics museum a child might see that a hunk of glass and metal is in the case. You might see that there is an old-fashioned X-ray tube in the case. The crucial point is that what we see depends on our theories. So we have no theory-neutral seeing. One might be tempted to fall back on sense-data at this point. Remember even if in some sense they are neutral, they are of no use to science until interpreted—until one brings theory to them.

We believe that Hanson's arguments destroy the sensory core theory in particular and quests for theory-neutral observation in general. In so far as the observational-theoretical distinction was motivated by finding a set of theory-independent statements, it is motivated by something that cannot be had.

A serious question remains: Given that observation is theory-laden, can theories be rationally compared? If theory-neutral observations are required, we are in trouble. Perhaps we can grant theory-ladenness and still show how theory comparison is rationally possible.

For instance, the fact that our observational terms are theory-laden does not mean that they are laden in a significant way with the higher-level theories to which we attach them. Why should it not be possible for two very different fundamental theories of the world to *share* their observational subtheories? We might be able to think of our theories of the observable world as modules that can be transported whole from one theory to another. If two theories share the same observational subtheory, then their observational claims can be compared. If they are not (or cannot be) united with the same observational subtheory, then they cannot be compared. Thus, it may be that theory-comparison is possible but not guaranteed. In other words:

(i) Because of the theory-ladenness of observation, it may be impossible *in certain cases* to compare different scientific theories on the basis of observation.

However, according to certain philosophers, the situation is much worse than (i) might indicate. According to Paul Feyerabend, the definition of a term is fixed by its role in a theory. Thus, the definition of "mass" as it occurs in Newtonian mechanics is given by *everything* that Newtonian mechanics has to say about mass. If a theory were to give "mass" even a slightly different role, then that theory would be working with a different definition of mass. For example, according to Newtonian mechanics, the mass of an object is a constant that does not depend on the object's speed; however, mass is dependent on velocity in special relativity. Therefore, Feyerabend infers that "mass" differs in meaning across the two theories. It follows from this that terms related to "mass," like "force" and "energy," will differ in meaning across the two theories as well. In fact, given the all-pervasive interconnections between terms in theories, it becomes at least plausible to say that any change in a theory will lead to a change in meaning of all of that theory's terms. However, if the terms of the earlier theory do not mean the same as the terms in the later theory, then there is no sense in which the two theories agree or disagree, either observationally or not. And, if there is no sense in which they can be said to agree or disagree, then there seems to be no sense in which they can be rationally compared. Hence, if Feyerabend's theory is true, it seems that theoretical comparison is impossible.[13] Instead of (i), we are left with:

(ii) Because of the theory-ladenness of observation, it is impossible to compare different scientific theories on the basis of observation.

Both (i) and (ii) pose problems for the philosopher of science. The very idea of science seems to depend on the idea of objectivity, the idea that we can rationally compare our scientific theories. If objectivity and rational theory choice are impossible, then the adoption of a scientific theory threatens to be no more than an arbitrary whim, reflecting no more than the prejudices of a moment, culture, or political viewpoint. The term "scientific knowledge" becomes a cruel irony and a badge of naivete for those who persist in using it. Clearly the threat is much worse given (ii) than (i). If we can never compare theories, then no rational theory choice seems possible. However, the threat is not completely dispelled by (i); the possibility of important but inadjudicable disputes remains. Regardless of whether we accept (i) or (ii), our conception of scientific progress will be challenged. And, it seems that Hanson's arguments for the theory-ladenness of observation leave us no other alternatives.

IV

Do Hanson's claims also have effects on the realism debate? The best answer is that they might. They do have one definite negative effect. With his arguments for the theory-ladenness of observation, Hanson lays to rest the idea that the observational-theoretical distinction is metaphysically or ontologically important. We had already rejected the idea that we could build an adequate theory of meaning that placed observational terms at its foundations. We had also given up the idea that observational statements provided a crucial link between truth and knowledge once we found that we could be wrong about them, too. Hanson's arguments lead us to believe that the mere existence of observational terms does not allow us to take the comparability of our scientific theories for granted. Therefore, there seems to be absolutely no motivation remaining for worrying about unobservable objects and theoretical statements in particular. If there is nothing particularly special about observable objects and observational terms, then we should not ask questions like "Do unobservables exist in the same way as observables?" and "Should we (and can we) build a theory of meaning according to which the meaning of nonobservational terms is parasitic on observational ones?" We might as well ask questions like "Do gophers exist in the same sense as whales?" and "Can we interpret our cutlery talk in terms of our major kitchen appliance vocabulary?" Divisions without an underlying metaphysical motive usually don't have any metaphysical import.

It may seem that, taken together, the attacks on the observational-theoretical distinction send antirealism into bankruptcy. Appearances can be misleading. If anything, they have caused a metaphysical boom. To see how this could be, notice that the realist's task was to make everything else as safe as observation. Since we found out that observation was not all that safe, in a sense, the realist succeeded: everything else *was* as safe as observation. But instead of securing all of our theories on the firm rock of observation, we found out that the "rock" was made of cork and was placidly floating along with us. All of the questions that we believed to apply to the nonobservational realm turned out to apply to everything. Where we used to ask about the sense in which nonobservational sentences were true, we must now ask about the nature of truth in general. Where we used to ask the ways in which unobservables could be said to exist, we must now ask about the way in which anything exists.

Hanson's arguments add some spice to these problems. Suppose that you endorse one theory and I another. Further, suppose that observational terms are theory-laden to a degree that makes rational theory comparison impossible. In such a situation we could draw one of the following morals about the nature of truth:

> RELATIVISM: There is no such thing as *the* truth. Instead, the notion of truth should be regarded as relative to one's basic theoretical standpoint. Where "Waves exist" may be true-for-you, it may not be true-for-me; and, there is no final standpoint outside of our way of looking at things that can transcend our theoretical prisons.

Relativism is plausible as long as we believe that truth is something that we can find out. Once we grant this, if you and I have an irreconcilable difference, one for which there is no objective and rational solution, then neither of us could be either simply right or simply wrong. However, if we reject the claim that the truth must be discoverable, we will not be forced into relativism. Instead we might embrace something else.

> SKEPTICISM: Whatever our rational practices do, there is not much reason to think that they take us toward the truth.

Since there are rationally inadjudicable disputes between ways of thinking, we may be locked into a way of thinking from which there is no rational road to the truth. In a way, this position saves the notion of objective truth only to make it useless. Because of this, the skeptic who wishes to maintain that we do science rationally cannot base rationality on truth. For the skeptic, the concept of truth would eventually drop out of his account of the practice and goals of science.

It seems that even limited theory-ladenness—i.e., claim (i) of the previous section—generates some interesting possibilities for the nature of truth and existence. But notice how strange things would be if theory-ladenness were complete and all-pervasive, that is, if (ii) were true rather than (i). The resulting skepticism would be total, for there would never be any way to move rationally from one theory to a successor theory. Furthermore, if we took the other route, the form of relativism we would end up with would be startling:

CONVENTIONALISM: What is true is entirely a matter of stipulation. Different theories are just different decisions on how to use theoretical terms. We can never rationally compare theories.

Conventionalim represents relativism taken to its logical limit. Where the relativist says that there are limits to what rationality can do, the conventionalist says that there is nothing that rationality can do. The conventionalist might look for other mechanisms that underly theory change. Perhaps it reflects rationally unmotivated shifts in the attitude of the population as a whole. For instance, it has been argued that the shift to quantum mechanics should not be understood as a rational scientific transition so much as an effect of Weimar Germany's disaffection with a mechanistic and deterministic world view, a view that they credited for taking Germany into the First World War.[14] Perhaps scientific change reflects something more political than cognitive: Michel Foucault has argued that one can best understand the birth of psychiatry and the modern concept of insanity in terms of the state's need to control a huge population.[15] Whatever, theoretical change neither reflects a convergence upon universal truth nor the embodiment of a universal rationality.

The debates concerning these positions are sometimes maddening but they are always fascinating. However, we could not begin to follow them out in this paper. Instead we have just tried to show the sorts of realism debate that arise from the weakening of the observational-theoretical distinction. There is no doubt that these are very different from the relatively tame one about the status of theoretical terms. The varieties of antirealism described above attack a much more cherished realism, the belief that there is one reality, one truth, where that truth is to be discovered by human beings, not invented by them.

V

We have seen that the observational-theoretical distinction fails to do the things that it was intended to do. It does not give us a way of explicating all scientific language. It does not provide us with a totally solid ground for all of our knowledge. And, it does not provide us with a sure way of objectively comparing all of our scientific theories. The distinction seems to be no more than a measure of our physiological limitations. The boundary of the distinction represents no more than the limits of our sensations: what is observable is what we can see or hear without the aid of instruments. What possible strength or importance could such a distinction have?

First, now that we realize what the distinction has turned out to be, it is difficult to argue that the distinction has any bearing on the question of what exists or what the concept of truth is: that our sensory limitations should delimit the class of things that exist is simply ludicrous. But, that does not mean that the distinction cannot play a very important role in our analysis of how we know things. In other words, although the distinction is almost certainly metaphysically barren, it may still retain some epistemological strength. None of the arguments that we have reviewed have attacked the position that observational statements are epistemically privileged. There are successful arguments against the position that observational statements are so privileged that they are certain. However, we can reject the certainty of observational statements without also rejecting the claim that they are somehow special.

What could this specialness amount to, given that certainty is ruled out? It would have to amount to our being more justified in our observational beliefs than in our nonobservational ones. In the final section of this paper we are going to determine whether there are good arguments for saying that observational statements are special. For the moment, let us suppose that they are, so that we can discuss the sort of antirealism that can arise out of a purely epistemological distinction. It is a purely epistemological form of antirealism, but it makes strong claims nevertheless:

EPISTEMOLOGICAL INSTRUMENTALISM: Although we can have justified beliefs (and know) about the observable realm, we cannot come to know about the unobservable realm.

Note the great difference between epistemological instrumentalism and the form of instrumentalism we discussed earlier. Where, according to the old instrumentalism, nonobservational terms do not refer and nonobservational statements do not say anything, the epistemological instrumentalist says that they can and do have truth values: it is just that we can never find out what those truth values are. The two forms of instrumentalism agree about the scientific role of nonobservational statements, which are held to be important only insofar as they lead to theories that are true at the observational level. It follows from this that the two forms of instrumentalism also agree on the goal of science. We do not do science in order to find the fundamental unobservable truth that will explain the world of experimental phenomena. We do science in order to control and predict the observable world. Thus, the realist that the epistemological instrumentalists take themselves to be opposing is someone who not only believes that nonobservational statements have truth values but also someone who believes both that these statements are knowable and that coming to know them constitutes a main goal of science.[16]

Again, it is not our goal to determine whether epistemological empiricism is true, but only to ascertain the relevance of the observational-nonobservational distinction for it. Although epistemological instrumentalism does not require the strongest of distinctions, it does require one for which observational statements are much more privileged than nonobservational ones. After all, epistemological instrumentalists do not merely hold that, in general, our observational beliefs are more justified than our nonobservational ones. They hold that our nonobservational beliefs are *never* justified. What feature of our observational beliefs (short of certainty) could serve as a foundation for this claim? In short, what epistemological strength does the distinction really have?

Despite the above brand of antirealism receiving enormous attention, even its proponents have not drawn the needed distinction in a perspicuous way. We want to provide the best argument we can for a distinction that could form the basis for epistemological instrumentalism. But the distinction will be interesting in its own right if it illuminates the preanalytical epistemological difference between the observational and the theoretical.

We could try this as a distinction: a state (i.e., the ball being red) is observable just in case there is a reliable way that could be used for finding out whether the state obtains without the use of measuring instruments. For example, if there is a red ball around us we can set up certain conditions; we can open our eyes and turn the lights on. Given those conditions, there is a reliable connection between the ball being red and our being in a certain sort of sensory state. In other words, the ball being red registers on our sensory apparatus.[17]

From this it follows that not all states are observable. However, some that are not *observable* may be *detectable*: a state (the electron having an energy of –13.6 ev) is detectable just in case we have a reliable way of finding out whether the state obtains, where this may involve the use of measuring instruments. If the electron has an energy of –13.6 ev, then we can set up our experimental apparatus so that there is a reliable connection between the electron's having that energy and our apparatus' being in a certain observable state. If the electron has a certain energy, then our experimental apparatus will show a reading of –13.6 in nice big red numbers. This reading is observable; it will register on our sensory apparatus.[18]

But, if this is the sense of observable that we are left with, then what motivation could there be for epistemological instrumentalism? It seems that observation and detection both provide us with reliable ways of finding out about the world. As such, we should conclude that if we know about observable states then we know about unobservable but detectable states as well. To answer this question we have to give a further specification of what's supposed to be reliable. Take the electron case as an example. We set up our experimental apparatus. Given the fact that our experimental apparatus is set up as a background condition, it seems that the electron will have an energy of –13.6 ev in exactly those cases where our apparatus reads –13.6. But there are other conditions that have to be met in order for our test to be foolproof.

The one that we want to concentrate on is the condition that basic electron theory is true. We can't rule out the possibility of another globally different view of the world that talks about completely different sorts of entities.[19] Someone who holds that theory could describe conditions that didn't involve electrons but which, if met, would entail that the apparatus would read −13.6. Thus although the combination of electron theory and experimental set up guarantees our meter reading, our meter reading does not guarantee the existence of electrons. The fact that electrons may be detectable doesn't necessarily give us good reasons to believe in the existence of electrons.

Perhaps defenders of epistemological instrumentalism believe that the situation is different in the case of observable states. Observing that my dog has fleas does give me good reasons for believing in fleas. This raises a question: can there be different accounts of the observable world *alone* each of which by its own lights is completely observationally accurate? Epistemological instrumentalism requires that observing be immune to the underdetermination that infects detection.

We are not completely convinced that the alleged asymmetry of underdetermination exists. The issue is presently an area of active discussion.

CONCLUSION

Here is a summary of our results: (1) The observational-theoretical distinction does not do the jobs for which it was originally intended. It does not guarantee us a sound philosophy of language, the possibility of scientific comparison, or a body of epistemologically transparent statements. As such, even if there is a significant distinction, it will not be one that will make the metaphysical status of unobservables problematic.

(2) Given the breakdown of the unobservable as a metaphysical haven, the metaphysical problems that were originally thought to concern only unobservables and statements about them can now be seen to concern all entities and statements. Our worries are now global ones about truth and existence in general. There is no special reason to worry about the existence of theoretical objects.

(3) If there is any significance to the observational-theoretical distinction, it will be epistemological. However, no definitive arguments have been given for its significance; and if it is significant, it may still not be strong enough to support the claim that we do know about observables but not about unobservables.[20]

NOTES

1. For a detailed exposition and critique of the two-tier view see Frederick Suppe, *The Structure of Scientific Theories* (Urbana: University of Illinois Press, 1974). The essay by Carnap in the present volume is an example of the view.

2. Carl Hempel, "The Theoretician's Dilemma" in *Aspects of Scientific Explanation* (New York: Macmillan, 1965), pp. 178–79.

3. This seems to be the position of Rudolf Carnap in his paper "Testability and Meaning," *Philosophy of Science* 3 (1936):420–71.

4. The example is from Israel Scheffler, *The Anatomy of Inquiry* (New York: Alfred A. Knopf. 1969), p. 127.

5. George Berkeley, *De Motu in Berkeley's Philosophical Writings*, ed. David Armstrong (New York: Macmillan, 1965), p. 251.

6. Carl Hempel, "Empiricist Criteria of Cognitive Significance: Problems and Changes," in *Aspects of Scientific Explanation*, p. 101.

7. Observational statements are statements whose nonlogical terms are observation terms.

8. Frederick Suppe, *The Structure of Scientific Theories*, p. 30.

9. See Ernest Nagel, *The Structure of Science* (New York: Harcourt, Brace, and World, 1961), pp. 129–140.

10. Peter Achinstein, *Concepts of Science* (Baltimore: The Johns Hopkins Press, 1968), p. 160.

11. See the essays by Putnam and Maxwell in this volume and Peter Achinstein, *Concepts of Science*, pp. 157–78.

12. See the essay by Hanson in this volume.

13. P. K. Feyerabend, "Explanation, Reduction, and Empiricism" in *Scientific Explanation, Space, and Time, Minnesota Studies in the Philosophy of Science*, vol. 3, ed. H. Feigl and G. Maxwell (Minneapolis: University of Minnesota Press, 1962), pp. 28–97.

14. P. Forman, "Weimar Culture, Causality, and Quantum Theory" in *Historical Studies in the Physical Sciences*, ed. R. McCormmach (Philadelphia: University of Pennsylvania Press, 1971), pp. 1–115.

15. Michel Foucault, *The Birth of the Clinic* (New York: Vintage Books, 1975).

16. For an example of what we are calling epistemological instrumentalism see Bas C. van Fraassen, *The Scientific Image* (Oxford: Clarendon Press, 1980).

17. This way of making the distinction is only meant to be suggestive. One needs to put what counts as the use of a measuring device in a precise way. On the present view, seeing that an electron is moving to the left by seeing a jet of smoke in a cloud chamber would be by using a measuring device.

18. A rigorous way of making the distinction between observation and detection in an alternative vocabulary is Fred Dretske's distinction between primary and secondary seeing. Fred Dretske, *Seeing and Knowing* (Chicago: University of Chicago Press, 1969).

19. For an introductory discussion of this possibility see Rudolf Carnap, *An Introduction to the Philosophy of Science*, ed. M. Gardner (New York: Basic Books, 1966), chap. 15.

20. More material on the relevance of the history of the observational-theoretical debate for the evolution of types of realism can be found in Carl Matheson, *Does Scientific Realism Matter?* dissertation, Syracuse University, 1986.

Case Study for Part 4

Imagine you are alone at home late one night in your second-floor bedroom quietly enjoying your philosophy of science anthology. Suddenly you hear a crash downstairs. Recognizing the possibility of a break-in, you call 911. You then creep slowly downstairs from your room, trusty textbook in hand to ward off any intruders. When you get to the living room you discover your stereo is missing and a window has been broken. When the police arrive you describe what you heard and saw, from which they conclude a burglar has broken into your house and made off with your stereo.

1. Is the police officers' conclusion warranted? Why or why not?

2. Mindful that neither you nor the police saw the intruder, what is the difference between the inference they made and the sorts of examples Stace has in mind?

Now imagine further that after the police leave, you notice several large indentations in your back yard, indentations that your vast experience with science fiction movies convinces you are consistent with a recent visit by an alien spacecraft.

3. If you were to consider the possibility that the theft of the stereo was due to an alien from outer space, would your conclusion be warranted?

4. Again, remembering that you did not see the intruder, what is the difference between this new inference and the sorts of examples Stace has in mind?

5. If you were to call the police back to your house to examine the indentations and the police officer who arrives does not share your interest in science fiction, would you and the officer "see" the same thing according to Hanson?

6. Is there a difference in the two existence claims made by the hypothesis that a burglar broke into your house and the hypothesis that an alien broke into your house according to Toulmin?

Study Questions for Part 4

1. Why should we distinguish observational from theoretical terms according to Carnap?

2. Why does Carnap believe that only part of the meaning of a theoretical term can be specified with reference to observational vocabulary? Where does the rest of its meaning come from?

3. What are Putnam's reasons for disputing whether it is possible to maintain the observational-theoretical distinction Carnap has drawn?

4. Hanson points out that in certain respects Kepler and Brahe see the same thing when they watch the sun arise from atop a hill, but in other respects they do not. Briefly identify these different respects and Hanson's reasons for distinguishing them.

5. What is wrong, according to Hanson, with the claim that so long as two observers look at the same object, they see the same thing?

6. What reasons does Stace provide for disputing the reality of unobservables? Do you agree with them? Why or why not?

7. Does Stace's argument apply with equal force against macroscopic examples like black holes, which are strictly speaking not directly observable?

8. What does Stace's argument imply with regard to historical events?

9. Toulmin appears to reach the same conclusion as Stace, but for different reasons. Briefly compare and contrast their reasons. Who presents the better argument and why?

10. In what ways does an analysis of how people use scientific terms a legitimate technique for resolving philosophical disputes?

11. Maxwell and Carnap both seem to recognize a continuum between observational terms and theoretical terms, yet Carnap persists in arguing the observational-theoretical dichotomy is significant. Are they talking about the same continuum? Why isn't Maxwell persuaded by Carnap's argument?

12. In what ways does a fictitious story like the "crobes" example developed by Maxwell represent a legitimate means of philosophical reasoning? Why should an adequate account of scientific practice be constrained by fictitious examples? Is he just changing names to protect the guilty? Or is there something more going on in his example?

13. Maxwell and Kline and Matheson dispute the observational-theoretical dichotomy, but for different reasons. Compare and contrast their approaches.

14. Does the realist-instrumentalist debate have any significance with regard to the actual practice of science? Explain your answer.

SELECTED BIBLIOGRAPHY

(* indicates contains an extensive bibliography)

[1] Aronson, J., R. Harre, and E. Way. *Realism Rescued.* London: Ducksworth Press, 1994. [Collection of essays that discuss scientific progress as it relates to debates about realism.]*

[2] Churchland, P. *Scientific Realism and the Plasticity of Mind.* Cambridge: Cambridge University Press, 1979. [Argues for a version of scientific realism and against the importance of the observational-theoretical distinction.]

[3] Dretske, F. *Seeing and Knowing.* Chicago: University of Chicago Press, 1969. [Relevant to understanding observation; intermediate level.]

[4] Eddington, A. *The Nature of the Physical World.* New York: Cambridge University Press, 1929, pp. ix–xiii,, 282–289. [A superrealist.]

[5] Grandy, R., ed. *Theories and Observation in Science.* Englewood Cliffs, N.J.: Prentice-Hall, 1973. [Classical essays on the two-tier conception of scientific theories and its problems.]

[6] Gutting, G. "Philosophy of Science." In *The Synoptic Vision,* edited by F. Delaney et al., pp. 73–86. Notre Dame, Ind.: University of Notre Dame Press, 1977. [Overview of W. Sellars's super-realism; will guide one in the Sellars literature.)

[7] Harre, R. *The Philosophies of Science.* 2d ed. Oxford: Oxford University Press, 1985, chap. 3 [Nice examples of theories; elementary.]

[8] Harre, R. *Varieties of Realism.* Oxford: Basil Blackwell Inc., 1986. [Surveys different types of realism.]

[9] Hempel, C. *Philosophy of Natural Science.* Englewood Cliffs, N.J.: Prentice-Hall, 1966, chap 6. [Basic survey of field.]

[10] Holton, G., and D. Roller. *Foundations of Modern Physical Science.* Reading, Mass.: Addison-Wesley, 1958. [An introduction to physics written by a physicist and a historian of science; requires only high school algebra; good examples of theories.]

[11] Kuhn, T. *The Structure of Scientific Revolutions.* 3d ed. Chicago: University of Chicago Press, 1996 [1962]. [An exciting work that has received enormous attention; presents an account of theories, observation (similar to Hanson's), and theory change; views are developed by way of historical examples.]

[12] Leplin, J., ed. *Scientific Realism.* Berkeley: University of California Press, 1984. [Classic essays on scientific realism.]

[13] Leplin, J. *A Novel Defense of Scientific Realism.* New York: Oxford University Press, 1997. [The title says it all.]*

[14] Nagel, E. *The Structure of Science.* New York: Harcourt, Brace, and World, 1961, chaps. 5 and 6. [An excellent discussion of the status of unobservables.]

[15] Suppe, F. *The Structure of Theories.* Urbana, Ill.: University of Illinois Press, 1977. [Detailed description of the standard view of theories, plus a review of all the work critical of the standard view; intermediate level.]*

[16] Suppe, F. *The Semantic Conception of Theories and Scientific Realism.* Urbana, Ill.: University of Illinois Press, 1989. [An extensive discussion of an alternative account of theories.]*

[17] van Fraassen, B. *The Scientific Image.* Oxford: Oxford University (Clarendon) Press, 1980. [Presents a provocative antirealist alternative known as constructive empiricism.]

Part 5

Confirmation and Acceptance

Introduction

I. The Problem

In parts 3 and 4, we addressed the structure of explanations and theories. In this part, the concern will be with how we justify our belief in a particular explanation or theory. Since hypotheses or lawlike statements are at the heart of both explanations and theories, the discussion will center on their justifications.

We are often told that scientists justify their hypotheses by comparing them to the facts or seeing how they fare in the face of the facts. There is a certain truth in this, though as with most slogans it is a misleading oversimplification. A purpose of the orienting remarks that follow is to make and motivate some of the bold distinctions to which a sensitive theory of justification must attend.

II. Confirmation

The following are examples of hypotheses:
- (a) Lead melts at 327° C.
- (b) Myelinated A-fibers transmit signals at 100 meters/ second.
- (c) The price of goods is inversely proportional to the demand for the goods.
- (d) The pressure of a gas times the volume of the gas is a constant ($PV = c$ or $P_1/P_2 = V_2/V_1$)

How do scientists justify their belief (or disbelief in (a)-(d)? Of course, they give reasons for their belief.

A place to begin in order to understand this reason-giving process is a distinction between two basic kinds of reasons: (i) epistemic reasons, (ii) practical reasons. Epistemic reasons count toward the truth or falsity of a hypothesis. Practical reasons do not bear on truth or falsity but rather on the usefulness of believing the hypothesis.

Consider the hypothesis that your next-door neighbor is madly in love with you. That she gives you gifts and winks at you as you leave for classes is an epistemic reason for believing the hypothesis.

Suppose further that you have an incredible inferiority complex. Unless you believe that people will greet you with open arms, you simply cannot face them at all. Furthermore, suppose that you wish very much to meet your neighbor. Now these facts about your psychology and goals are reasons for believing the hypothesis, but not epistemic reasons. They do not count toward the truthfulness of the hypothesis, but they make it useful for you to believe the hypothesis.

Later in this essay, the distinction above will be put to work. But, at least at the outset, it is not unreasonable to think that scientists offer epistemic reasons for (a)-(d). But what reasons?

In example (a), intuitively "that this piece of lead melted at 327° C" is an epistemic reason for believing (a). Typically it is called a confirming instance. A hypothesis that has the form of our (a), viz., everything that is *A* is *B*, has four types of instances: this thing is *A* and *B*, this thing is *A* and not *B*, this thing is not *A* and *B*, this thing is not *A* and not *B*. Members of the first type are confirming instances; members of the second type are disconfirming instances.

So far then, a hypothesis is justified by appeal to epistemic reasons where epistemic reasons are confirming instances of the hypothesis. Even though a certain amount of jargon has been introduced, what has been said is fairly commonsensical. In fact it seems to be what is behind the slogans that initiated the discussion. But notice that, even supposing what has been said is beyond doubt, it is vague at some crucial points.

We spelled out the meaning of "instance" for a hypothesis of the form "Everything that is *A* is *B*." But what about hypotheses of other forms such as (c) or (d)? A similar account of "instance" is needed for these and hypotheses of other forms. Though this will require a little formal sophistication there are no obvious theoretical problems in the way.

The second vague spot belies a deeper issue. Though we gave examples of *confirming* and *disconfirming* instances we did not define these notions. A careful reader might have noticed that we were conspicuously silent on how to understand "this thing is not *A* and *B*" and "this thing is not *A* and not *B*."

One worry can be simply put. A hypothesis of the form "Everything that is *A* is *B*" is logically equivalent to a hypothesis of the form "Everything that is not *B* is not *A*." If you think that "this *A* is a *B*" confirms the first hypothesis, as surely you must, then by parity of reasoning "this thing is not *B* and not *A*" ought to confirm the latter. Since the two hypotheses are logically equivalent it seems reasonable that "This thing is not *B* and not *A*" should also confirm "Everything that is *A* is *B*." But then "that this wax melts at 100° C" would confirm all lead melts at 327° C."[1]

There are further paradoxes but we need not pursue any of these.[2] The purpose of our reconnaissance is to spot pockets of resistance, not necessarily to eliminate them.

Suppose we could, with a little philosophical tenacity, define confirming (disconfirming) instances. Our theory of justification is far from complete. Just as one birdie does not a good round of golf make, though it surely helps, one confirming instance does not a hypothesis justify, though it counts toward the justification. We have been discussing what philosophers call qualitative features of confirmation—that is, defining a confirming instance. There are also quantitative features of confirmation.

An example will illustrate these features. The eighteenth-century chemist Robert Boyle is credited with discovering the law expressed in (d). He justified the hypothesis by providing the reader with the following data. In the first column, *A*, are values for the volume of a gas. In the second column, *B*, are the corresponding experimentally determined values for the pressure.

A	*B*	*C*
48	29 2/16	29 2/16
46	30 9/16	30 6/16
44	31 15/16	31 12/16
42	33 8/16	33 1/7
40	35 5/16	35
38	37	36 15/19
36	39 5/16	38 7/8
34	41 10/16	41 2/17
32	44 3/16	43 11/16
30	47 1/16	46 3/5
28	50 5/16	50
26	54 5/16	53 10/13
24	58 13/16	58 2/8
23	61 5/16	60 18/23
22	64 1/16	63 6/11
21	67 1/16	66 4/7
20	70 11/16	70
19	74 2/16	73 11/19
18	77 14/16	77 2/3
17	82 12/16	82 4/17
16	87 14/16	87 3/8
15	93 1/16	93 1/5
14	100 7/16	99 6/7
13	107 13/16	107 7/13
12	117 9/16	116 4/8

Boyle justifies his hypothesis not by a single confirming instance but by a set of confirming instances. Furthermore, he does not give us repeated measurements corresponding to a row in the chart but measures the pressure and volume over a range of values. Column C is also significant. If one plugs in V_1 and V_2, and P_1 and computes P_2 according to (d), he will discover a difference between the computed values for P_2 and the experimentally determined values. C gives the expected values. The differences are so slight that Boyle was willing to chalk them up to human error and defects in the experimental apparatus.[3]

This example brings out clearly that certain quantitative features of confirmation are relevant to the justification of a hypothesis. As a rule of thumb—the larger the quantity, the more diverse; and the more accurate the confirming instances, the greater the justification for the hypothesis.

Though this rule is surely correct, at least as a rough rule, it does not pass without inducing philosophical perplexity. How accurately must the "confirming instances" conform to the expected data in order to be *confirming* rather than disconfirming? How many instances are required before one can reasonably assert the truth of the hypothesis?

So far we have seen that a theory of justification must contain a theory of confirmation. Furthermore, one must distinguish qualitative aspects of confirmation from the quantitative aspects. There is one final feature of confirmation that needs to be sketched.

The distinction between observational and theoretical hypotheses, putting aside the difficulties in precisely drawing the distinction, is well known. The sample lawlike statements (a)-(d) are a subclass of observation statements. One important consequence of this is that the instances of these hypotheses can be known by observation in a fairly direct way. One can simply see if this piece of lead melts at 327° C. But with theoretical hypotheses this will not be true.

Some simple examples of theoretical hypotheses follow:

(e) All electrons have a charge of minus one.

(f) The weight of the "sea of air" surrounding the earth exerts a pressure on objects within the sea.

(g) The current density for any point within a conductor equals the reciprocal of the resistance of the conductor times the applied electrical field or $j = \frac{1}{p} E$.

The problem presented by theoretical hypotheses is not that they do not have instances, but rather since their instances cannot be directly known by observation one cannot in any straightforward way "compare the hypothesis to the facts."

Theoretical hypotheses are tested by deducing a test implication from the hypothesis and observing whether the test implication is true or not. If it is true, it is said to confirm the hypothesis; otherwise disconfirm. To distinguish these "instances" from the confirming and disconfirming instances let us call them confirming or disconfirming test implications.

Continuing our pneumatic examples, consider (f). Torricelli, a precursor of Boyle, deduced a number of test implications from the hypothesis. On one particularly impressive case he reasoned as follows: If (f) is true, then the pressure on objects at the top of a mountain ought to be less than on objects in the mountain's valley, since the column of air pushing on them would be less. So if (f) is true, then a barometer should read less on the top of a mountain than on the bottom. He tried this out and found a spectacular confirming test implication. The general structure of Torricelli's reasoning or the successful justification of a theoretical hypothesis can be represented by the following schema: (H = hypothesis; TI = test implication)

(s) if H then TI

TI

Therefore, H is confirmed.

The disconfirmation schema looks like this:

(t) if H then TI

not TI

Therefore, H is disconfirmed.

Just as quantitative considerations are important in justifying observational hypotheses, so are they in justifying theoretical hypotheses. The quantity and diversity and accuracy of test implications are again relevant. Torricelli strengthened his case by just such considerations. One final point about confirmation must be made. There is a dissimilarity between (s) and (t). (S) is not a valid argument form. Hence from the premises of (s) it is fallacious to conclude that H is true. This point is marked by claiming only that H is supported or confirmed. *H could* turn out to be false, though in so far as TI is the case, one has some reason to believe that H is true.

Since schema (t) is valid, it looks like one ought to be able to conclude not just that H is disconfirmed but that H is false. Given what has been said, that would be the correct conclusion. But the assumption that (t) captures the structure of disconfirmation is an oversimplification.

The actual form of the first premise of (t), and for that matter (s), is (n)

if H and A_1 and A_2 and . . . A_n then TI.

A_1, A_2, et cetera are auxiliary hypotheses—hypotheses assumed when it is claimed that H implies TI. For example, Torricelli was at least assuming that the pressure on the surface of a liquid is transferred through the liquid and that one can veridically observe the height of mercury through a glass barrier, i.e., that glass does not have some systematic distorting effects that will make it look as though the mercury has not risen when it really has. Contemporary philosophers believe that the first premise will always contain auxiliary hypotheses.

If one is more realistic about the first premise of (t), then formally the conclusion is: Therefore, H is false or A_1 is false or A_2 is false or . . . A_n is false. In other words, if we substitute (n) for the first premise of (t), then as a matter of pure logic alone one has a reason to conclude that at least one of H-A_n is false, but he has no reason to pick out a particular false statement from this set.

This point of this discussion of schema (t) is to show that just as confirmation does not prove that H is true, disconfirmation does not prove that H is false. On the assumption that the needed auxiliary hypotheses are highly confirmed the appropriate conclusion from (t) is that H has been disconfirmed, not found false.

II. ACCEPTANCE

There is a difference between having reasons that count toward the truth or falsity of a hypothesis and the acceptance of a hypothesis as true or the rejection of a hypothesis as false. During the Watergate hearings, most people would admit that testimony was given that counted toward the truth of the statement, "Nixon is a liar." Yet many of these people would not accept the statement as true.

Acceptance as usually understood is a broader notion than confirmation. Part of the reason for the acceptance or rejection of a hypothesis is the extent to which it has been confirmed or disconfirmed. But many philosophers have thought that there are additional criteria for acceptance or rejection. The following are often mentioned candidates: (a) the simplicity of a hypothesis, (b) the theoretical fecundity of a hypothesis, (c) the compatibility of a hypothesis with certain moral, political, or religious views, (d) the harm incurred if one's decision to accept or reject a hypothesis is mistaken.

We have already seen some hints as to what motivates the need for additional criteria. By way of confirmation and disconfirmation one cannot *prove* that a hypothesis is true or false. Of course, quantitative features are relevant, but when has one observed enough test implications or when are the test results accurate enough? There are further problems. When evaluating a given hypothesis, various quantitative features may be at odds, for example, as the variety of test implications increases the accuracy may decrease. It is also often the case that for a given hypothesis that is to explain certain phenomena there will be alternative hypotheses that are equally well confirmed and explain the same phenomena.

We shall briefly summarize the readings momentarily, but first some general questions to keep in mind when evaluating the importance of additional criteria such as (a)-(d). (1) From the scientific point of view, should *practical reasons* have anything to do with the acceptance of a hypothesis? (2) Of the additional criteria suggested, which are epistemic and which are practical? (3) Is it really the business of the scientist as a scientist to *accept* hypotheses? (The paper by R. Rudner in part 6 will help here.)

III. THE READINGS IN PART 5

In the essay "Hypothesis," W. V. Quine and J. S. Ullian list six virtues of a hypothesis. It is not clear whether some of these are fancy ways of talking about confirmation or whether they are additional criteria. You will have to evaluate each one individually.

Quine and Ullian have a lengthy discussion of simplicity. Simplicity is the additional criterion most often mentioned by philosophers. Try to decide whether it is an epistemic or a practical reason for acceptance. This essay has some arguments on the matter.

Giere's essay is a careful treatment of the confirmation of theoretical hypotheses. He provides some nice examples of theories which should help make sense of this somewhat formal topic. It might prove useful to compare Giere's treatment to that mentioned briefly in the introduction to this part.

The essays by Kuhn and Frank deal with what we have called additional criteria of acceptance. Kuhn wishes to emphasize the difference between argumentation in mathematics and argumentation in science. A consequence of the differences is, according to Kuhn, that two scientists can disagree about which of two competing hypotheses to accept, after a "complete" effort at confirmation, yet both scientists have good reasons for their views. Supposing Kuhn is correct, what does it tell us about the nature of the additional criteria?

Frank insists that, historically, such factors as compatibility with religious views have been decisive in which hypothesis is accepted. Furthermore, given how he understands theoretical talk, viz., instrumentally, he does not find such factors improper, or unscientific. In fact, Frank seems to support the use of these criteria. In Carl Hempel's essay some deep issues are raised about the philosophy of science itself. Is the philosopher describing scientific practice or giving a normative account of how science should ideally be done? Kuhn claims that in his work he is doing both. Hempel finds this joint task puzzling and is trying to see in what sense, if any, it is coherent.

A. D. K.

NOTES

1. See Hempel's "Studies in the Logic of Confirmation" in [1] of the bibliography for part 4.

2. See Goodman's "The New Riddle of Induction" in [2] of the bibliography for part 4.

3. For the historical and logical details of Boyle's work, see "Robert Boyle's Experiments in Pneumatics" in *Harvard Case Histories in Experimental Science*, vol. 1, ed. J. Conant and L. Nash. Cambridge, Mass.: Harvard University Press, 1948, pp. 1–64.

24

Hypothesis

W. V. Quine and J. S. Ullian

Some philosophers once held that whatever was true could in principle be proved from self-evident beginnings by self-evident steps. The trait of absolute demonstrability, which we attributed to the truths of logic in a narrow sense and to relatively little else, was believed by those philosophers to pervade all truth. They thought that but for our intellectual limitations we could find proofs for any truths, and so, in particular, predict the future to any desired extent. These philosophers were the rationalists. Other philosophers, a little less sanguine, had it that whatever was true could be proved by self-evident steps from two-fold beginnings: self-evident truths and observations. Philosophers of both schools, the rationalists and the somewhat less sanguine ones as well, strained toward their ideals by construing self-evidence every bit as broadly as they in conscience might, or somewhat more so.

Actually even the truths of elementary number theory are presumably not in general derivable, we noted, by self-evident steps from self-evident truths. We owe this insight to Godel's theorem, which was not known to the old-time philosophers.

What then of the truths of nature? Might these be derivable still by self-evident steps from self-evident truths together with observations? Surely not. Take the humblest generalization from observation: that giraffes are mute, that sea water tastes of salt. We infer these from our observations of giraffes and sea water because we expect instinctively that what is true of all observed samples is true of the rest. The principle involved here, far from being self-evident, does not always lead to true generalizations. It worked for the giraffes and the sea water, but it would have let us down if we had inferred from a hundred observations of swans that all swans are white.

Such generalizations already exceed what can be proved from observations and self-evident truths by self-evident steps. Yet such generalizations are still only a small part of natural science. Theories of molecules and atoms

404

are not related to any observations in the direct way in which the generalizations about giraffes and sea water are related to observations of mute giraffes and salty sea water.

It is now recognized that deduction from self-evident truths and observation is not the sole avenue to truth nor even to reasonable belief. A dominant further factor, in solid science as in daily life, is *hypothesis*. In a word, hypothesis is guesswork; but it can be enlightened guesswork.

It is the part of scientific rigor to recognize hypothesis as hypothesis and then to make the most of it. Having accepted the fact that our observations and our self-evident truths do not together suffice to predict the future, we frame hypotheses to make up the shortage.

Calling a belief a hypothesis says nothing as to what the belief is about, how firmly it is held, or how well founded it is. Calling it a hypothesis suggests rather what sort of reason we have for adopting or entertaining it. People adopt or entertain a hypothesis because it would explain, if it were true, some things that they already believe. Its evidence is seen in its consequences. . . .

Hypothesis, where successful, is a two-way street, extending back to explain the past and forward to predict the future. What we try to do in framing hypotheses is to explain some otherwise unexplained happenings by inventing a plausible story, a plausible description or history of relevant portions of the world. What counts in favor of a hypothesis is a question not to be lightly answered. We may note five virtues that a hypothesis may enjoy in varying degrees.

Virtue I is *conservatism*. In order to explain the happenings that we are inventing it to explain, the hypothesis may have to conflict with some of our previous beliefs; but the fewer the better. Acceptance of a hypothesis is of course like acceptance of any belief in that it demands rejection of whatever conflicts with it. The less rejection of prior beliefs required, the more plausible the hypothesis—other things being equal.

Often some hypothesis is available that conflicts with no prior beliefs. Thus we may attribute a click at the door to arrival of mail through the slot. Conservatism usually prevails in such a case; one is not apt to be tempted by a hypothesis that upsets prior beliefs when there is no need to resort to one. When the virtue of conservatism deserves notice, rather, is when something happens that cannot evidently be reconciled with our prior beliefs.

There could be such a case when our friend the amateur magician tells us what card we have drawn. How did he do it? Perhaps by luck, one chance in fifty-two; but this conflicts with our reasonable belief, if all unstated, that he would not have volunteered a performance that depended on that kind of luck. Perhaps the cards were marked; but this conflicts with our belief that he had had no access to them, they being ours. Perhaps he peeked or pushed, with help of a sleight-of-hand; but this conflicts with our belief in our perceptiveness. Perhaps he resorted to telepathy or clairvoyance; but this would wreak havoc with our whole web of belief. The counsel of conservatism is the sleight-of-hand.

Conservatism is rather effortless on the whole, having inertia in its favor. But it is sound strategy too, since at each step it sacrifices as little as possible of the evidential support, whatever that may have been, that our overall system of beliefs has hitherto been enjoying. The truth may indeed be radically remote from our present system of beliefs, so that we may need a long series of conservative steps to attain what might have been attained in one rash leap. The longer the leap, however, the more serious an angular error in the direction. For a leap in the dark the likelihood of a happy landing is severely limited. Conservatism holds out the advantages of limited liability and a maximum of live options for each next move.

Virtue II, closely akin to conservatism, is *modesty*. One hypothesis is more modest than another if it is weaker in a logical sense: if it is implied by the other, without implying it. A hypothesis A is more modest than A and B as a joint hypothesis. Also, one hypothesis is more modest than another if it is more humdrum: that is, if the events that it assumes to have happened are of a more usual and familiar sort, hence more to be expected. Thus suppose a man rings our telephone and ends by apologizing for dialing the wrong number. We will guess that he slipped, rather than that he was a burglar checking to see if anyone was home. It is the more modest of the two hypotheses, butterfingers being rife. We could be wrong, for crime is rife too. But still the butterfingers hypothesis scores better on modesty than the burglar hypothesis, butterfingers being rifer.

We habitually practice modesty, all unawares, when we identify recurrent objects. Unhesitatingly we recognize our car off there where we parked it, though it may have been towed away and another car of the same model may have happened to pull in at that spot. Ours is the more modest hypothesis, because staying put is a more usual and familiar phenomenon than the alternative combination.

It tends to be the counsel of modesty that the lazy world is the likely world. We are to assume as little activity as will suffice to account for appearances. This is not all there is to modesty. It does not apply to the preferred hypothesis in the telephone example, since Mr. Butterfingers is not assumed to be a less active man than one who might have plotted burglary. Modesty figured there merely in keeping the assumptions down, rather than in actually assuming inactivity. In the example of the parked car, however, the modest hypothesis does expressly assume there to be less activity than otherwise. This is a policy that guides science as well as common sense. It is even erected into an explicit principle of mechanics under the name of the law of least action.

Between modesty and conservatism there is no call to draw a sharp line. But by Virtue I we meant conservatism only in a literal sense—conservation of past beliefs. Thus there remain grades of modesty still to choose among even when Virtue I—compatibility with previous beliefs—is achieved to perfection; for both a slight hypothesis and an extravagant one might be compatible with all previous beliefs.

Modesty grades off in turn into Virtue III, *simplicity*. Where simplicity considerations become especially vivid is in drawing curves through plotted points on a graph. Consider the familiar practice of plotting measurements. Distance up the page represents altitude above sea level, for instance, and distance across represents the temperature of boiling water. We plot our measurements on the graph, one dot for each pair. However many points we plot, there remain infinitely many curves that may be drawn through them. Whatever curve we draw represents our generalization from the data, our prediction of what boiling temperatures would be found at altitudes as yet untested. And the curve we will choose to draw is the simplest curve that passes through or reasonably close to all the plotted points.

There is a premium on simplicity in any hypothesis, but the highest premium is on simplicity in the giant joint hypothesis that is science, or the particular science, as a whole. We cheerfully sacrifice simplicity of a part for greater simplicity of the whole when we see a way of doing so. Thus consider gravity. Heavy objects tend downward: here is an exceedingly simple hypothesis, or even a mere definition. However, we complicate matters by accepting rather the hypothesis that the heavy objects around us are slightly attracted also by one another, and by the neighboring mountains, and by the moon, and that all these competing forces detract slightly from the downward one. Newton propounded this more complicated hypothesis even though, aside from tidal effects of the moon, he had no means of detecting the competing forces; for it meant a great gain in the simplicity of physics as a whole. His hypothesis of universal gravitation, which has each body attracting each in proportion to mass and inversely as the square of the distance, was what enabled him to make a single neat system of celestial and terrestrial mechanics.

A modest hypothesis that was long supported both by theoretical considerations and by observation is that the trajectory of a projectile is a parabola. A contrary hypothesis is that the trajectory deviates imperceptibly from a parabola, constituting rather one end of an ellipse whose other end extends beyond the center of the earth. This hypothesis is less modest, but again it conduces to a higher simplicity: Newton's laws of motion and, again, of gravitation. The trajectories are brought into harmony with Kepler's law of the elliptical orbits of the planets.

Another famous triumph of this kind was achieved by Count Rumford and later physicists when they showed how the relation of gas pressure to temperature could be accounted for by the impact of oscillating particles, for in this way they reduced the theory of gases to the general laws of motion. Such was the kinetic theory of gases. In order to achieve it they had to add the hypothesis, by no means a modest one, that gas consists of oscillating particles or molecules; but the addition is made up for, and much more, by the gain in simplicity accruing to physics as a whole.

What is simplicity? For curves we can make good sense of it in geometrical terms. A simple curve is continuous, and among continuous curves the simplest are perhaps those whose curvature changes most gradually from point to point. When scientific laws are expressed in equations, as they so often are, we can make good sense of simplicity in terms of what mathematicians call the degree of an equation, or the order of a differential equation. This line was taken by Sir Harold Jeffreys. The lower the degree, the lower the order, and the fewer the terms, the simpler the equation. Such simplicity ratings of equations agree with the simplicity ratings of curves when the equations are plotted as in analytical geometry.

Simplicity is hard to define when we turn away from curves and equations. Sometimes in such cases it is not to be distinguished from modesty. Commonly a hypothesis A will count as simpler than A and B together; thus far simplicity and modesty coincide. On the other hand the simplicity gained by Newton's hypothesis of universal gravitation was not modesty, in the sense that we have assigned to that term; for the hypothesis was not logically implied by its predecessors, nor was it more humdrum in respect of the events that it assumed. Newton's hypothesis was simpler than its predecessors in that it covered in a brief unified story what had previously been covered only by two unrelated accounts. Similar remarks apply to the kinetic theory of gases.

In the notion of simplicity there is a nagging subjectivity. What makes for a brief unified story depends on the structure of our language, after all, and on our available vocabulary, which need not reflect the structure of nature. This subjectivity of simplicity is puzzling, if simplicity in hypotheses is to make for plausibility. Why should the subjectively simpler of two hypotheses stand a better chance of predicting objective events? Why should we expect nature to submit to our subjective standard of simplicity?

That would be too much to expect. Physicists and others are continually finding that they have to complicate their theories to accommodate new data. At each stage, however, when choosing a hypothesis subject to subsequent correction, it is still best to choose the simplest that is not yet excluded. This strategy recommends itself on much the same grounds as the strategies of conservatism and modesty. The longer the leap, we reflected, the more and wilder ways of going wrong. But likewise, the more complex the hypothesis, the more and wilder ways of going wrong; for how can we tell which complexities to adopt? Simplicity, like conservatism and modesty, limits liability. Conservatism can be good strategy even though one's present theory be ever so far from the truth, and simplicity can be good strategy even though the world be ever so complicated. Our steps toward the complicated truth can usually be laid out most dependably if the simplest hypothesis that is still tenable is chosen at each step. It has even been argued that this policy will lead us at least asymptotically toward a theory that is true.

There is more, however, to be said for simplicity: the simplest hypothesis often just is the likeliest, apparently, quite apart from questions of cagy strategy. Why should this be? There is a partial explanation in our ways of keeping score on predictions. The predictions based on the simpler hypotheses tend to be scored more leniently. Thus consider curves, where simplicity comparisons are so clear. If a curve is kinky and complex, and if some measurement predicted from the curve turns out to miss the mark by a distance as sizable as some of the kinks of the curve itself, we will count the prediction a failure. We will feel that so kinky a curve, if correct, would have had a kink to catch this wayward point. On the other hand, a miss of the same magnitude might be excused if the curve were smooth and simple. It might be excused as due to inaccuracy of measurement or to some unexplained local interference. This cynical doctrine of selective leniency is very plausible in the case of the curves. And we may reasonably expect a somewhat similar but less easily pictured selectivity to be at work in the interest of the simple hypotheses where curves are not concerned.

Considering how subjective our standards of simplicity are, we wondered why we should expect nature to submit to them. Our first answer was that we need not expect it; the strategy of favoring the simple at each step is good anyway. Now we have noted further that some of nature's seeming simplicity is an effect of our bookkeeping. Are we to conclude that the favoring of simplicity is entirely our doing, and that nature is neutral in the matter? Not quite. Darwin's theory of natural selection offers a causal connection between subjective simplicity and objective truth in the following way. Innate subjective standards of simplicity that make people prefer some hypotheses to others will have survival value insofar as they favor successful prediction. Those who predict best are likeliest to survive and reproduce their kind, in a state of nature anyway, and so their innate standards of simplicity are handed down. Such standards will also change in the light of experience, becoming still better adapted to the growing body of science in the course of the individual's lifetime. (But these improvements do not get handed down genetically.)

Virtue IV is *generality*. The wider the range of application of a hypothesis, the more general it is. When we find electricity conducted by a piece of copper wire, we leap to the hypothesis that all copper, not just long thin copper, conducts electricity.

The plausibility of a hypothesis depends largely on how compatible the hypothesis is with our being observers placed at random in the world. Funny coincidences often occur, but they are not the stuff that plausible hypotheses are made of. The more general the hypothesis is by which we account for our present observation, the less of a coincidence it is that our present observation should fall under it. Hence, in part, the power of Virtue IV to confer plausibility.

The possibility of testing a hypothesis by repeatable experiment presupposes that the hypothesis has at least some share of Virtue IV. For in a repetition of an experiment the test situation can never be exactly what it was for the earlier run of the experiment; and so, if both runs are to be relevant to the hypothesis, the hypothesis must be at least general enough to apply to both test situations.[1] One would of course like to have it much more general still.

Virtues I, II, and III made for plausibility. So does Virtue IV to some degree, we see, but that is not its main claim; indeed generality conflicts with modesty. But generality is desirable in that it makes a hypothesis interesting and important if true.

We lately noted a celebrated example of generality in Newton's hypothesis of universal gravitation, and another in the kinetic theory of gases. It is no accident that the same illustrations should serve for both simplicity and generality. Generality without simplicity is cold comfort. Thus take celestial mechanics with its elliptical orbits, and take also terrestrial mechanics with its parabolic trajectories, just take them in tandem as a bipartite theory of motion. If the two together cover everything covered by Newton's unified laws of motion, then generality is no ground for preferring Newton's theory to the two taken together. But Virtue II, simplicity, is. When a way is seen of gaining great generality with little loss of simplicity, or great simplicity with no loss of generality, the conservatism and modesty give way to scientific revolution.

The aftermath of the famous Michelson-Morley experiment of 1887 is a case in point. The purpose of this delicate and ingenious experiment was to measure the speed with which the earth travels through the ether. For two centuries, from Newton onward, it had been a well-entrenched tenet that something called the ether pervaded all of what we think of as empty space. The great physicist Lorentz (1853–1928) had hypothesized that the ether itself was stationary. What the experiment revealed was that the method that was expected to enable measurement of the earth's speed through the ether was totally inadequate to that task. Supplementary hypotheses multiplied in an attempt to explain the failure without seriously disrupting the accepted physics. Lorentz, in an effort to save the hypothesis of stationary ether, shifted to a new and more complicated set of formulas in his mathematical physics. Einstein soon cut through all this, propounding what is called the special theory of relativity.

This was a simplification of physical theory. Not that Einstein's theory is as simple as Newton's had been; but Newton's physics had been shown untenable by the Michelson-Morley experiment. The point is that Einstein's theory is simpler than Newton's as corrected and supplemented and complicated by Lorentz and others. It was a glorious case of gaining simplicity at the sacrifice of conservatism; for the time-honored ether went by the board, and far older and more fundamental tenets went by the board too. Drastic changes were made in our conception of the very structure of space and time. . . .

Yet let the glory not blind us to Virtue I. When our estrangement from the past is excessive, the imagination boggles; genius is needed to devise the new theory, and high talent is needed to find one's way about in it. Even Einstein's revolution, moreover, had its conservative strain; Virtue I was not wholly sacrificed. The old physics of Newton's classical mechanics is, in a way, preserved after all. For the situations in which the old and the new theories would predict contrary observations are situations that we are not apt to encounter without sophisticated experiment—because of their dependence on exorbitant velocities or exorbitant distances. This is why classical mechanics held the field so long. Whenever, even having switched to Einstein's relativity theory, we dismiss those exorbitant velocities and distances for the purpose of some practical problem, promptly the discrepancy between Einstein's theory and Newton's becomes too small to matter. Looked at from this angle, Einstein's theory takes on the aspect not of a simplification but a generalization. We might say that the sphere of applicability of Newtonian mechanics in its original simplicity was shown, by the Michelson-Morley experiment and related results, to be less than universal; and then Einstein's theory comes as a generalization, presumed to hold universally. Within its newly limited sphere, Newtonian mechanics retains its old utility. What is more, the evidence of past centuries for Newtonian mechanics even carries over, within these limits, as evidence for Einstein's physics; for, as far as it goes, it fits both.

What is thus illustrated by Einstein's relativity is more modestly exemplified elsewhere, and generally aspired to: the retention, in some sense, of old theories in new ones. If the new theory can be so fashioned as to diverge from the old only in ways that are undetectable in most ordinary circumstances, then it inherits the evidence of the old theory rather than having to overcome it. Such is the force of conservatism even in the context of revolution.

Virtues I through IV may be further illustrated by considering Neptune. That Neptune is among the planets is readily checked by anyone with reference material; indeed it passes as common knowledge, and there is for most of us no need to check it. But only through extensive application of optics and geometry was it possible to determine, in the first instance, that the body we call Neptune exists, and that it revolves around the sun. This required not only much accumulated science and mathematics, but also powerful telescopes and cooperation among scientists.

In fact it happens that Neptune's existence and planethood were strongly suspected even before that planet was observed. Physical theory made possible the calculation of what the orbit of the planet Uranus should be, but Uranus' path differed measurably from its calculated course. Now the theory on which the calculations were based was, like all theories, open to revision or refutation. But here conservatism operates: one is loath to revise extensively a well-established set of beliefs, especially a set so deeply entrenched as a basic portion of physics. And one is even more loath to abandon as spurious immense numbers of observation reports made by serious scientists. Given that Uranus had been observed to be as much as two minutes of arc from its calculated position, what was sought was a discovery that would render this deviation explicable within the framework of accepted theory. Then the theory and its generality would be unimpaired, and the new complexity would be minimal.

It would have been possible in principle to speculate that some special characteristic of Uranus exempted that planet from the physical laws that are followed by other planets. If such a hypothesis had been resorted to, Neptune would not have been discovered; not then, at any rate. There was a reason, however, for not resorting to such a hypothesis. It would have been what is called an *ad hoc hypothesis,* and ad hoc hypotheses are bad air; for they are wanting in Virtues III and IV. Ad hoc hypotheses are hypotheses that purport to account for some particular observations by supposing some very special forces to be at work in the particular cases at hand, and not generalizing sufficiently beyond those cases. The vice of an ad hoc hypothesis admits of degrees. The extreme case is where the hypothesis covers only the observations it was invented to account for, so that it is totally useless in prediction. Then also it is insusceptible of confirmation, which would come of our verifying its predictions.

Another example that has something of the implausibility of an ad hoc hypothesis is the water-diviner's belief that a willow wand held above the ground can be attracted by underground water. The force alleged is too special. One feels, most decidedly, the lack of an intelligible mechanism to explain the attraction. And what counts as intelligible mechanism? A hypothesis strikes us as giving an intelligible mechanism when the hypothesis rates well in familiarity, generality, simplicity. We attain the ultimate in intelligibility of mechanism, no doubt, when we see how to explain something in terms of physical impact or the familiar and general laws of motion.

There is an especially notorious sort of hypothesis which, whether or not properly classified also as ad hoc, shares the traits of insusceptibility of confirmation and uselessness in prediction. This is the sort of hypothesis that seeks to save some other hypothesis from refutation by systematically excusing the failures of its predictions. When the Voice from Beyond is silent despite the incantations of the medium, we may be urged to suppose that "someone in the room is interfering with the communication." In an effort to save the prior hypothesis that certain incantations will summon forth the Voice, the auxiliary hypothesis that untoward thoughts can thwart audible signals is advanced. This auxiliary hypothesis is no wilder than the hypothesis that it was invoked to save, and thus an uncritical person may find the newly wrinkled theory no harder to accept than its predecessor had been. On the other hand the critical observer sees that evidence has ceased altogether to figure. Experimental failure is being milked to fatten up theory.

These reflections bring a fifth virtue to the fore: *refutability,* Virtue V. It seems faint praise of a hypothesis to call it refutable. But the point, we have now seen, is approximately this: some imaginable event, recognizable if it occurs, must suffice to refute the hypothesis. Otherwise the hypothesis predicts nothing, is confirmed by nothing, and confers upon us no earthly good beyond perhaps a mistaken peace of mind.

This is too simple a statement of the matter. Just about any hypothesis, after all, can be held unrefuted no matter what, by making enough adjustments in other beliefs—though sometimes doing so requires madness. We think loosely of a hypothesis as implying predictions when, strictly speaking, the implying is done by the hypothesis together with a supporting chorus of ill-distinguished background beliefs. It is done by the whole relevant theory taken together.

Properly viewed, therefore, Virtue V is a matter of degree, as are its four predecessors. The degree to which a hypothesis partakes of Virtue V is measured by the cost of retaining the hypothesis in the face of imaginable events. The degree is measured by how dearly we cherish the previous beliefs that would have to be sacrificed to save the hypothesis. The greater the sacrifice, the more refutable the hypothesis.

A prime example of deficiency in respect of Virtue V is astrology. Astrologers can so hedge their predictions that they are devoid of genuine content. We may be told that a person will "tend to be creative" or "tend to be outgoing," where the evasiveness of a verb and the fuzziness of adjectives serve to insulate the claim from repudiation. But even if a prediction should be regarded as a failure, astrological devotees can go on believing that the stars rule our destinies; for there is always some item of information, perhaps as to a planet's location at a long gone time, that may be alleged to have been overlooked. Conflict with other beliefs thus need not arise.

All our contemplating of special virtues of hypotheses will not, we trust, becloud the fact that the heart of the matter is observation. Virtues I through V are guides to the framing of hypotheses that, besides conforming to past observations, may plausibly be expected to conform to future ones. When they fail on the latter score, questions are reopened. Thus it was that the Michelson-Morley experiment led to modifications, however inelegant, of Newton's physics at the hands of Lorentz. When Einstein came out with a simpler way of accommodating past observations, moreover, his theory was no mere reformulation of the Newton-Lorentz system; it was yet a third theory, different in some of its predicted observations and answerable to them. Its superior simplicity brought plausibility to its distinctive consequences.

Hypotheses were to serve two purposes: to explain the past and predict the future. Roughly and elliptically speaking, the hypothesis serves these purposes by implying the past events that it was supposed to explain, and by implying future ones. More accurately speaking, as we saw, what does the implying is the whole relevant theory taken together, as newly revised by adoption of the hypothesis in question. Moreover, the predictions that are implied are mostly not just simple predictions of future observations or other events; more often they are conditional predictions. The hypothesis will imply that we will make these further observations if we look in such and such a place, or take other feasible steps. If the predictions come out right, we can win bets or gain other practical advantages. Also, when they come out right, we gain confirmatory evidence for our hypotheses. When they come out wrong, we go back and tinker with our hypotheses and try to make them better.

What we called limiting principles in chapter 4 are, when intelligible, best seen as hypotheses—some good, some bad. Similarly, of course, for scientific laws generally. And similarly for laws of geometry, set theory, and other parts of mathematics. All these laws—those of physics and those of mathematics equally—are among the component hypotheses that fit together to constitute our inclusive scientific theory of the world. The most general hypotheses tend to be the least answerable to any particular observation, since subsidiary hypotheses can commonly be juggled and adjusted to accommodate conflicts; and on this score of aloofness there is no clear boundary between theoretical physics and mathematics. Of course hypotheses in various fields of inquiry may tend to receive their confirmation from different kinds of investigation, but this should in no way conflict with our seeing them all as hypotheses.

We talk of framing hypotheses. Actually we inherit the main ones, growing up as we do in a going culture. The continuity of belief is due to the retention, at each particular time, of most beliefs. In this retentiveness science even at its most progressive is notably conservative. Virtue I looms large. A reasonable person will look upon some of his or her retained beliefs as self-evident, on others as common knowledge though not self-evident, on others as vouched for by authority in varying degree, and on others as hypotheses that have worked all right so far.

But the going culture goes on, and each of us participates in adding and dropping hypotheses. Continuity makes the changes manageable. Disruptions that are at all sizable are the work of scientists, but we all modify the fabric in our small way, as when we conclude on indirect evidence that the schools will be closed and the planes grounded or that an umbrella thought to have been forgotten by one person was really forgotten by another.

NOTE

1. We are indebted to Nell E. Scroggins for suggesting this point.

25

Justifying Scientific Theories

Ronald Giere

Scientists sometimes talk about *justifying* theories, but more often they talk about *testing* models, hypotheses, or theories. To understand how theories are justified, then, we must first understand what constitutes a test of a model, hypothesis, or theory. Then we must learn how the results of tests are related to the *justification* of a theory.

Since *theories* are generalized *theoretical hypotheses,* justifying a theory is justifying a theoretical hypothesis. . . . But justifying a theoretical hypothesis itself requires that *it* be the *conclusion* of an argument in which some other statements are premises. We already know that the appropriate argument in this case will be an *inductive* argument. Now we must learn about the detailed structure of these arguments. This involves discovering what kinds of statements may serve as premises. . . .

I. TESTING NEWTON'S THEORY: HALLEY'S COMET

Newton's work on mechanics was first published in 1687. Around 1695, an English astronomer, Edmond Halley, began applying Newton's ideas to the motions of comets. He was probably acting on Newton's own suggestion that comets may be like small planets with very large elliptical orbits. In any case, comets were very interesting phenomena because they had always been viewed as mysterious or even ominous. Their appearances certainly exhibited no apparent regularity. If Newton's suggestion was correct, however, the behavior of comets would exhibit a great deal of underlying regularity.

Halley began investigating a comet that he himself had observed in 1682. The observations of 1682 gave a quite precise location, relative to the sun and the background stars, of the comet's path while it was within sight. However, since only a small part of the total orbit could be observed, it was not possible to determine how big it was and thus how long it would take for the comet to return. Indeed, it was impossible to determine whether the orbit was an ellipse, as Newton suggested, or a parabola. Newton's theory allowed the possibility of a parabolic orbit, but such an orbit would mean that the comet would come by only once and then leave the solar system forever. If, however, the orbit was elliptical, as Newton suggested, the comet should have traveled that same path many times before.

Halley began digging into the records of observations for previous comets. He found 24 cases, going back roughly 150 years, for which the observations were precise enough to compare with the observations of 1682. For two of these, one in 1606–1607 and one in 1530–1531, the recorded orbits were very close to that of the 1682 comet. The 1606 comet had in fact been observed by Kepler himself. Halley argued that it was extremely unlikely that three *different* comets should have such similar orbits, and concluded that these were three appearances of the *same* comet in an elliptical orbit with a period of roughly 76 years. He speculated, but could not prove, that the slight discrepancies in the three orbits were due to gravitational influences from the planets, particularly Jupiter.

But Halley didn't stop here. Using the data from all three cases, together with the hypothesis that he was dealing with a Newtonian system (or approximately Newtonian, since the observed orbits were not perfectly identical), Halley calculated the time of the next return. He boldly predicted that the comet should be seen again sometime in December, 1758. Figure 1 will help you to keep in mind the relevant details of this example. It is obviously not drawn to scale.

Halley published his work on comets in 1705. It was well received by Newton and the growing band of English Newtonians. It did little, however,

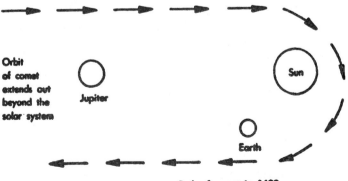

Figure 1. Halley's comet

to convince the French, who continued in the older theoretical tradition of Descartes. Halley himself died, a respected scientist, in 1743, fifteen years before the predicted return of the comet. By this time even the French were coming around to the Newtonian way of thinking. And Halley's prediction was remembered. In 1756 the French Academy of Science offered a prize for the best calculation of the return of the comet. The comet reappeared, as predicted, on Christmas Day, 1758, and was officially named "Halley's comet." This date provides a convenient point to mark the final triumph of the Newtonian theoretical tradition over the Cartesian.

Intuitively it seems quite clear that the success of Halley's *prediction* did provide substantial *justification* for Newton's *theory*. Why is this so? What made this a *good test* of Newton's theory? How does the experimental verification of Halley's prediction provide an *argument* that justifies the theory?

II. THE BASIC ELEMENTS OF A TEST

The preceding section is fairly typical of reports describing the justification of theoretical hypotheses. Such reports do not generally distinguish the *experiment* itself, the *result* of the experiment, and the *justification* of the hypothesis. These things must be distinguished, however, if we are to develop a general account that can be applied to a variety of cases. Because it is simpler and more typical, we will take Halley's second test, the predicted return of the comet in 1758, as our standard example. Later we will return to consider the first test, that based on earlier observations of the comet.

Experiments, Results, and Justification

Before we examine the details, it will be helpful to see some of the basic distinctions set out in a preliminary way. This will give you some idea of how the various pieces are supposed to fit together so that you can better appreciate the details as they are presented.

The basic idea behind any *experiment* is to arrange things so that in the end you will get the information you need to argue either that the hypothesis is true or that it is false. You ought to be able to tell at the beginning, *before* the experiment is actually performed, whether or not things have been so arranged. It was clear in 1705 that the predicted return of the comet in 1758 constituted a good test of Halley's hypothesis.

When the experiment terminates, something will have happened, or failed to happen. This is the *result* of the experiment. You cannot know what this is, of course, until the experiment is over. In Halley's case the results were not in until 1758, fifteen years after Halley himself had died. In this example the experiment was simple. Just wait until 1758 and look to see whether the comet reappears. The result of this experiment, of course, was that it did reappear. But it might *not* have. "The comet will reappear in 1758," is a contingent statement.

Now, to *justify* either the hypothesis or its negation, one constructs an *argument*. The premises of this argument will be statements that relate both to the experiment and to its results. Indeed, in our account, such arguments have only two premises, one describing the experiment and one describing its results. The conclusion of the argument will be either the hypothesis or its negation.

The Basic Elements

The Hypothesis. The most important element in a test is the *hypothesis* whose truth or falsity is at issue. The whole point of designing an experiment is to be able to come up with a justification either for the hypothesis or for its negation.

In Halley's case, the theoretical hypothesis at issue is that the comet and the sun together form a particular type of Newtonian Particle System, one in which a small mass (the comet) is in elliptical orbit around a large mass (the sun). Or, in light of the possible effects of Jupiter, the hypothesis is that they are *approximately* a Newtonian system of this type. Of course Halley was not only concerned with this one comet. He was concerned to show that all comets are Newtonian systems. Halley was also concerned with what is often called Newton's *celestial mechanics*—the *theory* that all celestial objects, including the Earth, form a Newtonian Particle System. We will later consider the status of these broader hypotheses.

The Prediction. The word "prediction" can be used to refer either to the possible event whose occurrence is predicted or to the *statement* that describes this possible event. It will generally be clear from the context which reference is intended.

In an experimental test of some hypothesis, the *prediction* always describes the occurrence of a possible state of some real system under investigation. To assert the prediction, then, is to say that the state will occur at the indicated time. This is clearly a contingent statement—that is, a statement that might be true or might be false. In our example, the prediction was that the comet would again be visible from the earth in December 1758. It turned out to be true.

Typically, as in Halley's second test, the prediction describes some possible *future* state of the system under study. That is why we use the word "prediction." However, it is not necessary for the prediction to describe a state of the system that lies in the future relative to the experiment. In Halley's first test, the "predicted" events were the *past* appearances of the comet. Similarly, in archaeology or evolutionary biology one studies systems that operated long ago. All the events in the history of these systems are in the past relative to the present investigation. In such investigations, then, the "prediction" would still describe some state of those systems, but a state that occurred long ago. The important thing, as you will soon learn, is whether the investigators knew about the predicted event and whether this knowledge played any part in the construction of the theory.

Initial Conditions. The term "initial conditions" may also be used to refer either to the occurrence of a state of the system or to a *statement* describing this occurrence. Typically, the initial conditions describe the occurrence of some state at the beginning of the experiment—thus the term "initial" conditions. In Halley's second test, the initial conditions described the positions and velocities of the comet that were recorded prior to 1705. Halley used this information in his calculation of the shape of the orbit and of the time of the comet's next return. In general, as illustrated by Halley's first test, it is not necessary that the initial conditions describe some state that occurred *before* the event described by the prediction. But they must describe some *other* state.

This brief rundown of the basic elements of a test leaves many important questions unanswered. Most of these questions, however, will be covered in the following discussion of the way these elements fit together in a *good* test of a scientific theory. . . .

III. WHAT MAKES A GOOD TEST?

We will now see how the elements of a test fit together to make a *good* test, that is, a test that provides the basis for a *justification* either of the hypothesis or of its negation. A test of a theoretical hypothesis is obviously a very complicated thing. It is nevertheless possible to isolate several general features of tests that are sufficient to determine the justified conclusion.

The Prediction Is Deducible

Halley began with the *theoretical hypothesis* that the comet and the sun together are (approximately) a Newtonian system of a particular type. Using Newton's theoretical model, together with the observed *initial conditions* (the partial orbits observed during previous appearances of the comet), he calculated, that is, *deduced*, the *prediction* that the comet would again be visible at the end of 1758. The first general feature of a *good test* of a theoretical hypothesis, then, is:

> The prediction is logically deducible from the hypothesis together with the initial conditions.

Note that whether a test has this feature is not a contingent matter of fact. It is a purely *logical* issue. It is equivalent to the question whether the *argument* from the hypothesis and the initial conditions to the prediction is *deductively valid.* . . . So this feature of a good test can be ascertained without assuming that the hypothesis is true and without knowing whether the prediction will be true or not. In general, being told that the prediction was calculated mathematically is sufficient for you to conclude that the prediction is deducible

That there must be *some* connections between the hypothesis and the prediction is obvious, because otherwise there would be no reason to regard the prediction as part of a test of the hypothesis. What we have found is that *one* connection is a *deductive* connection. This is a logical connection that goes *from* the hypothesis (plus initial conditions) *to* the prediction. The relationships in this first characteristic of a good test are pictured in Figure 2.

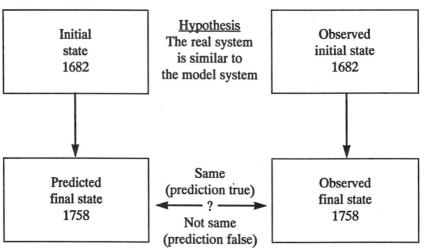

Figure 2. Theoretical hypothesis, initial condition, and prediction.

It is important to realize why the *initial conditions* must be included among the premises of the deduction. A theoretical hypothesis asserts only that some system has a specified *general structure*. It therefore tells you all the kinds of things that the system is *capable* of doing. But it does not tell you what the system is *in fact* doing at any particular moment. Thus it is impossible to deduce from the hypothesis alone any prediction that specified the actual state of the system at some definite time. The initial conditions, however, pin down the state of the system at some particular time. If the system is deterministic, as Newtonian systems are, that is sufficient to pin it down for all time.

The Prediction Is Improbable

To have a good test of a theoretical hypothesis, it is not enough that the prediction be implied. This is *necessary,* but *not sufficient.* To take a simple example, consider the different, and quite vague, prediction that some comet with some orbit or other would be observed after 1705. Now, this vague prediction also follows logically from Halley's hypothesis and observed data. Indeed, it follows logically from Halley's actual prediction that his comet would return in 1758. And as a matter of fact, a number of comets were observed between 1705 and Halley's death in 1743. Yet neither Halley nor anyone else ever claimed that these events justified his hypothesis. Why not? What was special about the prediction Halley made?

The trouble with the vague prediction of observing some comet or other after 1705 is that anyone with a basic knowledge of seventeenth-century astronomy could have made that prediction without knowing anything about Newton—and been quite confident of being correct. Good-sized comets had been appearing at least once every five years, on the average. Indeed, Halley himself had observed comets in both 1680 and 1681. Thus, given such frequent appearances by various comets, one wouldn't have needed Newton's ideas at all in order to make the vague prediction. In fact, then, even if the comet and sun were nothing at all like a simple two-particle Newtonian system, the prediction that some comet would be observed after 1705 was known to be highly likely to be true.

Halley's actual prediction was quite different. He predicted the appearance of a comet in a definite thirty-day period 53 years in the future (relative to when the prediction was published in 1705). Moreover, he did not predict any old comet. He predicted a comet with a precisely specified orbit. Anyone in 1705, knowledgeable about astronomy but ignorant of Newton's work, would have to have concluded that the appearance of a comet in December 1758 with that precisely specified orbit was *highly unlikely.* But not impossible! Nothing known in 1705 ruled out the *possibility* that such a comet would be observed. But everything known at the time (apart from Newton's hypotheses) counted against the prediction coming true. Without Newton's model, anyone would have to have regarded the success of Halley's prediction as a cosmic coincidence.

Whether the prediction is unlikely to be true is also in principle a *logical* matter that can be judged when the experiment is proposed without knowing whether the hypothesis or prediction are true. It is a matter of considering everything (other than the hypothesis at issue) then known to be relevant to the truth or falsity of the prediction, and determining whether this knowledge makes the truth of the prediction likely or not. In *principle* this could be a matter of deducing the improbability of the prediction from all current knowledge. In *practice* we must rely on scientists' informal judgments. Indeed, in Halley's time, probability theory was in its infancy. No one could have even pretended to calculate a definite probability for the appearance of a comet with the prescribed orbit. But any knowledgeable person could judge quite correctly that, leaving Newton's ideas aside, the prediction was not likely to come true.

The second general feature of a good test, then, can be summarized as follows:

> Relative to everything else known at the time (excluding the hypothesis being tested), it must be improbable that the prediction will turn out to be true.

*   *   *

The Prediction Is Verifiable

There is one important aspect of tests that we have thus far simply assumed. Before proceeding, we should make this assumption explicit.

> It must be possible, at the appropriate time, to verify whether the prediction is in fact true or false.

This is required because the truth or falsity of the prediction is going to provide part of the *justification* for our conclusion concerning the hypothesis. Thus it must be something we can justify *independently* of the truth or falsity of the hypothesis. Since a prediction is generally a description of the state of some system at a specified time, the predicted state must be one that can be observed or otherwise reliably detected.

As an illustration of the importance of this requirement, imagine that Halley had calculated the time and place at which the comet would be *farthest* from the Earth. He could have deduced such a prediction and it would have been very unlikely, apart from Newton's theory, that there should be a comet at that exact time and place. And this event would have happened well within Halley's lifetime so he could have been around to collect the glory. But no one had any way of telling whether this prediction was fulfilled or not. The comet is so small and the place so far away that even the best telescopes of the day could not have located it. The return to our solar system, however, was something which almost anyone could see with their naked eyes.

There is no necessity, of course, that the truth of the prediction be so directly observable as the return of Halley's comet. In modern science, the predictions that justify current hypotheses are often quite complex and require millions of dollars worth of equipment to verify. That doesn't matter. All that matters is that it can be done. . . .

IV. JUSTIFYING THEORETICAL HYPOTHESES

Good *experimental design* lies at the heart of good scientific research. An experiment is the scientists' way of asking Nature a question. The question has the form: Is this theoretical hypothesis true or false? Nature answers by exhibiting the prediction as true or false. If the prediction is true, the scientist, justifiably, concludes that the hypothesis is true. If the prediction is false, the hypothesis is refuted.

We will now look more closely at this form of reasoning, beginning with a general overview and then moving on to consider the *arguments* involved.

A Simple Inductive Rule

In justifying theoretical hypotheses, scientists follow a *simple inductive rule*.

> If the prediction is successful, the hypothesis is justified.
> if the prediction fails, the hypothesis is refuted.

The rule assumes, of course, that the experiment was well designed, which means that our conditions for a *good test* are satisfied. Figure 3 will help you at least to begin to see why this rule is a good rule.

Since the hypothesis is a contingent statement, either it is (approximately) *true* or it is *not* (approximately) *true*. Similarly, the prediction is a contingent statement that was chosen partly because it could be verified as true or false. Thus there are only two possible outcomes of the experiment: either the prediction is true or it is false. Applying the simple inductive rule, then, leads to one of two possible conclusions: the hypothesis is *justified* or it is *refuted*.

There are *four* possibilities shown in Figure 3. The hypothesis might be (approximately) true and end up either justified or refuted. Likewise, it might be false (not approximately true) and end up either justified or refuted. However, if the experiment satisfies the requirements for a *good test,* then two of these possibilities will be very likely and the other two very unlikely.

	If hypothesis approximately true	If hypothesis NOT approximately true
Hypothesis justified	Likely	Unlikely
Hypothesis refuted	Unlikely	Likely

Figure 3. Likelihood of results using the simple inductive rule.

Suppose first that the hypothesis is (approximately) *true.* Then, so long as the initial conditions are true, the prediction should also be true. Following the simple inductive rule, we shall conclude that the hypothesis is justified—which is what we should conclude. That is, we will justifiably claim that the hypothesis is (approximately) true when in fact it is. Of course, it is possible that something goes wrong somewhere along the line and that we end up concluding that the hypothesis is false when it is in fact true. But that should be highly unlikely.

Now suppose that the hypothesis is *false* (not approximately true). The prediction has been chosen so that it would be very unlikely to come out true unless the hypothesis is (approximately) true. That was one of our requirements for a good test. So in this case the prediction is likely to fail. Applying the simple inductive rule, we conclude that the hypothesis is refuted, that is, false (or not approximately true). This is what we should conclude. It is possible, of course, that the prediction succeeds even though the hypothesis is not even approximately true. It was possible, for example, that some other comet should have appeared in December 1758. It is even possible that the same comet, tossed about by other forces far outside the solar system, nevertheless, by accident, appeared as predicted. In these cases we would conclude, mistakenly, that the hypothesis is (approximately) true. Given everything known when the prediction was made, however, this would be an unlikely circumstance.

The simple inductive rule, then, satisfies our earlier definition of good *inductive* reasoning. It does not guarantee a true conclusion, but it does make it very likely that the conclusion reached is in fact true. What we have yet to see is the reasoning behind the rule set out explicitly in argument form with premises and conclusion. That comes next.

The Premises

We will begin by setting out the *general form* of the arguments for and against theoretical hypotheses. Then we will wrap up the Halley case, which is an example of an argument *for* a hypothesis. For a look at an argument against a theoretical hypothesis we shall need a new example.

As you might have guessed, arguments for or against theoretical hypotheses have the general form of *conditional arguments*. Their first premise will be a *conditional statement*. The second premise will include the *result* of the experiment, that is, either the prediction or its negation. And the conclusion will be either the hypothesis or its negation. So our main task is to formulate the conditional first premises of these arguments.

If the experiment is a *good test* of the hypothesis, then the prediction can be deduced from the hypothesis together with appropriate initial conditions. This means that the following *argument* form is *valid*.

H. IC. Thus, P.

where "H" stands for the Hypothesis, "IC" for the Initial Conditions, and "P" for the Prediction. This means . . . that the following *conditional statement* is a tautology:

Condition 1: If (H and IC), then P.

Since Condition 1 is justified by the corresponding deduction, it might be called the *deductibility condition* (or deducibility requirement). Being a tautology, this statement needs no further justification. It must be true.

Our second major requirement for a *good test* was that the prediction be something known to be *improbable* when the test is designed. In order to capture this requirement in a conditional statement, we must explicitly take account of the knowledge used to justify the claim that the prediction is indeed improbable. Such knowledge is often quite diverse. Let us therefore abbreviate all this knowledge by the letter "B," for *background* knowledge. The required conditional statement, then, may be written as:

Condition 2: If (Not H and IC and B), then very probably Not P.

This might be called the *improbability* condition (or improbability requirement).

If Condition 2 is true, its truth is in principle also a matter of logic. That is, in principle we can suppose that the statement "very probably Not P" can be deduced from the conjunction (Not H and IC and B). That such judgments are generally made informally, not by explicit calculation, does not change the logical status of the conditional statement.

This way of representing the second condition for a good test requires that theoretical hypotheses not be taken as asserting that the model *exactly* fits the real system. If we were to interpret H as asserting an exact fit, then it would no longer follow from (Not H and IC and B) that P is improbable. To see why this is so we need only imagine a model of mechanical motion that differs only slightly from Newton's. For example, instead of having the gravitational force vary inversely as the second power of the distance, it might be said to vary inversely as the 2.00001 power. This modified model provides us with a version of Not H, where H represents the exact Newtonian hypothesis. However, if we were to use this version of Not H, it would no longer follow that Not P is probable. Indeed, the difference between the two models is so slight that P would be very probable. Thus the "Not H" in Condition 2 must be read as saying that H is *not* even *approximately* true. This rules out all models that are sufficiently similar in structure to the chosen model that they would make P a likely outcome. So whether the prediction is judged likely or not then depends primarily on the rest of our knowledge at the time. Does everything else that is known imply that the prediction is improbable?

If Not H means that H is *not* approximately true, then H must mean that H is approximately true. Since no prediction is ever made with perfect accuracy, if only because there is always some uncertainty in the initial conditions, the degree of approximation in H need only be assumed to introduce less variation than the initial conditions. Then P, including its range of variation, follows deductively as required for Condition 1 to be justified. Recall that Halley's prediction was only accurate to within one month.

Assuming that the truth or falsity of the prediction can be verified at the appropriate time, Conditions 1 and 2 provide an alternate, if somewhat abstract, way of characterizing a *good* test of a theoretical hypothesis.

> A test of a theoretical hypothesis is a good test if and only if both Condition 1 and Condition 2 are true.

Alternatively, we are *justified* in regarding the test as a good test if and only if we are *justified* in claiming that both conditions are true. The virtue of defining a good test this way is that these same conditions will be the conditional premises for any *argument* used to justify or refute the hypothesis.

Refuting Argument

The theoretical hypothesis cannot be justified or refuted until the result of the experiment is in—that is, until the truth or falsity of the prediction has been determined. Once this is known, the appropriate argument can be constructed. Since the argument that would refute a hypothesis is simpler than one that would justify it, we will begin with refuting arguments.

Suppose, then, that the specified hypothesis, H, has been subjected to a good test. This means that both Condition 1 and Condition 2 have been justified. Now suppose it turns out that the *prediction* is *false*. That is, Not P is true. We now have all the ingredients to construct an argument justifying the conclusion Not H. We will set out this argument in full detail. You should work through it carefully, step by step, at least once, to convince yourself that it is indeed a valid argument. Later on we will use a condensed, short form of the argument.

The argument proceeds as follows:

First Premise:	If (H and IC), then P.
Second Premise:	Not P.
Preliminary Conclusion:	Not (H and IC).
Preliminary Conclusion:	Not H or Not IC.
Additional Premise:	IC.
Final Conclusion:	Not H.

Every step in this argument is deductively valid. The first premise is just Condition 1. The second premise is the result of the experiment; that is, the prediction failed. By the valid step of *denying the consequent* we reach the first preliminary conclusion. This conclusion is the denial of a conjunction, which we know leads validly to the next preliminary conclusion, a disjunction of the negations of the original component statements. . . . Finally, we must explicitly assert the Initial Conditions if we are to reach the conclusion that H is false. The final step is a valid *disjunctive syllogism*. Since every premise has been justified in advance, and every step in the argument is deductively valid, the conclusion is justified.

Once you understand the details of the above argument, it is much easier to collapse it into two premises and the conclusion as follows:

If (H and IC), then P.
Not P and IC.
Thus, Not H.

Justifying Argument

As before, we assume that some theoretical hypothesis, H, has been subjected to a *good test*. This time, however, the *prediction* comes true. In this case the argument justifying the hypothesis has the following form:

First Premise:	If (Not H and IC and B), then very probably Not P.
Second Premise:	P.
Preliminary Conclusion:	Not (Not H and IC and B).
Preliminary Conclusion:	H or Not IC or Not B.
Additional Premise:	IC and B.
Conclusion:	H.

Here again, we will go through this argument in detail before collapsing it into a shorter form.

The first premise is merely Condition 2. The second is the result of the test. The argument from these two premises to the first preliminary conclusion is an *inductive* argument. This is the *only* inductive step in the argument. The second preliminary conclusion is reached by *denying the conjunction*, a valid deductive inference. Only the step, "Not Not H, Thus H," has been left out. Finally, the initial conditions and the background knowledge are explicitly acknowledged as premises. This leads to the conclusion by a valid *disjunctive syllogism*.

Now we can collapse the argument into two premises and the conclusion:

If (Not H and IC and B), then very probably Not P.
P and IC and B.
Thus (inductively), H.

We will always use this "short form" in later discussions.

At this point, one warning is in order. You should recognize that the inductive step in the above argument looks a lot like *denying the consequent*, which is a valid argument. You might therefore be tempted to think that *any* argument of this form is a good inductive argument—a kind of inductive version of denying the consequent. Unfortunately, things are not so simple. Not all arguments of this form are good inductive arguments. What makes this one good is the assumption that we started out with a *good test* of H. This means that it is also true that H and IC imply P. Only with this additional assumption is it correct that the truth of the premises makes it very probable that the conclusion is also true.

*   *   *

V. CELESTIAL MECHANICS AND THE PHLOGISTON THEORY

The previous section was pretty abstract. Let us now put some flesh on these abstractions by examining two examples.

Justifying a Theory: Celestial Mechanics

It is about time we wrapped up the Halley episode. The *elements* of the test are:

H: The comet/sun system is (approximately) a Newtonian Particle System.
IC: The previously observed orbits of the comet.
P: The return of the comet in December 1758.

The prediction is easily verifiable when the time comes. The prediction was calculated (deduced) from the hypothesis and the initial conditions. And all prior information, B, indicated that, apart from Newton's hypothesis, the appearance of such a comet, as predicted, would be very improbable. So the test was a *good test* of the hypothesis.

The prediction turned out to be *true*. Applying the *simple inductive rule*, we conclude, with ample justification, that H is true. In order to exhibit the *argument* that shows in detail how the conclusion is justified, we formulate *Condition 2* and add the required additional premises. The argument, then, is,

If (Not H and IC and B), then very probably Not P.
P and IC and B.
Thus (inductively), H.

That is, the sun/comet system is (at least to a good approximation) a Newtonian Particle System of the type in question.

There is only one further part of the story to complete. In 1758 the conclusion everyone drew was much broader than Halley's hypothesis. In the first place, they appear to have concluded not only that Halley's comet is a Newtonian system, but that *all comets* are Newtonian systems. The justification for this conclusion might seem to be fairly weak since it amounts to generalizing on a *single* case. Yet they were by that time pretty well justified in regarding comets as being fundamentally alike. So any fundamental characteristic true of one is likely to be true of all. And its motions in space and time are surely a fundamental property. But the details of this type of reasoning are not easy to set out.

In the second place, Halley's success was taken to justify Newton's *theory* of *celestial mechanics*. This is the generalization that all celestial bodies fit Newtonian models. Here again the justification for this broad generalization is better than it might at first appear. By 1758 it was well established that the moon and planets obey Newtonian laws. Apart from the distant stars, comets were the only observable celestial motions not yet explicitly tested. By showing that comets are also Newtonian systems, Halley completed the generalization over known types of celestial bodies. Moreover, comets were correctly judged to be the most difficult case for Newtonian models because they exhibited the least apparent regularity of any celestial objects. If even comets are Newtonian systems, what celestial objects could not be?

Finally, Halley's success helped solidify the by then already strong Newtonian *theoretical tradition.* It would be 150 years before anyone seriously questioned the applicability of Newtonian models to celestial motions.

Why Independent Tests Are Desirable

Halley's prediction of the comet's return in 1758 was the basis for the *second* of two tests of the hypothesis that the 1682 comet is in an elliptical Newtonian orbit around the sun. The first test utilized the prediction that previous records of comets would show evidence of the same comet at earlier times. This test is complicated by the fact that it requires assumptions about the record-keeping system of earlier astronomers. To deduce the existence of appropriate records requires that there be record keepers. Halley was lucky that the period of his comet was short enough for it to have been observed by people such as Kepler. Moreover, the records kept have to be precise enough to distinguish comets with different orbits. To the extent that the records are imprecise, the probability of mistaking some other comet for the predicted comet increases. Halley's confidence in the precision of the more recent observations was sufficiently great for him to claim that it would be "next to a miracle" if the comets of 1530–1531, 1606–1607, and 1682 were three different comets with similar orbits. So he, at least, thought that his first prediction was deducible, improbable, and verifiable.

If Halley thought the first test was so good, what is the value of the second test? It makes the justification for the hypothesis even stronger. Roughly speaking, the more *improbable* the prediction relative to previous background knowledge, the stronger the justification if the prediction is successful. But that two improbable predictions should be successful is even more improbable than that either one by itself should be successful. This judgment requires no technical knowledge of probability theory. Being dealt a hand with four aces in an evening of bridge is improbable. Being dealt two such hands in a single evening is twice as improbable. So, even leaving aside the fact that the observations in 1682 and 1758 were more precise than previously, four matching orbits by four different comets would be more improbable than three. So the justification for the hypothesis had to be better in 1758 than it had been in 1705.

In general, then, the more independent tests, the stronger the justification for the hypothesis. But here, as everywhere, one soon reaches a point of diminishing returns. The appearance of Halley's comet in 1985–1986 adds little to the justification of Halley's original hypothesis.

Refuting a Theory: Phlogiston

To call a test of some hypothesis a *good test* does not mean that it ended up providing justification for the hypothesis being tested. It is not necessarily good from the standpoint of someone who wants to justify the hypothesis. A test is good in the more neutral sense that it is designed to provide *an* answer. The answer may be that the hypothesis is justified. But it may also be the reverse. The hypothesis may be refuted. It all depends on whether the prediction was true or not. That a test is a good test also does not mean that the prediction came out true. It only means that there was a clear prediction whose truth or falsity became known at the end of the experiment.

In Halley's case, of course, the prediction came true and the hypothesis was justified. In science, as in everyday life, one tends to remember only the winners. And history is the same. The losers tend to be forgotten. Some losers, however, are remembered, if only because they had a period of considerable success before their downfall. A famous loser is the phlogiston theory.

Fire, like motion and heredity, has always fascinated people. In the

Western world, recorded speculation about the nature of fire also goes back to the Greeks. To them we owe both the myth of Prometheus and the view that the world is made up of a few separate "elements": earth, air, fire, and water. All four are present in the process of combustion. The common-sense view of combustion is that something is driven out of the burning object, leaving only ashes behind. By the eighteenth century, this "something" had a well-established name, *phlogiston*—the fire stuff. Assuming that combustible material contains phlogiston explains most of the obvious facts about combustion. Heating drives off the phlogiston into the air; cooling makes it less volatile; smothering holds it in. The well-known fact that a burning candle placed in an enclosed container soon goes out was explained by saying that the enclosed air gets saturated with phlogiston so that the phlogiston remaining in the wax has nowhere to go.

Phlogiston accounts not only for combustion but also for the very important process of smelting. This is the process by which crude ores are turned into more refined metals. Generally this is done by carefully heating the ores, together with a measured amount of charcoal, to a controlled temperature. It was claimed that the charcoal contains an excess of phlogiston which, at moderately high temperatures, leaves the charcoal and combines with the ore to form the metal. This hypothesis was substantiated by the fact that further heating at higher temperatures returns the metal to its original state. The phlogiston is driven out of the metal by the higher temperature. Even rusting was explained as the result of the phlogiston slowly escaping from the metal.

These claims may be taken as the "laws" that define *phlogiston models*. Such models lay behind many hypotheses about systems undergoing combustion, rusting, or the process of smelting. The *phlogiston theory* was the general hypothesis that this sort of model fits most cases of combustion, smelting, rusting, and so on. In what follows we shall concentrate on combustion.

Combustion is very difficult to study. Most things we commonly burn are made of many different substances and give off many different gasses when burned. Moreover, combustion generally is rapid and violent. Progress in such studies required finding some simple, well-controlled subjects for experimentation. (Remember Mendel's pea plants.) In the 1770s, chemists developed a number of techniques for performing such experiments. The leaders were Joseph Priestley in England and Antoine Lavoisier in France. Priestly supported the phlogiston theory; Lavoisier led the revolution that overthrew it.

In the 1770s, utilizing techniques first developed by Priestley, Lavoisier performed a number of careful experiments with mercury. In one of these experiments he floated a precisely measured amount of mercury on a liquid and covered it with a glass jar, thus enclosing a known amount of air (see figure 4). The mercury was then heated using the rays of the sun focused by

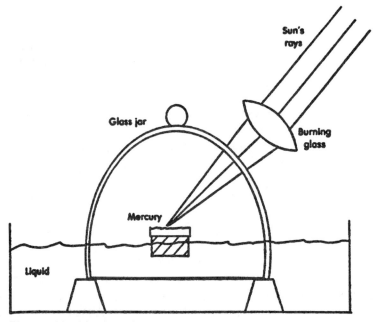

Figure 4. Lavoisier's experiment

a powerful magnifying glass (a burning glass). In such circumstances, as Lavoisier well knew, a red powder, or ash, forms on the surface of the mercury. Some of the mercury undergoes a controlled burning.

Applying the phlogiston model to this experiment, one would expect two things. First, the resulting mercury plus red ash should weigh *less* than the original sample of mercury alone. This is because some phlogiston must be driven off, leaving the ash behind. And the volume of air inside the jar should *increase* since it now contains the phlogiston that was driven out of the mercury. This means that the level of the liquid inside the jar would drop to make room for the additional "air."

Regarding this experiment as a test of the *phlogiston theory*, the elements of the test are these:

H: The experimental setup roughly fits a phlogiston model of combustion.
IC: The various facts describing the experiment are as outlined above.
P: The remaining mercury and red powder together weigh *less* than the original sample of mercury alone. And the level of the liquid inside the jar should go *down*.

Now, does this experiment constitute a *good test* of the hypothesis? There seems little reason whether the prediction *logically* follows from the hypothesis and initial conditions. This follows if anything does. So Condition 1 is satisfied. But is the prediction all that improbable given everything else known at the time? In other words, is Condition 2 justified? This is not so clear. The predicted results are, after all, the common-sense results. The phlogiston model is the common-sense model. Let's withhold judgment on Condition 2 and proceed with the story.

When Lavoisier completed the experiment, the water level had gone *up*, not down, and the mercury/ash combination weighed *more* than the original mercury alone. The prediction was *false*. Applying the simple inductive rule we conclude immediately that the hypothesis is false.

To construct the *argument* justifying our conclusion, we combine Condition 1 with the result of the experiment as follows:

If (H and IC), then P.
Not P and IC.
Thus, Not H.

And this argument is deductively valid.

So in the end the fact that Condition 2 was questionable did not matter. We didn't need it. Of course, if things had gone the other way, then Lavoisier would not have been in a strong position to argue that H is true. But then again, Lavoisier was not out to justify the phlogiston hypothesis. He was trying to refute it.

If the *phlogiston theory* includes the general hypothesis that *all* combustion-like processes fit phlogiston models, then Lavoisier's experiment refuted the theory as well. But no defender of the phlogiston theory interpreted it so broadly. What in fact happened is that members of the *theoretical tradition* based on phlogiston models *modified* their models to accommodate Lavoisier's results. This is a scientifically sound strategy for dealing with unwelcome facts. But the strategy does not pay off unless these new models yield justifiable hypotheses. Merely coming up with a revised model that fits the known results is not enough. Justification requires full-fledged tests that satisfy Condition 2 as well as Condition 1. Phlogiston theorists were not able to do this successfully.

Some of the revised phlogiston models attributed a *negative mass* to phlogiston. This put phlogiston chemists at odds with Newtonian physicists, for whom all particles have *positive* mass. Some models postulated that phlogiston is lighter than air and thus exhibits a buoyancy effect—like the bladder in a fish or the hot air balloons that were then popular in France. Other models specified that the escaping phlogiston is replaced by water vapor, which has greater mass and greater volume than phlogiston. This accounts both for the increased mass of the mercury sample and the decreased volume of the air. But none of these models were justified in applications to further experiments.

They kept on being refuted. By 1789 the phlogiston tradition was effectively dead, even though Priestley himself defended it in a text published as late as 1796. By that time Lavoisier was also dead, a victim of "the terror" that followed the French Revolution. But the theoretical tradition he founded continues. We know it as oxygen chemistry.

This example illustrates an important new (for us) feature of the scientific enterprise. It is easier to refute hypotheses than to justify them. To *refute* a hypothesis all you need is a single logical consequence that is false. To *justify* a hypothesis you need a logical consequence that is initially improbable but nevertheless turns out to be true. That is a much more difficult trick.

By the same token, trying to justify hypotheses is a much riskier business than trying to refute them. To justify a hypothesis you must find a prediction that all previous knowledge indicates is likely to be false. Thus, if the hypothesis in question is indeed false, this is very likely to be discovered in the very process of conducting an experiment that would justify the hypothesis.

It is sometimes thought that any objectivity there is in science is due to scientists being objective people. Some claim that this means being dispassionate regarding the truth or falsity of one's hypotheses. Others have even held that scientists should actively try to refute their own hypotheses! But this is a mistake. The objectivity of *science,* imperfect as it is, is not a function of the objectivity of *scientists.* It is a function of the "logical" rules of the game. These are embodied in the specification of a good test, and thus in Conditions 1 and 2. So there is no reason why scientists should not try to justify their hypotheses and be very disappointed if they are refuted. The rules of the game ensure that the harder one tries to get a good justification, the greater the risk of refutation—unless the hypothesis is indeed on the right track.

26

Objectivity, Value Judgment, and Theory Choice

Thomas S. Kuhn

In the penultimate chapter of a controversial book first published fifteen years ago, I considered the ways scientists are brought to abandon one time-honored theory or paradigm in favor of another. Such decision problems, I wrote, "cannot be resolved by proof." To discuss their mechanism is, therefore, to talk "about techniques of persuasion, or about argument and counterargument in a situation in which there can be no proof." Under these circumstances, I continued, "lifelong resistance [to a new theory] . . . is not a violation of scientific standards. . . . Though the historian can always find men—Priestley, for instance—who were unreasonable to resist for as long as they did, he will not find a point at which resistance becomes illogical or unscientific."[1] Statements of that sort obviously raise the question of why, in the absence of binding criteria for scientific choice, both the number of solved scientific problems and the precision of individual problem solutions should increase so markedly with the passage of time. Confronting that issue, I sketched in my closing chapter a number of characteristics that scientists share by virtue of the training which licenses their membership in one or another community of specialists. In the absence of criteria able to dictate the choice of each individual, I argued, we do well to trust the collective judgment of scientists trained in this way. "What better criterion could there be," I asked rhetorically, "than the decision of the scientific group?"[2]

A number of philosophers have greeted remarks like these in a way that continues to surprise me. My views, it is said, make of theory choice "a matter of mob psychology."[3] Kuhn believes, I am told, that "the decision of a scientific group to adopt a new paradigm cannot be based on good reasons of any kind, factual or otherwise."[4] The debates surrounding such choices must, my critics claim, be for me "mere persuasive displays without deliberative substance."[5] Reports of this sort manifest total misunderstanding, and I have occasionally said as much in papers directed primarily to other ends.

But those passing protestations have had negligible effect, and the misunderstandings continue to be important. I conclude that it is past time for me to describe, at greater length and with greater precision, what has been on my mind when I have uttered statements like the ones with which I just began. If I have been reluctant to do so in the past, that is largely because I have preferred to devote attention to areas in which my views diverge more sharply from those currently received than they do with respect to theory choice.

What, I ask to begin with, are the characteristics of a good scientific theory? Among a number of quite usual answers I select five, not because they are exhaustive, but because they are individually important and collectively sufficiently varied to indicate what is at stake. First, a theory should be accurate: within its domain, that is, consequences deducible from a theory should be in demonstrated agreement with the results of existing experiments and observations. Second, a theory should be consistent, not only internally or with itself, but also with other currently accepted theories applicable to related aspects of nature. Third, it should have broad scope: in particular, a theory's consequences should extend far beyond the particular observations, laws, or subtheories it was initially designed to explain. Fourth, and closely related, it should be simple, bringing order to phenomena that in its absence would be individually isolated and, as a set, confused. Fifth—a somewhat less standard item, but one of special importance to actual scientific decisions—a theory should be fruitful of new research findings: it should, that is, disclose new phenomena or previously unnoted relationships among those already known.[6] These five characteristics—accuracy, consistency, scope, simplicity, and fruitfulness—are all standard criteria for evaluating the adequacy of a theory. If they had not been, I would have devoted far more space to them in my book, for I agree entirely with the traditional view that they play a vital role when scientists must choose between an established theory and an upstart competitor. Together with others of much the same sort, they provide the shared basis for theory choice.

Nevertheless, two sorts of difficulties are regularly encountered by the men who must use these criteria in choosing, say, between Ptolemy's astronomical theory and Copernicus's, between the oxygen and phlogiston theories of combustion, or between Newtonian mechanics and the quantum theory. Individually the criteria are imprecise: individuals may legitimately differ about their application to concrete cases. In addition, when deployed together, they repeatedly prove to conflict with one another; accuracy may, for example, dictate the choice of one theory, scope the choice of its competitor. Since these difficulties, especially the first, are also relatively familiar, I shall devote little time to their elaboration. Though my argument does demand that I illustrate them briefly, my views will begin to depart from those long current only after I have done so.

Begin with accuracy, which for present purposes I take to include not only quantitative agreement but qualitative as well. Ultimately it proves the most nearly decisive of all the criteria, partly because it is less equivocal than the others but especially because predictive and explanatory powers, which depend on it, are characteristics that scientists are particularly unwilling to give up. Unfortunately, however, theories cannot always be discriminated in terms of accuracy. Copernicus's system, for example, was not more accurate than Ptolemy's until drastically revised by Kepler more than sixty years after Copernicus's death. If Kepler or someone else had not found other reasons to choose heliocentric astronomy, those improvements in accuracy would never have been made, and Copernicus's work might have been forgotten. More typically, of course, accuracy does permit discriminations, but not the sort that lead regularly to unequivocal choice. The oxygen theory, for example, was universally acknowledged to account for observed weight relations in chemical reactions, something the phlogiston theory had previously scarcely attempted to do. But the phlogiston theory, unlike its rival, could account for the metals' being much more alike than the ores from which they were formed. One theory thus matched experience better in one area, the other in another. To choose between them on the basis of accuracy, a scientist would need to decide the area in which accuracy was more significant. About the matter chemists could and did differ without violating any of the criteria outlined above, or any others yet to be suggested.

However important it may be, therefore, accuracy by itself is seldom or never a sufficient criterion for theory choice. Other criteria must function as well, but they do not eliminate problems. To illustrate I select just two—consistency and simplicity—asking how they functioned in the choice between the heliocentric and geocentric systems. As astronomical theories both Ptolemy's and Copernicus's were internally consistent, but their relation to related theories in other fields was very different. The stationary central earth was an essential ingredient of received physical theory, a tight-knit body of doctrine which explained, among other things, how stones fall, how water pumps function, and why the clouds move slowly across the skies. Heliocentric astronomy, which required the earth's motion, was inconsistent with the existing scientific explanation of these and other terrestrial phenomena. The consistency criterion, by itself, therefore, spoke unequivocally for the geocentric tradition.

Simplicity, however, favored Copernicus, but only when evaluated in a quite special way. If, on the one hand, the two systems were compared in terms of the actual computational labor required to predict the position of a planet at a particular time, then they proved substantially equivalent. Such computations were what astronomers did, and Copernicus's system offered them no labor-saving techniques; in that sense it was not simpler then Ptolemy's. If, on the other hand, one asked about the amount of mathematical apparatus required to explain, not the detailed quantitative features—

limited elongation, retrograde motion, and the like—then, as every school-child knows, Copernicus required only one circle per planet, Ptolemy two. In that sense the Copernican theory was the simpler, a fact vitally important to the choices made by both Kepler and Galileo and thus essential to the ultimate triumph of Copernicanism. But that sense of simplicity was not the only one available, nor even the one most natural to professional astronomers, men whose task was the actual computation of planetary position.

Because time is short and I have multiplied examples elsewhere, I shall here simply assert that these difficulties in applying standard criteria of choice are typical and that they arise no less forcefully in twentieth-century situations than in the earlier and better-known examples I have just sketched. When scientists must choose between competing theories, two men fully committed to the same list of criteria for choice may nevertheless reach different conclusions. Perhaps they interpret simplicity differently or have different convictions about the range of fields within which the consistency criterion must be met. Or perhaps they agree about these matters but differ about the relative weights to be accorded to these or to other criteria when several are deployed together. With respect to divergences of this sort, no set of choice criteria yet proposed is of any use. One can explain, as the historian characteristically does, why particular men made particular choices at particular times. But for that purpose one must go beyond the list of shared criteria to characteristics of the individuals who make the choice. One must, that is, deal with characteristics which vary from one scientist to another without thereby in the least jeopardizing their adherence to the canons that make science scientific. Though such canons do exist and should be discoverable (doubtless the criteria of choice with which I began are among them), they are not by themselves sufficient to determine the decisions of individual scientists. For that purpose the shared canons must be fleshed out in ways that differ from one individual to another.

Some of the differences I have in mind result from the individual's previous experience as a scientist. In what part of the field was he at work when confronted by the need to choose? How long had he worked there; how successful had he been; and how much of his work depended on concepts and techniques challenged by the new theory? Other factors relevant to choice lie outside the sciences. Kepler's early election of Copernicanism was due in part to his immersion in the Neoplatonic and Hermetic movements of his day; German Romanticism predisposed those it affected toward both recognition and acceptance of energy conservation; nineteenth-century British social thought had a similar influence on the availability and acceptability of Darwin's concept of the struggle for existence. Still other significant differences are functions of personality. Some scientists place more premium than others on originality and are correspondingly more willing to take risks; some scientists prefer comprehensive, unified theories to precise and detailed problem solutions of apparently narrower scope. Differentiating factors like these are described by my critics as subjective and are contrasted with the shared or objective criteria from which I began. Though I shall later question that use of terms, let me for the moment accept it. My point is, then, that every individual choice between competing theories depends on a mixture of objective and subjective factors, or of shared and individual criteria. Since the latter have not ordinarily figured in the philosophy of science, my emphasis upon them has made my belief in the former hard for my critics to see.

What I have said so far is primarily simply descriptive of what goes on in the sciences at times of theory choice. As description, furthermore, it has not been challenged by my critics, who reject instead my claim that these facts of scientific life have philosophic import. Taking up that issue, I shall begin to isolate some, though I think not vast, differences of opinion. Let me begin by asking how philosophers of science can for so long have neglected the subjective elements which, they freely grant, enter regularly into the actual theory choices made by individual scientists? Why have these elements seemed to them an index only of human weakness, not at all of the nature of scientific knowledge?

One answer to that question is, of course, that few philosophers, if any, have claimed to possess either a complete or an entirely well-articulated list of criteria. For some time, therefore, they could reasonably expect that further research would eliminate residual imperfections and produce an algorithm able to dictate rational, unanimous choice. Pending that achievement, scientists would have no alternative but to supply subjectively what the best current list of objective criteria still lacked. That some of them might still do so even with a perfected list at hand would then be an index only of the inevitable imperfection of human nature.

That sort of answer may still prove to be correct, but I think no philosopher still expects that it will. The search for algorithmic decision procedures has continued for some time and produced both powerful and illuminating results. But those results all presuppose that individual criteria of choice can be unambiguously stated and also that, if more than one proves relevant, an appropriate weight function is at hand for their joint application. Unfortunately, where the choice at issue is between scientific theories, little progress has been made toward the first of these desiderata and none toward the second. Most philosophers of science would, therefore, I think, now regard the sort of algorithm which has traditionally been sought as a not quite attainable ideal. I entirely agree and shall henceforth take that much for granted.

Even an ideal, however, if it is to remain credible, requires some demonstrated relevance to the situations in which it is supposed to apply. Claiming that such demonstration requires no recourse to subjective factors, my critics seem to appeal, implicitly or explicitly, to the well-known distinction between the contexts of discovery and of justification.[7] They concede, that is, that the subjective factors I invoke play a significant role in the discovery or invention of new theories, but they also insist that that inevitably intuitive process lies outside of the bounds of philosophy of science and is irrelevant to the question of scientific objectivity. Objectivity enters science, they continue, through the processes by which theories are tested, justified, or judged. Those processes do not, or at least need not, involve subjective factors at all. They can be governed by a set of (objective) criteria shared by the entire group competent to judge.

I have already argued that that position does not fit observations of scientific life and shall now assume that that much has been conceded. What is now at issue is a different point: whether or not this invocation of the distinction between contexts of discovery and of justification provides even a plausible and useful idealization. I think it does not and can best make my point by suggesting first a likely source of its apparent cogency. I suspect that my critics have been misled by science pedagogy or what I have elsewhere called textbook science. In science teaching, theories are presented together with exemplary applications, and those applications may be viewed as evidence. But that is not their primary pedagogic function (science students are distressingly willing to receive the word from professors and texts). Doubtless *some* of them were *part* of the evidence at the time actual decisions were being made, but they represent only a fraction of the considerations relevant to the decision process. The context of pedagogy differs almost as much from the context of justification as it does from that of discovery.

Full documentation of that point would require longer argument than is appropriate here, but two aspects of the way in which philosophers ordinarily demonstrate the relevance of choice criteria are worth noting. Like the science textbooks on which they are often modelled, books and articles on the philosophy of science refer again and again to the famous crucial experiments: Foucault's pendulum, which demonstrates the motion of the earth; Cavendish's demonstration of gravitational attraction; or Fizeau's measurement of the relative speed of sound in water and air. These experiments are paradigms of good reason for scientific choice; they illustrate the most effective of all the sorts of argument which could be available to a scientist uncertain which of two theories to follow; they are vehicles for the transmission of criteria of choice. But they also have another characteristc in common. By the time they were performed no scientist still needed to be convinced of the validity of the theory their outcome is now used to demonstrate. Those decisions had long since been made on the basis of significantly more equivocal evidence. The exemplary crucial experiments to which philosophers again and again refer would have been historically relevent to theory choice only if they had yielded unexpected results. Their use as illustrations provides needed economy to science pedagogy, but they scarcely illuminate the character of the choices that scientists are called upon to make.

Standard philosophical illustrations of scientific choice have another troublesome characteristic. The only arguments discussed are, as I have previously indicated, the ones favorable to the theory that, in fact, ultimately triumphed. Oxygen, we read, could explain weight relations, phlogiston could not, but nothing is said about the phlogiston theory's power or about the oxygen theory's limitations. Comparisons of Ptolemy's theory with Copernicus's proceed in the same way. Perhaps these examples should not be given since they contrast a developed theory with one still in its infancy. But philosophers regularly use them nonetheless. If the only result of their doing so were to simplify the decision situation, one could not object. Even historians do not claim to deal with the full factual complexity of the situations they describe. But these simplifications emasculate by making choice totally unproblematic. They eliminate, that is, one essential element of the decision situations that scientists must resolve if their field is to move ahead. In those situations there are always at least some good reasons for each possible choice. Considerations relevant to the context of discovery are then relevant to justification as well; scientists who share the concerns and sensibilities of the individual who discovers a new theory are ipso facto likely to appear disproportionately frequently among that theory's first supporters. That is why it has been difficult to construct algorithms for theory choice, and also why such difficulties have seemed so thoroughly worth resolving. Choices that present problems are the ones philosophers of science need to understand. Philosophically interesting decision procedures must function where, in their absence, the decision might still be in doubt.

That much I have said before, if only briefly. Recently, however, I have recognized another, subtler source for the apparent plausibility of my critics' position. To present it, I shall briefly describe a hypothetical dialogue with one of them. Both of us agree that each scientist chooses between competing theories by deploying some Bayesian algorithm which permits him to compute a value for $p(T,E)$, i.e., for the probability of a theory T on the evidence E available both to him and to the other members of his professional group at a particular period of time. "Evidence," furthermore, we both interpret broadly to include such considerations as simplicity and fruitfulness. My critic asserts, however, that there is only one such value of p, that corresponding to objective choice, and he believes that all rational members of the group must arrive at it. I assert, on the other hand, for reasons previously given, that the factors he calls objective are insufficient to determine in full any algorithm at all. For the sake of the discussion I have conceded that each individual has an algorithm and that all their algorithms have much in common. Nevertheless, I continue to hold that the algorithms of individuals are all ultimately different by virtue of the subjective considerations with which each must complete the objective criteria before any computations can be done. If my hypothetical critic is liberal, he may now grant that these subjective differences do play a role in determining the hypothetical algorithm on which each individual relies during the early stages of the competition between rival theories. But he is also likely to claim that, as evidence increases with the passage of time, the algorithms of different individuals converge to the algorithm of objective choice with which his presentation began. For him the increasing unanimity of individual choices is evidence for their increasing objectivity and thus for the elimination of subjective elements from the decision process.

So much for the dialogue, which I have, of course, contrived to disclose the non sequitur underlying an apparently plausible position. What converges as the evidence changes over time need only be the values of p that individuals compute from their individual algorithms. Conceivably those algorithms themselves also become more alike with time, but the ultimate unanimity of theory choice provides no evidence whatsoever that they do so. If subjective factors are required to account for the decisions that initially divide the profession, they may still be present later when the profession agrees. Though I shall not here argue the point, consideration of the occasions on which a scientific community divides suggests that they actually do so.

My argument has so far been directed to two points. It first provided evidence that the choices scientists make between competing theories depend not only on shared criteria—those my critics call objective—but also on idiosyncratic factors dependent on individual biography and personality. The latter are, in my critics' vocabulary, subjective, and the second part of my argument has attempted to bar some likely ways of denying their philosophic import. Let me now shift to a more positive approach, returning briefly to the list of shared criteria—accuracy, simplicity, and the like—with which I began. The considerable effectiveness of such criteria does not, I now wish to suggest, depend on their being sufficiently articulated to dictate the choice of each individual who subscribes to them. Indeed, if they were articulated to that extent, a behavior mechanism fundamental to scientific advance would cease to function. What the tradition sees as eliminable imperfections in its rules of choice I take to be in part responses to the essential nature of science.

As so often, I begin with the obvious. Criteria that influence decisions without specifying what those decisions must be are familiar in many aspects of human life. Ordinarily, however, they are called, not criteria or rules, but maxims, norms, or values. Consider maxims first. The individual who invokes them when choice is urgent usually finds them frustratingly vague and often also in conflict one with another. Contrast "He who hesitates is lost" with "Look before you leap," or compare "Many hands make light work" with "Too many cooks spoil the broth." Individually maxims dictate different choices, collectively none at all. Yet no one suggests that supplying children with contradictory tags like these is irrelevant to their education. Opposing maxims alter the nature of the decision to be made, highlight the essential issues it presents, and point to those remaining aspects of the decision for which each individual must take responsibility himself. Once invoked, maxims like these alter the nature of the decision process and can thus change its outcome.

Values and norms provide even clearer examples of effective guidance in the presence of conflict and equivocation. Improving the quality of life is a value, and a car in every garage once followed from it as a norm. But quality of life has other aspects, and the old norm has become problematic. Or again, freedom of speech is a value, but so is preservation of life and property. In application, the two often conflict, so that judicial soul-searching, which still continues, has been required to prohibit such behavior as inciting to riot or shouting fire in a crowded theater. Difficulties like these are an appropriate source for frustration, but they rarely result in charges that values have no function or in calls for their abandonment. That response is barred to most of us by an acute consciousness that there are societies with other values and that these value differences result in other ways of life, other decisions about what may and what may not be done.

I am suggesting, of course, that the criteria of choice with which I began function not as rules, which determine choice, but as values, which influence it. Two men deeply committed to the same values may nevertheless, in particular situations, make different choices as, in fact, they do. But that difference in outcome ought not to suggest that the values scientists share are less than critically important either to their decisions or to the development of the enterprise in which they participate. Values like accuracy, consistency, and scope may prove ambiguous in application, both individually and collectively; they may, that is, be an insufficient basis for a *shared* algorithm of choice. But they do specify a great deal: what each scientist must consider in reaching a decision, what he may and may not consider relevant, and what he can legitimately be required to report as the basis for the choice he has made. Change the list, for example by adding social utility as a criterion, and some particular choices will be different, more like those one expects from an engineer. Subtract accuracy of fit to nature from the list, and the enterprise that results may not resemble science at all, but perhaps philosophy instead. Different creative disciplines are characterized, among other things, by different sets of shared values. If philosophy and engineering lie too close to the sciences, think of literature of the plastic arts. Milton's failure to set *Paradise Lost* in a Copernican universe does not indicate that he agreed with Ptolemy but that he had things other than science to do.

Recognizing that criteria of choice can function as values when incomplete as rules has, I think, a number of striking advantages. First, as I have already argued at length, it accounts in detail for aspects of scientific behavior which the tradition has seen as anomalous or even irrational. More important, it allows the standard criteria to function fully in the earliest stages of theory choice, the period when they are most needed but when, on the traditional view, they function badly or not at all. Copernicus was responding to them during the years required to convert heliocentric astronomy from a global conceptual scheme to mathematical machinery for predicting planetary position. Such predictions were what astronomers valued; in their absence, Copernicus would scarely have been heard, something which had happened to the idea of a moving earth before. That his own version convinced very few is less important than his acknowledgment of the basis on which judgments would have to be reached if heliocentricism were to survive. Though idiosyncrasy must be invoked to explain why Kepler and Galileo were early converts to Copernicus's system, the gaps filled by their efforts to perfect it were specified by shared values alone.

That point has a corollary which may be more important still. Most newly suggested theories do not survive. Usually the difficulties that evoked them are accounted for by more traditional means. Even when this does not occur, much work, both theoretical and experimental, is ordinarily required before the new theory can display sufficient accuracy and scope to generate widespread conviction. In short, before the group accepts it, a new theory has been tested over time by the research of a number of men, some working within it, others within its traditional rival. Such a mode of development, however, *requires* a decision process which permits rational men to disagree, and such disagreement would be barred by the shared algorithm which philosophers have generally sought. If it were at hand, all conforming scientists would make the same decision at the same time. With standards for acceptance set too low, they would move from one attractive global viewpoint to another, never giving traditional theory an opportunity to supply equivalent attractions. With standards set higher, no one satisfying the criterion of rationality would be inclined to try out the new theory, to articulate it in ways which showed its fruitfulness or displayed its accuracy and scope. I doubt that science would survive the change. What from one viewpoint may seem the looseness and imperfection of choice criteria conceived as rules may, when the same criteria are seen as values, appear an indispensable means of spreading the risk which the introduction of support of novelty always entails.

Even those who have followed me this far will want to know how a value-based enterprise of the sort I have described can develop as a science does, repeatedly producing powerful new techniques for prediction and control. To that question, unfortunately, I have no answer at all, but that is only another way of saying that I make no claim to have solved the problem of induction. If science did progress by virtue of some shared and binding algorithm of choice, I would be equally at a loss to explain its success. The lacuna is one I feel acutely, but its presence does not differentiate my position from the tradition.

It is, after all, no accident that my list of the values guiding scientific choice is, as nearly as makes any difference, identical with the tradition's list of rules dictating choice. Given any concrete situation to which the philosopher's rules could be applied, my values would function like his rules, producing the same choice. Any justification of induction, any explanation of why the rules worked, would apply equally to my values. Now consider a situation in which choice by shared rules proves impossible, not because the rules are wrong but because they are, as rules, intrinsically incomplete. Individuals must then still choose and be guided by the rules (now values) when they do so. For that purpose, however, each must first flesh out the rules, and each will do so in a somewhat different way even though the decision dictated by the variously completed rules may prove unanimous. If I now assume, in addition, that the group is large enough so that individual differences distribute on some normal curve, then any argument that justifies the philosopher's choice by rule should be immediately adaptable to my choice by value. A group too small, or a distribution excessively skewed by external historical pressures, would, of course, prevent the argument's transfers.[8] But those are just the circumstances under which scientific progress is itself problematic. The transfer is not then to be expected.

I shall be glad if these references to a normal distribution of individual differences and to the problem of induction make my position appear very close to more traditional views. With respect to theory choice, I have never thought my departures large and have been correspondingly startled by such charges as "mob psychology," quoted at the start. It is worth noting, however, that the positions are not quite identical, and for that purpose an analogy may be helpful. Many properties of liquids and gases can be accounted for on the kinetic theory by supposing that all molecules travel at the same speed. Among such properties are the regularities known as Boyle's and Charles's law. Other characteristics, most obviously evaporation, cannot be explained in so simple a way. To deal with them one must assume that molecular speeds differ, that they are distributed at random, governed by the laws of chance. What I have been suggesting here is that theory choice, too, can be explained only in part by a theory which attributes the same properties to all the scientists who must do the choosing. Essential aspects of the process generally known as verification will be understood only by recourse to the features with respect to which men may differ while still remaining scientists. The tradition takes it for granted that such features are vital to the process of discovery, which it at once and for that reason rules out of philosophical bounds. That they may have significant functions also in the philosophically central problem of justifying theory choice is what philosophers of science have to date categorically denied.

What remains to be said can be grouped in a somewhat miscellaneous epilogue. For the sake of clarity and to avoid writing a book, I have throughout this paper utilized some traditional concepts and locutions about the viability of which I have elsewhere expressed serious doubts. For those who know the work in which I have done so, I close by indicating three aspects of what I have said which would better represent my views if cast in other terms, simultaneously indicating the main directions in which such recasting should proceed. The areas I have in mind are: value invariance, subjectivity, and partial communication. If my views of scientific development are novel—a matter about which there is legitimate room for doubt—it is in areas such as these, rather than theory choice, that my main departures from tradition should be sought.

SKIP

Throughout this paper I have implicitly assumed that, whatever their initial source, the criteria or values deployed in theory choice are fixed once and for all, unaffected by their participation in transitions from one theory to another. Roughly speaking, but only very roughly, I take that to be the case. If the list of relevant values is kept short (I have mentioned five, not all independent) and if their specification is left vague, then such values as accuracy, scope, and fruitfulness are permanent attributes of science. But little knowledge of history is required to suggest that both the application of these values and, more obviously, the relative weights attached to them have varied markedly with time and also with field of application. Furthermore, many of these variations in value have been associated with particular changes in scientific theory. Though the experience of scientists provides no philosophical justification for the values they deploy (such justification would solve the problem of induction), those values are in part learned from that experience, and they evolve with it.

The whole subject needs more study (historians have usually taken scientific values, though not scientific methods, for granted), but a few remarks will illustrate the sort of variations I have in mind. Accuracy, as a value, has with time increasingly denoted quantitative or numerical agreement, sometimes at the expense of qualitative. Before early modern times, however, accuracy in that sense was a criterion only for astronomy, the science of the celestial region. Elsewhere it was neither expected nor sought. During the seventeenth century, however, the criterion of numerical agreement was extended to mechanics, during the late eighteenth and early nineteenth centuries to chemistry and such other subjects as electricity and heat, and in this century to many parts of biology. Or think of utility, an item of value not on my initial list. It too has figured significantly in scientific development, but far more strongly and steadily for chemists than for, say, mathematicians and physicists. Or consider scope. It is still an important scientific value, but important scientific advances have repeatedly been achieved at its expense, and the weight attributed to it at times of choice has diminished correspondingly.

What may seem particularly troublesome about changes like these is, of course, that they ordinarily occur in the aftermath of a theory change. One of the objections to Lavoisier's new chemistry was the roadblocks with which it confronted the achievement of what had previously been one of chemistry's traditional goals; the explanation of qualities, such as color and texture, as well as of their changes. With the acceptance of Lavoisier's theory such explanations ceased for some time to be a value for chemists; the ability to explain qualitative variation was no longer a criterion relevant to the evaluation of chemical theory. Clearly, if such value changes had occurred as rapidly or been as complete as the theory changes to which they related, then theory choice would be value choice, and neither could provide justification for the other. But, historically, value change is ordinarily a belated and largely unconscious concomitant of theory choice, and the former's magnitude is regularly smaller than the latter's. For the functions I have here ascribed to values, such relative stability provides a sufficient basis. The existence of a feedback loop through which theory change affects the values which led to that change does not make the decision process circular in any damaging sense.

About a second respect in which my resort to tradition may be misleading, I must be far more tentative. It demands the skills of an ordinary language philosopher, which I do not possess. Still, no very acute ear for language is required to generate discomfort with the ways in which the terms "objectivity" and, more especially, "subjectivity" have functioned in this paper. Let me briefly suggest the respects in which I believe language has gone astray. "Subjective" is a term with several established uses: in one of these it is opposed to "objective," in another to "judgmental." When my critics describe the idiosyncratic features to which I appeal as subjective, they resort, erroneously I think, to the second of these senses. When they complain that I deprive science of objectivity, they conflate that second sense of subjective with the first.

A standard application of the term "subjective" is to matters of taste, and my critics appear to suppose that that is what I have made of theory choice. But they are missing a distinction standard since Kant when they do so. Like sensation reports, which are also subjective in the sense now at issue, matters of taste are undiscussable. Suppose that, leaving a movie theater with a friend after seeing a western, I exclaim: "How I liked that terrible potboiler!" My friend, if he disliked the film, may tell me I have low tastes, a matter about which, in these circumstances, I would readily agree. But, short of saying that I lied, he cannot disagree with my report that I liked the film or try to persuade me that what I said about my reaction was wrong. What is discussable in my remark is not my characterization of my internal state, my exemplification of taste, but rather my *judgment* that the film was a potboiler. Should my friend disagree on that point, we may argue most of the night, each comparing the film with good or great ones we have seen, each revealing, implicitly or explicitly, something about how he *judges* cinematic merit, about his aesthetic. Though one of us may, before retiring, have persuaded the other,

he need not have done so to demonstrate that our difference is one of judgment, not taste.

Evaluations or choices of theory have, I think exactly this character. Not
that scientists never say merely, I like such and such a theory, or I do not.
After 1926 Einstein said little more than that about his opposition to the
quantum theory. But scientists may always be asked to explain their choices,
to exhibit the bases for their judgments. Such judgments are eminently discussable, and the man who refuses to discuss his own cannot expect to be
taken seriously. Though there are, very occasionally, leaders of scientific
taste, their existence tends to prove the rule. Einstein was one of the few, and
his increasing isolation from the scientific community in later life shows how
very limited a role taste alone can play in theory choice. Bohr, unlike Einstein, did discuss the bases for his judgment, and he carried the day. If my
critics introduce the term "subjective" in a sense that opposes it to judgmental—thus suggesting that I make theory choice undiscussable, a matter of
taste—they have seriously mistaken my position.

Turn now to the sense in which "subjectivity" is opposed to "objectivity," and note first that it raises issues quite separate from those just discussed. Whether my taste is low or refined, my report that I liked the film is
objective unless I have lied. To my judgment that the film was a potboiler,
however, the objective-subjective distinction does not apply at all, at least not
obviously and directly. When my critics say I deprive theory choice of objectivity, they must, therefore, have recourse to some very different sense of
subjective, presumably the one in which bias and personal likes or dislikes
function instead of, or in the face of, the actual facts. But that sense of subjective does not fit the process I have been describing any better than the first.
Where factors dependent on individual biography or personality must be
introduced to make values applicable, no standards of factuality or actuality
are being set aside. Conceivably my discussion of theory choice indicates
some limitations of objectivity, but not by isolating elements properly called
subjective. Nor am I even quite content with the notion that what I have been
displaying are limitations. Objectivity ought to be analyzable in terms of criteria like accuracy and consistency. If these criteria do not supply all the
guidance that we have customarily expected of them, then it may be the
meaning rather than the limits of objectivity that my argument shows.

Turn, in conclusion, to a third respect, or set of respects, in which this
paper needs to be recast. I have assumed throughout that the discussions surrounding theory choice are unproblematic, that the facts appealed to in such
discussions are independent of theory, and that the discussions' outcome is
appropriately called a choice. Elsewhere I have challenged all three of these
assumptions, arguing that communication between proponents of different
theories is inevitably partial, that what each takes to be facts depends in part
on the theory he espouses, and that an individual's transfer of allegiance from

theory to theory is often better described as conversion than as choice. Though all these theses are problematic as well as controversial, my commitment to them is undiminished. I shall not now defend them, but must at least attempt to indicate how what I have said here can be adjusted to conform with these more central aspects of my view of scientific development.

For that purpose I resort to an analogy I have developed in other places. Proponents of different theories are, I have claimed, like native speakers of different languages. Communication between them goes on by translation, and it raises all translation's familiar difficulties. That analogy is, of course, incomplete, for the vocabulary of the two theories may be identical, and most words function in the same ways in both. But some words in the basic as well as in the theoretical vocabularies of the two theories—words like "star" and "planet," "mixture" and "compound," or "force" and "matter"—do function differently. Those differences are unexpected and will be discovered and localized, if at all, only by repeated experience of communication breakdown. Without pursuing the matter further, I simply assert the existence of significant limits to what the proponents of different theories can communicate to one another. The same limits make it difficult or, more likely, impossible for an individual to hold both theories in mind together and compare them point by point with each other and with nature. That sort of comparison is, however, the process on which the appropriateness of any word like "choice" depends.

Nevertheless, despite the incompleteness of their communication, proponents of different theories can exhibit to each other, not always easily, the concrete technical results achievable by those who practice within each theory. Little or no translation is required to apply at least some value criteria to those results. (Accuracy and fruitfulness are most immediately applicable, perhaps followed by scope. Consistency and simplicity are far more problematic.) However incomprehensible the new theory may be to the proponents of tradition, the exhibit of impressive concrete results will persuade at least a few of them that they must discover how such results are achieved. For that purpose they must learn to translate, perhaps by treating already published papers as a Rosetta stone or, often more effective, by visiting the innovator, talking with him, watching him and his students at work. Those exposures may not result in the adoption of the theory; some advocates of the tradition may return home and attempt to adjust the old theory to produce equivalent results. But others, if the new theory is to survive, will find that at some point in the language-learning process they have ceased to translate and begun instead to speak the language like a native. No process quite like choice has occurred, but they are practicing the new theory nonetheless. Furthermore, the factors that have led them to risk the conversion they have undergone are just the ones this paper has underscored in discussing a somewhat different process, one which, following the philosophical tradition, it has labelled theory choice.

NOTES

1. *The Structure of Scientific Revolutions,* 2d ed. (Chicago, 1970), pp. 148, 151–52, 159. All the passages from which these fragments are taken appeared in the same form in the first edition, published in 1962.

2. Ibid., p. 170.

3. Imre Lakatos. "Falsification and the Methodology of Scientific Research Programmes," in I. Lakatos and A. Musgrave, eds., *Criticism and the Growth of Knowledge* (Cambridge, 1970), 91–195. The quoted phrase, which appears on p. 178, is italicized in the original.

4. Dudley Shapere, "Meaning and Scientific Change," in R. G. Colodny, ed., *Mind and Cosmos: Essays in Contemporary Science and Philosophy,* University of Pittsburgh Series in the Philosophy of Science, vol. 3 (Pittsburgh, 1966), pp. 41–85. The quotation will be found on p. 67.

5. Israel Scheffler, *Science and Subjectivity* (Indianapolis, 1967), 81.

6. The last criterion, fruitfulness, deserves more emphasis than it has yet received. A scientist choosing between two theories ordinarily knows that his decision will have a bearing on his subsequent research career. Of course he is especially attracted by a theory that promises the concrete successes for which scientists are ordinarily rewarded.

7. The least equivocal example of this position is probably the one developed in Scheffler, *Science and Subjectivity,* chap. 4.

8. If the group is small, it is more likely that random fluctuations will result in its members' sharing an atypical set of values and therefore making choices different from those that would be made by a larger and more representative group. External environment—intellectual, ideological, or economic—must systematically affect the value system of much larger groups, and the consequences can include difficulties in introducing the scientific enterprise to societies with inimical values or perhaps even the end of that enterprise within societies where it had once flourished. In this area, however, great caution is required. Changes in the environment where science is practiced can also have fruitful effects on research. Historians often resort, for example, to differences between national environments to explain why particular innovations were initiated and at first disproportionately pursued in particular countries, e.g., Darwinism in Britain, energy conservation in Germany. At present we know substantially nothing about the minimum requisites of the social milieux within which a sciencelike enterprise might flourish.

27

Scientific Rationality:
Analytic vs. Pragmatic Perspectives

Carl G. Hempel

I. INTRODUCTION

Of all the current topics of inquiry in the philosophy of science, none has provoked a more intensive and fruitful discussion than the conflicting conceptions of science and its methodology that have been developed by analytic empiricists on one hand and by historically and sociologically oriented thinkers on the other.

By analytic empiricism, I understand here a body of ideas that has its roots in the logical empiricism of the Vienna Circle and the Berlin group, as well as in the work of kindred thinkers, such as Popper, Braithwaite, and Nagel. Among the protagonists of the more recent historic-sociological approach, I have in mind particularly Thomas Kuhn, Paul Feyerabend, and the late R. N. Hanson.

The considerations I propose to put before you concern two fundamental problems that have been at the center of the controversy between the two schools: namely

(1) *The problem of the rationality of science:* in what sense, on what grounds, and to what extent can scientific inquiry be qualified as a rational enterprise?

and

(2) *The problem of cognitive status of the methodology and philosophy of science:* are the principles set forth by these disciplines intended to provide norms for rational scientific inquiry, or are they meant to give an explanatory account of scientific research as a human activity?

Let me broadly and somewhat schematically sketch the background for this discussion. Analytic empiricism views the philosophy of science as a discipline which by "logical analysis" and "rational reconstruction" seeks to "explicate" the meanings of scientific terms and sentences and to exhibit the logical structure and the rationale of scientific theorizing. The philosophy of science is regarded as concerned exclusively with the logical and systematic aspects of sound scientific theorizing and of the knowledge claims it yields. On this view, the psychological, sociological, and historical facets of science as a human enterprise are irrelevant to the philosophy of science, much as the genetic and psychological aspects of human reasoning are held to be irrelevant for pure logic, which is concerned only with questions of the deductive validity of inferences, logical truth and falsity, consistency, provability, definability, and the like. The principles of logical theory clearly do not, and are not meant to, give a descriptive account of "how we think," of how human beings actually reason; but insofar as they provide criteria of logical validity and truth, they can be employed normatively, as standards for the critical appraisal of particular inferences and logical claims.

In view of analytic empiricism, the methodology of science similarly has the task of explicating standards for rational modes of formulating, appraising, and changing scientific knowledge claims; such explications may then serve as norms for the guidance of inquiry and for the critical appraisal of particular research procedures and theoretical claims: the logical empiricists' rejection of neovitalism as a pseudotheory and Popper's exclusion of psychoanalysis from the realm of scientific theories illustrate this point.

The historic-sociological school, on the other hand, rejects the idea of methodological principles arrived at, as it seems, by purely philosophical analysis; it insists that an adequate methodology must be based on a close study of the practice of scientific inquiry and must be capable of explaining certain characteristics of past and present research and theory change in science.

The contrasting views of the two schools have been thrown into sharp relief in their debate over the extent to which and the grounds on which it is possible to formulate and justify methodological principles for a rational choice between competing scientific theories.

According to analytic empiricism, such methodological principles have the character to generate criteria for rational theory choice; they might determine, for example, which of two competing theories possesses the higher rational credibility or acceptability in consideration of all the pertinent information available at the time. In fact, analytic attempts at explicating the notions of confirmation and of inductive probability and at formulating "rules of acceptance" for scientific hypotheses were efforts in just that direction.

Kuhn, in whose account of scientific revolutions the problem of choice between conflicting paradigmatic theories plays a central role, regards the search for such analytic criteria as basically misguided and as doomed to failure. He does acknowledge that there are certain considerations, noted also in the earlier philosophy of science, which influence the decisions scientists make in the context of theory choice; he speaks of them as professionally *5 criteria* shared preferences or values. Among them are a preference for theories of quantitative form whose predictions show a close fit with experimental or observational findings; for theories of wide scope; for theories that correctly predict novel phenomena, and some others. All these characteristics, he says, provide "good reasons for theory choice."[1]

But Kuhn rightly notes that those desiderata do not remotely suffice to determine a unique choice between competing theories, and he further insists that there are no methodological criteria which have that power and which would command the assent of the scientific community: there are no generally binding rules that compel the choice between theories on the basis of logic and experiment alone.[2] Theory choice is emphatically presented as the *?* result of group decisions which are not determined by rules of rational procedure of the sort analytic empiricism would envision.

Yet despite this naturalistic emphasis, Kuhn views science as a rational enterprise. Thus, he says: "scientific behavior, taken as a whole, is the best example we have of rationality" and "if history or any other empirical discipline leads us to believe that the development of science depends essentially on behavior that we have previously thought to be irrational, then we should conclude not that science is irrational, but that our notion of rationality needs adjustment here and there."[3] Thus, Kuhn regards his account of science as both empirical and normative. In response to the question, raised by Feyerabend, whether his pronouncements are to be read as descriptive or as prescriptive, he has said quite explicitly: "they should be read in both ways at once."[4]

II. THE JANUS-HEAD OF RATIONAL EXPLANATION AND THE IDEA OF AN E-N METHODOLOGY

But surely, the principles of a given methodology cannot literally be both explanatory and normative. For in the former case, they would form an empirical theory, supported by pertinent evidence, which would serve to explain certain significant aspects of the research behavior of scientists; whereas in the latter case, they would express conditions of rationality for scientific inquiry.

Yet the conception of a methodology, or some methodological principles, having this double aspect can be given a plausible construal, as is suggested by the following consideration: An explanation of a particular case of theory-choice or paradigm-change is clearly an explanation of a human decision. Now, human decisions and actions are typically explained by reference to the agent's motivating reasons; and in the formulation and even in the philosophical analysis of such accounts two distinct senses of accounting are not infrequently confounded or fused, namely, accounting for a decision in the sense of explaining it and accounting for a decision in the sense of justifying it. This is illustrated by formulations of the following type:

> Agent A intended to attain goal G and had a set B of beliefs about the circumstances in which he had to act—particularly about alternative means available to him for attaining his goal. But given A's goal and his beliefs, the appropriate or rational thing for him to do was X. That is why A decided on course of action X.

But surely, an account of this kind cannot explain why A did in fact choose X. For the norm it invokes about the proper thing to do under the specified circumstances has no explanatory force at all. Even if the norm be granted, it is entirely possible that A may have been aware of, or not committed to, the standard of appropriateness or rationality it expresses, and that he might, therefore, in fact not have chosen X. The argument invoking the normative principle offers no explanation of A's choice; but it does offer a *justification* for it by showing that what A chose to do was appropriate in the sense of the cited norm.

A corresponding *explanation* requires, instead of that norm, a psychological hypothesis to the effect that A had adopted the norm and thus had acquired a disposition to act in accordance with it. Let us note in passing that whereas a justificatory account will be acceptable only if the norm in question is acknowledged as sound, the acceptability of the corresponding explanatory account is subject to no such condition. If indeed A was disposed to act in accordance with the cited norm, then this fact can be invoked to explain A's decision, no matter whether the norm is deemed to be sound.

The tendency to fuse explanation with justification in accounting for human decisions is thus closely linked to the fact that when a decision is explained by attributing to the agent certain motivating considerations, then these considerations, if suitably spelled out, can be evaluated in their own right as to their "rationality" and such evaluation may indeed afford a justification of the given decision. An account which does possess this double aspect of explanation and justification might be called an *ideal rational account.*

There are some kinds of human decision which seem, in fairly close approximation, to admit of such a two-faced account. Consider, for example, the case of an engineer in charge of quality control who has to decide, on the basis of sample tests, whether a given batch of hormone tablets or of ball bearings manufactured by his firm is to be released for sale or rather to be reprocessed or destroyed because of excessive flaws. His decisions may well be explainable, and indeed predictable, on the basis of the sample findings, by the assumption that he has acquired the disposition to follow such and such specific decision criteria; while on the other hand, the criteria—or some more general decision-theoretical principles from which they can be derived—provide a justification for his particular decisions.

These considerations suggest a plausible construal (I am inclined to think: the only plausible construal) of the idea that certain principles in the methodology of science can play both an explanatory and a normative role. Such a two-faced, Janus-head methodological theory would have to present certain aspects of scientific research, such as theory testing and theory choice, as activities aimed at certain scientific goals, and carried out in accordance with specified rules which can be justified by showing that the modes of procedure they prescribe are rational means of pursuing the given goals.

In order to fulfill both its explanatory and its justificatory function, such a methodological theory for science or for some segment of it would have to specify certain goals of inquiry for the given field and certain procedural norms or rules for the pursuit of those goals; and it would then have to make two quite different claims:

(i) an explanatory claim, briefly to the effect that the practitioners of the science in question share a commitment, and thus a general disposition, to pursue the specified goals in accordance with the specified rules; and that, as a consequence, certain professional decisions made by the scientists can be explained by reference to those commitments;

(ii) a justificatory claim to the effect that the specified procedural rules determine appropriate, or "rational," ways of pursuing the given goals. (How the notion of rationality might be construed here is a question that will be considered shortly.)

I shall refer to a methodological theory of the kind here adumbrated as an explanatory-normative methodology, or as an *E-N methodology* for short. This is clearly an idealized notion: it is not to be expected that an *E-N methodology* can be formulated which would afford a satisfactory explanation even for the major turns in the development of science. Thus, for example, Lakatos, whose conception of a methodological theory has a certain affinity to that of an *E-N methodology,* emphasizes that for an explanation of actual occurrences in the history of science, "external" factors must often be invoked in addition to "internal," methodological, considerations.[5]

III. On the Notion of Scientific Rationality

Let me now turn to the idea of rationality of scientific inquiry, which has been the subject of extensive controversy in the recent philosophy of science. I think it interesting, but also somewhat disturbing, that Popper, Lakatos, Kuhn, Feyerabend, and others have made diverse pronouncements concerning the rationality or irrationality of science or of certain modes of inquiry without, however, as far as I am aware, giving a reasonably explicit characterization of the concept of rationality which they have in mind and which they seek to illuminate or to disparage in their methodological investigations. Thus we find Lakatos bringing the charge of irrationalism against Kuhn's account of scientific theorizing, whereas Kuhn, as noted earlier, claims that his account takes scientific research behavior, as a whole, to be the best example we have of rationality.

What is to be understood here by "rationality," and what kinds of consideration could be properly adduced in a critical appraisal of attributions of rationality to science as a whole or to certain modes of inquiry? I have no satisfactory general answers to these questions, but I would like to put before you some tentative reflections on the subject.

First, a mode of procedure or a rule calling for that procedure surely can be called rational or irrational only relative to the goals the procedure is meant to attain. Insofar as a methodological theory does propose rules or norms, these norms have to be regarded as instrumental norms: their appropriateness must be judged by reference to the objectives of the inquiry to which they pertain or, more ambitiously, by reference to the goals of pure scientific research in general.

Any attempt to give a characterization of "the goals of science" faces serious difficulties. But I hope you will permit me to begin with an avowedly oversimplified construal of those objectives; for even this schematic conception can serve quite well as a background against which to formulate some reflections concerning rationality which should be relevant also to more complex construals of the goals of scientific inquiry.

Let us assume, then, that science aims at establishing a sequence of increasingly comprehensive and accurate systems of empirical knowledge. Each such system might be thought of as represented by a set K of sentences which has been adopted, and is bound to be modified, in accordance with certain procedural rules.

See p 461

Given this goal, there are certain methodological norms which can be qualified as requirements of rational procedure on the ground that—speaking somewhat charitably—they are necessary conditions for any one of the changing systems K which science might accept at some time or another.

Among these necessary conditions are the requirement of intersubjective testability and the requirement of actual test with satisfactory outcome. Another such condition would require that any of the classes K be deductively closed, since the logical consequences of sentences accepted as presumably true must be presumed to be true as well. Another necessary condition would be that of logical consistency; and there may well be some others.

The goal-dependence of these conditions of rationality is clear. If, for example, it should be our goal to form a set of beliefs about the world that is emotionally reassuring or esthetically satisfying, then different procedural principles would qualify as rational. To further the attainment of our objectives, we might do well not to acknowledge all the logical consequences of accepted sentences and not to judge proposed beliefs in the light of all the relevant evidence obtainable—or better yet, perhaps, simply to forego empirical testing altogether.

In his plea for methodological anarchy, Feyerabend raises the question whether instead of "science as we know it today (the science of critical rationalism that has been freed from all inductive elements)," we would not prefer a science that is "more anarchistic and more subjective."[6] One might surely prefer the pursuit of the goals suggested by Feyerabend, and one might be convinced, as he seems to be, that it would be better for humanity to abandon "science as we know it"; but this does not change the fact that the modes of procedure appropriate for the pursuit of Feyerabend's objectives are not rational for science as it is commonly understood, and certainly not for the pursuit of the goals of science envisaged in our simple construal.

The modest norms of scientific rationality we have noted so far reflect necessary conditions for the attainment of the objectives of science. Could there be procedural norms reflecting sufficient conditions? Surely not; for such norms would, in effect, afford a solution to the problem of induction.

Let us look next at the familiar features—also mentioned by Kuhn—which scientists agree in regarding as desirable characteristics of scientific theories, and which therefore provide some basis for choosing between competing theories: closeness of quantitative fit between theoretical predictions and scientific data, large scope, prediction of novel effects, and some others. It seems to me that most of these desiderata are best regarded as providing a fuller characterization of the goals of scientific theorizing rather than as instrumental norms aimed at enhancing the prospects of achieving those goals. (Simplicity might similarly be regarded as a feature characterizing the goals of scientific theorizing; but alternatively, it might be viewed as an instrumental norm on the ground of the belief that the simpler theory is more likely to be true.)

But regardless of what role is assigned to the desiderata, it is clear, as Kuhn has stressed, that even if they are generally acknowledged by (belong to the "shared values" of) a scientific community, different scientists may, and do, understand them in somewhat different ways and may therefore differ in their judgments as to which of two theories satisfies a particular desideratum more fully. Moreover, scientists may differ in the priorities or relative weights they assign to the various desiderata. In sum, commitment to those norms does not ensure a uniform decision as to which of two theories outranks the other, by way of satisfying the entire set of desiderata.

Now, these last considerations are psychological and sociological and cannot, of course, prove it impossible to formulate precise general criteria of theory choice embodying those desiderata; but the difficulties encountered by analytic efforts to explicate such notions as the simplicity of theories or the degree of variety of the empirical phenomena covered by a theory (and thus, perhaps, its scope) do not augur well for the attainability of those analytic objectives. Indeed, Ernest Nagel argued long ago, on essentially the grounds just surveyed, that efforts to construct a general criterion of this kind were futile.[7]

IV. ON THE RELEVANCE OF DECISION THEORY TO PROBLEMS OF THEORY CHOICE

The preceding considerations also cast serious doubt on the promise of another idea that might suggest itself here, namely, an appeal to mathematical decision theory as a possible source of rational standards for a choice between competing theories, or knowledge systems, or paradigms.

Decision theory does provide objective criteria for choosing between alternative hypotheses; but these criteria presuppose that in addition to the hypotheses in question, a precise comparative or quantitative specification is given of the advantages that would result from adopting a given hypothesis if it is true, and the disadvantages that would result from adopting it if it should be false; and similarly, a specification of advantages or disadvantages connected with the rejection of a hypothesis which may in fact be true or false. "Utilities" and "disutilities" of the requisite kind can plausibly be specified in certain cases where the adoption of a hypothesis amounts to its application for some particular practical purpose, as in the case of industrial quality control. For theory choice is pure science, the requisite utilities or disutilities would have to express the gain or loss that would come from the adoption or the rejection of a true theory or a false one, in the light of the objectives of scientific research. But among those objectives are the desiderata considered a moment ago; and as far as these defy precise explication, decision theory is inapplicable.

Considerations of the kind outlined in the preceding section concerning the widely acknowledged desiderata for scientific theories led Kuhn to the claim that there are no generally binding rules of scientific procedure which unambiguously determine theory choice in the light of logic and experience alone. Kuhn therefore presented theory choice as the result of group decisions made by the scientists in the field, which are not fully governed by general rules, and which have to be accounted for in terms of the common training of the members of the group. This characterization was received by some critics with the charge of irrationalism and of appeal to mob psychology—a charge that may have seemed to be substantiated by Kuhn's comparison of theory change in science to a religious conversion experience involving a leap of faith. This may have seemed to license the adoption of any theory one might have faith in. But the requirements of testability, predictivity, evidential fit, and large scope, which Kuhn acknowledges, would suffice, despite their vagueness, to rule out, say, astrology or chiromancy, and in a competition with currently accepted theories.

And as for Kuhn's invocation, in the context of theory choice, of group decisions not governed by general rules of procedure, it may be of interest to note that in the first decade of this century, Pierre Duhem expressed quite a similar view. He argued, on purely logical grounds, that the outcome of a scientific experiment cannot refute a theoretical assumption in isolation, but only a comprehensive set of assumptions. If the experimental findings conflict with predictions deducible from the set, then some change has to be made in the total theory; but no objective logical criteria determine uniquely what changes should be made. That decision, Duhem says, must be left to "the good sense" of the scientists in the field.[8] Here, then, normative methodology leaves off, and only socio-psychological considerations are invoked, with perhaps a vaguely explanatory intent.

As for the charge of irrationalism that has been brought against Kuhn's account, I have to say that I cannot point to any justifiable canons of rational inquiry that Kuhn could be accused of having denied or slighted.

V. RATIONALITY AND INCOMMENSURABILITY

I want to take issue, however, with one argument, offered especially by Feyerabend, for the thesis that there can be no general criteria of choice between competing comprehensive theories. The argument is based on the idea that such theories do not share a single statement, that they are "completely disjointed, or incommensurable"; he cites classical and relativistic physics as an example.[9] I have strong doubts about this idea, largely because the notion of incommensurability is not made sufficiently clear. Feyerabend seems to hold that in the transition from one paradigm to another, all the terms taken over from the old theory into the new one come to function in a new set of theoretical principles and therefore change their meanings,—that this is the case even for the terms used to describe instrument readings and observational findings, and that therefore there is no possibility at all of comparing the two theories or the bodies of evidence relevant to them. In fact, Feyerabend concludes, concerning the possible bases of theory choice: "What remains are esthetic judgments, judgments of taste, and our subjective wishes."[10] Kuhn, too, has placed considerable emphasis on the incommensurability of paradigms, but he has not drawn quite such extreme and subjective conclusions.

Now, the transition from one paradigm to another clearly does bring with it considerable changes in the use of the terms taken over into the new theory: the paradigms differ in many of the sentences containing those terms. But why should a change of this kind be taken to signal a change in the meanings of the terms concerned rather than a change in the claims made about the entities they refer to? To illustrate: In *The Structure of Scientific Revolutions,* Kuhn remarks that the transition from the Ptolemaic to the Copernican System effected a change in the meanings of such terms as "planet," "earth," and "motion."[11] But on what grounds can it be claimed that the change in the statements containing those terms reflects a change in the meanings of the words rather than a change in what is assumed or asserted about the earth, the planets, and their motions?

In the absence of a more explicit account of meaning change for theoretical terms, claims of incommensurability cannot, I think, establish the impossibility of formulating criteria for a comparative evaluation of competing paradigms. It seems clear, for example, that the results of parallax measurements constitute evidence that is relevant to the Copernican as well as to the Ptolemaic theory, and that they thus form some basis for a comparison between the two.

VI. ON THE "TIMELESSNESS" OF METHODOLOGICAL NORMS

The methodological efforts of logical empiricism were not very much concerned with problems of theory change in science; but they did yield theories of confirmation and of probability for scientific hypotheses, which propounded explicit and precise standards for the comparison or appraisal of hypotheses in regard to the evidential support conferred upon them by the available empirical data. Carnap referred to his quantitative theory of confirmation as "inductive logic"; as the term suggests, the principles aimed at by theories of inductive support were thought of as being analogous to the principles of valid deductive inference formulated by deductive logic. At least implicitly, I think, it was assumed that methodological principles for appraising the rational credibility of hypotheses would have to hold a priori as it were, independently of any empirical matters, and immutably at all times—all this in analogy to the standard conception of the timelessness and immutable validity of the principles of deductive logic.

But it seems to me open to question whether this conception can be reasonably applied to all methodological principles of rational theory choice. For what kinds of theory can be reasonably entertained, or be regarded as worth serious consideration, will depend on the general view of the world—especially of its deep and pervasive features—that prevails at the time.

Thus, for example, Maxwell held that space and time coordinates should enter into scientific laws and theories, not absolutely or explicitly, but only in the form of differences; his reason being that the mere spatio-temporal location of an event could exert no causal influence on the course taken by that event: what occurs, under given physical conditions, at one spatio-temporal location will under the same physical conditions, equally happen at any other spatio-temporal location. Maxwell regarded this principle as providing a more precise expression of the idea that "the same causes will always produce the same effects."[12] Actually, we might note, his condition is satisfied not only by strictly causal or deterministic laws, but also by the probabilistic laws of contemporary science. It may be said to impose a constraint on rational theory choice, reflecting the assumption that in our world spatio-temporal locations have no "nomic efficacy," whether of a causal or of a probabilistic kind. But it is conceivable that that assumption may eventually come to be abandoned in the light of new scientific findings: and in this case, the methodological norm would have to be abandoned as well.

Similarly, 150 years ago, a theory propounding quantization of energy or perhaps even of space and time, or postulating the creation of electrons from empty space, would presumably not have been regarded as a rationally open theoretical option.

And Carnap himself, the leading figure among logical analysts, provided in his theory of inductive probability for a "continuum of inductive methods," each characterized by a real-number value of a parameter lambda, with the explicit understanding that the value of that parameter, and thereby the principles of the inductive methodology, could be changed in consideration of certain very general features of the empirical knowledge available at a given time.

That the standards of choice among the hypotheses or theories should depend on such general assumptions about the causal structure or about the basic articulation and the degree of orderliness of the world seems quite plausible and reasonable: but then, the conception of timelessness for all standards of rational theory choice must be abandoned. On this point, I find myself in agreement even with Feyerabend, though, as indicated earlier, I cannot follow him the rest of the way to anarchy.

VII. THE CONCEPTION OF RATIONALITY IN KUHN'S PERSPECTIVE ON SCIENCE

I have argued that to the extent that a methodological theory of science is to provide standards of rationality for scientific inquiry as well as a basis for explaining certain aspects of the actual research behavior of scientists, the theory must have the character of an *E-N* methodology: its norms must express goal-dependent, instrumental conditions of rationality; and its explanatory potential must lie in its ability to "rationalize" significant aspects of scientific research, i.e., to explain them by the assumption that the scientists in question are disposed to act in accordance with the specified norms.

Consider from this point of view Kuhn's remark, cited earlier, that his methodological pronouncements should be read as prescriptive and as descriptive at once.

Let us look first at their prescriptive import, which Kuhn characterizes schematically as follows: "The structure of my argument is simple and, I think, unexceptionable: scientists behave in the following ways; those modes of behavior have (here theory enters) the following essential functions; in the absence of an alternate mode *that would serve similar functions,* scientists should behave essentially as they do if their concern is to improve scientific knowledge."[13] This statement does seem to reflect an instrumental conception of the rationality of scientific behavior, the "essential function" of such behavior being the improvement of scientific knowledge. But surely, a methodological theory can yield prescriptions for scientific inquiry only insofar as it formulates explicit rules or norms for the conduct of inquiry: and it seems very questionable to me whether a theory in Kuhn's style and spirit can provide the requisite rules or norms.

I referred earlier to Kuhn's remark that scientific behavior is the best example we have of rationality, and that we had better change our standards of rationality if science seems to violate them. But any given piece of behavior can be described in many different ways; yet it can be qualified as rational or as irrational, if at all, then only under certain particular kinds of description, which represent it as a rule-following activity conforming to certain explicitly specified rules, in the pursuit of specified goals.

To put the point somewhat differently: Kuhn's remark that "scientists should behave essentially as they do if their concern is to improve scientific knowledge" leaves open the question: in what respects? What aspects, what characteristics of the actual behavior of scientists are the ones that matter, the ones to be emulated in a rational pursuit of the improvement of scientific knowledge? And in what sense, and on what grounds, can those characteristics be held to serve the function of improving scientific knowledge?

Kuhn's construal of the explanatory import of his ideas for actual cases of theory choice seems to me open to similar questions. Kuhn describes that import as follows: "take a *group* of the ablest available people with the most appropriate motivation; train them in some science and in the special ties relevant to the choice at hand; imbue them with the value system, the ideology, current in their discipline (and to a great extent in other scientific fields as well); and, finally, *let them make their choice.* If that technique does not account for scientific development as we know it, then no other will."[14]

But, literally speaking, a technique for training or programming people does not constitute an explanation of any part of their subsequent behavior. For that, we need an empirical theory which yields explicit specifications of those beliefs, values, ideological principles, et cetera whose acceptance will account for the theory choices in question. The claim that persons who are motivated and trained in the ways in which scientists are will make the choices actually encountered in science fails to specify just what aspects of the behavior of scientists are relevant to the explanation of theory choice; and insofar as those aspects are not made explicit, no explanatory claim can be made for the methodological theory.

set such impossibly high standards

Hempel believes in The Moral Person

The Right Thing

VIII. CONCLUSION

The conception of science as the exemplar of the rational pursuit of knowledge was emphatically held also by logical empiricism, which frequently supported and revised its explications of scientific concepts and procedures by appeal to scientific usage. One familiar example of this process is the gradual expansion of the ways in which Carnap explicated scientific modes of introducing new terms: from explicit definition to the use of reduction sentences and chains of such sentences on to the global specification of a set of theoretical concepts by means of theoretical postulates and correspondence rules: ever closer attention to the modes and needs of scientific theorizing led here to a far-reaching liberalization of what were regarded as essential conditions for the introduction of "meaningful" new terms. In this sense, the rational reconstructions formulated by logical empiricists did have a descriptive facet. But it must be acknowledged that the analytic empiricist concern with clarity and rigor and with systematic comprehensiveness and formal simplicity encouraged the formulation of some technically very impressive explicatory systems which are rather far removed from the objectives and modes of thinking that prompt scientific investigations; those systems, accordingly, possess only very limited explanatory potential.

Reconsideration and reorientation are thus clearly needed, and the critical and constructive ideas of the historic-pragmatic school have opened highly illuminating and promising new perspectives on issues of central concern to the methodology of science. But insofar as a proposed methodological theory of science is to afford an account of scientific inquiry as a rational pursuit, it will have to specify certain goals of scientific inquiry as well as some methodological principles observed in their pursuit; finally, it will have to exhibit the instrumental rationality of the principles in relation to the goals. Only to the extent that this can be done does the conception of science as the exemplar of rationality appear to be viable.

NOTES

1. See Kuhn (1970a), pp. 155ff.: (1970b), pp. 261f.
2. Kuhn (1971), pp. 144–145.

3. Kuhn (1971), p. 144.
4. Kuhn (1970b), p. 237.
5. See, for example, Lakatos (1971), pp. 105–108.
6. Feyerabend (1970), p. 76.
7. Nagel (1939), chapter 11, sec. 8.
8. Cf. Duhem (1962), part 2, chapter 6, section 10.
9. Feyerabend (1970), p. 280.
10. Feyerabend (1970), p. 90.
11. Kuhn (1970a), pp. 128, 149–150.
12. Maxwell (1887), pp. 31–32.
13. Kuhn (1970b), p. 237 (italics in original).
14. Kuhn (1970b), pp. 237–238 (italics in original).

REFERENCES

Duhem, P. 1962. *The Aim and Structure of Physical Theory.* New York: Atheneum. (French original first published in 1906).

Feyerabend, P. 1970. "Against Method: Outline of an Anarchistic Theory of Knowledge." In *Minnesota Studies in the Philosophy of Science,* vol. 4, edited by M. Radner and S. Winokur, 17–130.

Kuhn, T. S. 1970a. *The Structure of Scientific Revolutions.* 2d ed. Chicago: University of Chicago Press.

———— 1970b. "Reflections on My Critics." In *Criticism and the Growth of Knowledge,* edited by I. Lakatos and A. Musgrave, 231–278. New York: Cambridge University Press.

————. 1971. "Notes on Lakatos." In *P.S.A. 1970, Boston Studies in the Philosophy of Science,* vol. 8, edited by R. S. Cohen and R. C. Buck, 137–146.

Lakatos, I. 1971. "History of Science and Its Rational Reconstruction," In *P.S.A. 1970, Boston Studies in the Philosophy of Science,* vol. 8, edited by R. S. Cohen and R. C. Buck, 91–136.

Maxwell, J. C. 1878. *Matter and Motion.* New York: D. Van Nostrand.

Nagel, E. 1939. *Principles of the Theory of Probability.* University of Chicago Press.

Popper, K. R. 1959. *The Logic of Scientific Discovery.* London: Hutchinson.

28

The Variety of Reasons for the Acceptance of Scientific Theories

Philipp G. Frank

Among scientists it is taken for granted that a theory "should be" accepted if and only if it is "true"; to be true means in this context to be in agreement with the observable facts that can be logically derived from the theory. Every influence of moral, religious, or political considerations upon the acceptance of a theory is regarded as "illegitimate" by the so-called community of scientists. This view certainly has had a highly salutary effect upon the evolution of science as a human activity. It tells the truth—but not the whole truth. It has never happened that all the conclusions drawn from a theory have agreed with the observable facts. The scientific community has accepted theories only when a vast number of facts has been derived from few and simple principles. A familiar example is the derivation of the immensely complex motions of celestial bodies from the simple Newtonian formula of gravitation, or the large variety of electromagnetic phenomena from Maxwell's field equations.

If we restrict our attention to the two criterions that are called "agreement with observations" and "simplicity," we remain completely within the domain of activities that are cultivated and approved by the community of scientists. But, if we have to choose a certain theory for acceptance, we do not know what respective weight should be attributed to these two criterions. There is obviously no theory that agrees with *all* observations and no theory that has "perfect" simplicity. Therefore, in every individual case, one has to make a choice of a theory by a compromise between both criterions. However, when we try to specify the degree of "simplicity" in different theories, we soon notice that attempts of this kind lead us far beyond the limits of physical science. Everybody would agree that a linear function is simpler than a function of the second or higher degree; everybody would also admit that a circle is simpler than an ellipse. For this reason, physics is filled with laws that express proportionality, such as Hooke's law in elasticity or Ohm's law in electrodynamics. In all these cases, there is no doubt that a nonlinear relationship would describe the facts in a more accurate way, but one tries to get along with a linear law as much as possible.

There was a time when, in physics, laws that could be expressed without using differential calculus were preferred, and in the long struggle between the corpuscular and the wave theories of light, the argument was rife that the corpuscular theory was mathematically simpler, while the wave theory required the solution of boundary problems of partial differential equations, a highly complex matter. We note that even a purely mathematical estimation of simplicity depends upon the state of culture of a certain period. People who have grown up in a mathematical atmosphere—that is, saturated with ideas about invariants—will find that Einstein's theory of gravitation is of incredible beauty and simplicity; but to people for whom ordinary calculus is the center of interest, Einstein's theory will be of immense complexity, and this low degree of simplicity will not be compensated by a great number of observed facts.

However, the situation becomes much more complex, if we mean by *simplicity* not only simplicity of the mathematical scheme but also simplicity of the whole discourse by which the theory is formulated. We may start from the most familiar instance, the decision between the Copernican (heliocentric) and the Ptolemaic (geocentric) theories. Both parties, the Roman Church and the followers of Copernicus, agreed that Copernicus's system, from the purely mathematical angle, was simpler than Ptolemy's. In the first one, the orbits of planets were plotted as a system of concentric circles with the sun as center, whereas in the geocentric system, the planetary orbits were sequences of loops. The observed facts covered by these systems were approximately the same ones. The criterions of acceptance that are applied in the community of scientists today are, according to the usual way of speaking, in agreement with observed facts and mathematical simplicity. According to them, the Copernican system had to be accepted unhesitatingly. Since this acceptance did not happen before a long period of doubt, we see clearly that the criterions "agreement with observed facts" and "mathematical simplicity" were not the only criterions that were considered as reasons for the acceptance of a theory.

As a matter of fact, there were three types of reasons against the acceptance of the Copernican theory that remained unchallenged at the time when all "scientific" reasons were in favor of that theory. First, there was the incompatibility of the Copernican system with the traditional interpretation of the Bible. Second, there was the disagreement between the Copernican system and the prevailing philosophy of that period, the philosophy of Aristotle as it was interpreted by the Catholic schoolmen. Third, there was the objection that the mobility of the earth, as a real physical fact, is incompatible with the common-sense interpretation of nature. Let us consider these three types of reasons more closely. In the Book of Joshua this leader prays to God to stop the sun in its motion in order to prolong the day and to enable the people of Israel to win a decisive victory. God indeed "stopped the sun."

If interpreted verbally, according to the usage of words in our daily language, this means that the sun is moving, in flagrant contradiction with the Copernican theory. One could, of course, give a more sophisticated interpretation and say that "God stopped the sun" means that he stopped it in its motion relative to the earth. This is no longer contradictory to the Copernican system. But now the question arises: Should we adopt a simple mathematical description and a complicated, rather "unnatural" interpretation of the Bible or a more complicated mathematical description (motion in loops) and a simple "natural" interpretation of the biblical text? The decision certainly cannot be achieved by any argument taken from physical science.

If one believes that all questions raised by science must be solved by the "methods" of this special science, one must say: Every astronomer who lived in the period between Copernicus and Galileo was "free" to accept either the Copernican or the Ptolemaic doctrine; he could make an "arbitrary" decision. However, the situation is quite different if one takes into consideration that physical science is only a part of science in general. Building up astronomical theories is a particular act of human behavior. If we consider human behavior in general, we look at physical science as a part of a much more general endeavor that embraces also psychology and sociology. It is called by some authors "behavioristics." From this more general angle, the effect of a simplification in the mathematical formula and the simplification in biblical interpretation are quite comparable with each other. There is meaning in asking by which act the happiness of human individuals and groups is more favorably influenced. This means that, from the viewpoint of a general science of human behavior, the decision between the Copernican and Ptolemaic systems was never a matter of arbitrary decision.

The compatibility of a physical theory with a certain interpretation of the Bible is a special case of a much more general criterion: the compatibility of a physical theory with theories that have been advanced to account for observable phenomena outside the domain of a physical science. The most important reason for the acceptance of a theory beyond the "scientific criterions" in the narrower sense (agreement with observation and simplicity of the mathematical pattern) is the fitness of a theory to be generalized, to be the basis of a new theory that does not logically follow from the original one, and to allow prediction of more observable facts. This property is often called the "dynamical" character or the "fertility" of a theory. In this sense, the Copernican theory is much superior to the geocentric one. Newtonian laws of motion have a simple form only if the sun is taken as a system of reference and not the earth. But the decision in favor of the Copernican theory on this basis could be made only when Newton's laws came into existence. This act requires, however, creative imagination or, to speak more flippantly, a happy guessing that leads far beyond the Copernican and Ptolemaic systems.

However, long before the "dynamical" character of the Copernican system was recognized, the objection was raised that the system was incompatible with "the general laws of motion" that could be derived from principles regarded as "immediately intelligible" or, in other words, "self-evident" without physical experiment or observations. From such "self-evident" principles there followed, for example, the physical law that only celestial bodies (like sun or moon) moved "naturally" in circular orbits, while terrestrial bodies (like our earth) moved naturally along straight lines as a falling stone does. Copernicus's theory of a "motion of the earth in a circular orbit" was, therefore, incompatible with "self-evident" laws of nature.

Medieval scientists were faced with the alternatives: Should they accept the Copernican theory with its simple mathematical formulas and drop the self-evident laws of motion, or should they accept the complicated mathematics of the Ptolemaic system along with the intelligible and self-evident general laws of motion. Acceptance of Copernicus's theory would imply dropping the laws of motion that had been regarded as self-evident and looking for radically new laws. This would also mean dropping the contention that a physical law can be derived from "intelligible" principles. Again, from the viewpoint of physical science, this decision cannot be made. Although an arbitrary decision may seem to be required, if one looks at the situation from the viewpoint of human behavior it is clear that the decision, by which the derivation of physical laws from self-evident principles is abandoned, would alter the situation of man in the world fundamentally. For example, an important argument for the existence of spiritual beings would lose its validity. Thus, social science had to decide whether the life of man would become happier or unhappier by the acceptance of the Copernican system.

The objections to this system, on the basis of self-evident principles, have also been formulated in a way that looks quite different but may eventually, when the chips are down, not be so very different. Francis Bacon, the most conspicuous adversary of Aristotelianism in the period of Galileo, fought the acceptance of the Copernican theory on the basis of commonsense experience. He took it for granted that the principles of science should be as analogous as possible to the experience of our daily life. Then, the principles could be presented in the language that has been shaped for the purpose of describing, in the most convenient way, the experience of our daily existence—the language that everyone has learned as a child and that is called "common-sense language." From this daily experience, we have learned that the behavior of the sun and the planets is very different from that of the earth. While the earth does not emit any light, the sun and the planets are brilliant; while every earthly object that becomes separated from the main body will tend to fall back toward the center and stop there, the celestial objects undergo circular motion eternally around the center.

To separate the sun from the company of the planets and put the earth among these brilliant and mobile creatures, as Copernicus suggested, would have been not only unnatural but a serious violation of the rule to keep the principles of science as close to common sense as possible. We see by this example that one of the reasons for the acceptance of a theory has frequently been the compatibility of this theory with daily life experience or, in other words, the possibility of expressing the theory in common-sense language. Here is, of course, the source of another conflict between the "scientific" reasons for the acceptance of a theory and other requirements that are not "scientific" in the narrower sense. Francis Bacon rejected the Copernican system because it was not compatible with common sense.

In the eighteenth and nineteenth centuries, Newton's mechanics not only had become compatible with common sense but had even been identified with common-sense judgment. As a result, in twentieth-century physics, the theory of relativity and the quantum theory were regarded by many as incompatible with common sense. These theories were regarded as "absurd" or, at least, "unnatural." Lenard in Germany, Bouasse in France, O'Rahilly in Ireland, and Timiryaseff in Russia rejected the theory of relativity, as Francis Bacon had rejected the Copernican system. Looking at the historical record, we notice that the requirement of compatibility with common sense and the rejection of "unnatural theories" have been advocated with a highly emotional undertone, and it is reasonable to raise the question: What was the source of heat in those fights against new and absurd theories? Surveying these battles, we easily find one common feature, the apprehension that a disagreement with common sense may deprive scientific theories of their value as incentives for a desirable human behavior. In other words, by becoming incompatible with common sense, scientific theories lose their fitness to support desirable attitudes in the domain of ethics, politics, and religion.

Examples are abundant from all periods of theory-building. According to an old theory that was prevalent in ancient Greece and was accepted by such men as Plato and Aristotle, the sun, planets, and other celestial bodies were made of a material that was completely different from the material of which our earth consists. The great gap between the celestial and the terrestrial bodies was regarded as required by our common-sense experience. There were men—for example, the followers of Epicurus—who rejected this view and assumed that all bodies in the universe, earth and stars, consist of the same material. Nevertheless, many educators and political leaders were afraid that denial of the exceptional status of the celestial bodies in physical science would make it more difficult to teach the belief in the existence of spiritual beings as distinct from material bodies; and since it was their general conviction that the belief in spiritual beings is a powerful instrument to bring about a desirable conduct among citizens, a physical theory that supported this belief seemed to be highly desirable.

Plato, in his famous book *Laws*, suggested that people in his ideal state who taught the "materialistic" doctrine about the constitution of sun and stars should be jailed. He even suggested that people who knew about teachers of that theory and did not report them to the authorities should also be jailed. We learn from this ancient example how scientific theories have served as instruments of indoctrination. Obviously, fitness to support a desirable conduct of citizens or, briefly, to support moral behavior, has served through the ages as a reason for acceptance of a theory. When the "scientific criterions" did not uniquely determine a theory, its fitness to support moral or political indoctrination became an important factor for its acceptance. It is important to learn that the interpretation of a scientific theory as a support of moral rules is not a rare case but has played a role in all periods of history.

This role probably can be traced back to a fact that is well known to modern students of anthropology and sociology. The conduct of man has always been shaped according to the example of an ideal society; on the other hand, this ideal has been represented by the "behavior" of the universe, which is, in turn, determined by the laws of nature, in particular, by the physical laws. In this sense, the physical laws have always been interpreted as examples for the conduct of man or, briefly speaking, as moral laws. Ralph Waldo Emerson wrote in his essay *Nature* that "the laws of physics are also the laws of ethics." He used as an example the law of the lever, according to which "the smallest weight may be made to lift the greatest, the difference of weight being compensated by time."

We see the connection of the laws of desirable human conduct with the physical laws of the universe when we glance at the Book of Genesis. The first chapter presents a physical theory about the creation of the world. But the story of the Creation serves also as an example for the moral behavior of men; for instance, because the Creation took seven days, we all feel obliged to rest on each seventh day. Perhaps the history of the Great Flood is even more instructive. When the Flood abated, God established a Covenant with the human race: "Never again shall all flesh be cut off by the waters of a flood; neither shall there any more be a flood to destroy the earth." As a sign of the Covenant the rainbow appeared: "When I bring clouds over the earth and the bow is seen in the clouds, I will remember the Covenant which is between me and you, and the waters shall never again become a flood to destroy all flesh." If we read the biblical text carefully, we understand that what God actually pledged was to maintain, without exception, the validity of the physical laws or, in other words, of the causal law. God pledged: "While the earth remains, seedtime and harvest, cold and heat, summer and winter, day and night shall not cease."

All the physical laws, including the law of causality, were given to mankind as a reward for moral behavior and can be canceled if mankind does not behave well. So even the belief in the validity of causal laws in the physical world has supported the belief in God as the supreme moral authority who would punish every departure from moral behavior by abolishing causality. We have seen that Epicurean physics and Copernican astronomy were rejected on moral grounds. We know that Newton's physics was accepted as supporting the belief in a God who was an extremely able engineer and who created the world as a machine that performed its motions according to its plans. Even the generalization of Newtonian science that was advanced by eighteenth-century materialism claimed to serve as a support for the moral behavior of man. In his famous book *Man a Machine,* which has often been called an "infamous book," La Mettrie stresses the point that by regarding men, animals, and plants as beings of the same kind, man is taught to regard them all as his brothers and to treat them kindly.

It would be a great mistake to believe that this situation has changed in the nineteenth and twentieth centuries. A great many authors have rejected the biological theory that organisms have arisen from inanimate matter (spontaneous generation), because such a theory would weaken the belief in the dignity of man and in the existence of a soul and would, therefore, be harmful to moral conduct. In twentieth century physics, we have observed that Einstein's theory of relativity has been interpreted as advocating an "idealistic" philosophy, which, in turn, would be useful as a support of moral conduct. Similarly, the quantum theory is interpreted as supporting a weakening of mechanical determinism and, along with it, the introduction of "indeterminism" into physics. In turn, a great many educators, theologians, and politicians have enthusiastically acclaimed this "new physics" as providing a strong argument for the acceptance of "indeterminism" as a basic principle of science.

The special mechanism by which social powers bring about a tendency to accept or reject a certain theory depends upon the structure of the society within which the scientist operates. It may vary from a mild influence on the scientist by friendly reviews in political or educational dailies to promotion of his book as a best seller, to ostracism as an author and as a person, to loss of his job, or, under some social circumstances, even to imprisonment, torture, and execution. The honest scientist who works hard in his laboratory or computation room would obviously be inclined to say that all this is nonsense—that his energy should be directed toward finding out whether, say, a certain theory is "true" and that he "should not" pay any attention to the fitness of a theory to serve as an instrument in the fight for educational or political goals. This is certainly the way in which the situation presents itself to most active scientists. However, scientists are also human beings and are definitely inclined toward some moral, religious, or political creed. Those who deny emphatically that there is any connection between scientific theories and religious or political creeds believe in these creeds on the basis of indoctrination that has been provided by organizations such as churches or political parties. This attitude leads to the conception of a "double truth" that is not only logically confusing but morally dangerous. It can lead to the practice of serving God on Sunday and the devil on weekdays.

The conviction that science is independent of all moral and political influences arises when we regard science either as a collection of facts or as a picture of objective reality. But today, everyone who has attentively studied the logic of science will know that science actually is an instrument that serves the purpose of connecting present events with future events and deliberately utilizes this knowledge to shape future physical events as they are desired. This instrument consists of a system of propositions—principles—and the operational definitions of their terms. These propositions certainly cannot be derived from the facts of our experience and are not uniquely determined by these facts. Rather they are hypotheses from which the facts can be logically derived. If the principles or hypotheses are not determined by the physical facts, by what are they determined? We have learned by now that, besides the agreement with observed facts, there are other reasons for the acceptance of a theory: simplicity, agreement with common sense, fitness for supporting a desirable human conduct, and so forth. All these factors participate in the making of a scientific theory. We remember, however, that according to the opinion of the majority of active scientists, these extrascientific factors "should not" have any influence on the acceptance of a scientific theory. But who has claimed and who can claim that they "should not"?

This firm conviction of the scientists comes from the philosophy that they have absorbed since their early childhood. The theories that are built up by "scientific" methods, in the narrower sense, are "pictures" of physical reality. Presumably they tell us the "truth" about the world. If a theory built up exclusively on the ground of its agreement with observable facts tells the "truth" about the world, it would be nonsense to assume seriously that a scientific theory can be influenced by moral or political reasons. However, we learned that "agreement with observed facts" does not single out one individual theory. We never have one theory that is in full agreement but several theories that are in partial agreement, and we have to determine the final theory by a compromise. The final theory has to be in fair agreement with observations and also has to be sufficiently simple to be usable. If we consider this point, it is obvious that such a theory cannot be "the truth." In modern science, a theory is regarded as an instrument that serves toward some definite purpose. It has to be helpful in predicting future observable facts on the basis of facts that have been observed in the past and the present. It should also be helpful in the construction of machines and devices that can save us time and labor. A scientific theory is, in a sense, a tool that produces other tools according to a practical scheme.

In the same way that we enjoy the beauty and elegance of an airplane, we also enjoy the "elegance" of the theory that makes the construction of the plane possible. In speaking about any actual machine, it is meaningless to ask whether the machine is "true" in the sense of its being "perfect." We can ask only whether it is "good" or sufficiently "perfect" for a certain purpose. If we require speed as our purpose, the "perfect" airplane will differ from one that is "perfect" for the purpose of endurance. The result will be different again if we choose safety, or fun, or convenience for reading and sleeping as our purpose. It is impossible to design an airplane that fulfills all these purposes in a maximal way. We have to make some compromises. But then, there is the question: Which is more important, speed or safety, or fun or endurance? These questions cannot be answered by any agreement taken from physical science. From the viewpoint of "science proper" the purpose is arbitrary, and science can teach us only how to construct a plane that will achieve a specified speed with a specified degree of safety. There will be a debate, according to moral, political, and even religious lines, by which it will be determined how to produce the compromise. The policymaking authorities are, from the logical viewpoint, "free" to make their choice of which type of plane should be put into production. However, if we look at the situation from the viewpoint of a unified science that includes both physical and social science, we shall understand how the compromise between speed and safety, between fun and endurance is determined by the social conditions that produce the conditioned reflexes of the policymakers. The conditioning may be achieved, for example, by letters written to congressmen. As a matter of fact, the building of a scientific theory is not essentially different from the building of an airplane.

If we look for an answer to the question of whether a certain theory, say the Copernican system or the theory of relativity, is perfect or true, we have to ask the preliminary questions: What purpose is the theory to serve? Is it only the purely technical purpose of predicting observable facts? Or is it to obtain a simple and elegant theory that allows us to derive a great many facts from simple principles? We choose the theory according to our purpose. For some groups, the main purpose of a theory may be to serve as a support in teaching a desirable way of life or to discourage an undesirable way of life. Then, we would prefer theories that give a rather clumsy picture of observed facts, provided that we can get from the theory a broad view of the universe in which man plays the role that we desire to give him. If we wish to speak in a more brief and general way, we may distinguish just two purposes of a theory: the usage for the construction of devices (technological purpose) and the usage for guiding human conduct (sociological purpose).

The actual acceptance of theories by man has always been a compromise between the technological and the sociological usage of science. Human conduct has been influenced directly by the latter, by supporting specific religious or political creeds, while the technological influence has been rather indirect. Technological changes have to produce social changes that manifest themselves in changing human conduct. Everybody knows of the Industrial Revolution of the nineteenth century and the accompanying changes in human life from a rural into an urban pattern. Probably the rise of atomic power will produce analogous changes in man's way of life.

The conflict between the technological and the sociological aims of science is the central factor in the history of science as a human enterprise. If thoroughly investigated, it will throw light upon a factor that some thinkers, Marxist as well as religious thinkers, regard as responsible for the social crisis of our time: the backwardness of social progress compared with technological progress. To cure this illness of our time, an English bishop recommended, some years ago, the establishment of a "truce" in the advancement in technology, in order to give social progress some time to keep up with technological advancement. We have seen examples of this conflict in Plato's indictment of astrophysical theories that could be used as a support of "materialism." We note the same purpose in the fight against the Copernican system and, in our own century, against the Darwinian theory of evolution, against Mendel's laws of heredity, and so forth.

A great many scientists and educators believe that such a conflict no longer exists in our time, because now it is completely resolved by "the scientific method," which theory is the only valid one. This opinion is certainly wrong if we consider theories of high generality. In twentieth-century physics, we note clearly that a formulation of the general principles of subatomic physics (quantum theory) is accepted or rejected according to whether we believe that introduction of "indeterminism" into physics gives comfort to desirable ethical postulates or not. Some educators and politicians have been firmly convinced that the belief in "free will" is necessary for ethics and that "free will" is not compatible with Newtonian physics but is compatible with quantum physics. The situation in biology is similar. If we consider the attitude of biologists toward the question whether living organisms have developed from inanimate matter, we shall find that the conflict between the technological and the sociological purposes of theories is in full bloom. Some prominent biologists say that the existence of "spontaneous generation" is highly probable, while others of equal prominence claim that it is highly improbable. If we investigate the reasons for these conflicting attitudes, we shall easily discover that, for one group of scientists, a theory that claims the origin of man not merely from the "apes" but also from "dead matter" undermines their belief in the dignity of man, which is the indispensable basis for all human morality. We should note in turn that, for another group, desirable human behavior is based on the belief that there is a unity in nature that embraces all things.

In truth, many scientists would say that scientific theories "should" be based only on purely scientific grounds. But, exactly speaking, what does the word *should* mean in this context? With all the preceding arguments it can mean only: If we consider exclusively the technological purpose of scientific theories, we could exclude all criterions such as agreement with common sense or fitness for supporting desirable conduct. But even if we have firmly decided to do away with all "nonsense," there still remains the criterion of "simplicity," which is necessary for technological purposes and also contains, as we learned previously, a certain sociological judgment. Here, restriction to the purely technological purpose does not actually lead unambiguously to a scientific theory. The only way to include theory-building in the general science of human behavior is to refrain from ordering around scientists by telling them what they "should" do and to find how each special purpose can be achieved by a theory. Only in this way can science as a human activity be "scientifically" understood and the gap between the scientific and the humanistic aspect be bridged.

CASE STUDY FOR PART 5

Suppose a social scientist or a medical researcher justified using deception on patients or research subjects that had agreed to be part of a research effort. Suppose, further, that the researcher justified the lying, deception (maybe even harm) on the grounds that the knowledge acquired was both very important (it might save lives) and not obtainable unless the "volunteers" were deceived.

1. Suppose the "subjects" had live cancer cells implanted in them. How would you respond to the researcher?

2. Suppose people were made to feel like two cents about themselves because they did things they regretted (even at the time) but went along anyway.

3. Are there any situations in which it might be justified to lie, deceive, even harm innocent people to acquire valuable scientific knowledge?

4. What sorts of value judgments are at issue here, and how might they be justified?

5. Are the values used to defend such activities scientific or extrascientific? Explain.

STUDY QUESTIONS FOR PART 5

1. Are Quine and Ullian realists or antirealists?

2. Does the virtue of conservatism have any adverse effects on science?

3. Why is "modesty" a methodological virtue?

4. Why are ad hoc hypotheses bad for science?

5. Can a refutable hypothesis still be true? Explain.

6. Can Giere's proposals work if scientists are wrong about the initial conditions?

7. How can theory choice be rational if, as Kuhn claims, scientists use the same criteria but make different choices?

8. What's the distinction between the context of discovery and the context of justification? Would Kuhn or Feyerabend accept this distinction? Explain.

9. What is Kuhn's point in denying that there are rules for making theory choices?

10. What is the difference between normative and explanatory theories of inquiry, according to Hempel?

11. In what sense is it impossible to compare theories, according to Feyerabend?

12. What reasons, besides agreement with observation, does Frank claim play a role in scientific acceptance?

13. Does Frank hold the view that theory choice is subjective? Explain.

14 Must science agree with common sense to be acceptable? Give an example to refute this claim.

15. How would a "rational account" of science function as an ideal, according to Hempel?

16. What are instrumental norms? What's their role in science, if Hempel is right?

SELECTED BIBLIOGRAPHY

1. Achinstein, Peter, ed. *The Concept of Evidence.* New York: Oxford University Press, 1983. [Good essays.]

2. Bloor, David, and Edge, David, eds. *Science in Context.* Cambridge, Mass.: The MIT Press, 1982. [Sociology of science.]

3. Boyd, Richard, Philip Gasper, and J. D. Trout, eds. *The Philosophy of Science.* Cambridge, Mass.: The MIT Press, 1991. [Essays in part I.]

4. Brody B., ed. *Readings in Philosophy of Science*, part 3. Englewood Cliffs, N.J.: Prentice-Hall, 1970.

5. Earman, John, ed. *Testing Scientific Theories.* Minneapolis: University of Minnesota Press, 1983. [Advanced essays.]

6. Goodman, Nelson. *Fact, Fiction, and Forecast.* New York: Bobbs-Merrill, 1965. [A paradox in confirmation theory.]

7. Gutting, Gary, ed. *Paradigms and Revolutions.* Notre Dame, Ind.: University of Notre Dame Press, 1980. [Extends and criticizes Kuhn.]

8. Hacking, Ian. *Representing and Intervening.* Cambridge: Cambridge University Press, 1983. [Interesting approach to realism and antirealism.]

9. Hacking, Ian, ed. *Scientific Revolutions.* Oxford: Oxford University Press, 1981. [Postpositivism.]

10. Hull, David. *Science as Process.* Chicago: University of Chicago Press, 1988. [An approach to science emphasizing scientific activity.]

11. Kitcher, Philip. *The Advancement of Science.* Oxford: Oxford University Press, 1993. [A major synthesis.]

12. Kuhn, Thomas. *The Structure of Scientific Revolutions.* Chicago: University of Chicago Press, 1962. [A path-breaking essay.]

13. Lauden, Larry. *Progress and Its Problems.* Berkeley: University of California Press, 1977. [An important criticism of Kuhn and the sociology of science.]

14. McMullin, Ernan, ed. *Construction and Constraint.* Notre Dame, Ind.: University of Notre Dame Press, 1994. [Recent essays on scientific rationality.]

15. Pitt, Joseph, ed. *Theories of Explanation.* Oxford: Oxford University Press, 1988. [Good essays; advanced.]

16. Popper, Karl. *Conjectures and Refutations.* New York: Harper and Row, 1963. [Important reading.]

17. Savage, C. Wade, ed. *Scientific Theories.* Minneapolis: University of Minnesota Press, 1990. [Good surveys; difficult.]

18. Scheffler, Israel. *The Anatomy of Inquiry,* part 3. New York: Knopf, 1969. [A major discussion of positivist theories of confirmation.]

19. ———. *Science and Subjectivity.* New York: Bobbs-Merrill, 1967. [An early critic of Kuhn, Feyerabend, and Hanson, construed as relativists.]

Part 6

Science and Values

Introduction

I. The Problems

Problems about the role of value judgments within *science* and ethical disputes about *the uses* of science (e.g., in technology and social policy) are commonplace. Recent discussion of genetic engineering, behavior control, safety in nuclear power plants, human experimentation (e.g., deception and manipulation in research), medical ethics, and the IQ controversy are only the most dramatic instances of such problems. The basic problems are by no means new: just think of the controversy between Galileo and the Catholic Church about the nature of the solar system, or controversies between Darwin and his critics over the question of evolution.

However, underlying the many disputes covered by the phrase "science and values," there are several fundamental and persisting issues. Among the most obvious are these:

1. Can (is/should) science be "value free" or "neutral"? What do "value free" and "value neutral" mean in such contexts? What kinds of values are at issue? (This raises the question "What is value?" which must be bypassed here.) For instance, not all values are *moral* values, so that science may be *morally* neutral even though not *value* neutral in some other (or in a wider) sense. As we shall see, it is open to question whether science is or can be even morally neutral, much less value neutral.

2. If science *is* value free (in either the extended or the narrower sense of morally neutral: unless otherwise specified I shall mean the former whenever "value free" is used) what implications does this have for our conception of science, knowledge, and values, and for our views about the nature and aims of science and the social uses of science and technology? If science is *not* value free, what does *this* imply about the foregoing?

3. What are the best or most defensible concepts or theories about knowl-

481

edge, values (moral or nonmoral), science, and the best ways of conceptualizing their interconnections (or the lack thereof)? Here we must eventually deal with the concepts of rationality, objectivity, subjectivity, pure and applied science, and so on. (See part 1 of this volume.) And we must eventually decide what sorts of views will or will not be plausible candidates for helping us understand science, values, and their connections. For instance, will any theory according to which science and values are, and should be, totally unrelated be acceptable to us? This is a question that is discussed later on in this introduction.

It should come as no surprise that these and other issues have been given a variety of answers, and have generated a number of complex and often conflicting theories about science, values, and their interrelations. There is no point in attempting to even list, much less discuss, all or even many of these views in an introductory essay. Instead, some historical and philosophical backdrop for the selections in part 6 will be provided. The emphasis will be on those developments which get to the very heart of the issues, and which bear most directly on the readings.

It will prove instructive to begin with the question of *why* there is, or even should be, any problems about the relationships between science and values at all. This question is by no means rhetorical, but rather gets to the nub of the issue, which involves the relationship between science (and, more generally, knowledge) and values which finds expression in modern science and philosophy. It is here, after all, that we must look for the ideas that continue to dominate our culture's general outlooks on these questions.

II. WHY IS THERE A PROBLEM ABOUT "SCIENCE AND VALUES"?

The problems briefly outlined above come into existence with the advent of modern philosophy and science, especially the scientific revolution of the seventeenth century. In order to appreciate why, and how, this happens, it will be helpful to briefly sketch the views of the ancient Greeks—the molders of Western culture—on the issues of knowledge and values (or, at any rate, on the Greek approximations to these issues as they have since come to be understood). Then we can go on to sketch the development of the modern problems as they arise at the beginning of "modern history," which shall be dated here from the seventeenth century.

For the ancient Greeks, there are no distinctions between (a) science or knowledge and "values" or "the good" or between (b) science and philosophy or between (c) the objective and the subjective (as these concepts are understood within modern science and philosophy) or between (d) a "factual" or "descriptive" account of the world (e.g., in terms of the structural properties of things and the laws which govern them) and a "normative" or "evaluative" or (even) a "moral" interpretation of the world, as embodying a certain order, pattern, beauty, purpose, and even "goodness." (What is natural is also good in this view.) The most influential, and most forceful, presentation of the Greek view of the cosmos is articulated by Plato in his *Republic* (bks. iv–vii). For Plato, "objective" reality is characterized in terms of the idea or form of "the good." Reality is a unified, patterned, or ordered whole. In order to understand experience, we have to arrive at a knowledge of the laws and structures governing everything, as well as the order, patterning, or "purpose" which pervades all experience and which unifies it in a coherent and "meaningful" fashion. Such knowledge, which Plato calls "Dialectic," is not to be equated with what is today called "knowledge," since our term "knowledge" is often used as a synonym for "science" or "scientific knowledge." On Plato's view, the aims and methods of the empirical sciences are designed to give us at best only a limited insight into reality; more specifically, an insight into a certain kind of experience, a certain level of reality (the level of objects of experience like trees, rocks, and so on). They must be complemented by an insight into the more basic principles and patterns which govern everything. What for Plato is the "most real" is also the most abstract and the least accessible to ordinary experience. (Modern theoretical science, e.g., atomic theory, quantum physics, genetics, and chemistry, embody this Platonic ideal to some extent.) An "objective account" of things is not complete until everything is ordered into a unified picture, which involves the idea that purpose and norms (i.e., ordering principles) are not eliminable from such an account. It also involves the idea that science cannot give us either a complete account of everything, or an adequate account of even the objects of its legitimate concern, since these objects must be ultimately understood in terms of the principles which govern everything, including themselves and their place in the whole scheme.

One of the chief features of modern science and modern philosophy is the attempt to deny or else to truncate this platonic vision of the universe, and of the nature of knowledge and the good. In what follows, a brief sketch of these developments will be given.

III. ORIGIN AND NATURE OF THE PROBLEM

The issues concerning the relationship between science and values, including the role of values *in* science (the issue of value-neutrality) comes into modern Western history with the advent of the so-called mechanical picture of the world (especially classical Newtonian science), and the scientific revolution, most especially the epistemological and methodological revolution in science and philosophy inspired by its main architect, Rene Descartes (1596–1650).

According to this view, we must make a sharp distinction between what is objective and what is subjective in order to acquire reliable (i.e., "objective") knowledge of the world (including knowledge of human beings) by the use of reliable ("rational") methods of inquiry. Since, according to the mechanical world picture, nature is a vast machine governed by quantitative laws and relationships (nature is written in the language of mathematics), the objective features of the world turn out to be those features—matter, motion, and physical magnitudes—which constitute the nuts and bolts of the machine, together with the laws governing it. Only such features of experience are truly objective. Thus a rational methodology for inquiring about the machine's working—i.e., for acquiring knowledge—must take into account only those features which can be quantified, i.e., written in nature's language. The very essence of the world is given by the objective properties just mentioned, together with the mechanical laws which govern them. (These essential features of the world are dubbed "primary qualities" by Galileo and Locke.) Everything else, e.g., colors, values, interpretations, purpose, and theories, is not "objective" and thus does not belong in an objective account of the world, unless it can be "reduced" to objective terms, or explained away as illusory phenomena by such an account. (Later thinkers expanded the province of science to include those "subjective" items—called "secondary qualities"—that had an autonomous status for Descartes [e.g., mental phenomena and values] so that these came to be "reduced" to objective features or explained away entirely, as in modern behaviorism and materialism.)

In sum, objectivity means both: (a) objective in the sense of being about what is objective, and (b) objective in the sense of arriving at objective truths by methods which themselves take no account of anything "subjective," i.e., which are unbiased. (The search for mechanical "fool-proof" methods, e.g., computer algorithms, cost benefit decisions, is the ultimate outcome of this ideal of rational, objective method.)

IV. INITIAL OBJECTIONS TO "OBJECTIVISM"

This view already has insuperable difficulties: at least we can now see this (which is not to make the anachronistic claim that its classical proponents were flawed for not seeing it: it is just as easy to be a "Monday-morning quarterback" in history as in football).

First of all, to paraphrase Woody Allen (*Love and Death*): Objectivity is subjective, and subjectivity is objective; at least the latter point is certainly true: the fact that I am (say) in pain is no less objective a fact about me than the fact that I weigh 160 pounds.

Second, the view being considered is, paradoxically, rooted in the notion that "objective knowledge" is a "rational reconstruction" of the private, i.e., "subjective," experiences of a collection of knowers (viz., out of those subjective experiences which represent the essence of the world). Modern epistemology is rooted in this conundrum.

Third, as M. Polanyi and others show, if we merely wanted objective truths, we (as a species) would devote virtually all of our intellectual energy to studying interstellar dust, and only a fraction of a microsecond studying ourselves (or anything else, for that matter) since, objectively speaking, human beings are of no cosmic significance in the objective order of things! Obviously no one would take this requirement on objectivity seriously (this concept of objectivity is theological—God sees the world objectively as an outside omniscient observer). What we are seeking are truths which are interesting, which are useful and valuable to us. In a word, knowledge, truth, and objectivity are (or are rooted in) values and human purposes.

Fourth, not only do knowledge, objectivity, and truth—and thus methodology—turn out to be, or at least be grounded in, values; on some views they are, or are grounded in, moral ideals. In any event, the search for knowledge expresses a value; and thus distinctions between reliable and unreliable knowledge claims, between good and bad methods, and so on, are partly normative judgments. At this point it may be worth citing the words of N. I. Bukharin, who says: "The idea of the self-sufficient character of science . . . is naive; it confuses the *subjective passions* of the professional scientist . . . with the *objective social role* of this kind of activity, as an activity of vast practical importance.")

V. SOME IMPLICATIONS OF "OBJECTIVISM"

Despite these "obvious" difficulties, the fact remains that the distinction between objectivity and subjectivity, as regards claims governing methodology and the content of an objective world picture, dominates modern science, philosophy, and Western culture from Descartes and Galileo up to the present. The ideas of value neutrality, the uses of cost-benefit analyses to arrive at "rational" decisions in politics, science, and technology, the attempt to use knowledge for social and political ends (e.g., behavior control techniques), and so on, are just sophisticated outgrowths of this cluster of ideas. So, too, is the idea that scientific method affords the only rational methods for solving problems, so that value judgments are either not rational, or else are concerned merely with problems about calculating efficiency, or about decision-making under conditions of uncertainty. Both proponents and opponents of the classical picture of the world and the attendant ideas of objectivity and rationality (e.g., behaviorists, on the one hand, and so-called "neoromantics" and existentialists, on the other) share this conception that values are essentially subjective and irrational if they are anything more than predictions or calculations about means to an end.

But it is becoming glaringly obvious that these assumptions are connected with an inadequate conception of *both* science *and* values, and of the interrelationships between the two areas. No one is reluctant to distinguish between "good" and "bad" science, or between science and nonscience or pseudoscience (e.g., astronomy vs. astrology). (See part 1.) Yet on the view being discussed we are not supposed to be able to distinguish between good and bad moralities, or between acceptable vs. unacceptable value judgments. But this combination of "normative" science and "positive" ethics is internally incoherent and grossly inadequate.

VI. SOME COROLLARIES OF "OBJECTIVISM"

Two clusters of ideas conspire to produce this result. (i) Many advocates of the view being discussed either attempt to turn ethics into a science, or to explain value judgments scientifically (e.g., in terms of conditioning or historical or economic determinism). The latter kind of strategy usually leads to some form of ethical nihilism or extreme ethical relativism: All we can do is explain the origins of ethical behavior in terms of some objective scientific theory. On this view the point or content (i.e., the "autonomy" of value judgments is either lost or obscured. But this strategy explains why, for an advocate of this approach, "positive ethics" is the inevitable result. Moreover, it is just because "normative science" shows that objectivism is the only correct scientific approach that "positive ethics" turns out to be compatible, indeed, required by, objectivism. (ii) The idea of value neutrality, especially as this idea shows up in the social and policy sciences, is greatly influenced by views which attempt to reconcile "objective" science and "subjective" morality, by drawing theoretical limits to science, in order to save morality and human freedom. On this view, (a) science and morality cannot conflict (they are "complementary") since (b) they have nothing to do with each other: they govern different spheres of experience (these relate to the differences between "man as object" and as actor). But the price to be paid for this move is just the idea that science is, and must be, value neutral and that value judgments are merely subjective and irrational acts of the will.

Connected with this view is the idea that rational justification is essentially (hypothetico) deductive. The ultimate principles of a system, whether a moral system, a scientific theory, or a formal system, such as geometry, or the "brute facts" or "data" are beyond rational dispute. They must either be arbitrarily stipulated and accepted or be taken as self-evident, and then used to define what a rational proof or justification is *within* the system. The "ultimates" must be accepted as given. Relativism is the view that there are different, equally rational or acceptable (incompatible) ultimates. When one reaches these ultimates, be it a body of "hard facts," or axioms or moral principles, one has reached bedrock. One can then either accept or reject them. If the former, one can then show that the principles which follow from them are rational. In the latter case, one is free to adopt different "ultimates." At the same time, finding rational principles, or making rational decisions within the system, becomes a matter of, say, finding the best means of optimizing the ultimate principles or ends postulated by the system.

Ultimately, this theory of justification is part and parcel of the idea that science is value free, that value judgments are really objective judgments about the best means of optimizing goals which cannot be rationally assessed, and of the view that value judgments can and must be explained objectively (e.g., by deterministic explanations) or else are merely arbitrary fiats of individuals or cultures (which view amounts to nihilism or relativism). This is the most dramatic way in which the theory of objectivity we are discussing is already pregnant with modern nihilism; for this view already structures our view of values as either just objective items of a social system or an individual's behavioral repertoire or else as just "subjective" reactions to the objective facts, which do not belong in an objective scientific picture of the world.

It turns out, that *both* of these approaches to values amount to a kind of relativism, which often embodies a very conservative ideology, i.e., supports the idea that existing views of morality cannot be challenged, and thus that value judgments are really judgments about the best means to those ends and values which are in existence. (This is why cost-benefit analyses embody the view that the optimization of a given end is the only standard for making value judgments.) It thus turns out, on the "C-B" view, that the idea of value neutrality really amounts to the idea that value judgments (to the degree that they are rational and objective) are just judgments of efficiency concerning the best means to a given end. The ends (e.g., purposes) are determined scientifically, and this means (ultimately) they are either given or explained by some deterministic theory as being inevitable, ultimate facts. In any event, objectivism certainly is not "value-free."

VII. RECENT DEVELOPMENTS

In recent years new light has been shed on the nature of science, values, rationality and the sorts of issues discussed in parts 1–5 of this book. Much of the impetus behind these (often controversial) developments stems from the work of feminists, postmodernists, and sociologists of science, as well as from writers exploring the sociopolitical context of modern science.

These analyses usually move further away from "objectivist" analyses of values. But they do more than this. They shed new light on a range of issues that involve questions of values in science, and have produced some astonishingly complex and worthwhile insights into the processes of scientific discovery and justification. At the same time, they have produced strong reactions from advocates of more orthodox approaches to science, values, rationality and the rest of the topics covered so far in the volume. The result of this has been the so-called science wars, in which advocates of these new approaches are charged with being "antiscience," with blurring the distinction between science and nonscience, and with seeking to give pride of place to irrationality and to forces such as ethnic and gender identity and political ideology, not just in society, but within the very core of modern science itself. Advocates of these approaches, at a minimum, insist that it is important to study science with reference to the issues they bring out. This may very well make science better, and will surely increase our understanding of science, which is always a good thing. If these analyses require a more nuanced and complex account of science, so be it.

This is not the place to rehearse these issues in detail. It may be helpful instead to sketch the main lines of argument of each of the three approaches mentioned above (feminism, postmodernism, sociology of science) with special emphasis on controversies about values.

First, feminism. Many feminists, especially scholars with working familiarity with one or another science, have claimed that the assumptions, methods, and guiding values of many sciences, e.g, medicine, biology, psychology, have been gender biased. When, for example, male medical researchers draw inferences about premenstrual syndrome either without studying women, or by coloring their views of women with male biases; when male psychologists make generalizations about human moral development after using only male subjects, many scholars (and not only feminists or women) raise questions about what is going on. Is there something about science, or its guiding values, assumptions, and methods, that produces such biased results? Are there "women's ways of knowing" that modern science neglects? In some cases questions about the distinction between science and nonscience is connected to the so-called male bias in favor of treating nature as an object of domination, or the so-called values of domination and control characteristic of Western civilization. These issues and others have been seriously debated. In the reading by Giere, the topics covered relate more narrowly to questions about feminism, methodology, and the issue of scientific reliability.

The sociology of science, a discipline founded in the late nineteenth century as an outgrowth of some of the debates discussed in the Introduction to part 2, has grown exponentially in recent years. Initially spurred on by Kuhn, whom many interpreted as giving science a sociological account, the sociology (and psychology) of science has taken many forms and raised many issues. Can scientific realism be defended if science is viewed as a social practice? Is science better than voodoo? Do the activities of human beings create the world that scientists study? How do social and personal values, which may not always be rational, as well as personal idiosyncracies and human foibles, including the desire for fame and power, influence the activities of scientists in labs, and how does this relate to the way scientists tell stories about what they do and theorize about the results? Is science a practice like any other, in which case the same human foibles play a role everywhere, so that science is no longer special, even different from a game? If so, what becomes of scientific realism and rationality? Are there any special scientifc values or standards of rationality? Are they to be merely identified and explained by sociologists and psychologists? Does an "acceptable" or "rational" theory require just as much sociological or psychological (causal?) explanation or reduction as a "failed" or "irrational" one? Indeed, can "acceptable," "rational," "failed," and "irrational" be given anything but sociological or psychological interpretations? Do we then get another form of objectivism, in which whatever is done is right or wrong just because the social standards, practices, and values do or do not endorse it?

There is now a growth industry called Science and Technology Studies (STS) which discusses these sorts of issues and many others. Some of their main advocates are right in the center of the science wars, which so far seem to have only just begun.

Finally, postmodernism is a general phenomenon that has pervaded all aspects of culture and society, even though there is little, if any, agreement about what the term means, or what, if any, its main claims and arguments are. For purposes of this discussion, postmodernism calls into question the validity of the ideas of truth, knowledge, reality, objectivity, rationality, and progress that both underlie and are taken for granted by modern science, indeed, by modern, post-Enlightenment culture in the West. The very distinction between "facts" and "values," "knowledge" and "power," "interpretation" and "reality," "discovery" and "invention" have been challenged in various ways. Some writers consider science a "social construct," no more or less objective than stories, novels, or folk tales. Some postmodernists have called for a "blurring of the genres": physics and history (for instance) are just different types of texts or forms of writing. As the reader can imagine, the current science wars involve basic questions that go way beyond the issues discussed in this book, as well as beyond academic and social debates about the nature and value of science in our world. Only time will tell how these debates, which take us back to part 1 of this volume, evolve, and what their consequences will be.

Without even attempting to define "postmodernism" here, the thrust of many writings given this label is to challenge the concepts "reality," "knowledge," "truth," "objectivity," "meaning," even "the world." Such terms are indeterminate. They have the status of "social constructs" fabricated by some ruling class or powerful class (white men, Europeans, elitists) who impose them on everybody else, or at least use rhetoric to destroy all other ideas and render their own constructs as "objective reality" and "truth." At the very least, skepticism and doubts about these terms challenge assumptions that both realists and antirealists share about the very possibility of scientific knowledge and progress. Another line of reasoning connects issues of knowledge to power, male dominance, and eurocentrism. Science is just another story, neither more nor less privileged than any other. It is only for political and cultural reasons that science, rationality, evidence, and argumentation are "privileged." Finally, science has become dependent upon success, measured in terms of power and wealth; science is now an economic commodity, and neither has nor requires any kind of justification in terms of its benefits to humanity. Growing skepticism about the social, economic, and cultural benefits of science, combined with misgivings about the uses or misuses of science and technology for destructive purposes, point in the same general directions, according to many postmodernists.

Postmodernism is more radical than antirealism or even relativism, simce it is based upon Nietzsche's view that the universe is a meaningless, chaotic flux, which has no meaning. It is we humans who impose various orders on the world for purposes of survival, cultural flourishing, or whatever. Postmodernists are even more skeptical than Nietzsche, since they suspect that modern science is rooted in sexism, racism, eurocentrism, male bias, and dominance, and is in fact "privileged" for reasons that have little, if anything, to do with the greater rationality or evidential support of science over myths. In a way, postmodernism does radicalize various antirealist and relativist tendencies in modern thought, e.g, the writings of Kuhn, Feyerabend, and radical feminists. However, although some postmodernist themes are not novel, they are more extreme and potentially more destructive of many of the assumptions that give modern science and technology pride of place in our world today. The reader must decide whether or not this result makes postmodernism worth further study.

VIII. THE READINGS IN PART 6

In his essay Rudner argues that the need for scientists to decide when and if the available evidence is strong enough or good enough to warrant the acceptance of a hypothesis is a normative or evaluative activity; hence, science is essentially value-laden.

Hempel discusses a number of basic relations between science and values. "Categorical" value judgments—e.g., "X is good/right or bad/wrong," are not part of science or provable/disprovable by science. Science can help us make "hypothetical" value judgments, however. Thus, "if we want X, then we should do A" can be assessed by science, since scientific

knowledge can help us predict the outcome of decisions, find the most efficient means to our ends, et cetera. Science can also explain how and why
individuals and groups have the values they do.

McMullin looks at the issues discussed by Rudner and Hempel in a
broad historical and analytic framework. His views about the normative
nature of science and values are broader and more radical than either
Rudner's or Hempel's, although he seems to be less radical in his views about
science, values, and theory choice than Kuhn.

Hollinger traces some of the debates about science and values in the
twentieth century. He then sketches some ideas of the contemporary German
theorist Jurgen Habermas, who tries to give us a more adequate view of the
relation between science, values, and politics.

Giere considers whether standard views in the philosophy of science can
avoid the dangers of gender bias. He proposes a version of scientific realism
that he thinks may deal with this problem.

R. H.

29

The Scientist *Qua* Scientist
Makes Value Judgments

Richard Rudner

The question of the relationship of the making of value judgments in a typically ethical sense to the methods and procedures of science has been discussed in the literature at least to that point which e. e. cummings somewhere refers to as "The Mystical Moment of Dullness." Nevertheless, albeit with some trepidation, I feel that something more may fruitfully be said on the subject.

In particular the problem has once more been raised in an interesting and poignant fashion by recently published discussions between Carnap[1] and Quine[2] on the question of the ontological commitments which one may make in the choosing of language systems.

I shall refer to this discussion in more detail in the sequel; for the present, however, let us briefly examine the current status of what is somewhat loosely called the "fact-value dichotomy."

I have not found the arguments which are usually offered, by those who believe that scientists do essentially make value judgments, satisfactory. On the other hand the rebuttals of some of those with opposing viewpoints seem to have had at least a *prima facie* cogency although they too may in the final analysis prove to have been subtly perverse.

Those who contend that scientists do essentially make value judgments generally support their contentions by either

(a) pointing to the fact that our having a science at all somehow "involves" a value judgment; or

(b) by pointing out that in order to select, say, among alternative problems, the scientist must make a value judgment; or (perhaps most frequently)

(c) by pointing to the fact that the scientist cannot escape his quite human self—he is a "mass of predilections," and these predilections must inevitably influence all of his activities not excepting his scientific ones.

To such arguments, a great many empirically oriented philosophers and scientists have responded that the value judgments involved in our decisions to have a science, or to select problem (a) for attention rather than problem (b) are, *of course*, extrascientific. If (they say) it is necessary to make a decision to have a science before we can have one, then this decision is literally prescientific and the act has thereby certainly not been shown to be any part of the *procedures* of science. Similarly the decision to focus attention on one problem rather than another is extraproblematic and forms no part of the procedures involved in dealing with the problem *decided* upon. Since it is *these* procedures which constitute the method of science, value judgments, so they respond, have not been shown to be involved in the scientific method as such. Again, with respect to the inevitable presence of our predilections in the laboratory, most empirically oriented philosophers and scientists agree that this is "unfortunately" the case; but, they hasten to add, if science is to progress toward objectivity the influence of our personal feelings or biases on experimental results must be minimized. We must try not to let our personal idiosyncrasies affect our scientific work. The perfect scientist—the scientist *qua* scientist does not allow this kind of value judgment to influence his work. However much he may find doing so unavoidable *qua* father, *qua* lover, *qua* member of society, *qua* grouch, *when* he does so he is not behaving *qua* scientist.

As I indicated at the outset, the arguments of neither of the protagonists in this issue appear quite satisfactory to me. The empiricists' rebuttals, telling *prima facie* as they may against the specific arguments that evoke them, nonetheless do not appear ultimately to stand up, but perhaps even more importantly, *the original arguments* seem utterly too frail.

I believe that a much stronger case may be made for the contention that value judgments are essentially involved in the procedures of science. And what I now propose to show is that scientists as scientists *do* make value judgments.

Now I take it that no analysis of what constitutes the method of science would be satisfactory unless it comprised some assertion to the effect that the scientist as scientist accepts or rejects hypotheses.

But if this is so then clearly the scientist as scientist does make value judgments. For, since no scientific hypothesis is ever completely verified, in accepting a hypothesis the scientist must make the decision that the evidence is *sufficiently* strong or that the probability is *sufficiently* high to warrant the acceptance of the hypothesis. Obviously our decision regarding the evidence and respecting how strong is "strong enough," is going to be a function of the *importance*, in the typically ethical sense, of making a mistake in accepting or rejecting the hypothesis. Thus, to take a crude but easily manageable example, if the hypothesis under consideration were to the effect that a toxic ingredient of a drug was not present in lethal quantity, we would require a relatively high degree of confirmation or confidence before accepting the hypothesis—for the consequences of making a mistake here are exceedingly grave by our moral standards. On the other hand, if, say, *our* hypothesis stated that, on the basis of a sample, a certain lot of machine, stamped belt buckles was not defective, the degree of confidence we should require would be relatively not so high. *How sure we need to be before we accept a hypothesis will depend on how serious a mistake would be.*

The examples I have chosen are from scientific inferences in industrial quality control. But the point is clearly quite general in application. It would be interesting and instructive, for example, to know just how high a degree of probability the Manhattan Project scientists demanded for the hypothesis that no uncontrollable pervasive chain reaction would occur, before they proceeded with the first atomic bomb detonation or first activated the Chicago pile above a critical level. It would be equally interesting and instructive to know why they decided that *that* probability value (if one was decided upon) was high enough rather than one which was higher; and perhaps most interesting of all to learn whether the problem in this form was brought to consciousness at all.

In general then, before we can accept any hypothesis, the value decision must be made in the light of the seriousness of a mistake, that the probability is *high enough* or that the evidence is *strong enough*, to warrant its acceptance.

Before going further, it will perhaps be well to clear up two points which might otherwise prove troublesome below. First I have obviously used the term "probability" up to this point in a quite loose and preanalytic sense. But my point can be given a more rigorous formulation in terms of a description of the process of making statistical inference and of the acceptance or rejection of hypotheses in statistics. As is well known, the acceptance or rejection of such a hypothesis presupposes that a certain level of significance or level of confidence or critical region be selected.[3]

It is with respect at least to the *necessary* selection of a confidence level or interval that the necessary value judgment in the inquiry occurs. For, "the size of the critical region (one selects) is related to *the risk one wants to accept* in testing a statistical hypothesis."[3*, p. 435]

And clearly how great a risk one is willing to take of being wrong in accepting or rejecting the hypothesis will depend upon how seriously in the typically ethical sense one views the consequences of making a mistake.

I believe, of course, that an adequate rational reconstruction of the procedures of science would show that every scientific inference is properly constructable as a statistical inference (i.e., as an inference from a set of characteristics of a sample of a population to a set of characteristics of the total population) and that such an inference would be scientifically in control only insofar as it is statistically in control. But it is not necessary to argue this point, for even if one believes that what is involved in some scientific inferences is not statistical probability but rather a concept like strength of evidence or degree of confirmation, one would still be concerned with making the decision that the evidence was *strong enough* or the degree of confirmation *high enough* to warrant acceptance of the hypothesis. Now, many empiricists who reflect on the foregoing considerations agree that acceptances or rejections of hypotheses do essentially involve value judgments, but they are nonetheless loathe to accept the conclusion. And one objection which has been raised against this line of argument by those of them who are suspicious of the intrusion of value questions into the "objective realm of science," is that actually the scientist's task is only to *determine* the degree of confirmation or the strength of the evidence which exists for a hypothesis. In short, they object that while it may be a function of the scientist *qua member of society* to decide whether a degree of probability associated with the hypothesis is high enough to warrant its acceptance, *still* the task of the scientist *qua* scientist is *just the determination* of the degree of probability or the strength of the evidence for a hypothesis and not the acceptance or rejection of that hypothesis.

But a little reflection will show that the plausibility of this objection is merely apparent. For the determination that the degree of confirmation is say, *p*, or that the strength of evidence is such and such, which is on this view being held to be the indispensable task of the scientist *qua* scientist, is clearly nothing more than *the acceptance by the scientist of the hypothesis that the degree of confidence is p or that the strength of the evidence is such and such*; and as these men have conceded, acceptance of hypotheses does require value decisions. The second point which it may be well to consider before finally turning our attention to the Quine-Carnap discussion has to do with the nature of the suggestions which have thus far been made in this essay. In this connection, it is important to point out that the preceding remarks do *not* have as their import that an empirical description of every present day scientist ostensibly going about his business would include the statement that he made a value judgment at such and such a juncture. This is no doubt the case; but it is a hypothesis which can only be confirmed by a discipline which cannot be said to have gotten extremely far along as yet; namely, the Sociology and Psychology of Science, whether such an empirical description is warranted, cannot be settled from the armchair.

My remarks have, rather, amounted to this: Any adequate analysis or (if I may use the term) rational reconstruction of the method of science must comprise the statement that the scientist *qua* scientist accepts or rejects hypotheses; and further that an analysis of that statement would reveal it to entail that the scientist *qua* scientist makes value judgments.

I think that it is in the light of the foregoing arguments, the substance of which has, in one form or another, been alluded to in past years by a number of inquirers (notably C. W. Churchman, R. L. Ackoff, and A. Wald), that the Quine-Carnap discussion takes on heightened interest. For, if I understand that discussion and its outcome correctly, although it apparently begins a good distance away from any consideration of the fact-value dichotomy, and although all the way through it both men touch on the matter in a way which indicates that they believe that questions concerning the dichotomy are, if anything, merely tangential to their main issue, yet it eventuates with Quine by an independent argument apparently in agreement with at least the conclusion here reached and also apparently having forced Carnap to that conclusion. (Carnap, however, is expected to reply to Quine's article and I may be too sanguine here.)

The issue of ontological commitment between Carnap and Quine has been one of relatively long standing. In this recent article,[1] Carnap maintains that we are concerned with two *kinds* of questions of existence relative to a given language system. One is what kinds of entities it would be permissible to speak about as existing when that language system is used, i.e., what kind of *framework* for speaking of entities should our system comprise. This, according to Carnap, is an external question. It is the practical question of what sort of linguistic system we want to choose. Such questions as "Are there abstract entities?" or "Are there physical entities?" thus are held to belong to the category of external questions. On the other hand, having made the decision regarding which linguistic framework to adopt, we can then raise questions like "Are there any black swans?" "What are the factors of 544?" et cetera. Such questions are *internal* questions.

For our present purposes, the important thing about all of this is that while for Carnap *internal* questions are theoretical ones, i.e., ones whose answers have cognitive content, external questions are not theoretical at all. They are *practical questions*—they concern our decisions to employ one language structure or another. They are of the kind that face us when for example we have to decide whether we ought to have a Democratic or a Republican administration for the next four years. In short, though neither Carnap nor Quine employ the epithet, they are *value questions*.

Now if this dichotomy of existence questions is accepted Carnap can still deny the essential involvement of the making of value judgments in the procedures of science by insisting that concern with *external* questions, admittedly necessary and admittedly axiological, is nevertheless in some sense a prescientific concern. But most interestingly, what Quine then proceeds to do is to show that the dichotomy, as Carnap holds it, is untenable. This is not the appropriate place to repeat Quine's arguments which are brilliantly presented in the article referred to. They are in line with the views he has expressed in his "Two Dogmas of Empiricism" essay and especially with his introduction to his recent book, *Methods of Logic*. Nonetheless the final paragraph of the Quine article I'm presently considering sums up his conclusions neatly:

Within natural science there is a continuum of gradations, from the statements which report observations to those which reflect basic features say of quantum theory or the theory of relativity. The view which I end up with, in the paper last cited, is that statements of ontology or even of mathematics and logic form a continuation of this continuum, a continuation which is perhaps yet more remote from observation than are the central principles of quantum theory or relativity. The differences here are in my view differences only in degree and not in kind. Science is a unified structure, and in principle it is the structure as a whole, and not its component statements one by one, that experience confirms or shows to be imperfect. Carnap maintains that ontological questions, and likewise questions of logical or mathematical principle, are questions not of fact but of choosing a convenient conceptual scheme or framework for science; and with this I agree only if the same be conceded for every scientific hypothesis. (n. 2, pp. 71–72.)

✳ In the light of all this I think that the statement that *Scientists qua Scientists* make value judgments is also a consequence of Quine's position.

Now, if the major point I have here undertaken to establish is correct, then clearly we are confronted with a first-order crisis in science and methodology. The positive horror which most scientists and philosophers of science have of the intrusion of value considerations into science is wholly understandable. Memories of the (now diminished but a certain extent still continuing) conflict between science and, e.g., the dominant religions over the intrusion of religious value considerations into the domain of scientific inquiry, are strong in many reflective scientists. The traditional search for objectivity exemplifies science's pursuit of one of its most precious ideals. But for the scientist to close his eyes to the fact that scientific method *intrinsically* requires the making of value decisions, for him to push out of his consciousness the fact that he does make them, can in no way bring him closer to the ideal of objectivity. To refuse to pay attention to the value decisions which *must* be made, to make them intuitively, unconsciously, haphazardly, is to leave an essential aspect of scientific method scientifically out of control.

What seems called for (and here no more than the sketchiest indications of the problem can be given) is nothing less than a radical reworking of the ideal of scientific objectivity. The slightly juvenile conception of the cold-blooded, emotionless, impersonal, passive scientist mirroring the world perfectly in the highly polished lenses of his steel-rimmed glasses—this stereotype—is no longer, if it ever was, adequate.

What is being proposed here is that objectivity for science lies at least in becoming precise about what value judgments are being and might have been made in a given inquiry—and even, to put it in its most challenging form, what value decisions ought to be made; in short that a science of ethics is a necessary requirement if science's progress toward objectivity is to be continuous.

Of course the establishment of such a science of ethics is a task of stupendous magnitude and it will probably not even be well launched for many generations. But a first step is surely comprised of the reflective self-awareness of the scientist in making the value judgments he must make.

NOTES

1. R. Carnap, "Empiricism, Semantics, and Ontology," *Revue Internationale de Philosophie*, II (1950): 20–40.

2. W. V. Quine, "On Carnap's Views on Ontology," *Philosophical Studies*, 11, no. 5, (1951).

3. "In practice three levels are commonly used: 1 per cent, 5 per cent and 0.3 of one per cent. There is nothing sacred about these three values; *they have become established in practice without any rigid theoretical justification*." (my italics) (subnote 3*, p. 435). To establish significance at the 5 percent level means that one is willing to take the risk of accepting a hypothesis as true when one will be thus making a mistake, one time in twenty. Or in other words, that one will be wrong (over the long run) once every twenty times if one employed a .05 level of significance. See also (subnote 3† chap. v) for such statements as "which of these two errors is most important to avoid (it being necessary to make such a decision in order to accept or reject the given hypothesis) is a *subjective matter . . .*" (my italics) (subnote 3†, p. 262).

*A. C. Rosander, *Elementary Principles of Statistics* (New York: D. Van Nostrand Co., 1951).
†J. Neyman, *First Course in Probability and Statistics* (New York: Henry Holt & Co., 1950).

30

Science and Human Values

Carl G. Hempel

1. The Problem

Our age is often called an age of science and of scientific technology, and with good reason: the advances made during the past few centuries by the natural sciences, and more recently by the psychological and sociological disciplines, have enormously broadened our knowledge and deepened our understanding of the world we live in and of our fellow men; and the practical application of scientific insights is giving us an ever increasing measure of control over the forces of nature and the minds of men. As a result, we have grown quite accustomed, not only to the idea of a physico-chemical and biological technology based on the results of the natural sciences, but also to the concept, and indeed the practice, of a psychological and sociological technology that utilizes the theories and methods developed by behavioral research.

This growth of scientific knowledge and its applications has vastly reduced the threat of some of man's oldest and most formidable scourges, among them famine and pestilence; it has raised man's material level of living, and it has put within his reach the realization of visions which even a few decades ago would have appeared utterly fantastic, such as the active exploration of interplanetary space.

But in achieving these results, scientific technology has given rise to a host of new and profoundly disturbing problems: The control of nuclear fission has brought us not only the comforting prospect of a vast new reservoir of energy, but also the constant threat of the atom bomb and of grave damage, to the present and to future generations, from the radioactive by-products of the fission process, even in its peaceful uses. And the very progress in biological and medical knowledge and technology which has so strikingly reduced infant mortality and increased man's life expectancy in large areas of our globe has significantly contributed to the threat of the "population explosion," the rapid growth of the earth's population which we are facing today, and which, again, is a matter of grave concern to all those who have the welfare of future generations at heart.

Clearly, the advances of scientific technology on which we pride ourselves, and which have left their characteristic imprint on every aspect of this "age of science," have brought in their train many new and grave problems which urgently demand a solution. It is only natural that, in his desire to cope with these new issues, man should turn to science and scientific technology for further help. But a moment's reflection shows that the problems that need to be dealt with are not straightforward technological questions but intricate complexes of technological and moral issues. Take the case of the population explosion, for example, To be sure, it does pose specific technological problems. One of these is the task of satisfying at least the basic material needs of a rapidly growing population by means of limited resources; another is the question of means by which population growth itself may be kept under control. Yet these technical questions do not exhaust the problem. For after all, even now we have at our disposal various ways of counteracting population growth; but some of these, notably contraceptive methods, have been and continue to be the subject of intense controversy on moral and religious grounds, which shows that an adequate solution of the problem at hand requires, not only knowledge of technical means of control, but also standards for evaluating the alternative means at our disposal; and this second requirement clearly raises moral issues.

There is no need to extend the list of illustrations: any means of technical control that science makes available to us may be employed in many different ways, and a decision as to what use to make of it involves us in questions of moral valuation. And here arises a fundamental problem to which I would now like to turn: Can such valuational questions be answered by means of the objective methods of empirical science, which have been so successful in giving us reliable, and often practically applicable, knowledge of our world? Can those methods serve to establish objective criteria of right and wrong and thus to provide valid moral norms for the proper conduct of our individual and social affairs?

2. Scientific Testing

Let us approach this question by considering first, if only in brief and sketchy outline, the way in which objective scientific knowledge is arrived at. We may leave aside here the question of *ways of discovery*; i.e., the problem of how a new scientific idea arises, how a novel hypothesis or theory is first conceived; for our purposes it will suffice to consider the scientific *ways of validation*; i.e., the manner in which empirical science goes about examining a proposed new hypothesis and determines whether it is to be accepted or rejected. I will use the word "hypothesis" here to refer quite broadly to any statements or set of statements in empirical science, no matter whether it deals with some particular event or purports to set forth a general law or perhaps a more or less complex theory.

As is well known, empirical science decides upon the acceptability of a proposed hypothesis by means of suitable tests. Sometimes such a test may involve nothing more than what might be called direct observation of pertinent facts. This procedure may be used, for example, in testing such statements as "It is raining outside," "All the marbles in this urn are blue," "The needle of this ammeter will stop at the scale point marked 6," and so forth. Here a few direct observations will usually suffice to decide whether the hypothesis at hand is to be accepted as true or to be rejected as false.

But most of the important hypotheses in empirical science cannot be tested in this simple manner. Direct observation does not suffice to decide, for example, whether to accept or to reject the hypotheses that the earth is a sphere, that hereditary characteristics are transmitted by genes, that all Indo-European languages developed from one common ancestral language, that light is an electromagnetic wave process, and so forth. With hypotheses such as these, science resorts to indirect methods of test and validation. While these methods vary greatly in procedural detail, they all have the same basic structure and rationale. First, from the hypothesis under test, suitable other statements are inferred which describe certain directly observable phenomena that should be found to occur under specifiable circumstances if the hypothesis is true; then those inferred statements are tested directly; i.e., by checking whether the specified phenomena do in fact occur; finally, the proposed hypothesis is accepted or rejected in the light of the outcome of these tests. For example, the hypothesis that the earth is spherical in shape is not directly testable by observation, but it permits us to infer that a ship moving away from the observer should appear to be gradually dropping below the horizon; that circumnavigation of the earth should be possible by following a straight course; that high-altitude photographs should show the curving of the earth's surface; that certain geodetic and astronomical measurements should yield such and such results; and so forth. Inferred statements such as these can be tested more or less directly; and as an increasing number and variety of them are actually borne out, the hypothesis becomes increasingly confirmed. Eventually, a hypothesis may be so well confirmed by the available evidence that it is accepted as having been established beyond reasonable doubt. Yet no scientific hypothesis is ever proved completely and definitively; there is always at least the theoretical possibility that new evidence will be discovered which conflicts with some of the observational statements inferred from the hypothesis, and which thus leads to its rejection. The history of science records many instances in which a once accepted hypothesis was subsequently abandoned in the light of adverse evidence.

3. INSTRUMENTAL JUDGMENTS OF VALUE

We now turn to the question whether this method of test and validation may be used to establish moral judgments of value, and particularly judgments to the effect that a specified course of action is good or right or proper, or that it is better than certain alternative courses of action, or that we ought—or ought not—to act in certain specified ways.

By way of illustration, consider the view that it is good to raise children permissively and bad to bring them up in a restrictive manner. It might seem that, at least in principle, this view could be scientifically confirmed by appropriate empirical investigations. Suppose, for example, that careful research had established (1) that restrictive upbringing tends to generate resentment and aggression against parents and other persons exercising educational authority, and that this leads to guilt and anxiety and an eventual stunting of the child's initiative and creative potentialities; whereas (2) permissive upbringing avoids these consequences, makes for happier interpersonal relations, encourages resourcefulness and self-reliance, and enables the child to develop and enjoy his potentialities. These statements, especially when suitably amplified, come within the purview of scientific investigation; and though our knowledge in the matter is in fact quite limited, let us assume, for the sake of the argument, that they had actually been strongly confirmed by careful tests. Would not scientific research then have objectively shown that it is indeed better to raise children in a permissive rather than in a restrictive manner?

A moment's reflection shows that this is not so. What would have been established is rather a conditional statement; namely, that *if* our children are to become happy, emotionally secure, creative individuals rather than guilt-ridden and troubled souls *then* it is better to raise them in a permissive than in a restrictive fashion. A statement like this represents a *relative*, or *instrumental, judgment of value*. Generally, a relative judgment of value states that a certain kind of action, M, is good (or that it is better than a given alternative M_1) if a specified goal G is to be attained; or more accurately, that M is good, or appropriate, for the attainment of goal G. But to say that is tantamount to asserting either that, in the circumstances at hand, course of action M will definitely (or probably) lead to the attainment of G, or that failure to embark on course of action M will definitely (or probably) lead to the nonattainment of G. In other words, the instrumental value judgment asserts either that M is a (definitely or probably) sufficient means for attaining the end or goal G, or that it is a (definitely or probably) necessary means for attaining it. Thus, a relative, or instrumental, judgment of value can be reformulated as a statement which expresses a universal or a probabilistic kind of means-ends relationship, and which contains no terms of moral discourse—such as "good," "better," "ought to"—at all. And a statement of this kind surely is an empirical assertion capable of scientific test.

4. CATEGORICAL JUDGMENTS OF VALUE

Unfortunately, this does not completely solve our problem; for after a relative judgment of value referring to a certain goal G has been tested and, let us assume, well confirmed, we are still left with the question of whether the goal G ought to be pursued, or whether it would be better to aim at some alternative goal instead. Empirical science can establish the conditional statement, for example, that if we wish to deliver an incurably ill person from intolerable suffering, then a large dose of morphine affords a means of doing so; but it may also indicate ways of prolonging the patient's life, if also his suffering. This leaves us with the question whether it is right to give the goal of avoiding hopeless human suffering precedence over that of preserving human life. And this question calls, not for a relative but for an *absolute*, or *categorical, judgment of value* to the effect that a certain state of affairs (which may have been proposed as a goal or end) is good, or that it is better than some specified alternative. Are such categorical value judgments capable of empirical test and confirmation?

Consider, for example, the sentence "Killing is evil." It expresses a categorical judgment of value which, by implication, would also categorically qualify euthanasia as evil. Evidently, the sentence does not express an assertion that can be directly tested by observation; it does not purport to describe a directly observable fact. Can it be indirectly tested, then, by inferring from it statements to the effect that under specified test conditions such and such observable phenomena will occur? Again, the answer is clearly in the negative. Indeed, the sentence "Killing is evil" does not have the function of expressing an assertion that can be qualified as true or false; rather, it serves to express a standard for moral appraisal or a norm for conduct. A categorical judgment of value may have other functions as well; for example, it may serve to convey the utterer's approval or disapproval of a certain kind of action, or his commitment to the standards of conduct expressed by the value judgment. Descriptive empirical import, however, is absent; in this respect a sentence such as "Killing is evil" differs strongly from, say, "Killing is condemned as evil by many religions," which expresses a factual assertion capable of empirical test.

Categorical judgments of value, then, are not amenable to scientific test and confirmation or disconfirmation; for they do not express assertions but rather standards or norms for conduct. It was Max Weber, I believe, who expressed essentially the same idea by remarking that science is like a map: it can tell us how to get to a given place, but it cannot tell us where to go. Gunnar Myrdal, in his book *An American Dilemma* (p. 1052), stresses in a similar vein that "factual or theoretical studies alone cannot logically lead to a practical recommendation. A practical or valuational conclusion can be derived only when there is at least one valuation among the premises." Nevertheless, there have been many attempts to base systems of moral standards on the findings of empirical science; and it would be of interest to examine in some detail the reasoning which underlies those procedures. In the present context, however, there is room for only a few brief remarks on this subject.

It might seem promising, for example, to derive judgments of value from the results of an objective study of human needs. But no cogent derivation of this sort is possible. For this procedure would presuppose that it is right, or good, to satisfy human needs—and this presupposition is itself a categorical judgment of value: it would play the role of a valuational premise in the sense of Myrdal's statement. Furthermore, since there are a great many different, and partly conflicting, needs of individuals and of groups, we would require not just the general maxim that human needs ought to be satisfied, but a detailed set of rules as to the preferential order and degree in which different needs are to be met, and how conflicting claims are to be settled; thus, the valuational premise required for this undertaking would actually have to be a complex system of norms; hence, a derivation of valuational standards simply from a factual study of needs is out of the question.

Several systems of ethics have claimed the theory of evolution as their basis; but they are in serious conflict with each other even in regard to their most fundamental tenets. Some of the major variants are illuminatingly surveyed in a chapter of G. G. Simpson's book, *The Meaning of Evolution*. One type, which Simpson calls a "tooth-and-claw ethics," glorifies a struggle for existence that should lead to a survival of the fittest. A second urges the harmonious adjustment of groups or individuals to one another so as to enhance the probability of their survival, while still other systems hold up as an ultimate standard the increased aggregation of organic units into higher levels of organization, sometimes with the implication that the welfare of the state is to be placed above that of the individuals belonging to it. It is obvious that these conflicting principles could not have been validly inferred from the theory of evolution—unless indeed that theory were self-contradictory, which does not seem very likely.

But if science cannot provide us with categorical judgments of value, what then can serve as a source of unconditional valuations? This question may either be understood in a pragmatic sense, as concerned with the sources from which human beings do in fact obtain their basic values. Or it may be understood as concerned with a systematic aspect of valuation; namely, the question where a proper system of basic values is to be found on which all other valuations may then be grounded.

The pragmatic question comes within the purview of empirical science. Without entering into details, we may say here that a person's values—both those he professes to espouse and those he actually conforms to—are largely absorbed from the society in which he lives, and especially from certain influential subgroups to which he belongs, such as his family, his schoolmates, his associates on the job, his church, clubs, unions, and other groups. Indeed his values may vary from case to case depending on which of these groups dominates the situation in which he happens to find himself. In general, then, a person's basic valuations are no more the result of careful scrutiny and critical appraisal of possible alternatives than is his religious affiliation. Conformity to the standards of certain groups plays a very important role here, and only rarely are basic values seriously questioned. Indeed, in many situations, we decide and act unreflectively in an even stronger sense; namely, without any attempt to base our decisions on some set of explicit, consciously adopted, moral standards.

Now, it might be held that this answer to the pragmatic version of our question reflects a regrettable human inclination to intellectual and moral inertia; but that the really important side of our question is the systematic one: If we do want to justify our decisions, we need moral standards of conduct of the unconditional type—but how can such standards be established? If science cannot provide categorical value judgments, are there any other sources from which they might be obtained? Could we not, for example, validate a system of categorical judgments of value by pointing out that it represents the moral standards held up by the Bible, or by the Koran, or by some inspiring thinker or social leader? Clearly, this procedure must fail, for the factual information here adduced could serve to validate the value judgments in question only if we were to use, in addition, a valuational presupposition to the effect that the moral directives stemming from the source invoked *ought* to be complied with. Thus, if the process of justifying a given decision or a moral judgment is ever to be completed, certain judgments of value have to be accepted without any further justification, just as the proof of a theorem in geometry requires that some propositions be accepted as postulates, without proof. The quest for a justification of *all* our valuations overlooks this basic characteristic of the logic of validation and of justification. The value judgments accepted without further justification in a given context need not, however, be accepted once and for all, with a commitment never to question them again. This point will be elaborated further in the final section of this essay.

As will hardly be necessary to stress, in concluding the present phase of our discussion, the ideas set forth in the preceding pages do not imply or advocate moral anarchy; in particular, they do not imply that any system of values is just as good, or just as valid, as any other, or that everyone should adopt the moral principles that best suit his convenience. For all such maxims have the character of categorical value judgments and cannot, therefore, be implied by the preceding considerations, which are purely descriptive of certain logical, psychological, and social aspects of moral valuation.

5. RATIONAL CHOICE: EMPIRICAL AND VALUATIONAL COMPONENTS

To gain further insight into the relevance of scientific inquiry for categorical valuation let us ask what help we might receive, in dealing with a moral problem, from science in an ideal state such as that represented by Laplace's conception of a superior scientific intelligence, sometimes referred to as Laplace's demon. This fiction was used by Laplace, early in the nineteenth century, to give a vivid characterization of the idea of universal causal determinism. The demon is conceived as a perfect observer, capable of ascertaining with infinite speed and accuracy all that goes on in the universe at a given moment; he is also an ideal theoretician who knows all the laws of nature and has combined them into one universal formula; and finally, he is a perfect mathematician who, by means of that universal formula, is able to infer, from the observed state of the universe at the given moment, the total state of the universe at any other moment; thus past and future are present before his eyes. Surely, it is difficult to imagine that science could ever achieve a higher degree of perfection!

Let us assume, then, that, faced with a moral decision, we are able to call upon the Laplacean demon as a consultant. What help might we get from him? Suppose that we have to choose one of several alternative courses of action open to us, and that we want to know which of these we *ought* to follow. The demon would then be able to tell us, for any contemplated choice, what its consequences would be for the future course of the universe, down to the most minute detail, however remote in space and time. But, having done this for each of the alternative courses of action under consideration, the demon would have completed his task: he would have given us all the information that an ideal science might provide under the circumstances. And yet he would not have resolved our moral problem, for this requires a decision as to which of the several alternative sets of consequences mapped out by the demon as attainable to us is the best; which of them we ought to bring about. And the burden of this decision would still fall upon our shoulders: it is we who would have to commit ourselves to an unconditional judgment of value by singling out one of the sets of consequences as superior to its alternatives. Even Laplace's demon, or the ideal science he stands for, cannot relieve us of this responsibility.

In drawing this picture of the Laplacean demon as a consultant in decision-making, I have cheated a little; for if the world were as strictly deterministic as Laplace's fiction assumes, then the demon would know in advance what choice we were going to make, and he might disabuse us of the idea that there were several courses of action open to us. However that may be, contemporary physical theory has cast considerable doubt on the classical conception of the universe as a strictly deterministic system: the fundamental laws of nature are now assumed to have a statistical or probabilistic rather than a strictly universal, deterministic, character.

But whatever may be the form and the scope of the laws that hold in our universe, we will obviously never attain a perfect state of knowledge concerning them; confronted with a choice, we never have more than a very incomplete knowledge of the laws of nature and of the state of the world at the time when we must act. Our decisions must therefore always be made on the basis of incomplete information, a state which enables us to anticipate the consequences of alternative choices at best with probability. Science can render an indispensable service by providing us with increasingly extensive and reliable information relevant to our purpose; but again it remains for us to *evaluate* the various probable sets of consequences of the alternative choices under consideration. And this requires the adoption of pertinent valuational standards which are not objectively determined by the empirical facts.

This basic point is reflected also in the contemporary mathematical theories of decision-making. One of the objectives of these theories is the formulation of decision rules which will determine an optimal choice in situations where several courses of action are available. For the formulation of decision rules, these theories require that at least two conditions be met: (1) Factual information must be provided specifying the available courses of action and indicating for each of these its different possible outcomes—plus, if feasible, the probabilities of their occurrence; (2) there must be a specification of the values—often prosaically referred to as utilities—that are attached to the different possible outcomes. Only when these factual and valuational specifications have been provided does it make sense to ask which of the available choices is the best, considering the values attaching to their possible results.

In mathematical decision theory, several criteria of optimal choice have been proposed. In case the probabilities for the different outcomes of each action are given, one standard criterion qualifies a choice as optimal if the probabilistically expectable utility of its outcome is at least as great as that of any alternative choice. Other rules, such as the maximin and the maximax principles, provide criteria that are applicable even when the probabilities of the outcomes are not available. But interestingly, the various criteria conflict with each other in the sense that, for one and the same situation, they will often select different choices as optimal.

The policies expressed by the conflicting criteria may be regarded as reflecting different attitudes toward the world, different degrees of optimism or pessimism, of venturesomeness or caution. It may be said therefore that the analysis offered by current mathematical models indicates two points at which decision-making calls not solely for factual information, but for categorical valuation, namely, in the assignment of utilities to the different possible outcomes and in the adoption of one among many competing decision rules or criteria of optimal choice. . . .

6. VALUATIONAL "PRESUPPOSITIONS" OF SCIENCE

The preceding three sections have been concerned mainly with the question whether, or to what extent, valuation and decision presuppose scientific investigation and scientific knowledge. This problem has a counterpart which deserves some attention in a discussion of science and valuation; namely, the question whether scientific knowledge and method presuppose valuation.

The word "presuppose" may be understood in a number of different senses which require separate consideration here. First of all, when a person decides to devote himself to scientific work rather than to some other career, and again, when a scientist chooses some particular topic of investigation, these choices will presumably be determined to a large extent by his preferences, i.e., by how highly he values scientific research in comparison with the alternatives open to him, and by the importance he attaches to the problems he proposes to investigate. In this explanatory, quasi-causal sense the scientific *activities* of human beings may certainly be said to presuppose valuations.

Much more intriguing problems arise, however, when we ask whether judgments of value are presupposed by the body of scientific *knowledge*, which might be represented by a system of statements accepted in accordance with the rules of scientific inquiry. Here presupposing has to be understood in a systematic-logical sense. One such sense is invoked when we say, for example, that the statement "Henry's brother-in-law is an engineer" presupposes that Henry has a wife or a sister: in this sense, a statement presupposes whatever can be logically inferred from it. But, as we noted earlier, no set of scientific statements logically implies an unconditional judgment of value; hence, scientific knowledge does not, in this sense, presuppose valuation.

There is another logical sense of presupposing, however. We might say, for example, that in Euclidean geometry the angle-sum theorem for triangles presupposes the postulate of the parallels in the sense that that postulate is an essential part of the basic assumptions from which the theorem is deduced. Now, the hypotheses and theories of empirical science are not normally validated by deduction from supporting evidence (though it may happen that a scientific statement, such as a prediction, is established by deduction from a previously ascertained, more inclusive set of statements); rather, as was mentioned in section 2, they are usually accepted on the basis of evidence that lends them only partial, or "inductive," support. But in any event it might be asked whether the statements representing scientific knowledge presuppose valuation in the sense that the grounds on which they are accepted include, sometimes or always, certain unconditional judgments of value. Again the answer is in the negative. The grounds on which scientific hypotheses are accepted or rejected are provided by empirical evidence, which may include observational findings as well as previously established laws and theories, but surely no value judgments. Suppose for example that, in support of the hypothesis that a radiation belt of a specified kind surrounds the earth, a scientist were to adduce, first, certain observational data, obtained perhaps by rocket-borne instruments; second, certain previously accepted theories invoked in the interpretation of those data; and finally, certain judgments of value, such as "it is good to ascertain the truth." Clearly, the judgments of value would then be dismissed as lacking all logical relevance to the proposed hypothesis since they can contribute neither to its support nor to its disconfirmation.

But the question whether science presupposes valuation in a logical sense can be raised, and recently has been raised, in yet another way, referring more specifically to valuational presuppositions of scientific *method*. In the preceding considerations, scientific knowledge was represented by a system of statements which are sufficiently supported by available evidence to be accepted in accordance with the principles of scientific test and validation. We noted that as a rule the observational evidence on which a scientific hypothesis is accepted is far from sufficient to establish that hypothesis conclusively. For example, Galileo's law refers not only to past instances of free fall near the earth, but also to all future ones; and the latter surely are not covered by our present evidence. Hence, Galileo's law, and similarly any other law in empirical science, is accepted on the basis of incomplete evidence. Such acceptance carries with it the "inductive risk" that the presumptive law may not hold in full generality, and that future evidence may lead scientists to modify or abandon it.

A precise statement of this conception of scientific knowledge would require, among other things, the formulation of rules of two kinds: First, *rules of confirmation*, which would specify what kind of evidence is confirmatory, what kind disconfirmatory for a given hypothesis. Perhaps they would also determine a numerical *degree* of evidential support (or confirmation, or inductive probability) which a given body of evidence could be said to confer upon a proposed hypothesis. Secondly, there would have to be *rules of acceptance*: these would specify how strong the evidential support for a given hypothesis has to be if the hypothesis is to be accepted into the system of scientific knowledge; or, more generally, under what conditions a proposed hypothesis is to be accepted, under what conditions it is to be rejected by science on the basis of a given body of evidence.

Recent studies of inductive inference and statistical testing have devoted a great deal of effort to the formulation of adequate rules of either kind. In particular, rules of acceptance have been treated in many of these investigations as special instances of decision rules of the sort mentioned in the preceding section. The decisions in question are here either to accept or to reject a proposed hypothesis on the basis of given evidence. As was noted earlier, the formulation of "adequate" decision rules requires, in any case, the antecedent specification of valuations that can then serve as standards of adequacy. The requisite valuations, as will be recalled, concern the different possible outcomes of the choices which the decision rules are to govern. Now, when a scientific rule of acceptance is applied to a specified hypothesis on the basis of a given body of evidence, the possible "outcomes" of the resulting decision may be divided into four major types: (1) the hypothesis is accepted (as presumably true) in accordance with the rule and is in fact true; (2) the hypothesis is rejected (as presumably false) in accordance with the rule and is in fact false; (3) the hypothesis is accepted in accordance with the rule, but is in fact false; (4) the hypothesis is rejected in accordance with the rule, but is in fact true. The former two cases are what science aims to achieve; the possibility of the latter two represents the inductive risk that any acceptance rule must involve. And the problem of formulating adequate rules of acceptance and rejection has no clear meaning unless standards of adequacy have been provided by assigning definite values or disvalues to those different possible "outcomes" of acceptance or rejection. It is in this sense that the method of establishing scientific hypotheses "presupposes" valuation: the justification of the rules of acceptance and rejection requires reference to value judgments.

In the cases where the hypothesis under test, if accepted, is to be made the basis of a specific course of action, the possible outcomes may lead to success or failure of the intended practical application; in these cases, the values and disvalues at stake may well be expressible in terms of monetary gains or losses; and for situations of this sort, the theory of decision functions has developed various decision rules for use in practical contexts such as industrial quality control. But when it comes to decision rules for the acceptance of hypotheses in pure scientific research, where no practical applications are contemplated, the question of how to assign values to the four types of outcome mentioned earlier becomes considerably more problematic. But in a general way, it seems clear that the standards governing the inductive procedures of pure science reflect the objective of obtaining a certain goal, which might be described somewhat vaguely as the attainment of an increasingly reliable, extensive, and theoretically systematized body of information about the world. Note that if we were concerned, instead, to form a system of beliefs or a world view that is emotionally reassuring or esthetically satisfying to us, then it would not be reasonable at all to insist, as science does, on a close accord between the beliefs we accept and our empirical evidence; and the standards of objective testability and confirmation by publicly ascertainable evidence would have to be replaced by acceptance standards of an entirely different kind. The standards of procedure must in each case be formed in consideration of the goals to be attained; their justification must be relative to those goals and must, in this sense, presuppose them.

7. CONCLUDING COMPARISONS

If, as has been argued in section 4, science cannot provide a validation of categorical value judgments, can scientific method and knowledge play any role at all in clarifying and resolving problems of moral valuation and decision? The answer is emphatically in the affirmative. I will try to show this in a brief survey of the principal contributions science has to offer in this context.

First of all, science can provide factual information required for the resolution of moral issues. Such information will always be needed, for no matter what system of moral values we may espouse—whether it be egoistic or altruistic, hedonistic or utilitarian, or of any other kind—surely the specific course of action it enjoins us to follow in a given situation will depend upon the facts about that situation; and it is scientific knowledge and investigation that must provide the factual information which is needed for the application of our moral standards.

More specifically, factual information is needed, for example, to ascertain (a) whether a contemplated objective can be attained in a given situation; (b) if it can be attained, by what alternative means and with what probabilities; (c) what side effects and ulterior consequences the choice of a given means may have apart from probably yielding the desired end; (d) whether several proposed ends are jointly realizable, or whether they are incompatible in the sense that the realization of some of them will definitely or probably prevent the realization of others.

By thus giving us information which is indispensable as a factual basis for rational and responsible decision, scientific research may well motivate us to change some of our valuations. If we were to discover, for example, that a certain kind of goal which we had so far valued very highly could be attained only at the price of seriously undesirable side effects and ulterior consequences, we might well come to place a less high value upon that goal. Thus, more extensive scientific information may lead to a change in our basic valuations—not by "disconfirming" them, of course, but rather by motivating a change in our total appraisal of the issues in question.

Secondly, and in a quite different manner, science can illuminate certain problems of valuation by an objective psychological and sociological study of the factors that affect the values espoused by an individual or a group; of the ways in which such valuational commitments change; and perhaps of the manner in which the espousal of a given value system may contribute to the emotional security of an individual or the functional stability of a group.

Psychological, anthropological, and sociological studies of valuational behavior cannot, of course, "validate" any system of moral standards. But their results can psychologically effect changes in our outlook on moral issues by broadening our horizons, by making us aware of alternatives not envisaged, or not embraced, by our own group, and by thus providing some safeguard against moral dogmatism or parochialism.

Finally, a comparison with certain fundamental aspects of scientific knowledge may help to illuminate some further questions concerning valuation.

If we grant that scientific hypotheses and theories are always open to revision in the light of new empirical evidence, are we not obliged to assume that there is another class of scientific statements which cannot be open to doubt and reconsideration, namely, the observational statements describing experiential findings that serve to test scientific theories? Those simple, straightforward reports of what has been directly observed in the laboratory or in scientific field work, for example—must they not be regarded as immune from any conceivable revision, as irrevocable once they have been established by direct observation? Reports on directly observed phenomena have indeed often been considered as an unshakable bedrock foundation for all scientific hypotheses and theories. Yet this conception is untenable; even here, we find no definitive, unquestionable certainty.

For, first of all, accounts of what has been directly observed are subject to error that may spring from various physiological and psychological sources. Indeed, it is often possible to check on the accuracy of a given observation report by comparing it with the reports made by other observers, or with relevant data obtained by some indirect procedure, such as a motion picture taken of the finish of a horse race; and such comparison may lead to the rejection of what had previously been considered as a correct description of a directly observed phenomenon. We even have theories that enable us to explain and anticipate some types of observational error, and in such cases, there is no hesitation to question and to reject certain statements that purport simply to record what has been directly observed.

Sometimes relatively isolated experimental findings may conflict with a theory that is strongly supported by a large number and variety of other data; in this case, it may well happen that part of the conflicting data, rather than the theory, is refused admission into the system of accepted scientific statements—even if no satisfactory explanation of the presumptive error of observation is available. In such cases it is not the isolated observational finding which decides whether the theory is to remain in good standing, but it is the previously well-substantiated theory which determines whether a purported observation report is to be regarded as describing an actual empirical occurrence. For example, a report that during a spiritualistic seance, a piece of furniture freely floated above the floor would normally be rejected because of its conflict with extremely well-confirmed physical principles, even in the absence of some specific explanation of the report, say, in terms of deliberate fraud by the medium, or of high suggestibility on the part of the observer. Similarly, the experimental findings reported by the physicist Ehrenhaft, which were claimed to refute the principle that all electric charges are integral multiples of the charge of the electron, did not lead to the overthrow, nor even to a slight modification, of that principle, which is an integral part of a theory with extremely strong and diversified experimental support. Needless to say, such rejection of alleged observation reports by reason of their conflict with well-established theories requires considerable caution; otherwise, a theory, once accepted, could be used to reject all adverse evidence that might subsequently be found—a dogmatic procedure entirely irreconcilable with the objectives and the spirit of scientific inquiry.

Even reports on directly observed phenomena, then, are not irrevocable; they provide no bedrock foundation for the entire system of scientific knowledge. But this by no means precludes the possibility of testing scientific theories by reference to data obtained through direct observation. As we noted, the results obtained by such direct checking cannot be considered as absolutely unquestionable and irrevocable; they are themselves amenable to further tests which may be carried out if there is reason for doubt. But obviously if we are ever to form any beliefs about the world, if we are ever to accept or to reject, even provisionally, some hypothesis or theory, then we must stop the testing process somewhere; we must accept some evidential statements as sufficiently trustworthy not to require further investigation for the time being. And on the basis of such evidence, we can then decide what credence to give to the hypothesis under test, and whether to accept or to reject it.

This aspect of scientific investigation seems to me to have a parallel in the case of sound valuation and rational decision. In order to make a rational choice between several courses of action, we have to consider, first of all, what consequences each of the different alternative choices is likely to have. This affords a basis for certain relative judgments of value that are relevant to our problem. If *this* set of results is to be attained, this course of action ought to be chosen; if *that other* set of results is to be realized, we should choose such and such another course; and so forth. But in order to arrive at a decision, we still have to decide upon the relative values of the alternative sets of consequences attainable to us; and this, as was noted earlier, calls for the acceptance of an unconditional judgment of value, which will then determine our choice. But such acceptance need not be regarded as definitive and irrevocable, as forever binding for all our future decisions: an unconditional judgment of value, once accepted, still remains open to reconsideration and to change. Suppose, for example, that we have to choose, as voters or as members of a city administration, between several alternative social policies, some of which are designed to improve certain material conditions of living, whereas others aim at satisfying cultural needs of various kinds. If we are to arrive at a decision at all, we will have to commit ourselves to assigning a higher value to one or the other of those objectives. But while the judgment thus accepted serves as an unconditional and basic judgment of value for the decision at hand, we are not for that reason committed to it forever—we may well reconsider our standards and reverse our judgment later on; and though this cannot undo the earlier decision, it will lead to different decisions in the future. Thus, if we are to arrive at a decision concerning a moral issue, we have to accept some unconditional judgments of value; but these need not be regarded as ultimate in the absolute sense of being forever binding for all our decisions, any more than the evidence statements relied on in the test of a scientific hypothesis need to be regarded as forever irrevocable. All that is needed in either context are *relative* ultimates, as it were: a set of judgments—moral or descriptive—which are accepted at the time as not in need of further scrutiny. These relative ultimates permit us to keep an open mind in regard to the possibility of making changes in our heretofore unquestioned commitments and beliefs; and surely the experience of the past suggests that if we are to meet the challenge of the present and the future, we will more than ever need undogmatic, critical, and open minds.

31

Values in Science

Ernan McMullin

Thirty years ago, Richard Rudner argued in a brief essay in *Philosophy of Science* that the making of value-judgments is an essential part of the work of science. He fully realized how repugnant such a claim would be to the positivist orthodoxy of the day, so repugnant indeed that its acceptance (he prophesied) would bring about a "first-order crisis in science and methodology" (1953, p. 6). Carnap, in particular, has been emphatic in excluding values from any role in science proper. His theory of meaning had led him to conclude that "the objective validity of a value . . . cannot be asserted in a meaningful statement at all" (1932/1959, p. 77). The contrast between science, the paradigm of meaning, and all forms of value-judgment could scarcely have been more sharply drawn: "it is altogether impossible to make a statement that expresses a value-judgment." No wonder, then, that Rudner's thesis seemed so shocking.

Thirty years later, the claim that science is value-laden might no longer even seem controversial, among philosophers of science, at least, who have become accustomed to seeing the pillars of positivism fall, one by one. One might even characterize the recent deep shifts in theory of science as consequences (many of them, at least) of the growing realization of the part played by value-judgment in scientific work. If this way of describing the Kuhnian "revolution" seems unfamiliar, it is no doubt due in part to the uneasiness that the ambiguity of the terms "value" and "value-judgment" still engenders. There are other ways of describing what has happened since the 1950s on philosophy of science that do not require so much preliminary ground-clearing.

Nevertheless, I shall try to show that the watershed between "classic" philosophy of science (by this meaning, not just logical positivism but the logicist tradition in theory of science stretching back through Immanuel Kant and René Descartes to Aristotle) and the "new" philosophy of science can best be understood by analyzing the change in our perception of the role played by values in science. I shall begin with some general remarks about the nature of value, go on to explore some of the historical sources for the claim that judgment in science is value-laden, and conclude by reflecting on the implications of this claim for traditional views of the objectivity of scientific knowledge-claims. I will not address the problem of the social sciences, where these issues take on an added complexity. They are, as we shall soon see, already complicated enough in the context of natural sciences.

I. THE ANATOMY OF VALUE

"Value" is one of those weasel words that slip in and out of the nets of the philosopher. We shall have to try to catch it first, or else what we have to say about the role of values in science may be of small use. It is not much over a hundred years since the German philosopher, Hermann Lotze, tried to construct a single theory of value which would unite the varied value-aspects of human experience under a single discipline. The venture was, of course, not really new since Plato had attempted a similar project long before, using the cognate term "good" instead of "value." Aristotle's response to Plato's positing of the Good as a common element answering to one idea was to point to the great diversity of ways in which the term "good" might be used. In effect, our response to Lotze's project of a general axiology would likewise be to question the usefulness of trying to find a single notion of value that would apply to all contexts equally well.

Let us begin with the sense of "value" that the founders of value-theory seem to have preferred. They took it to correspond to such features of human experience as attraction, emotion, and feeling. They wanted to secure an experiential basis for value in order to give the realm of value an empirical status just as valid as that of the (scientific) realm of fact. The reality of *emotive* value (as it may be called) lies in the feelings of the subject, not primarily in a characteristic of the object. Value-differences amount, then, to differences of attitude or of emotional response in specific subjects.

If one takes "value" in this sense, value-decision becomes a matter of clarifying emotional responses. To speak of value-*judgment* here (as indeed is often done) is on the whole misleading since "judgment" could suggest a cognitive act, a weighing-up. When the value of something is determined by one's attitude to it, the declaration of this value is a matter of value-*clarification* rather than of judgment, strictly speaking. It was primarily from this sense of value that the popular positivist distinction between differences of belief and differences of attitude took its origin, though C. L. Stevenson (who, when specifying his own notion of attitude, recalls R. B. Perry's definition of "interest" as a psychological disposition to be for or against something) allows that value-differences may have components both of attitude *and* belief (1949, p. 591).

It seems plausible to hold that emotive values are alien to the work of natural science. There is no reason to think that human emotionality is a trustworthy guide to the structures of the natural world. Indeed, there is every reason, historically speaking, to view emotive values, as Francis Bacon did, as potentially distortive "Idols," projecting in anthropomorphic fashion the pattern of human wants, desires, and emotions on a world where they have no place. When "ideology" is understood as a systematization of such values, it automatically becomes a threat to the integrity of science. The notion of value which is implicit in much recent social history of science, as well as in many analyses of the science-ideology relationship, is clearly that of emotive value.

A second kind of "value" is more important for your quest. A property or set of properties may count as a value in an entity of a particular kind because it is desirable for an entity of that kind. (The same property in a different entity might *not* count as a value.) The property can be a desirable one for various sorts of reasons. Speed is a desirable trait in wild antelope because it aids survival. Sound heart action is desirable in an organism with a circulatory system because of the functional needs of the organism. A retentive memory is desirable for a lawyer because of the nature of the lawyer's task. Sharpness is desirable in a knife because of the way in which it functions as a utensil. Efficiency is desirable in a business firm if the firm is to accomplish the ordinary ends of business. . . .

Let us focus on what these examples have in common. (In another context, we might be more concerned about their differences.) In each case, the desirable property is an objective characteristic of the entity. We can thus call it a *characteristic* value. In some cases, it is relative to a pattern of human ends; in others, it is not. In some cases, a characteristic value is a means to an end served by the entity possessing it; in others, it is not. In all cases, it serves to make its possessor function better as an entity of that kind.

Assessment of characteristic values can take on two quite different forms. One can judge the extent to which a particular entity realizes the value. We may be said to *evaluate* when we judge the speediness of a particular antelope or the heartbeat of a particular patient. On the other hand, we may be asked to judge whether or not (or to what extent) this characteristic really is a value for this kind of entity. How much do we *value* the characteristic? Here we are dealing, not with particulars, but with the more abstract relation of characteristic and entity under a particular description. Why *ought* one value speed in an antelope, rather than strength, say? How important *is* a retentive memory to a lawyer?

The logical positivists stressed the distinction between these two types of value-judgment, what I have called evaluation and valuing.[1] Valuing they took to be subjective and thus foreign to science. Evaluation, however, may be permissible because it expresses an estimate of the degree to which some commonly recognized (and more or less clearly defined) type of action, object, or institution is embodied in a given instance" (Nagel 1961, p. 492).[2] Notice the presupposition here: clear definition of the characteristic is required in order that there be a standard against which an estimate may be made. It was already a large concession to allow a role for mere estimation (as against measurement proper) in science; no further concession would be allowed.

Value-judgment, in the sense of evaluation could thus fall on the side of the factual, and the old dichotomy between fact and value could still be maintained. Value-judgment in the sense either of valuing or of evaluating, where the characteristic value is *not* sharply defined, was still to be rigorously excluded from science. Such value-judgment (so the argument went) is necessarily subjective; it involves a decision which is not rule-guided, and therefore has an element of the arbitrary. It intrudes individual human norms into what should ideally (if it were to be properly scientific) be an impersonal mapping of propositions onto the world.

What was offensive about value-judgment, then, was *not* its concern with characteristic values. Indeed, when such values are *measured* (when, for example, human blood-pressure is measured as a means to determining any departure from "normality"), the results are obviously "scientific" in the most conservative sense. Not every judgment in regard to characteristic value counts therefore as a "value-judgment," as this term has come to be used. Such a judgment must not only be concerned with value, but must function, not as measurement does, but in a nonmechanical, individual way. Since it is a matter of experience and skill, individual differences in judgment can thus in the normal course be expected.

It is clear, therefore, where the tension arises between value-judgment and not only the positivist view of science but the entire classical theory of science back to Aristotle. Max Weber spoke for that long tradition when, in his effort to eliminate value-judgment from social science, he opposed any form of assessment which could not immediately be enforced on all. The objectivity of science (he insisted) requires public norms accessible to all, and interpreted by all in the same way (Weber 1917).

What I want to argue here is that value-judgment, in just the sense that Weber deplored, *does* play a central role in science. Both evaluation *and* value are involved. The attempt to construe all forms of scientific reasoning as forms of deductive or inductive inference fails. The sense of my claim that science is value-laden is that there are certain characteristic *epistemic* values which are integral to the entire process of assessment in science. Since my topic is "values in science," there are, however, some other construals of this title that ought to be briefly addressed first, in order to be laid aside.

II. OTHER CONSTRUALS

One value, namely truth itself, has always been recognized as permeating science. In the classic account, it was in fact the goal of the entire enterprise. Unlike other values, it was deemed to have nothing of the personal about it. On the contrary, it connoted an objective relation of proposition and world and thus was constitutive of the very category of fact itself. But this was not thought to weaken the maxim that science should be value-free, because the values that were thus being enjoined from intrusion into the work of science were the particular ones that would tend to compromise the objectivity of the effort and not the transcendental one which defined the tradition of science itself.

There has been much debate in recent philosophy of science about the sense in which truth can still be taken to be constitutive of science. The correspondence view of truth as a matching of language and mind-independent reality has been assailed by Ludwig Wittgenstein and many other more recent critics like Hilary Putnam and Richard Rorty. More to the point here, it seems clear that when a scientist "accepts" a theory, even a long-held theory, he is not claiming that it is *true*. The predicate in terms of which theory is valued is not truth, as the earlier account held it to be. We speak of a theory as being "well-supported," "rationally acceptable," or the like. To speak of it as *true* would suggest that a later anomaly that would force a revision or even abandonment of the theory can in principle be excluded. The recent history of science would make both scientists and philosophers wary of any such presumption, except perhaps in cases of very limited theories or ones which are vaguely stated

It can, however, be argued that truth is still a sort of horizon-concept or ideal of the scientific enterprise, even though we may not be able to assert truth in a definitive manner of any component of science along the way. There are many variations of this view, one which was clearly articulated a century ago by C. S. Peirce. I do not intend to discuss this issue further here (though I will return to it obliquely in my conclusion), because to argue that truth is at least in some sense a characteristic value admissible in science is hardly novel, and does not constitute the point of division with classic logicist theories of science that I am seeking to identify.

Nor am I concerned here with *ethical* values. Weber and the positivists of the last century and this one recognized that the work of science makes ethical demands on its practitioners, demands of honesty, openness, and integrity. Science is a communal work. It cannot succeed unless results are honestly reported, unless every reasonable precaution be taken to avoid experimental error, unless evidence running counter to one's own view is fairly handled, and so on. These are severe demands, and scientists do not always live up to them. Outright fraud, as we have been made uncomfortably aware in recent years, does occur. But so far as we can tell, it is rare and does not threaten the integrity of the research enterprise generally. In any event, there never has been any disagreement about the value-ladenness of science where moral values of this kind are concerned. If I am to make a claim about a change in regard to the recognition of the proper presence in science of value-judgment, it cannot be in regard to those moral values which have always been seen as essential to the success of communal inquiry.[3]

In support of his claim that "value-judgments are essentially involved in the procedures of science," Rudner argued that the acceptance of a scientific hypothesis:

> is going to be a function of the importance, in the typically ethical sense, of making a mistake in accepting or rejecting the hypothesis. Thus, to take a crude but easily manageable example, if the hypothesis under consideration were to the effect that a toxic ingredient of a drug was not present in lethal quantity, we would require a relatively high degree of confirmation or confidence before accepting the hypothesis, for the consequences of making a mistake here are exceedingly grave by our moral standards. (1953, p. 2)

This notion of hypothesis "acceptance" is dangerously ambiguous. Rudner takes it to mean: "approve as a basis for a specific kind of action." But acceptance in this sense is not part of theoretical science, strictly speaking. When a physicist "accepts" a particular theory, this can mean that he believes it to be the best-supported of the alternatives available or that he sees it as offering the most fruitful research-program for the immediate future. These are *epistemic* assessments; they attach no values to the theoretical alternatives other than those of likelihood or probable fertility. On the other hand, if theory is being applied to practical ends, and the theoretical alternatives carry with them outcomes of different value to the agents concerned, we have the typical decision-theoretic grid involving not only likelihood estimates but also "utilities" of one sort or another. Such utilities are irrelevant to theoretical science proper and the scientist is not called upon to make value-judgments in their regard as part of his scientific work. The values of life and death involved in a decision to use or not to use a possibly toxic drug in a case where it alone seems to offer a chance of recovery are not relevant to the much more limited question as to whether or not the drug *would* be toxic for this patient.

The utilities typically associated with the application of science to human ends in medicine, engineering and the like, cannot, therefore, be cited as a reason for holding natural science itself to be value-laden. The conclusion that Rudner draws from his analysis of hypothesis-"acceptance" is that "a science of ethics is a necessary requirement if science's progress toward objectivity is to be continuous." But scientists are (happily!) not called on to "accept" hypotheses in the sense he is presupposing,[4] and so his conclusion does not go through.[5] If we are to hold that the work of science is value-laden, it ought to be for another reason.

My argument for the effective presence of "values in science" does not, then, refer to the constitutive role in science of the value or truth; nor does it refer to the ethical values required for the success of science as a communal activity, or to the values implicit in decision-making in applied science. Rather, it is directed to showing that the appraisal of theory is in important respects closer in structure to value-judgment than it is to the rule-governed inference that the classic tradition in philosophy of science took for granted.

Not surprisingly, the recognition of this crucial epistemological shift has been slow and painful. Already there are intimations of it among the more perceptive nineteenth-century philosophers of science. William Whewell, for example, describes a process very like value-judgment in his influential account of the "consilience of inductions," though he draws back from the threatening subjectivism of this line of thought, asserting that consilience will amount to "demonstration" in the long run (Laudan 1981). The logical positivists, as already noted, resolutely turned the theory of science back into the older logicist channels once more. Yet as they (and their critics) tried to characterize the strategies of science in closer detail, doubts began to grow. To these earlier anticipations of our theme, I now briefly turn.

III. ANTICIPATIONS

The prevailing inductivism of the nineteenth century made it seem as though science ultimately consisted of *laws*, that is, statements of empirical regularities. These laws were arrived at by generalization from the facts of observation; the facts themselves were regarded as an unproblematic starting-point for the process of induction. It was, of course, realized that the laws were open to revision as measuring apparatus was improved, as the ranges of the variables were extended, as new relevant factors were discovered. There was no *logic*, strictly speaking, which would lead from a finite set of observation-statements to a universally valid law of nature.

Human decision had to enter in, therefore, by way of curve-fitting, extrapolation, and estimates of relevance. Such decision was not arbitrary; there were skills and techniques to be learnt which would aid the scientist in drawing the best generalizations from the data available. Was this not a matter of value-judgment rather than of a common logic of formal rules? We would say so today. But the point was not so evident then, or perhaps it would be more accurate to say that it seemed of little importance.

The reason was that the laws were taken to be true to a degree of approximation that could be improved indefinitely. Thus the influence of these decisional aspects, where the individual skills of curve-fitting, extrapolation, and estimation of relevance, entered into the process of formulating a law, would be progressively lessened, as the law came closer and closer to being an exact description of the real, that is, as the law gradually attained the status of fact. Thus, even though value-judgment did enter, in a number of ways, into the process of inductive generalization, its presence could in practice be ignored. It was, after all, no more than an accessory activity, of little significance to the ultimate deliverances of science, namely, the exact statements of the laws of nature.

The logical positivists still adhered to this nomothetic ideal. But from the beginning, they encountered difficulties as soon as they tried to spell out how an inductive method might work. The story is a familiar one. I am going to focus on only two episodes in it, one involving Karl Popper and the other Rudolf Carnap, in order to show how "value-uneasiness" was already in evidence among philosophers of science fifty years ago, though in neither of these episodes was it altogether satisfactorily characterized.

As we all know, Popper rejected the nomothetic ideal of science that the logical positivists took over from the nineteenth century. For him, science is a set of conjectures rather than a set of laws. The testing of conjectures is thus the central element in scientific method and it can work only by falsification, when a basic statement conflicts with a conjectured explanation, leading to the rejection of the conjecture. The entire logical weight of this operation is carried by the "basic statements," that is, reports of observable events at particular locations in space and time.

But now a difficulty arises. Could not the basic statements themselves be falsified? They could not consistently be held immune to the test-challenge that Popper saw as the criterion of demarcation between science and nonscience. But if the basic statements themselves are open to challenge, how is the whole procedure of falsification of conjecture to work? It sounds as if a destructive regress cannot be avoided.

Popper's answer is to say that:

> Every test of a theory, whether resulting in its corroboration or falsification, must stop at some basic statement or other which we decide to accept. . . . Considered from a logical point of view, the situation is never such that it compels us to stop at this particular basic statement rather than at that. . . . For any basic statement can again in its turn be subjected to tests, using as a touchstone any of the basic statements which can be deduced from it, with the help of some theory, either the one under test or another. This process has no natural end. Thus if the test is to lead anywhere, nothing remains but to stop at some point or other and say we are satisfied for the time being. (1934/1959, p. 104)

Thus the designation of a statement as a "basic" one is not definitive, and hence falsification is not quite the decisive logical step Popper would have liked it to be. He continues: "Basic statements are accepted as the result of a decision or convention; and to that extent they are conventions" (p. 106). His choice of the term "convention" here is a surprising one since it carries the overtone of arbitrariness, of *arbitrary* choice and not just choice. But Popper is explicit in excluding this suggestion. He criticizes Otto Neurath, in fact, who (he says) made a "notable advance" by recognizing that protocol statements are not irrevocable, but then failed to specify a method by which they might be evaluated. Such a move, he goes on,

> leads nowhere if it is not followed by another step; we need a set of rules to limit the arbitrariness of "deleting" (or else "accepting") a protocol sentence. Neurath fails to give any such rules and thus unwittingly throws empiricism overboard. For without such rules, empirical statements are no longer distinguished from any other sort of statements. (p. 97)

For Popper, the need for such a line of demarcation takes precedence over any other demand. So if there are to be decisions regarding the basic statements, these must (he says) be "reached in accordance with a procedure governed by rules" (p. 106). If there are *rules*, however, to guide the decision, it sounds as though a definite answer might be obtained by the application of these rules in any given case. And so the properly *decisional* element would be minimal, and value-judgment (as we have defined it) would not enter in.

But, in fact, we discover that the word, "rule," here (like the word, "convention") is not to be taken literally. When Popper specifies how these "rules" would operate, all he has to say is that we can arrive at:

> a procedure according to which we stop only at a kind of statement that is especially easy to test. For it means that we are stopping at statements about whose acceptance or rejection the various investigators are likely to reach agreement. (p. 104)

So that ease in testing is to guide the investigator in deciding which statements to designate as basic. But this clearly operates here as a *value* rather than as a *rule*. There could be differences in judgment as to the extent to which the value was realized in a given case. Popper himself says of his "rules" that though they are "based on certain fundamental principles" which aim at the discovery of objective truth, "they sometimes leave room not only for subjective convictions but even for subjective bias" (p. 110).

Thus what we have here is value-judgment, not the application of rule, strictly speaking. There is no *rule* as to where to stop the testing process. If some investigators prolong it further than others do, we would not be inclined to describe this as either "following" or "breaking" a rule. But we *would* call it the pursuing of a particular goal or value.

Popper's use of the term "convention" to describe the element of value-judgment in the designation of basic statements has proved misleading to later commentators, even though he explicitly rejected classical conventionalism, mainly because it was unable, in his view, to generate a proper criterion of demarcation between science and nonscience (McMullin 1978a, section 7). lmre Lakatos, for example, described Popper's view as a form of "revolutionary conventionalism" because of its explicit admission of the role of decisional elements in the scientific process. This led him to characterize his own MSRP as a way of "rationalizing classical conventionalism," rational because the criteria for distinguishing between "hard core" and "protective belt" can be partially specified, as can the criteria of theory-choice, but "conventional" because the process is not one of a mechanical application of rule, involving, as it does, individual judgment (1970, p. 134). Joseph Agassi likewise proposed that the most accurate label for Popper's theory of science is "modified conventionalism" (Agassi 1974, p. 693) to which suggestion Popper rather testily responded "I am not a conventionalist, whether modified or not" (Popper 1974, p. 1117).

Much of the confusion prompted by Popper's use of the term "convention" might have been avoided if he had used the notion of value-judgment instead. It has precisely the flexibility that he needed in order to distance himself, as he wished to do, from both positivism and conventionalism, from positivism because of his insistence upon the decisional elements in the selection of basic statements and from conventionalism because he believed that the values guiding judgment in this case are grounded in the "autonomous aim" of science, which is the pursuit of objective knowledge (Popper 1974, p. 1117).

Though the admission of value-judgment into science had moved Popper away from his rationalist moorings, it is significant that he never extended the range of value-judgment to theory-choice, which today to us would seem the much more likely locus. Even though he allows that "the choice of any theory is an act, a practical matter" (1935/1959, p. 109), his opposition to verification made him wary of allowing that theories might ever be "accepted." To the extent that they are, it is a provisional affair, he reminds us. But this sort of provisional acceptance is still, in his view, *decisively* influenced by the success of the theory in avoiding falsification (McMullin 1978a, p. 224). Rationalism is thus preserved at this level by the assumption of a more or less decisive method of choosing between theories at any given stage of development.

This is the assumption that Carnap helped, somewhat unwittingly perhaps, to undermine. In 1950, he drew his famous distinction between "internal" questions, which can be answered within a given linguistic framework and "external" questions, which bear on the acceptability of the framework itself (1950, p. 214). The point of the distinction was to clarify the debate about the existence of such abstract entities as classes or numbers to which Carnap assimilated the question of theoretical entities like electrons. To ask about the existence of such entities *within* a given linguistic framework is perfectly legitimate, he said, and an answer can be given along logical or empirical lines. But to ask about the *reality* of such a system of entities taken as a whole is to pose a metaphysical question to which only a pseudo-answer can be given. The question can, however, be framed in a different way and then it becomes perfectly legitimate. We can ask whether the linguistic framework itself is an appropriate one for our purposes, whatever they may be. This is the form in which external questions *should* be put in order to avoid idle philosopher's questions about the existence of numbers or electrons.

Once the question is put in this way, he goes on, it is seen to be a practical, not a theoretical matter. The decision to accept a particular framework:

> although itself not of a cognitive nature, will nevertheless usually be influenced by theoretical knowledge, just like any other deliberate decision concerning the acceptance of linguistic or other rules. The purposes for which the language is intended to be used, for instance, the purpose of communicating factual knowledge, will determine which factors are relevant for the decision. (1950, p. 208)

And he goes on to enumerate some of the factors that might influence a pragmatic decision of this sort: he mentions the "efficiency, fruitfulness and simplicity" of the language, relative to the purposes for which it is intended. These are *values*, of course, and so what he is talking about here (though he does not explicitly say so) is value-judgment.

In this essay, Carnap is worrying mainly about the challenge of the nominalists to such entities as classes, properties, and numbers. He wants to answer this challenge, not by asserting the existence of these entities directly —this would violate his deepest empiricist convictions—but by appealing to the practical utility of everyday language where terms corresponding to these entities play an indispensable role. And so he counters Occam's razor with a plea for "tolerance in permitting linguistic forms" (1950, p. 220). As long as the language is efficient as an instrument, he says, it would be foolish, indeed harmful, to impoverish it on abstract nominalist grounds.

But Carnap conceded much more than he may have realized by this maneuver. By equating the general semantical problem of abstract entities with the problem of theoretical entities in science, he implied that pragmatic "external" criteria are the appropriate ones for deciding on the acceptability of the linguistic frameworks of science, that is, of scientific theories. For the first time, he is implicitly admitting that the tight "internal" logicist criteria which he had labored so long to impose on the problems of confirmation and explanation are inappropriate when it is the very language of science itself, that is, the theory, that is in question.

It is the *theory* that leads us to speak of electrons; to assess this usage, we have to evaluate as a single unit the *theory* in which this concept occurs and by means of which it is defined. If more than one "linguistic framework" for theory is being defended in some domain, the decision as to which is the better one has to be resolved, not by inductive logic, but by these so-called external criteria.

The term, "external," was obviously an unhappy choice, as things would turn out. The questions Carnap dubs "external" would be external to science only if theory-decision is external to science. They were external to his logicist conception of how science ought to be carried on, of course. Only if science can be regarded as a "given" formal system can the enterprise of the logician get under way. The question of whether a particular theory *should* be "given" or whether another might not accomplish the theoretical ends of science better, cannot be properly (i.e., "internally") posed in the original positivist scheme of things.

Once Carnap *allowed* it to be posed, however "externally," it would not be long until theory-evaluation would be clearly recognized as the most "internal" of all scientific issues, defining as it does scientific rationality and scientific progress. After we have discarded his term "external," we still retain his insight that the structure of decision in regard to the acceptability of a theoretical language is not one of logical rule but of value-judgment.

IV. THEORY-CHOICE AS VALUE-JUDGMENT

This gets us up only to 1950, which seems like very long ago in philosophy of science. Yet the shape of things to come is already clear to us, even though it was by no means clear then. The watershed between classic theory of science and our as yet unnamed postlogicist age has been variously defined since then. But for our purposes here, it can best be laid out in four propositions, three of them familiar, the other (P3) a little less so.

P1: The goal of science is theoretical knowledge.

P2: The theories of science are underdetermined by the empirical evidence.

P3: The assessment of theories involves value-judgment in an essential way.

P4: Observation in science is theory-dependent.

P1 tells us that the basic explanatory form in science is theory, not law, and thus that retroduction, not induction, is the main form of scientific validation. Theories by their very nature are hypothetical and tentative; they remain open to revision or even to rejection. P2 reminds us that there is no direct logical link, of the sort that classical theories of science expected, between evidence and theory. Since one is not compelled, as one would be in a logical or mathematical demonstration, one has to rely on oblique modes of assessment. And P3 tells us that these take the form of value-judgments.

P4 serves to emphasize that a thesis in regard to theory-appraisal has broader scope. To the extent that scientific observation is theory-dependent, it is also indirectly value-impregnated. This last point is not stressed any further in what follows, but it is well that it should be kept in mind lest it be thought that only *one* element in science, theory-choice, is affected by the shift described here, and that the traditional logicist/empiricist picture might be sustained at all other points.

So much for the schema. It can be found, more or less in the form in which I have sketched it here, in the work of Kuhn, specifically in his 1973 essay "Objectivity, Value-judgment, and Theory Choice" (1977). He asks there what are the characteristic values of a good scientific theory and lists, as a start, five that would be pretty well agreed upon. I will rework his fist just a little, and add some comments.

Predictive accuracy is the desideratum that scientists would usually list first. But one has to be wary about the emphasis given it. As Lakatos and Feyerabend in particular have emphasized, scientists must often tolerate a certain degree of inaccuracy, especially in the early stages of theory-development. Nearly every theory is "born refuted"; there will inevitably be anomalies it cannot handle. There will be idealizations that have to be worked out in order to test the theory in complex concrete contexts. Were this demand to be enforced in a mechanical manner, the results for science could be disastrous. Nevertheless, a high degree of predictive accuracy is in the long run something a theory *must* have if it is to be acceptable.

A second criterion is *internal coherence*. The theory should hang together properly; there should be no logical inconsistencies, no unexplained coincidences. One recalls the primary motivating factor for many astronomers in abandoning Ptolemy in favor of Copernicus. There were too many features of the Ptolemaic orbits, particularly the incorporation in each of a one-year cycle and the handling of retrograde motions, that seemed to leave coincidence unexplained and thus, though predictively accurate, to appear as ad hoc.

A third is *external consistency*: consistency with other theories and with the general background of expectation. When steady-state cosmology was proposed as an alternative to the Big Bang hypothesis in the late 1940s, the criticism it first had to face was that it flatly violated the principle of conservation of energy, which long ago attained the status almost of an a priori in mechanics. Even if Fred Hoyle had managed to make his model satisfy the other demands laid on it, such as the demand that it yield testable predictions in advance and not just after the fact, it would always have had a negative rating on the score of external consistency.

A fourth feature that scientists value is *unifying power*, the ability to bring together hitherto disparate areas of inquiry. The standard illustration is James Maxwell's electromagnetic theory. A more limited, but quite striking, example would be the plate-tectonic model in geology. Over the past twenty years, it has successfully explained virtually all major features of the earth's surface. What has impressed geologists sufficiently to persuade most (not all) of them to overcome the scruples that derive, for example, from the lack of a mechanism to account for the plate-movements themselves, is not just its predictive accuracy but the way in which it has brought together previously unrelated domains of geology under a single explanatory roof.

A further, and quite crucial, criterion is *fertility*. This is rather a complex affair (McMullin 1976). The theory proves able to make novel predictions that were not part of the set of original explananda. More important, the theory proves to have the imaginative resources, functioning here rather as a metaphor might in literature, to enable anomalies to be overcome and new and powerful extensions to be made. Here it is the *long-term* proven ability of the theory or research program to generate fruitful additions and modifications that has to be taken into account.

One other, and more problematic, candidate as a theory-criterion is *simplicity*. It was a favorite among the logical positivists because it could be construed pragmatically as a matter of convenience or of aesthetic taste, and seemed like an optional extra which the scientist could decide to set aside, without affecting the properly epistemic character of the theory under evaluation (Hempel 1966, pp. 40–45). Efforts to express a criterion of "simplicity" in purely formal terms continue to be made, but have not been especially successful.

One could easily find other desiderata. And it would be important to supply some detailed case-histories in order to illustrate the operation of the ones I have just listed. But my concern here is rather to underline that these criteria clearly operate as *values* do, so that theory choice is basically a matter of value-judgment. Kuhn puts it this way:

> The criteria of [theory] choice function not as rules, which determine choice, but as values which influence it. Two men deeply committed to the same values may nevertheless, in particular situations, make different choices, as in fact they do. (1977, p. 331)

They correspond to the two types of value-judgement discussed above in section 1. First, different scientists may *evaluate* the fertility, say, of a particular theory differently. Since there is no algorithm for an assessment of this sort, it will depend on the individual scientist's training and experience. Though there is likely to be a very large measure of agreement, nonetheless the skills of evaluation here are in part personal ones, relating to the community consensus in complex ways.

Second, scientists may not attach the same relative weights to different characteristic values of theory, that is they may not *value* the characteristics in the same way, when, for example, consistency is to be weighed over against predictive accuracy. It is above all because theory has more than one criterion to satisfy, and because the "valuings" given these criteria by different scientists may greatly differ, that disagreement in regard to the merits of rival theories can on occasion be so intractable.

It would be easy to illustrate this by calling on the recent history of science. A single example will have to suffice. The notorious disagreement between Niels Bohr and Albert Einstein in regard to the acceptability of the quantum theory of matter did not bear on matters of predictive accuracy. Einstein regarded the new theory as lacking both in coherence and in consistency with the rest of physics. He also thought it failing in simplicity, the value that he tended to put first. Bohr admitted the lack of consistency with classical physics, but played down its importance. The predictive successes of the new theory obviously counted much more heavily with him than they did with Einstein. The differences between their assessments were not solely due to differences in the values they employed in theory-appraisal. Disagreement in substantive metaphysical belief about the nature of the world also played a part. But there can be no doubt from the abundant testimony of the two physicists themselves that they had very different views as to what constituted a "good" theory.

The fact that theory-appraisal is a sophisticated form of value-judgment explains one of the most obvious features of science, a feature that could only appear as a mystery in the positivist scheme of things. Controversy, far from being rare and wrong-headed, is a persistent and pervasive presence in science at all levels. Yet if the classical logicist view of science had been right, controversy would be easily resolvable. One would simply employ an algorithm, a "method," to decide which of the contending theories is best confirmed by the evidence available. At any given moment, there would then be a "best" theory, to which scientists properly versed in their craft ought to adhere.

But, of course, not only is this *not* the case, but it would be a disaster if it *were* to be the case (McMullin 1983). The clash of theories, Popper has convinced us, is needed in order that weak spots may be probed and potentialities fully developed. Popper's own theory of science made it difficult to see how such a pluralism of theories could be maintained. But once theory-appraisal is recognized to be a complex form of value-judgment, the persistence of competing theories immediately follows as a consequence.

Thomas Kuhn characteristically sees the importance of value-difference not so much in the clash of theories—such controversy is presumably not typical of his "normal science"—as in the period of incipient revolution when a new paradigm is struggling to be born:

Before the group accepts it, a new theory has been tested over time by the research of a number of men, some working within it, others within its traditional rival. Such a mode of development, however, *requires* a decision process which permits rational men to disagree, and such disagreement would be barred by the shared algorithm which philosophers generally have sought. If it were at hand, all conforming scientists would make the same decision at the same time. ... I doubt that science would survive the change. What from one point of view may seem the looseness and imperfection of choice criteria conceived as rules may, when the same criteria be seen as values, appear an indispensable means of spreading the risk which the introduction or support of novelty always entails. (1961, p. 220)

It almost seems as though the value-ladenness of theory-decision is specially designed to ensure the continuance of controversy and to protect endangered but potentially important new theoretical departures. A Hegelian might see in this, perhaps, the cunning of Reason in bringing about a desirable result in a humanly unpremeditated way. But, of course, these are just the fortunate consequences of the nature of theory-decision itself. It is not as though theories could be appraised in a different more rule-guided way. One is forced to recognize that the value-ladenness described above derives from the problematic and epistemologically complex way in which theory relates to the world. It is *only* through theory that the world is scientifically understood. There is no alternative mode of access which would allow the degree of "fit" between theory and world to be independently assessed, and the values appropriate to a good theory to be definitively established. And so there is no way to exchange the frustrating demands of value-judgment for the satisfying simplicities of logical rule.

V. EPISTEMIC VALUES

Even though we cannot *definitively* establish the values appropriate to the assessment of theory, we saw just a moment ago that we can provide a tentative list of criteria that have gradually been shaped over the experience of many centuries, the values that are implicit in contemporary scientific practice. Such characteristic values I will call *epistemic*, because they are presumed to promote the truthlike character of science, its character as the most secure knowledge available to us of the world we seek to understand. An epistemic value is one we have reason to believe will, if pursued, help toward the attainment of such knowledge. I have concentrated here on the values that one expects a good *theory* to embody. But there are, of course, many other epistemic values, like that of reproducibility in an experiment or accuracy in a measurement.

When I say that science is value-laden, I would not want it to be thought that these values derive from theory-appraisal only. Value-judgment permeates the work of science as a whole, from the decision to allow a particular experimental result to count as "basic" or "accepted" (the decisional element that Popper stressed), to the decision not to seek an alternative to a theory which so far has proved satisfactory. Such values as these may be pragmatic rather than epistemic; they may derive from the finiteness of the time or resources available to the experimenter, for example. And sometimes the borderline between the epistemic and the pragmatic may be hard to draw, since (as Pierre Duhem and Karl Popper among others have made clear) it is essential to the process of science that pragmatic decisions be made, on the temporary suspension of further testing for example.

Of course, it is not pragmatic values that pose the main challenge to the epistemic integrity of the appraisal process. If values are needed in order to close the gap between underdetermined theory and the evidence brought in its support, presumably all sorts of values can slip in: political, moral, social, and religious. The list is as long as the list of possible human goals. I shall lump these values together under the single blanket term, "nonepistemic." The decision as to whether a value is epistemic or nonepistemic in a particular context can sometimes be a difficult one. But the grounds on which it should be made are easy to specify in the abstract. When no sufficient case can be made for saying that the imposition of a particular value on the process of theory choice is likely to improve the *epistemic* status of the theory, that is, the conformity between theory and world, this value is held to be nonepistemic in the context in question. This decision is itself, of course, a value-judgment and there is an obvious danger of a vicious regress at this point. I hope it can be headed off, and will return to this task in a moment.

But first, one sort of factor that plays a role in theory-assessment can be hard to situate. Externalist historians of science have been accustomed to grouping under the elastic term "value" not only social and personal goals but also various elements of world-view, metaphysical, theological, and the like. Thus, for example, when Newton's theology or Bohr's metaphysics affected the choice each made of "best" theory in mechanics, such historians have commonly described this as an influence of "values" upon science (see, for example, Graham 1981).

Since I have been arguing so strongly here for the value-ladenness of science, it might seem that I should welcome this practice. But it is rooted, I think, in a sort of residual positivism that is often quite alien to the deepest convictions of the historians themselves who indulge in it (McMullin 1982). They would be the first to object to the label "externalist," but here they are assuming that a philosophical world-view is of its nature so "external" to science that it must be flagged as a "value," and consequently dealt with quite differently from the point of view of explanation.

Let me try to clarify the source of my opposition to this practice. A philosophical system can in certain contexts serve as a value, as a touchstone of decision. So for that matter can a scientific theory. But this does not convert it into a "value" in the sense in which social historians sometimes interpret this term, namely as something for which socio-psychological explanation is all-sufficient. The effect of calling metaphysics a "value" can be to shift it from the category of *belief* to be explained in terms of reasons adduced, in the way that science is ordinarily taken, to the category of *goal* to be explained, in terms of character, upbringing, community pressures, and the rest.

What I am arguing for is the potentially *epistemic* status that philosophical or theological world-view can have in science. From the standpoint of today, it would be inadmissible to use theological argumentation in mechanics. Yet Isaac Newton in effect did so on occasion. In describing this, it is important to note that theology functioned for him as an epistemic factor, as a set of reasons that *Newton* thought were truth-bearing (McMullin 1978b, p. 55). It did *not* primarily operate as a value if by "value" one were to mean a socio-psychological causal factor, superimposed upon scientific argument from the outside, to be understood basically as a reflection of underlying social or psychological structures.

Now, of course, the historian may find that someone's use of theological or philosophical considerations *did*, on a given occasion, reflect such structures. But this has to be historically *proven*. The question must not be begged by using the term "value" as externalist historians have too often done. Incidentally, the pervasive presence of nonstandard epistemic factors in the history of science is the main reason, to my mind, why the one-time popular internal-external dichotomy fails. Sociologists of science in the "strong program" tradition are more consistent in this respect. They *do* take metaphysics and theology to be a reflection of socio-psychological structure, but of course, they regard science itself in the same epistemically unsympathetic light. My point here has simply been that it is objectionable to single out nonstandard forms of argument in science by an epistemically pejorative use of the term "value" (McMullin 1983).

VI. THE PLACE OF ACT IN A WORLD OF VALUES

That being said, let me return to the question that must by now be uppermost in the reader's mind. What is left of the vaunted objectivity of science, the element of the factual, in all this welter of value-judgment? Once the camel's nose is inside, the tent rapidly becomes uncomfortable. Is there any reasoned way to stop short of a relativism that would see in science no more than the product of a contingent social consensus, bearing testimony to the historical particularity of culture and personality much more than to an objective truth about the world? I think there is, but I can at this stage only provide an outline of the argument needed. It requires two separate steps.

Step one is to examine the epistemic values employed in theory-appraisal, the values that lie at the heart of the claim that theory assessment in science is essentially value-laden, and to ask how they in turn are to be validated, and how in particular, circularity is to be avoided in doing so. First, let me recall how the skills of epistemic value-judgment are *learnt*. Apprentice scientists learn them not from a method book but from watching others exercise them. They learn what to expect in a "good" theory. They note what kinds of considerations carry weight, and why they do so. They get a feel for the relative weight given the different kinds of considerations, and may quickly come to realize that there are divergences here in practice. Their own value-judgments will gradually become more assured, and will be tested against the practice of their colleagues as well as against historical precedent (Polanyi 1958; Kuhn 1962).[6]

What is the epistemic worth of the consensus from which these skills derive? Kuhn is worried about the validity of invoking history as warrant in this case:

> Though the experience of scientists provides no philosophical justification for the values they deploy (such justification would solve the problem of induction), those values are in part learned from that experience and they evolve with it. (1977, p. 335)

This is to take the Hume-Popper challenge to induction far too seriously (unless, of course, "justification" were to be taken to mean definitive proof). The characteristic values guiding theory-choice are firmly rooted in the complex learning experience which is the history of science; this is their primary justification, and it is an adequate one.

We have gradually learnt from this experience that human beings have the ability to create those constructs we call "theories" which can provide a high degree of accuracy in predicting what will happen, as well as accounting for what has happened, in the world around us. It has been discovered, further, that these theories can embody other values too, such values as coherence and fertility, and that an insistence on these other values is likely to enhance the chances over the long run of the attainment of the first goal, that of empirical accuracy.

It was not always clear that these basic values *could* be pursued simultaneously.[7] In medieval astronomy, it seemed as though one had to choose between predictive accuracy and explanatory coherence, the Ptolemaic epicycles exemplifying one and Aristotelian cosmology the other. Since the two systems were clearly incompatible, philosophers like Aquinas reluctantly concluded that there were two sorts of astronomical science, one (the "mathematical") which simply "saved the appearances," and the other (the "physical") whose goal it was to explain the *truth* of things (Duhem, 1908/1969, chapter 3). Galileo's greatest accomplishment, perhaps, was to demonstrate the possibility of a single science in which the values of both the physical and the mathematico-predictive traditions could be simultaneously realized (Machamer 1978).

There was nothing *necessary* about this historical outcome. The world might well have turned out to be one in which our mental constructions would *not* have been able to combine these two ideals. What became clear in the course of the seventeenth century was that they *can* be very successfully combined, and that other plausible values can be worked in as well. When I say "plausible" here, I am suggesting that there is a second convergent mode of validation for these values of theory-appraisal (for "valuings" in the sense defined in section 1).

We can endeavor to *account for* their desirability in terms of a higher-order epistemological account of scientific knowing. This is to carry retroduction to the next level upwards. It is asking the philosopher to provide a theory in terms of which such values as fertility would be shown to be appropriate demands to lay on scientific theory. The philosopher's ability to provide just such a theory (and it is not difficult to do this) in turn then testifies to the reliability to taking these criteria to be proper values for theory-appraisal in the first place. This is only the outline of an argument, and much more remains to be filled in. But perhaps I have said enough to indicate how one could go about showing that the characteristic values scientists have come to expect a theory to embody are a testimony to the *objectivity* of the theory, as well as of the involvement of the subjectivity of the scientist in the effort to attain that objectivity.

There is a further argument I would use in support of this conclusion, but it is based on a premise that is not shared by all. That is the thesis of scientific realism. I think that there are good reasons to accept a cautious and carefully restricted form of scientific realism, prior to posing the further question of the objective basis of the values we use in theory-appraisal (McMullin 1983). The version of realism I have in mind would suggest that in many parts of science, like geology and cell-biology, we have good reasons to believe that the models postulated by our current theories gives us a reliable, though still incomplete, insight into the structures of the physical world.

Thus, for example, we would suppose that the success of certain sorts of theoretical model would give us strong reason to believe that the core of the earth *is* composed of iron, or that stars *are* glowing masses of gas. We have no direct testimony regarding either of these beliefs, of course. To claim that the world *does* resemble our theoretical models in these cases, is to claim that the method of retroduction on which they are based, and which rests finally on the values of theory-appraisal I have already discussed, is in fact (at least in certain sorts of cases) reliable in what it claims.[8] Obviously, the realist thesis will not hold, or will hold only in attenuated form, where theory is still extremely underdetermined (as in current elementary-particle theory) or where the ontological implications of the theory are themselves by no means clear (as in classical mechanics).

And so, to conclude step one, there is reason to trust in the values used commonly in current science for theory-appraisal as something much more than the contingent consensus of a peculiar social subgroup.

But a further step is needed, because these values do not of themselves *determine* theory-choice, a point I have stressed from the beginning. And so other values *can* and *do* enter in, the sorts of value that sociologists of science have so successfully been drawing to our attention of late, as they scrutinize particular episodes in the history of science. I am thinking of such values as the personal ambition of the scientist, the welfare of the social class to which he or she belongs, and so on. Has the camel not, then, poked its wet nose in beside us once again?

It has, of course, but perhaps we can find a way to push it out—or almost out—one final time. The process of science is one long series of test and tentative imaginative extensions. When a particular theory seems to have triumphed, when Louis Pasteur has overcome Felix Pouchet, to cite one nineteenth-century illustration that has recently come in for a lot of attention from social historians of science (Farley and Geison 1976), it is not as though the view that has prevailed is allowed to reign in peace. Other scientists attempt to duplicate experimental claims; theoreticians try to extend the theories involved in new and untried ways; various tests are devised for the more vulnerable theoretical moves involved, and so on. This is not just part of the mythology of science. It really *does* happen, and is easy to document.

To the extent that nonepistemic values and other nonepistemic factors have been instrumental in the original theory-decision (and sociologists of science have rendered a great service by revealing how much more pervasive these factors are than one might have expected), they are gradually sifted by the continued application of the sort of value-judgment we have been describing here. The nonepistemic, by very definition, will not in the long run survive this process. The process is designed to limit the effects not only of fraud and carelessness, but also of ideology, understood in its pejorative sense as distortive intrusion into the slow process of shaping our thought to the world.

NOTES

1. The terminology of "evaluation" and "valuing" is used by Kovesi (1967) in a somewhat different way. He supposes value-judgment to apply to things via their descriptions. Thus, we "evaluate" particulars insofar as they "fall under a certain description" (p. 151). Whereas we "value" things "insofar as they are such and such." We would "evaluate" a particular lawyer as a lawyer (being given a description of the qualities that make up a lawyer), whereas we would

"value" lawyers for what they are, as indispensable to the conduct of complex communities or however we might wish to describe their "value" in some broader context. (His aim is to contrast "evaluation" with moral judgment.) My focus is on specific characteristic values, on the Y-ness of X's, where his is on entity-descriptions, on X-ness itself as a subject for evaluation or valuing. The advantage of the former is that it makes the basis of the value-judgment specific. It focuses evaluation on the characteristic which can be present to a greater or lesser degree. And it provides a *context* for valuing which Kovesi's notion appears to lack, thus risking confusion with emotive value. Finally, Kovesi's emphasis on description could mislead, since the characteristic value need not be *described*, strictly speaking. Indeed, as we shall see below, the frequent inability to give explicit descriptions of characteristic values is an essential feature of evaluation as it occurs in science. The emphasis on X-ness (which does need describing) rather than on the Y-ness of X's (where the Y may be only summarily indicated) is the root of the difference. I am indebted to Carl Hempel and David Solomon for discussions of the topics of this section.

2. Nagel used the terms "characterize" and "appraise" instead of our "evaluate" and "value." The example he gives of "characterizing" is the evaluation of the degree of anemia a particular animal suffers from against a standard of "normality" in the red blood-corpuscle count. (See "The Value-Oriented Bias of Social Inquiry," Nagel 1961, pp. 485–502.)

3. Some philosophers assimilate epistemic values to moral values, so that for them the values implicit in theory-appraisal are broadly moral ones. Putnam, for example, takes adherence to these values on the part of scientists to be "part of our idea of human cognitive flourishing, and in hence part of our idea of total human flourishing, of Eudaimonia" (1981, p. 134). The analysis of characteristic value given in section 1, and even more the discussion of the warrant for epistemic value in section 6 below, would lead me to question this assimilation of the epistemic to the moral under the very vague notion of "flourishing." To pursue this further would, however, require further analysis of the nature of moral knowledge.

4. A point already made by Richard Jeffrey (1956) in a response to the Rudner article.

5. In fairness, it should be added that Rudner drew attention in the same paper to the value-implications of the new directions that Carnap and Quine were just beginning to chart. But these consequences were obscured by his emphasis on the ethical aspects of theory-acceptance. He evidently supposed that all of these considerations would converge, but in fact, they did not, and could not.

6. Polanyi and Kuhn relate such skills as that of theory-assessment (and pattern-recognition, which is ultimately theory-dependent) in rather different ways to the learning experience of the apprentice scientist. I would lean more to Kuhn's analysis in this case, but in the context of my argument here, it is sufficient to note the affinity between these two authors rather than to press their differences.

7. Kuhn attaches a higher degree of fixity to the epistemic values of theory-choice then I would. He takes the five he describes to be "permanent attributes of science," provided the specification be left vague (1977, p. 335).

8. This is where I would diverge from Putnam (1981), who otherwise defends a view of the role of value-judgment in science similar to the one outlined here. In the spirit of Kant, he wants to find a middle way between objectivism and subjectivism, between what he regards as the extremes of "metaphysical realism" and "cultural relativism." The former he defines as being based on "the notion of a transcendental match between our representation and the world" which he briskly characterizes as "nonsense" (1981, p. 134). Blocked from taking the epistemic values to be the means of gradually achieving such a correspondence, he is thus forced to make them in some sense ultimates. "Truth is not the bottom line; truth itself gets its life from our criteria of rational acceptability" (p. 130). What he wants to stress, he says, is "the dependence of the empirical world on our criteria of rational acceptability" (p. 134). Instead of merely holding that "our knowledge of the world presupposes values" (the thesis that I am

arguing for in this essay), he is led then to "the more radical claim that what *counts* as the real world depends upon our values" (p. 137).

But such a position leaves him (in my view) with no vantage-point from which it would be possible to correct, or gradually adjust, the epistemic values themselves. They constitute for him "part of our conception of human flourishing" (p. xi). But there can be many such conceptions; against Aristotle (whom he takes to defend a single ideal of human flourishing), he argues for a "diversity" of ways in which such flourishing might properly he construed (p. 148). But how then can he *also* reject some such ways as "wrong, as infantile, as sick, as one sided" (p. 198)? What grounds are available in his system for such a rejection? He says that "we revise our very criteria of rational acceptability 'in the light of our theoretical picture of the empirical world" (p. 134), but gives no hint as to how this is to be done in practice. He cites "coherence" as a sort of supercriterion which appears to be necessary to *any* ideal of human flourishing (p. 132). But what if someone were to *reject* such a criterion? Putnam says such a person is "sick." But are there *arguments* he could use to warrant this diagnosis?

I do not think that in the end this "middle way" works. The tilt to idealism is obvious. But it would take a more elaborate analysis to show this. (This footnote and footnote 3 were added in proof. Had I seen Putnam's book before I wrote this text, I would have attempted a fuller discussion of it.)

REFERENCES

Agassi, Joseph. (1974). "Modified Conventionalism Is More Comprehensive than Modified Essentialism." In Schilpp (1974), pp. 693–696.

Carnap, Rudolf. (1932). "Überwindung der Metaphysik durch logische Analyse der Sprache." *Erkenntnis* 2:219–241. (Reprinted as "The Elimination of Metaphysics Through Logical Analysis of Language." (trans.) Arthur Pap. In *Logical Positivism*, edited by A. J. Ayer, 60–81. Glencoe, Ill.: Free Press, 1959.

———. (1950). "Empiricism, Semantics, and Ontology." *Revue internationale de philosophie* 4: 20–40. (As reprinted in *Meaning and Necessity*. 2nd ed. Chicago: University of Chicago Press, 1956, pp. 205-221.)

Duhem, Pierre. (1908). Εψςειν τά φαι νόμενα: *Essai sur la notion de theorie physique de Platon a Galilée*. Paris: A. Hermann. (Reprinted as *To Save the Phenomena: An Essay on the Idea of Physical Theory from Plato to Galileo*. (trans.) E. Doland and C. Maschler. Chicago: University of Chicago Press, 1969.)

Farley, John and Gerald Geison. (1976). "Science, Politics, and Spontaneous Generation in 19th Century France: The Pasteur-Pouchet Debate." *Bulletin of the History of Medicine* 48: 161–198.

Graham, Loren. (1981). *Between Science and Values*. New York: Columbia.

Hempel, Carl G. (1966). *Philosophy of Natural Science*. Englewood Cliffs, N.J.: Prentice-Hall.

Jeffrey, Richard. (1956). "Valuation and Acceptance of Scientific Hypotheses." *Philosophy of Science* 23: 237–246.

Kovesi, Julius. (1967). *Moral Notions*. London: Routledge and Kegan Paul.

Kuhn, Thomas. (1961). "The Function of Measurement in Modern Physical Science." *Isis* 52:161–190. (As reprinted in Kuhn, Thomas. *The Essential Tension*, 178–224. Chicago: University of Chicago Press, 1977.)

——— (1962). *The Structure of Scientific Revolutions*. Chicago: University of Chicago Press.

———. (1977). "Objectivity, Value Judgment, and Theory Choice." In *The Essential Tension*. Chicago: University of Chicago Press, pp. 320–339.

Lakatos, Imre. (1970). "Falsification and the Methodology of Scientific Research Programmes." In *Criticism and the Growth of Knowledge,* edited by I. Lakatos and A. Musgrave, 91–195. Cambridge: Cambridge University Press.

Laudan, Larry. (1981). "William Whewell on the Consilience of Inductions." In *University of Western Ontario Series in Philosophy of Science.* Volume 19, *Science and Hypothesis,* 163–180. Dordrecht: Reidel.

Machamer, Peter. (1978). "Galileo and the Causes." In *University of Western Ontario Series in Philosophy of Science.* Volume 14, *New Perspectives on Galileo,* edited by R. Butts and J. Pitt, 161–180. Dordrecht: Reidel.

McMullin, Ernan. (1976). "The Fertility of Theory and the Unit for Appraisal in Science." In *Boston Studies in the Philosophy of Science.* Volume 39, *Essays in Honor of Imre Lakatos,* edited by P. Feyeraband et al., 395–432. Dordrecht: Reidel.

———. (1978a). "Philosophy of Science and its Rational Reconstructions." In *Boston Studies in the Philosophy of Science.* Volume 58, *Progress and Rationality in Science,* edited by G. Radnitzky and G. Andersson, 221–252. Dordrecht: Reidel.

———. (1978b). *Newton on Matter and Activity.* Notre Dame, Ind.: University of Notre Dame Press.

———. (1982). "The Role of 'Values' in Understanding Science." *Hastings Center Report* 12, no. 6: 38–40.

———. (1983). "Scientific Controversy and Its Termination." In *Scientific Controversies,* edited by A. Kaplan and H. T. Engelhardt. Cambridge: Cambridge University Press.

———. (1985). "The Case for Scientific Realism." In *Scientific Realism,* edited by J. Leplin. Los Angeles: University California Press.

Nagel, Ernest. (1961). *The Structure of Science.* New York: Harcourt, Brace.

Polyani, Michael. (1958). *Personal Knowledge.* London: Routledge and Kegan Paul.

Popper, Karl. (1934). *Logik der Forschung.* Vienna: Springer. (Reprinted as *The Logic of Scientific Discovery.* London: Hutchinson, 1959.)

———. (1974). "Replies to My Criticism." In Schilpp (1974), 961–1197.

Putnam, Hilary. (1981). *Reason, Truth, and History.* Cambridge: Cambridge University Press.

Rudner, Richard. (1953). "The Scientist *qua* Scientist Makes Value-judgments." *Philosophy of Science* 20: 1–6.

Schilpp, Paul A., ed. (1974). *The Library of Living Philosophers.* Volume 14, book 2: *The Philosophy of Karl Popper.* LaSalle, Ill.: Open Court.

Stevenson, Charles. (1949). "The Nature of Ethical Disagreement." In *Readings in Philosophical Analysis,* edited by H. Feigl and W. Sellars, 587–593. New York: Appleton-Century-Crofts.

Weber, Max. (1917). "Der Sinn der 'Wertfreiheit' der soziologischen und ökonomischen Wissenschraften." *Logos* 7: 40–88. (Reprinted as "The Meaning of 'Ethical Neutrality' in Sociology and Economics." In *The Methodology of the Social Sciences,* translated by E. A. Shils and H. A. Finch, 1–47. Glencoe, Ill: The Free Press, 1949.)

32

From Weber to Habermas

Robert Hollinger

1. THE DIALECTICS OF OBJECTIVISM

In many respects, all the issues about science and values discussed in part 6 represent the logical culmination of the dialectic of Objectivism that begins with Descartes and Galileo. It will be helpful to briefly summarize this view (which is spelled out in more detail elsewhere in this book).[1] According to Objectivism, the world, which is the subject of scientific investigation, is governed by laws and processes that are quite independent of human beliefs, values, desires, and interpretations. Knowledge must mirror or correspond to the objective world; and our methods must also be objective, i.e., unbiased and nonpersonal, so as to guarantee this result. But this requires that science and our methods must purge themselves of anything subjective, i.e., any biases, psychological factors, values, etc., that would prevent our methods from resulting in objective knowledge. On this view, claims not capable of being described in the language of science, and any method that is not totally person free, and interest neutral, must be discarded. Hence, value judgments must either (A) be capable of being expressed in the language of science or else (B) are merely subjective expressions of preferences, and not truth claims, and thus belong outside of science.

The first of these attitudes toward value judgments, i.e., (A), attempts to turn ethical and political questions into scientific or technological problems, especially within modern utilitarian and cost-benefit terms. Value judgments are divided into two types: "categorical" and "instrumental." A categorical value judgment is of the form "X is good/bad" or "X is right/ wrong." Some categorical value judgments are said to be "ultimate" or "basic," i.e., since they don't derive from any other values or from any facts. Instrumental value judgments are judgments about what is the best means of achieving some categorical (including basic) values. Categorical values are often called

539

"intrinsic" values, because they are valued for themselves; e.g., health, pleasure, happiness. Instrumental values are usually sought because they help us realize other valued things, especially intrinsic values. Thus, if health is an intrinsic value, going to the doctor is an instrumental one, relative to the value we place on health. Health, and some intrinsic values, may also be instrumentally valued with respect to other values. And some instrumental values may, in some situations, be regarded as intrinsic values. Those who talk about ultimate values often believe that some values are only intrinsic and that all other values are really instrumental to the realization of our ultimate values.

Thesis (A) can now be expressed as follows. All ultimate values must be expressible in the language of science. But since the language of science is a description of the Objective world, all ultimate value judgments are descriptions of some aspect of the objective world. E.g., the "naturalistic" view of values, according to which "good" means "whatever human beings ultimately value," implies that it is just a fact about human beings that if pleasure is our ultimate value, good just is pleasure. We cannot raise the question whether it is right or wrong, good or bad, that somebody values X because it gives him/her pleasure. Ultimate values cannot be proved right or wrong, good or bad. We can only prove that people do have them. More generally, no categorical value judgment can be proven or disproven as right or wrong, good or bad, by science. Instrumental values, which tell us how to maximize our categorical (and thus our ultimate) values, can be shown by science to either effectively maximize our categorical values or not. Utilitarianism and modern cost-benefit analysis are the expressions of these assumptions. Given any ultimate value (or categorical value judgment) X, all we need to do is to discover which means or instrumental value will get us X in the most cost-efficient way. (See Hempel's in part 6.) So ethics becomes a branch of applied science: we need to predict how to realize our ultimate and other categorical values in the most rational ways open to us. This is sometimes called the thesis of technological or instrumental rationality. (See Introduction to this part.)

The second way of dealing with values, i.e., (B), is the idea that all value judgments, perhaps even scientific knowledge itself, to the extent that it relies on value judgments, are either arbitrary, nonrational, or irrational. They are the result of sociological, cultural, and psychological factors that can explain the prevalence of certain ideas and values, but cannot justify them on "rational" grounds. Thus, many versions of Marxism, which take knowledge and value claims to be "ideological," and writings in the "Sociology of Knowledge," explain our scientific beliefs and moral values in cultural, economic, and historical terms, while presupposing thereby that questions about their truth and rationality are out of the question. On some versions of (B), basic values are just matters of taste, and therefore cannot be rationally discussed or shown to be right or wrong, good or bad. The basic argument in either case is that only science contains statements that can be true or false, and thus objective. Ethics does not contain such statements, and cannot be reformulated in the language of science. Therefore ethics does not express or deal with knowledge.

It seems as if (A) and (B) are in some ways different and even incompatible views. But some writers believe that in at least one crucial respect they amount to the same thing. The claim that these two approaches to values amount to the same thing was first articulated by Edmund Husserl (1859–1938), the founder of Phenomenology. In Kuhn's language, the claim is that (A) and (B) are variants of a single paradigm. The underlying assumptions of this paradigm: (1) Everything is divided into the "objective" and the "subjective." (2) Only scientific Knowledge is objective. (3) Ultimate value judgments are either scientific or subjective. (4) Ultimate values, whether they are taken to reflect basic facts about human nature, personal decisions, or a culture's dominant values, cannot be rationally debated, or proven right or wrong, good or bad. Husserl believed that both (A) and (B) are embodied in the logic of Objectivism. On this view, (A) and (B) lead to the view that there are no moral absolutes; that there is no rational way of assessing the great variety of ultimate value judgments, the unreflective acceptance of certain values; the growing inability of society to reach rational agreement about ethical issues. In short, they lead to wholesale acceptance of whatever ultimate values in fact exist, even if they rest upon prejudice.

But we should be able to be reflective about our basic moral judgments, discuss them, criticize them, and correct them in rational ways. We should also aspire to some sort of universal moral consensus or agreement. But (A) and (B), according to Husserl, do not let us do so. This is the force of his claim that they must accept basic values as "given."

All "naturalistic" attempts to reduce human values and mental life to the law of the natural and physical sciences actually presuppose, for Husserl, an allegiance to certain values, e.g., science is the best way of enhancing human life, that ultimate values cannot be proven right, or even questioned, within the framework of modern science. Such a judgment of ultimate value must be accepted without being provable. But in denying the possibility of rationally defending ultimate values, modern science simultaneously rests upon a set of values: those embedded in the dominant views about knowledge and values mentioned above under (A) and (B). Husserl claimed that this picture thus rested on a contradiction. Moreover, by being unable and unwilling to question its own methods and assumptions, modern science is unable to raise, let alone address, questions about its own significance. The value of modern science as the best way of enhancing human life is thus (on his view) an assumption. Yet the scientific views about knowledge and values have become the dominant views of our culture. But what if these views are just unfounded prejudices, asks Husserl? In particular, what if the view that only the modern sciences tell us what's real, true, or valuable is unjustified and unjustifiable? What would we be left with? Nothing, according to Husserl. That is, our culture would be pervaded by nihilism, the claim that everything is meaningless and without value, which is just what Husserl believed had happened to Western culture.

Husserl believed that the cultural and political implications of both of these versions of nihilism are very dangerous. Husserl wanted to be able to get behind the unexamined prejudices that underlie both of these views, to show that modern science and human values equally are rooted in the practical activities of human beings, which could eventually be shown to rest upon universal and rational criteria. He hoped that we could eventually undermine both objectivism and the various ramifications of it, discussed under (A) and (B) above.

II. WEBER

Max Weber's (1864–1920) views about knowledge, reality, and values manage to combine themes from both (A) and (B). Our world is a meaningless, infinite flux. In order to understand it, we have to adopt some point of view, which is determined by our personal interests and values, or by those of our culture. This is the thesis of "value relevance." Our values and interests, which are subjective, give us an angle of vision on a narrow range of data, which can then be explained scientifically, i.e., objectively. For example, if we decide to find out why there are more suicides in the United States during the Christmas season than at other times of the year, our interests and values determine what we'll study. We can still get some objective answers to the questions we pose, if we use scientific method in an unbiased way. But our quest for knowledge is always severely restricted, since the world, and therefore any chunk of it we may want to study, is tremendously complex, and since there are any number of equally valid, but equally arbitrary, points of view we can adopt. For instance, we can study nature in many ways and for many reasons, and from many points of view. Each point of view is limited: it lets us understand only very small bits of nature. We can never, according to Weber, explain everything within the framework of any one point of view or theory. The world is too complex and we are too ignorant to even hope for a unified account of everything. This is even more true of any attempt to study human history or society. We can, for example, study Western culture from an unlimited number of perspectives. Moreover, Weber believed that we cannot expect to piece all of them together to get a unified and coherent history of our culture. So there is a sense in which Weber does undermine the idea of objective knowledge, if this is construed as the search for a unified scientific world picture (as Objectivism aspires to be).

In addition, value judgments are subjective for Weber, because they cannot be proven by science. Thus, the sciences must be value free or value neutral. Categorical value judgments are subjective. Science can tell us what the consequences of subjectively based decisions are, so that we can be aware of the consequences of our actions. This is what Weber calls "the ethics of responsibility." Modern societies must be governed by this model: public (and private) actions and policies are a combination of arbitrary preferences and cost-effective means for achieving them.

Let us consider the political implications of this view. In modern liberal democracies there exists a plurality of interests, points of view, values, and goals. (One need only reflect upon the abortion controversy.) How can democratic societies prevent chaos and institute public policies? Weber's answer is roughly this: The values and goals of our political leaders are no more rational than those of anybody else. But leaders must be strong enough to convince legislators and bureaucrats to accept their goals and policies. They must, at the same time, exercise the ethics of responsibility by consulting scientific experts, who will advise them of the consequences, and thus the wisdom, of their values and decisions. (For Weber, scientists cannot make political decisions, because the sciences must be value-free, and because value judgments are subjective.) Responsible political leaders will also have some convictions, but will balance these moral ideals with an assessment of the consequences of acting upon them. Considering only consequences is dangerous, because it can lead to expediency at the cost of noble ideals. Attention to ideals regardless of the consequences leads to fanaticism, which is (at least) equally dangerous in a political leader.

Weber's views about science, values, and politics dominate the world today. Political and moral issues become technological problems, and political goals and values are the result of a so-called consensus of all the interests involved, or else are taken to be "obvious," "rational," or even scientifically proven (given the use of the utilitarian idea that it is rational to maximize benefits over costs). All value judgments are equally valid, all points of view must have their say, we must be tolerant of all points of view and use cost-benefit analysis, together with the democratic political process, to arrive at "rational," value-neutral, "scientific" solutions to our problems. Since the values of individuals and a culture must be accepted as "givens," yet are all equally subjective, politics takes the form of a power struggle among the various values and interests, with the "right" solution being determined by whoever has the most power or persuasive force and a cost-benefit argument that their interests will benefit the greatest number of people.

What are the practical implications of all this? On some views, modern life oscillates between two ideals: the view that all social and personal problems require some combination of scientific knowledge, technological expertise, and cost-benefit analysis; and the idea that politics is a continual power struggle among competing values and interests. On either alternative (in practice they are often combined), some fear, democracy is in danger (either by a growing technological form of public decision making, or by virtue of an increasing lack of basic consensus). Thus, many people seem to believe that our judgment (say) not to help starving people in another country can be justified in cost-benefit terms (promoting their survival will make things worse, given the world population problem and the scarcity of world resources) or else the judgment whether to help or not becomes almost a matter of taste or individual preference.

If these fears are well founded, we may well ask what has gone wrong. How can we explain how modern science, together with liberal, humane values, results in this state of affairs?

III. BETWEEN WEBER AND HABERMAS

There have been a number of answers to this question, beginning perhaps with the analysis of Western culture that one finds in Friedrich Nietzsche (1844–1900), Karl Marx (1818–1883), as well as Weber himself. Nietzsche claimed that our culture is dominated by nihilism, and that the modern scientific outlook is rooted in the belief in God and in an absolute, objective perspective on the universe. But this illusion is actually rooted in certain values and interests, which Nietzsche believed are unhealthy, because they are symptomatic of a hatred for life, of diversity and change (and ultimately a fear of death). The belief in moral absolutes is part of this same cluster of values. Nietzsche believed that individuals should learn to affirm life and create their own values. He also felt society needs strong creative leaders, who would be artists and philosophers, and not scientists or technicians. They would create a set of cultural ideals and values, and society would not be democratic, but guided by this elite class of creative people. Life requires the dominance of healthy artistic illusions, on Nietzsche's view, since only art affirms life.

While Marx did believe in objective truth, he also believed it was necessary to see that all knowledge is rooted in human interests. It is only the interest that the downtrodden have in unmasking the ideological distortions of dominant views of knowledge, truth, and values—which served the economic interests of the ruling class—that make objective knowledge possible, because this was the only universal interest. (Hence, objective knowledge is not interest free; it is rooted in universal human interests.) So for Marx most of our beliefs and values are not true, but ideological underpinnings of the existing social order. We need to analyze and unmask the economic sources of these forms of "false consciousness" until we achieve a socialist society, where the gap between appearance and reality will be overcome, thereby making it possible to dispense with many forms of theory, in favor of a science that would give a free people the information they need to carry on their cooperative life activities.

Weber, who was strongly influenced by Marx and Nietzsche, was quite pessimistic about the dominance of technological rationality and the subjectivity of values in the modern world. But he was also very fatalistic about the possibilities of overcoming "the iron cage" in which these two factors put us. More recently, Martin Heidegger (1889–1976) has claimed that modern science is a manifestion of the will to power, our drive to dominate and control the world. Modern science is essentially technological. Moreover, Heidegger (influenced by Husserl, whom he worked under) argues against the objectivist idea that science gives us a privileged access to truth, knowledge, rational methods, and values. This is a prejudice, according to Heidegger, and is really a manifestation of nihilism. Another group of recent authors, notably Michel Foucault (1927–1984) and Jacques Derrida (b. 1930) combine insights from Nietzsche, Marx, Freud, and Heidegger to argue that Western science is essentially rooted in the will to power, which leads to dominating political practices. Many Marxists, e.g., Herbert Marcuse (1898–1979), had already combined these insights in the thirties to show that the ideal of dominating nature inevitably leads to the domination of human beings, through the development of the human sciences and the transformation of ethics and politics into branches of technology, dominated by capitalism and the model of cost- benefit analysis that is controlled by "experts."[2]

The upshot of all these analyses was a growing cultural pessimism, a movement away from science and humanistic-liberal ideals, an emphasis on rebellion and on art as a source of liberation; in short, a distrust of the entire Enlightenment. All of these writers believed that the Enlightenment project of enhancing human life through the application of more and more scientific knowledge to society had in fact resulted in just the reverse of enlightenment and liberation. Moreover, knowledge itself—especially modern natural and human science—was rooted in power relations, capitalism, the will to power, and not (arguably) in value-free objective methods and results. Hence, they believed that the Enlightenment had to fail, and thus must be completely rejected. It is one of Habermas's goals to show that this analysis is a mistake, even though he agrees that the liberating potential of the Enlightenment has not yet been fully realized.

IV. HABERMAS

In the remainder of this paper I want to consider the views of Jurgen Habermas (b. 1929), a contemporary German philosopher and social theorist, because he not only provides a systematic interpretation and criticism of

Weber, but also develops an important alternative picture of science, culture, and values that provides a more positive assessment of science than do the views of Nietzsche, Heidegger, Foucault, and Derrida. To some extent he agrees with their criticisms of modern society, but thinks it is too one-sided and overly pessimistic. Since Habermas's views are very complex, I shall summarize his main thoughts in a very brief and general way.[3]

In his earlier writings[4] Habermas develops the following criticism of Weber: Weber's views combine "scientism" and "decisionism." "Scientism" is the doctrine that only the results and methods of the physical and natural sciences are valid and rational. There is only one type of knowledge and one rational method. All other disciplines and practices must either be assimilated to the doctrines of the natural and physical sciences and their methods (of prediction, control, and technological rationality) or else be dismissed as irrational and not knowledge. "Decisionism" is the view that since "categorical" value judgments can be neither reduced to scientific knowledge nor proved by science, and therefore not by scientific method, they are the expressions of personal and arbitrary decisions. Scientism and decisionism are two sides of the same coin, according to Habermas.

Habermas believes that Weber's views, which are representative of most of Western thinking since the time of Galileo, are an expression of how scientism and decisionism are related. What Habermas does, in effect, is to provide an alternative conception of knowledge and values that undermines these two views, while also providing an argument that the pessimism of Weber and others confuses the fact that the Enlightenment project has not been realized with the false claim that it cannot be realized and must be abandoned in favor of an aesthetic model of culture, politics, and human existence.

In his book, *Knowledge and Human Interests*,[5] Habermas, who shares the idea that knowledge is rooted in universal human interests, and is not interest-free, isolates three perennial human interests. (These are historically necessary features of human life; Habermas does not try to defend the Platonic vision outlined in the Introduction to this part.) These are the "technical," the "practical," and "the emancipatory" interests. The "technical" interest relates to the human need to control nature for survival. "Labor" is the form of activity that fulfills this need. Modern science and cost-benefit rationality best serve this interest by allowing us to predict and control nature in rational ways. The practical interest is the interest in human communication, interaction, and a common life. This interest in "interaction" requires us to understand human behavior by adopting something like the methods advocated by Taylor in his essay in part 2 of the present volume. The human sciences cannot model themselves on the physical and natural sciences, but must find ways of understanding and interpreting human communication and interaction. They must be "hermeneutical" or "interpretive" (more like a dialogue than an experiment) and must allow us to enhance our capacity for human communication. Public life must also revitalize these ideals, which involves abandoning the notion that ethics and policies are branches of technology, and that only the experts in society have the right to participate in public debate and decision-making.[6] Finally, the emancipatory interest is the idea that knowledge should enhance our freedom and improve human life by emancipating us from oppressive forces: material, political, psychological, and ideological. "Critical theory" is that type of knowledge which best furthers this goal. In particular, it makes it possible to reflect about our basic values and beliefs, and to debate them rationally. (More about this later.)

Ideally, all three interests, and each type of knowledge—natural science, human science, critical theory—should do their proper job and work in harmony with each other. But when scientism and decisionism dominate, this does not happen. Instead, practical and emancipatory interests are reduced to the technical interest, i.e., all problems become technological problems. And all forms of knowledge are reduced to the knowledge of nature that we get from the physical and natural sciences. For Habermas, it is only the abuse of the physical and natural sciences and their technological model of rationality that causes our social problems. If left to their proper business, they would be most helpful and not dangerous. So Habermas does not want to reject or even criticize science or technical rationality, but only put them in their place, and foster the development of the other two human interests and their type of knowledge.

He therefore thinks that attempts to substitute an aesthetic for a "rational" model of life and culture—or the attempt to reject the technical interest or reduce it to the practical—which, on his view, Marcuse, Heidegger, Nietzsche, Foucault, and Derrida advocate, is just as wrong as the attempt to reduce the practical sphere to the technical domain.

One direct outgrowth of Habermas's alternative to scientism/decisionism is an attempt to revitalize the classic conception of politics.[7] The Greeks, especially Aristotle, distinguished between "technical" and "practical" questions, and believed there were three types of knowledge: theoretical, practical and productive. Theoretical knowledge, e.g., science, had no relevance to ethical and political life, which was the realm of practical knowledge and included public discussion by citizens and common deliberation about goals. Politics is not a branch of technology, but involves genuine democracy. Productive knowledge, e.g., ship building, is governed by "techne," i.e., instrumental reason, that accepts goals, and (unlike politics) does not deliberate about them. For the Greeks, these are separate activities and forms of knowledge. Political consensus arrived at by dialogue is rational. It is only by requiring that political consensus be based on scientific or technological methods that decisionism can gain a foothold, according to Habermas. We tend to confuse the practical and the technical; we conceive of politics, human interaction, and liberation as resulting from the application of theoretical knowledge-cost-benefit analysis to the "applied" sphere. This is a dangerous mistake, according to Habermas.

To sum up, the dominant attitudes about science, values, and culture that Weber's views represent is the result of the illegitimate dominance of one type of knowledge, one human value or interest, and one type of decision procedure, on modern society. By articulating the practical and emancipatory interests/values, and indicating how they are best fulfilled by practical knowledge and critical theory, Habermas claims to have undermined scientism/decisionism, preserved what is worthwhile about science and technology, refuted the idea that politics is a branch of technology, and provided a vision of democracy as based upon rational dialogue rather than cost-benefit analysis.

In his more recent writings, Habermas recasts his earlier views.[8] He also tries to show that Weber and cultural pessimists such as Marcuse, Nietzsche, and Heidegger confuse the fact that science and technological rationality have been abused and misused (which he agrees with) from the claim that science and technology are inherently dangerous and must lead to all the ills of modern society. The Enlightenment still has a liberating potential.

This potential is for a society based upon the beneficial dimensions of science and technology, as well as the ideal of genuine democracy through citizen participation in dialogue and shared rational consensus. Politics is neither a branch of technology nor a tug of war between competing arbitrary values, but a rational dialogue that is rooted in the classic notions of deliberation and practical knowledge. A genuine dialogue can only take place when the material and ideological obstacles to human communication and cooperation are eliminated. In an ideal situation, whenever an ultimate or categorical moral or political value or issue arose, people would consider only the relevant arguments and evidence for and against, in terms of aiming at what can give rise to a rational, open consensus. Everybody would have an equal opportunity to participate in the dialogue, which should culminate in an agreement that is free, rational, and in the general interest.

Habermas now talks about the two main components of modern society: the "Systems Sphere" of science, technology, corporate capitalism, and bureaucracy, and the "Cultural Sphere"—the realm of public and private life, morality, culture, and human interaction. The systems sphere has "colonized" the cultural sphere (the "lifeworld"). We must learn how to harmonize and integrate developments in the systems sphere and the lifeworld in order to achieve a world in which our knowledge can be used for truly liberating purposes in the cultural sphere as well as the systems sphere. We must, in other words, learn to integrate science and technology with our best political and moral ideals, which are continuously evolving. Habermas develops a model of human evolution as a collective learning device, and postulates the idea of selective evolution to explain the disparity between our growing scientific and technological knowledge and our relative lack of progress in the social and political spheres. We have not yet evolved to the point where we have in place genuinely democratic ways of achieving rational consensus, owing to the dominance of the technological and subjectivist views of values and politics. This is his answer to Weber. So our present task involves a theory of the evolution of society—and the differential development of our scientific knowledge and moral ideals—which allows us to measure the disparity between developments in the systems sphere and the lifeworld, while providing a model for their integration and for moral/ political progress.

Habermas accepts the idea that moral progress involves the ability to take a universal and impartial view about moral and political issues. A dialogue about basic moral issues permits people to reflect about basic values, including their own, by subjecting them to criticism and correction. Only basic values that can be accepted by all partners to an open, rational discussion are acceptable. On this view, basic values can be rationally discussed, corrected, and moved toward more universal acceptance, provided the political factors necessary to promote genuine dialogue exist.

NOTES

1. A good deal of my discussion, especially in this section, summarizes the material in the Introduction to part 6 in this volume.

2. See here, William Leiss, *The Domination of Nature*, (Boston: Beacon, 1972).

3. See here, Thomas McCarthy, *The Critical Theory of Jürgen Habermas*, (Cambridge, Mass.: MIT Press, 1978).

4. Jürgen Habermas, *Knowledge and Human Interests*, (Boston: Beacon, 1971). *Toward a Rational Society*, (Boston: Beacon, 1970). *Theory and Practice*, (Boston: Beacon, 1973).

5. See especially *Knowledge and Human Interests*.

6. See especially *Toward a Rational Society*.

7. See especially *Theory and Practice*. Cf. McCarthy, op. cit., ch. 1.

8. See especially Jürgen Habermas. *Theory of Communicative Action*, vol. 1. (Boston: Beacon, 1984; vol. 2, 1987). Jürgen Habermas. *Communication and the Evolution of Society* (Boston: Beacon, 1979).

33

The Feminist Question in the Philosophy of Science

Ronald Giere

INTRODUCTION

My title is a reflection of Sandra Harding's *The Science Question in Feminism* (1986). Her science question in feminism is this: Feminist claims of masculine bias in science are often themselves based on scientific studies, particularly the findings of various social sciences. But if the claims or methods of science are in general as suspect as many feminists claim, then appeals to scientific findings to support charges of bias are undercut. In short, is it possible simultaneously to appeal to the authority of science while issuing general challenges to that same authority?

My feminism question in the philosophy of science is this: To what extent is it possible to incorporate feminist claims about science within the philosophy of science? Are feminist claims about science compatible with a philosophy of science that rejects relativism? Are they compatible with a philosophy of science that embraces realism? In short, how seriously should philosophers of science, in general, take the claims of feminists that the philosophy of science should incorporate feminist claims about science? The answer to my question, of course, depends both on what feminist claims one considers and on one's conception of the philosophy of science.

From the standpoint of the philosophy of science, the most significant claim of feminist scholars is that the very *content* of accepted theory in many areas of science reveals the gender bias of the mostly male scientists who created it. Moreover, the theories in question came to be accepted through the application of accepted methodological practices. So the sciences and scientists involved cannot be written off as obviously biased or otherwise marginal. Thus, gender bias in the content of accepted science is both possible and, in some cases, actual.

CASE STUDIES

An appropriate starting point for an examination of feminist critiques of science is with the many case studies of actual scientific research purporting to demonstrate masculine bias in the results of what had been regarded as clear cases of acceptable scientific practice. Investigating such cases, however, is much more difficult than one might think. Before explaining why, I will provide a rough taxonomy of cases and mention a few examples.

The most convincing cases are those in which the subject matter of the science consists either of real human beings or higher mammals, and the theories in question focus on aspects of life in which gender is obviously a variable. This includes parts of many sciences such as anthropology, sociology, ethology, and primate evolution. Standard examples of these sorts of cases include theories of human evolution based on a model of "man the hunter." According to these theories, the evolution from higher primates to humans was driven by selective forces operating in small groups of male hunters. The use of tools, the development of language, and particularly human forms of social organization, have all been claimed to have evolved in the context of hunting by males. This theory has been the standard theory in many fields for several generations. This approach was not seriously challenged until women entered these fields in more than token numbers and began developing an alternative model of "woman the gatherer." These women have argued that gathering and elementary agriculture likewise require complex skills, social organization, communication, and the development of basic tools. And, they argue, the evidence for this theory is at least as good as that for the standard "man the hunter" paradigm. The lesson drawn is that the "man the hunter" account was the accepted theory for so long at least in part because it was developed and sustained by scientific communities dominated by men with masculine values and experiences. Developing a plausible rival required women with female values and experiences.[1] The investigations of Longino and Doell (Longino, 1990, chs. 6 and 7) into theories of the biological origin of sex differences in humans provides another outstanding example of this type of case.

A second category consists of cases in which the subjects are humans or primates, but the theories are not directly about obviously gendered aspects of their lives. Here a good example comes from the field of psychological and moral development. The standard theories for most of the twentieth century were those developed by Freud, Erikson, and Kohlberg. These theories purported to be theories of "human development" but were in fact based primarily on studies of boys and men. When studies of girls and women were made, observed differences were treated as "deviations" from the established norm, or even as evidence of failure by girls to reach the higher stages of development. A contrary view emerged in the 1970s through the work of female psychologists such as Carol Gilligan as reported in her now classic book *In a Different Voice* (1982). Gilligan studied moral development in both men and women, but concentrated on women. Her conclusion was that women are neither deviant nor lagging in their moral development, just different. The lesson is the same as in the "man the hunter" model.

A third category of cases involves living, but nonmammalian subjects, and theories in which sex is not a salient variable. A good example here is Barbara McClintock's work on genetic transposition as interpreted by Evelyn Fox Keller in her 1983 book, *A Feeling for the Organism*. Keller argues that McClintock approached her subject with values and interests that were connected with the fact that she was not a man in a profession dominated by men. McClintock had an appreciation for complexity, diversity, and individuality, and an interest in functional organization and development, which was at variance with the desire for simple mechanical structures that motivated most of her male colleagues. That, according to Keller, explains both why McClintock was able to make the discoveries she did, and why her mostly male colleagues failed for so long to understand or appreciate what she had done.

The fourth and most difficult category for the feminist critique involves nonliving subjects, and theories that obviously do not explicitly incorporate gender as a relevant variable. This includes sciences from molecular biology to high-energy physics. Here Keller (1985, 1992, 1995) and a few others have argued that the influence of gender can be seen in the metaphors that, they claim, both motivate and give meaning to the theories that are generally accepted. DNA, for example, is thought of as a kind of genetic control center issuing orders along a hierarchical chain of command—a clearly male, military, or corporate, metaphor.

For any of these cases to be effective as a critique of science, one must maintain *both* that they exhibit a clear masculine bias and that they nevertheless constitute examples of acceptable scientific practice. To dismiss the cases, therefore, one can argue either that the case for masculine bias is not sufficiently substantiated, or that bias does exist, but the cases are not acceptable science. The power of the antifeminist position lies in the fact that one can use the argument for gender bias as itself grounds for concluding that the case is one of bad science, thus undercutting the feminist critique. And this strategy is likely to be most successful in the examples where the *prima facie* case for masculine bias seems strongest. Suspicion of the scientific credibility of such "soft" sciences as anthropology and cognitive development long antedated feminist critiques of theories in these fields.

I believe that a credible case for the feminist position has been made in at least some of these examples, but this claim can only be substantiated by a detailed examination of the cases themselves. So, rather than engage the debate at this level, I will shift my attention to the question whether it is *theoretically* possible that the feminist conclusion is correct. Could there be gender bias in what by all other criteria must count as good science? There is a rhetorical as well as a theoretical reason for raising this question. Many philosophers and philosophers of science simply do not regard it as theoretically possible that the feminist critique could be correct. For these philosophers, looking carefully at the cases is merely an academic exercise. To be convinced, therefore, that it is worth even considering the implications of the feminist critique for the philosophy of science, one must first be convinced that it is at least theoretically possible that the critique is correct. That is what I hope to do here—make a convincing case that it is theoretically possible.

SOME SOURCES OF THE ANTIFEMINIST POSITION

I will consider several sources of the assumption that the feminist position is theoretically impossible. If it can be shown that the antifeminist position rests on inadequate foundations, that would undercut the assumption that the feminist position is theoretically impossible.

One source is the Enlightenment ideal of science. The cornerstone of the Enlightenment ideal is the view that the ability to acquire genuine knowledge of the world is independent of personal virtue or social position. Popes and kings, bishops and knights, have no special access to genuine knowledge. What matters is the correct employment of natural reason, and that is, in principle, within the grasp of any normal person. The irrelevance of gender was presumed, although too often because women were deemed not capable of exercising the powers of natural reason. In the present-day philosophical canon, most of the thinkers between Descartes and Kant held an enlightenment picture of science, even if, like Descartes, they were precursors rather than participants in the Enlightenment as such. To a large extent, much of contemporary philosophy simply presupposes this Enlightenment ideal. And that at least partly explains why so many contemporary philosophers and philosophers of science find it simply impossible that gender might matter for what counts as legitimate scientific knowledge.

Feminists, not surprisingly, tend to take a dim view of the Enlightenment. I would urge a middle ground, insisting that the Enlightenment was a genuine advance over what came before, but recognizing that its presumption of the gender neutrality of human reason was *merely* a presumption, and not based on any firm grounds, particularly not the sorts of empirical investigations now common in the cognitive and social sciences. But I do not want to dwell on the enlightenment. There are sources much closer to our own time for the view that the feminist critique could not possibly be correct.

The current configuration of views within philosophy of science in the United States derives mainly from European sources transmitted by refugees displaced by World War II. For the most part, these influential refugees were

German speaking members of a loosely knit group advocating a scientific philosophy, a *"Wissenschaftliche Weltauffassung."* These thinkers were repelled by the various neo-Kantian idealisms then dominant within German philosophy, and in German intellectual life generally. And they were simultaneously inspired by the new physics associated above all others with the work of Einstein.

In a nutshell, the position of the scientific philosophers was that to understand the nature of fundamental categories like space and time, one should look to Einstein's relativity theory, not to the a priori theorizing of neo-Kantian philosophers. Similarly, to understand the nature of causality, one should look at the new quantum theory. Their program was a radical program, a program to *replace* much of philosophy as it was generally practiced in Germany with a new scientific philosophy. It is thus not surprising that none of these philosophers occupied positions of great influence, whether intellectual or institutional, within the German speaking philosophical world.

The most prominent at the time was Moritz Schlick, Professor of Philosophy at the University of Vienna. But he was not really part of the Viennese philosophical establishment. The chair he held had earlier belonged to Ernst Mach, a philosopher-scientist of radical empiricist persuasions. Schlick himself was murdered by a former student, under somewhat shadowy circumstances, in 1936.

Before his death, however, Schlick had provided both philosophical inspiration and institutional support for the Vienna Circle. It was he who, in 1926, brought the young Rudolf Carnap to Vienna as an instructor in philosophy. And it was Schlick who maintained contact with Wittgenstein, who had his own program for a philosophy to end all philosophies. But it was Carnap who became the intellectual leader of the Vienna Circle, a heterogeneous group of mathematicians, natural scientists, social scientists, and scientifically trained philosophers like himself.

The 1920s and early 1930s were disquieting times in Germany and Austria. Political life, often played out in the streets, was fractured left and right. The threat of anarchy ended abruptly on January 30, 1933 when Hitler came to power in Germany. The scientific philosophers were overwhelmingly internationalist in outlook; liberal, socialist, or even communist in political orientation; and many were Jews. For such people, life in Germany, and even in Austria, became increasingly difficult.

Most prominent among the scientific philosophers outside of Vienna was Hans Reichenbach in Berlin. While a student of physics and mathematics in the teens, Reichenbach was active in socialist student movements. That ended when he began teaching science and mathematics in various *Technische Hochschule*. He also began publishing logical-philosophical analyses of Einstein's theory of relativity. In 1927, Einstein, together with Planck and von Laue, arranged for Reichenbach to be offered an untenured position in the physics department at the University of Berlin, the pinnacle of German, and, at that time, world, physics. The philosophers in Berlin voted not to admit Reichenbach as a member of their department, but Einstein, at least initially, welcomed his help in carrying on his own intellectual battles with the neo-Kantians over the nature of space, time, and causality. Reichenbach relished the role.

With the imposition of the Nazi racial laws in the spring of 1933, Reichenbach, along with hundreds of other German professors, was dismissed from his post. Einstein, having resigned from abroad, found safe haven at the newly created Institute for Advanced Study in Princeton. Reichenbach was among fifty or so former German professors who accepted five-year contracts at the University of Istanbul. This was part of Kemal Ataturk's effort to bring Turkey into the modern world. Even before his call to Berlin, Reichenbach had been exploring the possibility of emigrating to the United States. Now he resumed these efforts in earnest. As part of his plan to find a position in the United States, he put aside his technical work both on relativity and on the theory of probability, and began writing, in English, a general work on scientific epistemology. That work, *Experience and Prediction*, was completed in 1937 and published by the University of Chicago Press in 1938—the year Reichenbach began his tenure at UCLA.

In the very first section of that book, titled "The Three Tasks of Epistemology," Reichenbach introduces his distinction between "the context of discovery" and "the context of justification," remarking that "epistemology is only occupied in constructing the context of justification" (p. 7). The introduction of the distinction is not the conclusion of any argument. It is a precondition for the analysis to follow. In fact, this distinction, though of course not in these words, had existed in German philosophy for half a century. But this seems to be the first time it appeared in Reichenbach's writings. It reappears only once in *Experience and Prediction*, near the end of the final chapter on probability and induction, where he writes (p. 382):

> What we wish to point out with our theory of induction is the logical relation of the new theory to the known facts. We do not insist that the discovery of the new theory is performed by a reflection of a kind similar to our expositions; we do not maintain anything about the question of how it is performed—what we maintain is nothing but a relation of a theory to facts, *independent of the man who found the theory*. There must be some definite relation of this kind, or there would be nothing to be discovered by the man of science. Why was Einstein's theory of gravitation a great discovery, even before it was confirmed by astronomical observations? Because Einstein saw—as his predecessors had not seen—that the known facts indicate such a theory. . . . (emphasis added).

Here I wish to indulge in a bit of historical speculation. The speculation is this: When Reichenbach writes of "a relation of theory to facts, independent of the man who found the theory," he is thinking primarily of Einstein, whose views were vilified in the Nazi press not because of any lack of a proper logical relation between Einstein's theories and the facts, but simply because of a personal fact about the man with whom those theories originated—he was a Jew. Reichenbach's own personal situation differed from Einstein's mainly in that his accomplishments and, consequently, his reputation, were less exalted.[2]

One can now see a clear connection between contemporary feminist critiques of science and Reichenbach's use of the distinction between discovery and justification. Reichenbach, I believe, made it a precondition for doing scientific epistemology that the very notion of "Jewish science" be philosophically inadmissible. The Nazi racial laws were not only a crime against humanity, they were a crime against philosophical principle. The feminist notion of "masculine science," or any sort of gendered science, is not a principle any different. It makes the epistemological status of a scientific theory dependent on facts about the scientists themselves, as historical persons, quite apart from internal, logical relations between fact and theory.

Even if I am mistaken about the personal motivation behind Reichenbach's use of a then well-known distinction in his first general epistemological work, there is no doubt that his understanding of the distinction rules out the relevance of gender to any philosophically correct understanding of legitimate scientific knowledge. Moreover, this understanding of the task of scientific epistemology was shared by most of the European scientific philosophers. And it was these philosophers who came to dominate philosophical thought about science in the United States in the postwar period.

One might object that this is all just so much history of the philosophy of science. Where are the arguments? I hope it is clear that this response begs the question at issue. The validity of the discovery-justification distinction was not established by argument. It was, as is clear in Reichenbach's book, part of the initial statement of the task of a scientific epistemology. It is part of that conception of scientific epistemology that gender or other cultural factors cannot possibly play any role in establishing the legitimacy of scientific claims. My "argument" has been that it is to a large extent due to the legacy of those whose conception of the philosophy of science was formed in the war against Nazi power and ideology that the idea of gendered science still seems to many as being simply impossible.

The point of my historical remarks can be put more sharply. The insistence on the irrelevance of origins which has characterized logical empiricism in America is refuted by the history of that movement itself. The prominence of many doctrines, like the discovery-justification distinction, was not the result of argument, but an assumption forming the conceptual context within which arguments were formulated. The only way to understand why those doctrines were held is to inquire into the historical origins of their role in that movement. Indeed, it is a revealing irony that later criticisms of the discovery-justification distinction focused exclusively on its validity or usefulness, not on its origins.

THE POSSIBILITY OF GENDER BIAS IN POSTPOSITIVIST PHILOSOPHY OF SCIENCE

The contemporary feminist movement in America has its own roots in the civil rights movement and the antiwar movement of the 1960s. That was a different war, a different generation, and a different set of political circumstances. The major influence on the philosophy of science of that decade was Thomas Kuhn's *The Structure of Scientific Revolutions*. Kuhn clearly did not set out to become a hero of the 1960s cultural revolution. Nor could one who wrote so unselfconsciously about "the man of science" have been promoting a feminist agenda. Yet his work has, I think correctly (e.g., by Keller, 1985), been seen as providing support for the possibility of gendered science.

In Kuhn's book, the distinction between "the context of discovery" and "the context of justification," in just those words, appears again in the very first chapter. Here, however, Kuhn himself remarks that the distinction seems not to have been the result of any investigation into the nature of science. Rather, he claims, it was part of a framework within which the study of science had been carried out. He makes clear that his own inquiry does not presuppose any such distinction. And, indeed, Kuhn's own theory of science, with its emphasis on the role of individual judgment exercised by scientists in communities, yields nothing that would rule out the possible influence of gender on the eventual beliefs of a typical scientific community.

In the philosophical profession at large, it is widely believed that Kuhn was part of a historical turn in the philosophy of science which superseded logical empiricism. That belief is mistaken on at least two counts. First, the historical tradition within the philosophy of science did not supersede logical empiricism. It was, rather, a rival philosophical tradition which emerged around 1960 and was in part stimulated by Kuhn's work. Logical empiricism continued to evolve both in terms of the study of particular scientific theories and in terms of general methodological inquiries. Both sorts of developments are exemplified, for example, in the works of Bas van Fraassen (1980, 1989, 1991). Second, Kuhn himself was only marginally a part of the historical tradition within the philosophy of science. Most of the philosophers of science associated with that tradition, including Paul Feyerabend, N. R. Hanson, Imre Lakatos, Larry Laudan, Ernan McMullin, Dudley Shapere, and Stephen Toulmin, shared Kuhn's rejection of logical empiricism. And they agreed with his focus on scientific development as the central notion for the study of science. But, for the most part, they also shared a rejection of Kuhn's own theory of science.

With the obvious exception of Feyerabend, these historically oriented philosophers of science sought not to reject the Logical Empiricist idea of an objective connection between data and theory, but to replace the idea of a *logical connection* between data and theory with that of *rational progress* within a research tradition. This shift is clearest in the case of Lakatos. For Lakatos, a research program is progressive to the extent that it generates successful novel predictions yielding new confirmed empirical content. There appears to be no room in this definition for any influence from cultural variables such as gender. I will now argue that the apparent impossibility of gender bias in postpositivist philosophical theories of rational progress is only apparent. It is possible even on Lakatos's hard-line account.

One of the many lessons Kuhn claimed to have learned from his study of the history of science was that scientists rarely abandon a research tradition unless they first can at least imagine a promising alternative. Both Lakatos and Laudan explicitly adopted this idea, arguing that the evaluation of a research tradition is not based on a two-place relationship between data and a theory, but on a three-place relationship between data and at least two rival research traditions.

There cannot be many examples in the history of science where the existing rival research programs exhaust all the logical possibilities. So it is typically possible that the theories making up the existing rival research programs are in fact all false. Nevertheless, as Kuhn argued, and almost everyone else agreed, it is rare to have a scientific field in which there is no clearly favored research program. There is typically an establishment position. It follows that, at any particular time, which research program is most progressive by any proposed criteria depends on which of the logically possible research programs are among the actually existing rival programs. Against other logically possible rivals, the current favorite might not have fared so well. Moreover, Lakatos and Laudan, but most others as well, retain a distinction between discovery and justification to the extent that their accounts of rational progress place few if any constraints on how a possible research program comes to be an active contender. There is little to rule out this process being driven by gender bias or any other cultural value.

So, for any leading research program, it is possible that its position as the current leading contender is in part a result of gender or some other cultural bias. If these biases had been different, other programs might have been considered, and a different program might have turned out to be comparatively more progressive at the time in question. In short, the fact that a given program is judged normatively most progressive by stated criteria might possibly be due, at least in part, to the operation of gender biases in the overall process of scientific inquiry. And that is enough to establish the possibility that the feminist critique is correct in at least some cases.

A POPPERIAN RESPONSE

My earlier survey of leading scientific philosophers omitted any mention of Karl Popper. That was deliberate, because, as I see it, Popper had little influence on what became logical empiricism, particularly in America, until after publication of the 1959 English edition of his 1935 monograph, *Logic der Forschung*, under the even more misleading title, *The Logic of Scientific Discovery*. Despite his own claims that it was he who killed positivism (1974), the accidental fact that the English edition of Popper's book appeared shortly before Kuhn's put him in a position to become a primary defender of the positivist faith against the Kuhnian heresy.

The titles of Popper's book are misleading because, on his account of science, there is no such thing as a "logic" of research or of scientific discovery. The main role for logic in science is the use of *modus tollens* in the refutation of a universal generalization by a statement describing a negative instance. This form of inference requires no reference to alternative hypotheses. So, apart from questions about how one establishes the truth of the required singular "observation statement," this form of inference would seem to be immune to gender or any other cultural influences. Popper's work thus shows that it is possible to construct a theory of science which maintains a strong enough distinction between the contexts of discovery and justification to eliminate the possibility of gender bias. But it also shows how very difficult it is to construct a *good* theory of science that fulfills this requirement. No one better exhibited the shortcomings, not to say the utter implausibility, of Popper's theory of science than his successor, Imre Lakatos—and Lakatos borrowed heavily from Kuhn. It should be noted that the approaches to scientific justification taken by both Carnap and Reichenbach would, if successful, also eliminate any possibility for gender or other cultural biases. For both, theory evaluation is not comparative, at least not in any obvious way. I will not elaborate this point further because these approaches have few defenders today.

The successor to Carnap's conception of inductive logic is a subjective probability logic, as championed, for example, by Carnap's associate, Richard Jeffrey (1965). Theories of subjective probability, however, place only minimal constraints on how an individual assigns initial probabilities to any theory. This leaves lots of room for individual scientists to assign high initial probabilities to theories reflecting their own particular gender biases. The best the probabilistic approach can offer is proof of the diminishing influence of the initial probability assignment in the face of increasing observational evidence. But there is no way of knowing, in this framework, how much the probability assigned a particular theory at any given time might be the product of some form of bias, including gender bias. That leaves feminist critiques as much room as they need.

In sum, there is little in current philosophical theories of science that supports the widespread opinion that gender bias is impossible within the legitimate practice of science. That opinion seems mainly the product of a traditional adherence to an enlightenment ideal of science strongly reinforced by the historical origins of twentieth-century scientific philosophy in Europe and its rebirth as logical empiricism in America. As disquieting as it may seem to many, we shall have to learn to live with the empirical possibility of "Jewish science." That is, for any particular scientific theory, it must be an *empirical* question whether its acceptance as the best available account of nature might be due at least in part to its having been created and developed by Jewish scientists rather than scientists embodying some other religious tradition. In another cultural context in which science as we know it is generally practiced, some other theory might now be the accepted theory. Whether or not this is true for any *particular* theory can only be determined empirically by looking in detail at the history of how that theory achieved its present status. The irrelevance of religious origins cannot be guaranteed a priori. The same holds for gender.

PERSPECTIVAL REALISM

In countenancing the relevance of cultural forces in the acceptance of scientific theories, have we not moved too far in the direction of *relativism*? In particular, is this position compatible with a reasonable scientific realism? I think it is, but the issue is complex. If we suppose that the world is organized in a way that might be mirrored in a humanly constructable linguistic system, then there is indeed a problem. For then realism seems to require that we could have reason to believe that our theories are literally *true* of the world. The objects in the world are grouped as our theories say they are and behave as our theories say they should behave. If, however, what we *take to be* true of the world is influenced by cultural factors, there is no reason to think that this influence would promote the development of actually true theories and considerable reason to suspect that it would do just the opposite. That sounds like relativism, not realism.

Radical though it may seem, I think the resolution of this problem is to reject the usefulness of the notion of *truth* in understanding scientific realism. I do not mean that we cannot use an everyday notion of truth, as when asserting that it is indeed true that the earth is round. Here truth may be understood as no more than a device for linguistic ascent. Rather, it is the *analysis* of truth developed in the foundations of logic and mathematics, and used in formal semantics, that we should reject in our attempts to understand modern science. But if we reject the standard analyses of truth and reference, what resources have we left with which even to formulate claims of realism for science? The answer is that the notion of linguistic truth is but one form of the more general notion of *representation*. What realism requires is only that our theories well *represent* the world, not that they be true in some technical sense. So we need a notion of representation for science that does not rest on the usual analyses of truth for linguistic entities. What might that be?

A first step is to reject the analysis of scientific theories as sets of statements in favor of a model-based account which makes nonlinguistic models the main vehicles for representing the world, and places language in a supporting role.[3] We may, of course, use language to characterize our models, and what we say of the models is true. But this is merely the truth of definition, and requires little analysis. The important representational relationship is something like *fit* between a model and the world. Unlike truth, fit is a more qualitative relationship, as clothes may be said to fit a person more or less well. Of course we can say it is true that the clothes fit, but this is again merely the everyday use of the notion of truth.

Here I can offer no general analysis of the notion of fit, only a further analogy—maps. There are many different kinds of maps: road maps, topological maps, subway maps, plat maps, et cetera. And it can hardly be denied that maps do genuinely represent at least some aspects of the world. How else can we explain their usefulness in finding one's way in otherwise unfamiliar territory? Moreover, the idea of mapping the world has long been present in science. There were star charts before there were world atlases, and scientists around the world are now busy "mapping" the human genome. Maps have many of the representational virtues we need for understanding how scientists represent the world. There is no such thing as a universal map. Neither does it make sense to question whether a map is true or false. The representational virtues of maps are different. A map may, for example, be more or less accurate, more or less detailed, of smaller or larger scale. Maps require a large background of human convention for their production and use. Without such they are no more than lines on paper. Nevertheless, maps do manage to correspond in various ways with the real world.

Since no map can include *every* feature of the terrain to be mapped, what determines which features are to be mapped, and to what degree of accuracy? Obviously these specifications cannot be read off the terrain itself. They must be imposed by the mapmakers. Presumably which set of specifications gets imposed is a function of the *interests* of the intended users of the maps.

Among cartographers, those whose job it is to make maps, it is assumed that constructing a map requires a prior selection of features to be mapped. Another aspect of mapmaking emphasized by cartographers is *scale*, particularly for linear dimensions. How many units of length in the actual terrain are represented by one unit on the map? These two aspects of mapmaking, feature selection and scale, are related. The greater the scale the more features that can be represented. The required trade-offs again typically would reflect the interests of the intended users.

It is not stretching an analogy too far to say that the selection of scale and of features to be mapped determines the *perspective* from which a particular map represents the intended terrain. Photographs taken from different locations provide more literal examples of different perspectives on a terrain or a building. In any case, given a perspective in this sense, it is an empirical question whether a particular map successfully represents the intended terrain. If it does, we can reasonably claim a form of *realism* for the relationship between the map and the terrain mapped. I will call this form of realism *perspectival realism*.[4]

Standard analyses of reference and truth suggest a metaphysics in which the domain of interest consists of discrete objects grouped into sets defined by necessary and sufficient conditions. Likewise, there is a metaphysics suggested by perspectival realism. Rather than thinking of the world as packaged into sets of objects sharing definite properties, perspectival realism presents it as highly complex and exhibiting many qualities that at least appear to vary continuously. One might then construct maps that depict this world from various perspectives. In such a world, even a fairly successful realistic science might well contain individual concepts and relationships inspired by various cultural interests. It is possible, therefore, that our currently acceptable scientific theories embody cultural values and nevertheless possess many genuine representational virtues.

FEMINIST REALISM

There is an unfortunate mismatch in terminology between feminist and general philosophers of science. Within the philosophy of science generally, the distinction between empiricists and realists concerns the sort of epistemic commitment one has toward "unobservable" or "theoretical" entities and properties. Empiricists would restrict our commitments to the observable phenomena; realists make no such restrictions. "Feminist empiricist," on the other hand, characterizes someone who thinks some theories may embody gender biases, but also thinks such biases can be detected using standard scientific methods. Moreover, better theories, which may embody other biases, can be proposed and validated. Feminist empiricism, therefore, is neutral regarding the general debate between empiricists and realists. Of course a feminist empiricist might also be an empiricist in the more general sense, but that would be an additional commitment beyond feminist empiricism. More significantly for my purposes, a feminist empiricist could be a *realist* in the more general sense. Thus, feminist realism is not inherently an incoherent doctrine. Given current usage, it turns out, misleadingly, to be a special case of feminist empiricism.

To my knowledge, no feminist philosopher of science has claimed to be a feminist realist. I expect this is because realists have often claimed to know *the truth* about many things, or at least to be *rationally justified* in claiming such knowledge. Feminists, quite naturally, are suspicious of any such claims. From my point of view, this suspicion presupposes the mistaken view that realism must be understood in terms of truth in the standard philosophical sense. Abandoning this presupposition, one is free to adopt a perspectival account of realism which is far more congenial to the interests of feminists. Moreover, adopting perspectival realism does not commit one to any form of special scientific rationality. Perspectival realism is perfectly compatible with a thoroughgoing naturalism which appeals only to the naturally evolved cognitive capacities of human agents together with their historically developed cultural artifacts. It is to be expected that such agents would typically project their cultural values, including gender values, into the models they develop to explain phenomena in the world. And some of these models could be expected to end up part of established science. That is just what feminist philosophers of science have been claiming all along.

NOTES

1. For an overview and references on this topic see Longino (1990, 106–111).

2. I have developed these and related themes at greater length in Giere, 1996.

3. For further elaboration and references on model-based accounts of scientific theories see Giere, 1988.

4. I find inspiration for both this terminology and the concept in some works of Donna Haraway, particularly her paper, "Situated Knowledges: The Science Question in Feminism and the Privilege of Partial Perspective," reprinted in Haraway, 1991.

REFERENCES

Giere, R. N. *Explaining Science*. Chicago: University of Chicago Press, 1988.

———. "From Wissenschaftliche Philosophie to Philosophy of Science." In *Origins of Logical Empiricism*, edited by R. Giere and A. Richardson. Minnesota Studies in the Philosophy of Science, Vol. 26. Minneapolis: University of Minnesota Press, 1996.

Gilligan, C. *In a Different Voice*. Cambridge, Mass.: Harvard University Press, 1982.

Haraway, D. J. *Simians, Cyborgs, and Women*. New York: Routledge, 1991.

Harding, S. *The Science Question in Feminism*. Ithaca, N.Y.: Cornell University Press, 1986.

Jeffrey. R. C. *The Logic of Decision*. New York: McGraw-Hill, 1965; 2nd ed. Chicago: University of Chicago Press, 1983.

Keller, E. F. *A Feeling for the Organism*. New York: W. H. Freeman, 1983.

———. *Reflections on Gender and Science*. New Haven, Conn.: Yale University Press, 1985.

———. *Secrets of Life, Secrets of Death: Essays on Language, Gender, and Science*. New York: Routledge, 1992.

———. *Refiguring Life: Metaphors of Twentieth-Century Biology*. New York: Columbia University Press, 1995.

Kuhn, T. S. *The Structure of Scientific Revolutions.* Chicago: University of Chicago Press, 1962. 2nd ed. 1970.

Longino, H. E. *Science as Social Knowledge.* Princeton, N.J.: Princeton University Press, 1990.

Popper, K. R. *Logic der Forschung: Zur Erkenntnistheorie der Modernen Naturwissenschaft.* Wien: Springer Verlag, 1935.

———. *The Logic of Scientific Discovery.* London: Hutchinson, 1959.

———. "Intellectual Autobiography." In *The Philosophy of Karl Popper.* 2 vols., edited by R. A. Schilpp. La Salle: Open Court, 1974.

Reichenbach, H. *Experience and Prediction.* Chicago: University of Chicago Press, 1938.

van Fraassen, B. C. *The Scientific Image.* Oxford: Oxford University Press, 1980.

———. *Laws and Symmetry.* Oxford: Oxford University Press, 1989.

———. *Quantum Mechanics.* Oxford: Oxford University Press, 1991.

Many claims made by scientists of all sorts—psychologists, medical researchers, and others—have involved making generalizations about human beings based solely on the use of male subjects, or else have involved claims about, say, women, people of color, and gays and lesbians that seem questionable on scientific grounds, and also appear to be motivated by a political agenda of some sort.

a. Suppose the claim were that women who suffer PMS have a serious psychological disorder.

b. Suppose the claim is that women are naturally less successful than men in our society because they produce less testosterone after puberty.

c. Suppose the claim is that African Americans have less intelligence than Caucasians or Asians because they are genetically inferior.

d. Suppose the claim is that gays and lesbians are "abnormal" because they suffer some genetic defect.

1. Must all claims of these sorts involve bad science, or immoral science?

2. What if the claims were true? Would that imply any particular value judgment, policy, or course of action?

3. Suppose female psychologists claimed to prove that men (or white men, or heterosexual white men) all suffered from a genetic defect that made them prone to clinical depression. Would people object that this was bad or immoral science? Should they?

4. How would one go about arguing, from examples like these, that science is, generally speaking, gender biased (or racist, or homophobic) and a tool for sexism, racism, et cetera? How can we tell if these sorts of cases are the exceptions rather than the rule? Must good science make such cases impossible (as Giere seems to argue)? Explain.

5. Do these sorts of cases—you might want to find some of your own if you don't already know of any— tell us anything significant about the connections between science, values, and politics in today's society?

STUDY QUESTIONS FOR PART 6

1. What is the role of values in science for Rudner?

2. Why does Hempel claim that "categorical" values are unsupportable by science?

3. How do scientists help society make value judgements?

4. How does McMullin respond to Hempel's proposals?

5. Why does McMullin claim that values are a more basic element of the world than "facts"?

6. What are "epistemic values"?

7. How do epistemic values influence theory choice?

8. What does Habermas mean by "selective evolution"?

9. How does the theory of selective evolution explain the conflict between science and values?

10. How does Habermas's view of deliberation and dialogue democratize technology?

11. How does Habermas's idea of critique respond to Weber?

12. Why must the charge that science is gender biased be limited only to examples of good or acceptable science, according to Giere?

13. Why doesn't the Enlightenment rule out the possibility of gender bias in science?

14. Why doesn't positivism rule it out?

15. Why is the distinction between the context of discovery and the context of justification inadequate to make gender bias in science impossible?

16. Why doesn't Popper's view of science help?

17. In what ways is Kuhn's theory unhelpful in ruling out the possibility of gender science?

18. What is perspectival realism? How does it differ from usual construals of realism?

19. How might feminists use perspectival realism to deal with the problem of gender bias in science?

20. Does Giere adequately distinguish perspectival realism from relativism? Is this a problem?

SELECTED BIBLIOGRAPHY

[* Indicates good bibliographies]

1. Brown, R., ed. *Scientific Rationality: The Sociological Turn*. Boston: Reidel, 1984. [Themes from Barnes, Lauden, and Kuhn.]*

2. Brubaker, Rogers. *The Limits of Rationality*. Boston: Unwin, 1984. [Weber's views on science, values, and politics.]

3. Bryant, Christopher. *Positivism in Social Theory and Research*. New York: St. Martin's, 1985. [Values and methods in twentieth-century social science.]*

4. Calhoun, Craig. *Critical Social Science*. Cambridge: Blackwell, 1995. [An advanced survey of critical theory by a sociologist.]*

5. Duran, Jane. *Philosophies of Science/Feminist Theory*. Boulder,

Colo.: Westview, 1998. [Feminist assessments of standard approaches (Popper, Kuhn, etc.).]*

6. Erwin, Edward, Sidney Gendin, and Lowell Kleiman, eds. *Ethical Issues in Scientific Research*. New York: Garland, 1994. [Good essays on topics such as fraud in science.]

7. Farnham, Christie, ed. *The Impact of Feminist Research in the Academy*. Bloomington: Indiana University Press, 1987. [Survey essays on sociology, psychology, political science, etc.]*

8. Fuller, Steve. *Philosophy of Science and Its Discontents*. Rev. ed. New York: Guilford Press, 1992. [Fuller's first programmatic essay.]

9. ———. *Philosophy, Rhetoric, and the End of Knowledge*. Madison: University of Wisconsin, 1993. [Fuller, a leading proponent of STS, develops his latest version of it.]*

10. Geuss, Raymond. *The Idea of Critical Theory*. Cambridge: Cambridge University Press, 1980. [Very important but advanced.]

11. Gross, Paul, and Norm Levitt. *The Higher Superstition*. Baltimore, Md.: Johns Hopkins Press, 1994. [The first round in the science wars: critical of feminism, postmodernism, science studies, etc.)

12. Gross, Paul, and Norm Levitt, eds. *The Flight from Science and Reason*. Baltimore, Md.: Johns Hopkins Press, 1997. [Key readings in the science wars, against feminism and postmodernism]*

13. Harding, Sandra, ed. *Feminism and Methodology*. Bloomington: Indiana University Press, 1987.[Essays by feminists about issues in social sciences.]

14. ———. *The "Racial" Economy of Science*. Bloomington: Indiana University Press, 1994. [Essays about race and gender in scientific practice and ideology.]

15. ———. *The Science Question in Feminism*. Ithaca, N.Y.: Cornell University Press, 1986. [Influential essays.]

16. ———. *Whose Science? Whose Values?* Ithaca, N.Y.: Cornell University Press, 1991. [More recent essays by Harding.]

17. Hollinger, Robert. *Postmodernism and the Social Sciences*. Thousand Oaks, Calif.: Sage, 1994. [A historical and thematic study.]*

18. Lauden, Larry. *Science and Values*. Berkeley: University of California Press, 1984. [A notable discussion, excellent for classroom.]

19. ———. *Beyond Positivism and Relativism*. Boulder, Colo.: Westview, 1996. [A lucid critique of recent versions of relativism.]*

20. ———. *Science and Relativism*. Chicago: University of Chicago Press, 1990. [A lucid study, good for use as a text.]*

21. Lawson, Hilary, and Lisa Appinanesi, eds. *Dismantling Truth: Reality in the Post-Modern World*. New York: St. Martin's, 1989. [Accessible essays on postmodernism, science, and truth.]

22. Longino, Helen. *Science as Social Knowledge*. Princeton, N.J.:

Princeton University Press, 1990. [An important interpretation of science from a feminist standpoint.]*

23. Megill, Alan, ed. *Rethinking Objectivity*. Durham, N.C.: Duke University Press, 1994. [Original essays by feminists and others.]

24. Nelson, Lynn, ed. *Feminist Methodologies*. Boulder, Colo.: Westview Press, 1986. [A collection of original and interesting essays.]*

25. Nelson, Lynn Hinkinson, and Jack Nelson, eds. *Feminism, Science, and Philosophy of Science*. Amsterdam: Kluwer, 1996. [Good essays.]*

26. Pickering, Andrew, ed. *Science as Practice and Culture*. Chicago: University of Chicago Press, 1992. [Recent essays in the sociology of science and science studies.]*

27. Proctor, Robert. *Value-Free Science?* Cambridge, Mass.: Harvard University Press, 1991. [A historical and sociological discussion.]*

28. Rouse, Joseph. *Knowledge and Power*. Ithaca, N.Y.: Cornell University Press, 1987. [An important book on science, knowledge, politics.]*

29. Tuana, Nancy, ed. *Feminism and Science*. Bloomington: Indiana University Press, 1991. [Important feminist essays on natural and social sciences, medicine, etc.]*

30. Weber, Max. *The Methodology of the Social Sciences*. Oxford: Oxford University Press, 1949. [A classic work. Very advanced.]

Appendix

Advice for Instructors

This book is intended for undergraduates who have at least a high school background in science, but little or no background in philosophy. The readings have been selected to (1) provide an introductory-level survey of some perennial issues in philosophy of science, (2) illustrate philosophical reasoning, and (3) serve as a catalyst for class discussion. Each section begins with a brief introduction, followed by a seminal article expressing a fundamental perspective and reactions to it. The sections conclude with a case study, study questions, and a select bibliography of further and more current literature on the same topic. The organization of this book reflects a deep-seated conviction on our part that an introductory philosophy text should present students with an accessible introductory-level discussion of one view on a particular topic followed by reactions to that viewpoint. It is an approach that impresses on the reader how intelligent authors engaged in the same topic can reach different conclusions, in contrast to other texts that focus on the development of one particular viewpoint. We believe that our approach is superior, because it illustrates the dialectic approach of philosophy while inviting the reader to reach his or her own conclusions about a particular topic.

Finding journal articles or book selections that are appropriate for an introductory course, are self-contained, refer to the same topic (and preferably each other), and espouse different views, while being brief at the same time is the difficult task successive editions of this book have attempted to improve upon. Part of this difficulty reflects the nature of scholarly writing: academics write for fellow academics, and generally on topics that are too specific or specialized for an introductory course. And while some instructors may lament our decision to focus on traditional topics of philosophy of science rather than trendier issues, our experience has been that undergraduates in an introductory course are more receptive to discussions about, say, how to distinguish science from pseudoscience than more faddish issues that presuppose a background in more fundamental topics.

For this book to be an effective teaching tool, there is no substitute for students reading the articles before they meet in class to discuss them. Learning is an active process, one that requires instructors to engage students rather than treat them as passive observers who take notes on an instructor's predigested version of the course material. Some of your students will recognize this from the start, but realistically speaking, most will not. Part of this may reflect the absence of proper incentives—if students do not see the rewards of coming prepared for class (i.e., how it will affect their grade), they will understandably direct their efforts to those activities that do. Part of this most assuredly also reflects contemporary undergraduate students' lack of basic reading skills. Many will find the task of reading the articles outside of class to be a daunting and utterly foreign activity. One way to confront these difficulties is to make reading the articles a more structured activity using handouts like the ones provided at the end of this essay. The handout clarifies what students should be getting out of the readings, and can be used as the basis for in-class discussions. We suggest that students be required to turn in completed handouts for each article at the end of class discussion for credit, and that they receive more than just a check or check plus for the work they do.

The background essays for each section are brief, giving readers only the basics they need to make sense of the selections. We have intentionally avoided summarizing the articles in the background essays, as doing so would invariably lead to students' relying on the background readings as a crutch rather than doing the work of reading and making sense of the articles. If you would like a more historically oriented course, consider using an additional text, such as Salmon et al. (1992).

Part of any philosophy course should involve practice writing and defending philosophical positions. In addition to the discussion handout mentioned above, we suggest instructors also have students write a term paper in connection with the course. Kuhn's *Structure of Scientific Revolutions* is an obvious candidate for study, as many of the issues developed in his famous work arise repeatedly and in different contexts throughout the text. Each part concludes with numerous additional references as well. Seech (1995) provides an excellent introduction on how to write philosophical papers for students with little background in philosophy.

Following the discussion handout are examples of a few suggested reading lists for instructors who wish to depart from the order of the articles presented in the text.

D. W. R.

REFERENCES

Kuhn, T. S. (1997). *The Structure of Scientific Revolutions*. 3d ed. Chicago: University of Chicago Press.

Salmon et al. (1992). *Introduction to the Philosophy of Science*. Englewood Cliffs, N.J.: Prentice-Hall. (S)

Seech, Z. (1995). *Writing Philosophy Papers*. 2d ed. Belmont, Calif.: Wadsworth Publishing Co.

SAMPLE HANDOUT AND ASSOCIATED WORKSHEET

Reading Philosophical Articles

Much, if not most, of the learning you do in connection with this course will be the product of the work you do alone and in groups on assignments outside of class. The following handout has been developed to structure your learning and help you make sense of the articles you will be asked to read in conjunction with this course. The handout also contains suggestions on how to complete part I of a companion worksheet containing specific questions about a particular article.

What is a philosophical article?

For the purpose of our course, philosophical articles are academic articles written by professionals formally addressing a philosophical topic, such as the morality of euthanasia. Many works in literature address philosophical topics as well. What makes an article distinctively philosophical, however, is the presence of a sustained argument in favor of a particular position (the author's thesis or main conclusion).

Some suggestions on how to read a philosophical article

The first goal of reading any philosophical article is to identify what the author's main conclusion is and what argument(s) he/she uses in favor of that conclusion. It's only when you understand the author's views that you are in a position to evaluate the argument and assess whether you agree or disagree with the author and determine which parts of the article are persuasive and which are not.

  Surprisingly, the best way to approach an academic article for the first time is to skim the whole before actually reading it. Quickly reviewing the text will give you an idea of the overall structure of the paper, the author's writing style, and often some idea of how the author argues for a particular point of view. The most important thing to do is locate the author's main conclusion, as it is (usually) the main thesis that each of the various parts of the article support.

  As you start to read the article, start keeping a list of terms and how the author defines or uses them. The strength of the author's argument as a whole often depends upon the extent to which he/she can win the assent of the reader to particular definitions of specific terms. Many arguments about the morality of abortion, to give you one example, depend heavily upon a specific definition of personhood. Sometimes during the course of the article the author will redefine a term, and if so, it's important to keep track of what the author's motivations are for adopting a different definition.

You should also keep a list of the evidence the author uses in support of the main conclusion. Making a list as you go along will greatly assist you in identifying the overall argument the author puts forth for his/her thesis. As you might expect, each piece of the author's argument in favor of the main conclusion must itself be justified and this often involves the use of subarguments for particular claims supporting the main conclusion.

Articles often contain other persuasive elements, which, while not part of the formal argument, can nevertheless work in the author's favor. An author's writing style, for instance, can serve to present the issues in such a way as to suggest that the author has truly considered all serious objections to his/her position. Vivid or striking examples can be so compelling as to win the reader's assent (e.g., photographs of fetuses at ten weeks are often used by antiabortionists). Almost all academic writing has some rhetorical elements—for our purposes, it's important to recognize and to distinguish these forms of persuasion from the author's formal argument.

Specific suggestions for filling out part I of the Discussion Sheet

The first part of your worksheet is meant to help you identify what the author said and how he/she argues for a particular claim, the thesis or main conclusion of the article. You should do this part in three steps: (1) skim the article to get a sense of what it is about; (2) read the article once or twice to identify what the author's evidence for the conclusion is; and (3) summarize and identify (reconstruct) the author's argument, rereading sections of the text as necessary.

1. You can find the *title* of the article and the *author*(s) name(s) on the very first page of the article. The *topic* or subject of the article may be contained in the title, but sometimes finding the topic of the article is less straightforwvard. The title "A modest proposal," for instance, doesn't tell you what the article is about. Authors also sometimes introduce the topic of discussion by pointing out its connections to broader issues; this being so, the first paragraph may be misleading. After skimming the entire article, ask yourself what the article *as a whole* is about. Identifying the *main conclusion* or thesis of the article is the very next step. You may find it lurking in an introductory paragraph, as when the author announces his/her intentions: "The purpose of this article is to demonstrate that capital punishment is wrong." Sometimes it will appear in a concluding paragraph. As noted above, authors often provide subarguments for particular pieces of evidence they use to support their main conclusion. So the pitfall you must vigilantly avoid is mistaking a subargument for the main argument of the paper. (Hint: consider whether the claim you think is the main conclusion is supported by the bulk

or only part of the article.) One of the best ways to identify the main conclusion is to consider, if you had to say it in *one* sentence, what the author(s) would like you to believe about the topic, (e.g., "The central claim the author of this article would like the reader to believe is. . . .")

2. To identify the important *terms* of the article a useful rule of thumb is that you should (1) write down any terms that are unfamiliar to you, (2) write down any terms the author defines or otherwise characterizes, and (3) consider writing down terms that appear to have important roles in the author's argument. There are several reasons why an author may not define a particular term—the term in question has an established usage in the literature, the author believes the term is intuitively grasped by all intelligent readers, or perhaps part of the purpose of the article is to reveal an ambiguity regarding how the term is used.

To identify the author's *evidence* for the main conclusion, consider the relevance of particular claims in the article with regard to the author's thesis as a whole. Does the statement support the main conclusion or some other claim in the paper? If a particular point seems tangential or only remotely connected to the main conclusion, chances are it is not part of the author's direct evidence for his/her thesis. The relevance of a particular line of evidence may also be revealed later in the paper. (The author may also use as evidence a claim he/she believes is so uncontroversial that it doesn't have to be explicitly stated. This is called a *suppressed or unstated premise* of the argument.)

You should be wary of ignoring whole sections of the author's article as unrelated to the argument for the main conclusion or some subargument. Peer-reviewed articles of the sort we will be reading have gone through multiple drafts, in which the author has repeatedly attempted to refine the prose of the article to clarify the connections between each of the points raised to the main conclusion and excise unrelated or tangential lines of thought.

3 . Your goal in this section is to articulate in just a few sentences what the author's argument is for his/her main conclusion. The discussion worksheet breaks this down into two steps. First, summarize the article, and second, with your summary in mind, reconstruct the author's argument. The summary should give you an overall perspective on how each of the various pieces of evidence you found earlier supports the conclusion. (As you gain more practice, you may find that your summary looks a great deal like the argument.) When you write the argument itself, you need to focus more on the logic of the argument (i.e., how the evidence supports the conclusion). State these connections explicitly.

DISCUSSION WORKSHEET

Name: _____ Philosophy of Science

This handout will help you make sense of the readings outside of class and get more out of our discussion. Think of it as a review sheet for the exam. There are *three* things you should do with it. (1) *Prior to class*, answer the questions under parts I and II of this handout in a color of ink or pencil different from that you will use in class. Read part III for suggestions on how to prepare for discussion. (2) *During discussion*, add to or modify what you have written. (3) *After discussion*, answer the questions on the last page and submit the completed handout (all four pages) at the end of class.

PART I. WHAT THE AUTHOR REALLY SAID

1.   Directions: Skim the article as a whole to answer the first few questions below and get a sense of the structure of the article and the author's writing style.

TITLE OF ARTICLE:
AUTHOR(S) OF ARTICLE:
TOPIC:
MAIN CONCLUSION:

2.   Directions: As you read the article, write down any important /unfamiliar terms and start a list of the reasons or evidence the author provides in favor of the main conclusion (next page).

TERMS: List any terms or concepts that are unfamiliar or appear to be important. If the author provides a definition, be sure to write that down, too. Circle any you feel need clarification or discussion.

Important terms   Definitions

EVIDENCE: List any evidence (reasons) the author provides for the main conclusion. Each of these may appear as a subconclusion of its own argument. If you spot evidence in favor of a subconclusion, list that as well and identify which subconclusion it supports. Circle any that you feel needs clarification or discussion.

1) EVIDENCE FOR MAIN CONCLUSIONS

2) EVIDENCE FOR SUBCONCLUSIONS   SUBCONCLUSION SUPPORTED

IDENTIFY OTHER PERSUASIVE ELEMENTS.

Were there any other aspects of this article, such as the way it was presented, its use of examples, the author's writing style, etc., that made the article persuasive or nonpersuasive? List any you find.

3. Directions: After reading the article, complete the following:

SUMMARIZE, using the evidence you found above, the author's argument for the main conclusion. State points directly rather than "he says" or "It's about." (Don't evaluate the argument here.)

IDENTIFY the author's argument for the main conclusion by rewriting the summary you just wrote *as an argument* using just three or four sentences. Evidence for the conclusion should be placed in the form of premises, and the very last sentence should state the main conclusion. Your focus this time should be on the logic of the argument, i.e., *how* the evidence supports the main conclusion. (Again, resist the temptation to evaluate the argument here.)

PART II.—WHAT I THINK ABOUT THIS. The questions below ask you to evaluate the article. Beyond the elements of the article itself, consider whether the author has considered alternative viewpoints and what the consequences of adopting his/her position might be.

FIRST REACTIONS. List or write up any reactions you have to the article. Do you agree with the author? Why or why not? (Don't comment on everything—just the things you either strongly agree or disagree with.)

WHERE DOES THE AUTHOR GO WRONG? Remembering the argument you found for the author under "Some Suggestions on How to Read a Philosophical Article," identify what part of the argument, either evidence or the logic linking the premises to the conclusion, you think is mistaken. (Even if you agree with the author, play the devil's advocate by identifying what you consider to be the weakest point of the argument.)

WHAT IS THE STRONGEST PART OF THE AUTHOR'S ARGUMENT? Again, identify one part of the argument you think works well.

DEVELOP YOUR OWN POSITION. State your own position on this issue and sketch how you might argue for it. (If you find the author's argument compelling, suggest another way one might argue for the same position.)

PART III. DISCUSSION. By the end of discussion, you should understand the author's position, its strengths and weaknesses, your own position, and how you might argue for it. Review your answers to the above sections just before class to figure out what you'd like to get out of discussion. Come prepared to volunteer your own answers and raise questions as we walk through Parts I and II of this handout. Don't let the discussion move on until you understand that part of the author's argument.

PART IV. EVALUATE THE DISCUSSION. (Please do this at the end of class.) Directions: For each incomplete sentence below, check (x) the column completing the thought that best describes your experience in discussion today.

A) Overall reaction: B) Identification of the author's argument:

	A lot	Some	None	Discussion helped me identify . . .	Yes	?	No	Found on own
I learned	___	___	___	the author's main conclusion	___	___	___	_____
				the logic of the argument	___	___	___	_____
I participated	___	___	___	a strength of the argument	___	___	___	_____
				a weakness of the argument	___	___	___	_____
I enjoyed	___	___	___	my own views on this topic	___	___	___	_____

C) Discussion Format. Did analyzing this article using a discussion format help you?

(circle) YES   NO

Write below any additional comments you have on how your discussion went and how it could be improved.

D) Handout. Did this handout help you prepare for discussion?

(circle) YES   NO

Write below any additional comments you have on this handout and how it could be improved.

(Modified from W. F. Hill, *Learning through Discussion* [Thousand Oaks, Calif.: Sage Publications Inc., 1969].)

SAMPLE READING LISTS

1. Set of readings for a class that meets once a week.

The book lends itself to a biweekly arrangement, with students reading a single article (or a single article and the associated background essay) for each class. For classes that meet once a week, instructors will need to consider having students come to class having prepared two readings. Here are several pairings of articles that will work well together when considered during a single class:

Part I: Science and Pseudoscience
  Popper and Ziman, Feyerabend and Thagard

Part II: The Relationship of the Hard and Soft Sciences
  Taylor and Kuhn, Machlup and Rosenberg, Fay and Moon

Part III: Explanation and Law
  Hempel, Lambert and Britten, and Cartwright, Salmon, and Kitcher

Part VI: Science and Values
  Rudner and Hempel, McMullin and Hollinger, Giere

2. Set of readings for courses devoted to a topic that transcends the organization presented in the text.

A. Philosophy of the Social Sciences

Part I: Science and Pseudoscience
  Popper, Ziman, Feyerabend

Part III: Explanation and Law
  Hempel

Part V: Confirmation and Acceptance
  Quine and Ullian, Giere, Kuhn, Hempel, Frank

Part VI: Science and Values
  Rudner, Hempel, McMullin, Hollinger, Giere

Part II: The Relationship of the Hard and Soft Sciences
  Taylor, Kuhn, Fay and Moon, and Rosenberg

B. *Approaches to Philosophy of Science*

Positivism

Hempel (Part III), Carnap (Part IV), Frank (Part V), Rudner (Part VI), Hempel (Part VI), Rosenberg (Part II), and Machlup (Part II)

Kuhn and His Critics

Kuhn (Part V), Kuhn (Part II), Popper (Part I), Feyerabend (Part I), Hempel (Part V), McMullin (Part VI)

Postpositivism

Putnam (Part IV), Hanson (Part IV), Toulmin (Part IV), Matheson and Kline (Part IV), van Fraassen (Part III), Fay and Moon (Part II)

C. *Science and Culture*

Science and Society

Feyerabend (Part I), Thagard (Part I), Kitcher (Part I), Taylor (Part II), Machlup (Part II), Fay and Moon (Part II), Hanson (Part III), Giere (Part VI), Hollinger (Part VI), Giere (Part VI)

Science and Values

Kuhn (Part V), Hempel (Part V), Frank (Part V), Rudner (Part V), Hempel (Part VI), McMullin (Part VI)